AGRICULTURAL WASTE MANAGEMENT

Problems, Processes, and Approaches

ENVIRONMENTAL SCIENCES

An Interdisciplinary Monograph Series

ARTHUR C. STERN, editor, AIR POLLUTION, Second Edition. In three volumes, 1968

L. FISHBEIN, W. G. FLAMM, and H. L. FALK, CHEMICAL MUTAGENS: Environmental Effects on Biological Systems, 1970

DOUGLAS H. K. LEE and DAVID MINARD, editors, PHYSIOLOGY, ENVIRONMENT, AND MAN, 1970

KARL D. KRYTER, THE EFFECTS OF NOISE ON MAN, 1970

R. E. MUNN, BIOMETEOROLOGICAL METHODS, 1970

M. M. KEY, L. E. KERR, and M. BUNDY, PULMONARY REACTIONS TO COAL DUST: "A Review of U. S. Experience," 1971

DOUGLAS H. K. LEE, editor, METALLIC CONTAMINANTS AND HUMAN HEALTH, 1972

DOUGLAS H. K. LEE, editor, ENVIRONMENTAL FACTORS IN RESPIRATORY DISEASE, 1972

H. ELDON SUTTON and MAUREEN I. HARRIS, editors, MUTAGENIC EFFECTS OF ENVIRONMENTAL CONTAMINANTS, 1972

RAY T. OGLESBY, CLARENCE A. CARLSON, and JAMES A. McCANN, editors, RIVER ECOLOGY AND MAN, 1972

LESTER V. CRALLEY, LEWIS T. CRALLEY, GEORGE D. CLAYTON, and JOHN A. JURGIEL, editors, INDUSTRIAL ENVIRONMENTAL HEALTH: The Worker and the Community, 1972

MOHAMED K. YOUSEF, STEVEN M. HORVATH, and ROBERT W. BULLARD, PHYSIOLOGICAL ADAPTATIONS: Desert and Mountain, 1972

DOUGLAS H. K. LEE and PAUL KOTIN, editors, MULTIPLE FACTORS IN THE CAUSATION OF ENVIRONMENTALLY INDUCED DISEASE, 1972

MERRIL EISENBUD, ENVIRONMENTAL RADIOACTIVITY, Second Edition, 1973

JAMES G. WILSON, ENVIRONMENT AND BIRTH DEFECTS, 1973

RAYMOND C. LOEHR, AGRICULTURAL WASTE MANAGEMENT: Problems, Processes, and Approaches, 1974

Agricultural Waste Management

PROBLEMS, PROCESSES, AND APPROACHES

Raymond C. Loehr
DEPARTMENT OF AGRICULTURAL ENGINEERING
CORNELL UNIVERSITY
ITHACA, NEW YORK

ACADEMIC PRESS New York and London 1974
A Subsidiary of Harcourt Brace Jovanovich, Publishers

ACADEMIC PRESS, INC.
111 Fifth Avenue, New York, New York 10003

United Kingdom Edition published by
ACADEMIC PRESS, INC. (LONDON) LTD.
24/28 Oval Road, London NW1

Library of Congress Cataloging in Publication Data

Loehr, Raymond C
Agricultural waste management.

(Environmental sciences series)
Includes bibliographical references.
1. Agricultural wastes. I. Title.
[DNLM: 1. Agriculture. 2. Air pollution–Prevention and control. 3. Refuse disposal. 4. Water pollution–Prevention and control.
WA788 L825a 1973]
TD930.L63 628.5 73-5296
ISBN 0–12–455250–1

PRINTED IN THE UNITED STATES OF AMERICA

ANDERSONIAN LIBRARY
WITHDRAWN
FROM
LIBRARY
STOCK
UNIVERSITY OF STRATHCLYDE

To the many individuals whose professional and personal interest and concern have made this book possible. Special recognition should be given to GEB, LWW, GAR, REM, NCB, and with greatest warmth and appreciation to JML.

CONTENTS

3 Environmental Impact

4 Waste Characteristics

FUNDAMENTALS AND PROCESSES

5 Biological Processes

6 Ponds and Lagoons

7 Aerobic Treatment

8 Anaerobic Treatment

9 Utilization of Agricultural Wastes

10 Land Disposal of Wastes

11 Nitrogen Control

12 Physical and Chemical Treatment

MANAGEMENT APPROACHES

13 Management

PREFACE

Enhancement of environmental quality is an accepted national goal. Historically, the major efforts to maintain and enhance environmental quality have focused upon problems caused by urban centers. This emphasis has been due to pressing problems in controlling industrial pollution, in treating domestic liquid wastes, in disposing of municipal solid wastes, and perhaps to an instinctive feeling that agriculturally related environmental quality problems were uncontrollable and/or minor. Only recently has attention been given to the waste problems of agriculture.

The specific role of agriculture as it affects the environmental quality of the nation is unclear. The contribution of agriculture to water, air, and nuisance problems compared to contributions from industrial and domestic sources is difficult to assess. The available information suggests that the contribution from agriculture may be significant at the regional or local level. Data on fish kills from feedlot runoff, nutrient problems due to runoff from cultivated lands, the quantities of animal and food-processing wastes produced nationally, the pollutional strength of these wastes, the possible contamination of groundwaters from crop production and land disposal of wastes, and the increasing size of agricultural production operations indicate that considerable attention must be given to the development of a number of alternative methods to handle, treat, and dispose of agricultural wastes with minimum contamination of the environment.

The past decade has seen increasing monies and manpower devoted to finding solutions to the management of agricultural wastes. While considerable knowledge of the magnitude of the problem and of possible technical solutions to specific problems exists, detailed management methods to prevent contamination of the environment have only recently become available.

Agriculture is being faced with a number of constraints as the nation attempts to improve the quality of the environment. Since the concept of a totally unimpaired or totally polluted environment is not meaningful, feasible compromises must be obtained between agricultural production and environmental quality control to assure adequate food for the nation, adequate profit for the producer, and an acceptable environment for the public.

Proper management approaches to achieve this compromise are difficult to determine. Part of the difficulty occurs because the effects of environmental quality constraints on the dynamics of agricultural production are unclear. In addition, we are not in a position to adequately assess the producer or public cost of imposing such constraints on agriculture.

Both long- and short-term solutions must be found. Certain solutions will be of a stopgap nature until more definitive and comprehensive agricultural waste management systems are developed. These systems will contain the fundamentals of all successful waste management systems which include being technically feasible and economically attractive, requiring minimum supervision and maintenance, being flexible enough to allow for the varied and seasonal nature of the waste and the agricultural operation, and producing consistent results.

A workable strategy to develop satisfactory agricultural waste management systems will include imaginative developments and concepts from sanitary engineering, soil science, agricultural engineering, agricultural economics, poultry and animal sciences, and food and crop sciences. The solution to specific problems and the development of feasible agricultural waste management systems will occur by combining the fundamentals of these disciplines with the practice of agricultural production.

This book presents a summary of the processes and approaches applicable to the solution of agricultural waste management problems. In the context of this book, agricultural wastes are defined as the excesses and residues from the growing and first processing of raw agricultural products, i.e., fruits, vegetables, meat, poultry, fish, and dairy products. Implications and possible management systems for crop production are also discussed. The book is intended as a bridge between theory and practice as well as between the many disciplines that are involved in agricultural waste management.

Emphasis is placed on those processes that appear most adaptable to the treatment, disposal, and management of agricultural wastes. Fundamental concepts are followed by details describing the use of processes and management approaches. Examples in which the processes or approaches were used with agricultural wastes are included to illustrate the fundamentals as well as the design and operational facets. The references cited are not intended to be exhaustive. They were chosen because their data either illustrated the scientific or design fundamentals of a process or illustrated the application of specific processes to agricultural wastes.

The book attempts to place the agricultural waste problem in reasonable perspective, to illustrate engineering and scientific fundamentals that can be applied to the management of these wastes, to illustrate the role of the land in waste management, and to discuss guidelines for the development of feasible waste management systems.

The book developed primarily as an introductory text for individuals interested in knowing and applying feasible agricultural waste management concepts and approaches. Included are agricultural producers who are being required to incorporate waste management systems in their operations and who require knowledge of the advantages and disadvantages of alternative systems; scientists and engineers who require an understanding of the characteristics of agricultural wastes and waste treatment, disposal, and management alternatives; governmental agencies that make and enforce environmental quality control regulations and have need for information on agriculturally produced problems, agricultural production, and feasible waste management methods and systems; and environmentalists who desire an appreciation of the problems of agricultural production as well as of feasible approaches to manage the wastes and residues that unavoidably occur.

Agricultural waste management is in its early developmental stages. New and better approaches will be developed and more basic studies will elucidate the fundamentals affecting processes and management systems. It is hoped that this work will stimulate further inquiry and that the material presented will minimize and avoid, if possible, many environmental quality problems that are associated with agricultural production.

Cooperation of my many colleagues at Cornell and in governmental and industrial organizations who took the time and interest to review portions of the book is gratefully acknowledged and appreciated.

Raymond C. Loehr

THE PROBLEM

1

Current Constraints

Introduction

Agriculturally caused pollution is but one part of the national environmental quality problem. All pollution sources, i.e., municipal, industrial, marine, agricultural, and mining sources, must be considered in an integrated manner to improve the quality of the environment. Recent changes in agricultural production methods have caused natural interest in agriculturally related pollution to escalate. Such pollution is no longer considered minor or uncontrollable.

Changing agricultural practices have altered the traditional view of agricultural production. All agricultural production is becoming more intensive. Had agricultural production practices remained static, food production and the standard of living of the American public would not have reached the high levels enjoyed today. However, remarkable changes in the efficiency of United States agriculture have occurred in recent decades. Farm size and productivity per farm worker have increased significantly. Intensive crop and animal production have taken on many aspects of industrial operations.

The increased efficiency of agricultural production has generated or been associated with a variety of environmental problems. Efficiency of agricultural production and quality of the environment are inescapably interrelated and frequently appear to be diametrically opposed. Methods of handling, treating, and disposing of agricultural wastes may adversely affect air, water, and soil quality, and may be a nuisance to those who dwell nearby. Encroachment of suburbia into rural areas has sharpened the awareness of the problems generated by the handling and disposing of agricultural wastes.

Within recent years it has been documented that agriculturally related pollution is not minor and deserves the increasing attention of scientists, engineers, and administrators interested in the enhancement of the environment. A number of reports and symposia have been developed to place the agricultural waste problem in perspective, to establish priorities, and to estimate the cost of needed research and control measures to retard possible deterioration of environmental quality by agricultural activities (1–5).

Examples of adverse environmental quality problems attributed to agricultural operations include: excessive nutrients from lands used for crop production or waste disposal that unbalance natural ecological systems and increase eutrophication; microorganisms in waste discharges that may impair the use of surface waters for recreational use; impurities in groundwater from land disposal of wastes; contaminants that complicate water treatment; depletion of dissolved oxygen in surface waters causing fish kills and septic conditions; and odors from concentrated waste storage and land disposal.

The causes and concerns of agricultural waste treatment and disposal are analogous to the environmental problems caused by people. When people were fewer in number, when agricultural production was less concentrated, and when both were better distributed throughout the land, their wastes could be absorbed without adversely affecting the environment. Aggregations of people in cities and the development of large-scale industrial operations have caused the air and water pollution as well as health problems of which we are increasingly aware. While the problem of municipal and industrial wastes has been increasing for decades, the agricultural waste problem has been more recent and dramatic being apparent to those most closely associated with the problem for only about a decade and to those less closely associated for only the past few years.

Perhaps the most dramatic changes in agricultural production have occurred in animal production which has changed from small, individual farm operations into an industry involving large-scale enterprises. Small animals, such as chickens and hogs, are confined within small areas and buildings in which the environmental conditions are controlled to produce the greatest weight gain in the shortest period of time. There is an increasing trend for cattle to be finished in similarly controlled areas, dry lot feedlots. Under such conditions it is no longer possible for these animals to drop their wastes on pastures where the wastes can be adsorbed by nature without adversely affecting the environment.

Concentrated animal feeding operations may be a prime pollution contributor in some areas of the country. For example, certain sections of the Midwest have feedlots in watersheds containing high animal concentrations. Intensive storms can flush heavy loadings of animal waste runoff to small streams where the volume of storm water may be insufficient to provide adequate dilution along the entire watercourse. Fish kills and lowered recreational values have resulted. Runoff from manured fields and effluents from animal waste disposal lagoons also can affect the quality of adjacent streams.

Odors can be another problem associated with animal production facilities especially during land spreading of the wastes. Some odors are inevitable near such facilities. Odors can be reduced by proper sanitation in production facilities and by proper waste management.

There is an increasing concern with the environmental effect of the disposal of wastes on the land. Land application of wastes has been a traditional method of agricultural waste disposal and remains the best approach in most locations. The high concentrations of animal wastes in a small area and the disposal of the wastes on the soil have raised questions about surface runoff and groundwater quality problems.

Because of inexpensive fertilizers and increased labor costs, use of animal wastes as fertilizers has been less economical. Transportation costs and the quantity of the wastes produced at confined animal operations have caused an interest in liquid handling and treatment systems and in drying and solids destruction systems. These in turn have caused other problems.

Food processing wastes are an agricultural waste problem. In contrast to animal wastes which are very high strength, low volume wastes, food processing wastes are lower strength, higher volume wastes. The discharge of untreated food processing wastes to streams has caused pollution problems and the improper use of land disposal by irrigation can cause resultant runoff, soil clogging, and odor problems.

In the last few years, nutrient budget studies have indicated another concern of agricultural production. The quantity of nitrogen in a number of stream and river basins has been inferred to be a result of agricultural practice, especially of crop fertilization and animal waste disposal in these areas. Undoubtedly, some of the excess nutrients from crop production and waste disposal are reaching surface waters and groundwaters and contributing to the eutrophication and nutrient concentrations in some areas. The magnitude of this contribution requires better information and will vary between basins.

Past and, to a large extent, current pollution control activities are being directed at domestic and industrial wastes. When pollution from these sources are controlled, wastes from agricultural operations still may impart considerable undesirable material to the waters of the nation unless a determined and continuing effort is made to control real and potential pollution from agricultural operations.

The primary focus of agricultural waste management is on the obvious problems such as odor control and feedlot runoff. There is, however, awareness of potential long-range problems associated with land runoff from rural and agricultural lands, contaminant leaching to groundwaters, and salt buildup where land is used for waste disposal or where water reuse is practiced. Odor problems are mainly social in nature and efforts to control this problem represent the minimum management needed. The time scale of emphasis on agricultural waste management problems is

Present: Odor control from confined animal operations; biochemical oxygen demand (BOD) and suspended solids control of liquid agricultural wastes

Near future: Nitrogen control; rural and agricultural runoff

Future: Inorganic salt control

Legal Constraints

The increase in environmental pollution has required protection of public interest through legal restraints in the form of state and federal statutes and regulations and by individual action seeking redress for damages. The available legislation covers pollution irrespective of its source. Certain types of pollution have been considered as "natural" or "background" pollution. "Natural" pollution originates from sources such as runoff from urban, rural, and forest lands, natural chloride seeps, decaying vegetation such as leaves and crop residue, and animals on pasture and grazing lands. Such pollution is difficult to control because of its diverse nature, lack of controllable point sources, and inadequate knowledge concerning feasible abatement techniques. It has been assumed that this type of pollution is of small significance compared to other sources of pollution.

Recent developments in agricultural production techniques have altered the traditional concept of including pollution from agricultural production as "background" or uncontrollable pollution. As a result, the available legal restraints are being applied increasingly to agricultural operations.

Federal

Federal pollution control legislation has been developed since the turn of the century but with increasing rapidity since 1948. The basic policy and philosophy of water pollution control in this country can be found in the Water Pollution Control Act of 1948 and subsequent legislation (6). The basic policy of federal water pollution control legislation includes the following: (a) Congress has the authority to exercise control of pollution in the waterways of the nation, (b) both health and welfare are benefited by the prevention and control of water pollution, and (c) a national policy for the prevention, control, and abatement of water pollution shall be established and implemented.

The Water Quality Act of 1965 greatly expanded the scope of federal activities and caused the federal government to assume a leading role in the control of water pollution. One of the far-reaching effects of the 1965 Act was the provision for establishing water quality standards. Each state was to develop water quality criteria and a plan for implementation and enforcement. The water quality criteria adopted by a state were to be the water quality standards applicable to the interstate waters or portions thereof within that state if it was determined by the Secretary of the Department of Interior that the criteria and plan were consistent with the purposes for which the standards were to be established. In establishing the standards, consideration was to be given to their use and value for public water supplies; recreation and aesthetics; fish, other aquatic life, and wildlife; agricultural uses; and industrial water supplies.

A change in the water pollution control philosophy of the nation has occurred with the establishment of the standards. The emphasis is now on the amount of wastes that can be kept out of the water rather than on the amount of wastes that can be accepted by the waters without causing serious pollution problems. This philosophy will guide acceptable waste treatment and disposal methods as well as legal actions in the future.

The standards were to protect the public health and welfare and enhance the quality of water. Enhancement of water quality was taken to mean that all existing water uses would be protected and all wastes amenable to control would be controlled. All wastes amenable to treatment were to receive secondary treatment, or its equivalent as a minimum requirement.

The criteria may vary within a state depending upon water use, i.e., public water supply, industrial water supply, recreation, agricultural purposes, or receipt of treated waters. A review of the criteria from various states reveals common parameter ranges pertinent to this discussion: dissolved oxygen levels between 4 and 8 mg/l; ammonia concentrations less than 2.0 mg/l as N; total soluble phosphorus concentrations less than 0.05 or 0.1 mg/l; coliform concentrations ranging from less than 1000/100 ml for recreational waters to less than 20,000/100 ml for most other classes; no solids which are readily visible and attributed to sewage or other wastes. These criteria refer to concentrations in the stream, not in the waste discharges.

The above criteria generally are for maintenance of higher quality waters. When a given parameter was not specified for certain streams, a statement such as "not in amounts as to be injurious or impair the waters for any best usage assigned to this class" or "below concentrations which will be detrimental for established beneficial use" was included. For a specific situation the actual state water quality criteria should be read for clarity and accuracy.

Many states now are assuming that wastes from agricultural operations are controllable, are including these wastes in surveillance and enforcement proceedings, and are requiring the wastes to be controlled to avoid pollution of the environment. As an example, New York has required removal of at least 85% of the BOD and suspended solids and has considered removal of phosphorus from the waste waters from the Long Island duck farms. Disinfection levels also have been specified for these wastes.

The wastes discharged from agricultural operations will be required to meet the water quality criteria established by the states for specific streams if the wastes are discharged to or reach these streams. Because of the high concentrations of many agricultural waste waters, the alternatives of waste management are obvious. The wastes can be treated to a high level or can be disposed of on the land taking due care to avoid subsequent runoff and/or groundwater pollution. Another alternative is to avoid locating agricultural operations discharging high strength wastes to streams where it is difficult to meet the water quality standards. Where wastes are discharged to surface waters, agricultural producers may find standards applied to specify the quality of effluent that can be discharged. Such effluent standards can be in terms of oxygen demanding material [BOD, chemical oxygen demand (COD)], solids, nutrients, bacteria, and flow.

One unmistakable implication of pollution control activities in the nation is that facilities producing agricultural wastes will have to consider the impact of existing legal restraints on both site selection and waste treatment and disposal in all expansion plans and the establishment of new facilities. The facilities should be located where treatment and disposal can be obtained at minimum cost and without adversely affecting the environment.

The National Environmental Policy Act of 1969 declared it national policy to "encourage productive and enjoyable harmony between man and his environment and to promote efforts which will prevent or eliminate damage to the environment . . . ." An important aspect of the Act was to require that federal agencies must include a report evaluating the environmental effects of all proposed actions and legislation. Consideration of the environmental impact of actions in the early planning stages escalates environmental quality concepts to a much higher level of public and private consciousness than ever before.

If the environmental impact concept were to pervade all of society, it might be possible to alter national and personal life styles and ethics concerning environmental quality. Having the public consider environmental consequences before action of any type is taken, rather than after, has the potential of eliminating many of the obvious pollution problems. The implications of NEPA extend beyond the specific provisions found in the Act. NEPA is a positive rather than punitive law that reflects public concern for environmental quality. All societal components including agriculture will be affected by the specific and implied provisions of the Act.

One of the difficulties in integrating pollution control activities to avoid pollution transferral, such as liquid to solid, liquid to gaseous, or land to water, has been the independency of pollution control agencies at the federal level. In 1970, the national Environmental Protection Agency was established. EPA contains the federal activities dealing with solid waste management, air and water pollution control, and water hygiene among others and has the responsibility to integrate such environmental quality activities to minimize pollution transferral. The federal government has the legislation to control agricultural pollution and to encourage states to take a more aggressive role in their abatement activities.

The 1972 Amendment to the Federal Water Pollution Act continued the emphasis on keeping wastes out of surface waters. The Amendment declared that it is the national goal that discharge of pollutants into the navigable waters of the United States be eliminated by 1985. To achieve this goal, effluent limitations on point waste sources were to be achieved by the application of best practicable control technology by 1977 and of best available technology by 1983. The technology shall be economically achievable for a category or class of point sources as well as being practicable and available. The burden will be on industry to show that their treatment facilities represent the maximum use of technology within its economic capability. A number of agricultural operations such as dairy product processing, fruit, vegetable, and seafood processing, feedlots, and fertilizer manufacturing are included among the point sources that will be subject to effluent limitations.

With the passage of the 1972 Amendments, national pollution control emphasis shifted from water quality standards to effluent limitations. This approach was taken because of difficulties in linking waste discharge quality with stream quality and in enforcing previous legislation. The basic assumption, constrained only by the availability of economical control technology, is that the nation will strive toward complete elimination of water pollution. For many industries, this philosophy will result in a shift from waste water treatment plants to closed-loop systems that recycle waste water.

The Environmental Protection Agency was authorized to issue permits for the discharge of treated wastes into inland and coastal waters if the wastes met the conditions of the Act. States may conduct their own discharge permit programs with the approval of the EPA.

Agriculturally caused pollution problems were officially recognized for the first time in the 1972 Amendments. Certain agricultural waste point sources were subject to effluent limitations, specific appropriation authorization for agricultural waste research and demonstration projects was included, and area and statewide waste water treatment management plans were to include a process to identify agriculturally related nonpoint sources of pollution and to set forth procedures to control such sources. The EPA Administrator was authorized to issue guidelines for identification and control of pollution from a variety of nonpoint sources, including those from field, crop, and forest land. The concept of recycling potential sewage pollutants through production of agriculture, silviculture, or aquaculture products was encouraged.

The goal of eliminating waste discharges to surface waters will require that liquid and solids residuals be disposed of by other means. When methods to manage agricultural wastes to avoid surface water pollution are contemplated, it is equally important to avoid other types of pollution. Air pollution can result from drying and incineration, inadequate land disposal methods can cause groundwater and runoff pollution, and odor nuisances can be generated by improperly timed land disposal operations. Transferral of pollution from one form to another will not be successful or tolerated.

State

Legal controls on water quality have been recognized as traditional state responsibilities. Considerable variation has existed among states in their activities. While most state legislation does not mention agricultural pollution specifically, such pollution is not excluded. Because of the increased interest in agricultural pollution control, some states have developed regulations for such control.

The control of water pollution from animal feedlot operations has come under specific regulations in a number of states. The Kansas regulations are an example (7). These regulations stipulate that the operation of existing and proposed confined feeding operations must register with the Kansas State Department of Health. Where a potential for water pollution exists, suitable water pollution control facilities must be constructed in accordance with plans and specifications approved by the Department of Health.

Approved facilities normally include (a) diversion of runoff from nonfeedlot areas, (b) retention ponds for all waste water and runoff contacting animal wastes, (c) application of liquid waste to agricultural land, and (d) application of solid wastes to agricultural land. An important aspect of the regulations is that if, in the judgment of the Department, a proposed or existing confined feeding operation does not constitute a water pollution problem, provision of water pollution control facilities will not be required. Modifications of these regulations, the use of permits to assume adherence to federal, state, and local regulations, and voluntary registration have been considered and adopted for feedlots in other states. Many states rely upon existing regulations and a review of each situation to control agricultural pollution problems.

Most of current state pollution control legislation has been written for municipal and traditional industrial wastes. Agricultural wastes have different volume and concentration relationships than municipal and industrial wastes. In addition, many agricultural wastes do not reach streams except as land runoff. Disposal on the land rather than to streams is common. As a result strict application of current regulations may be difficult. Use of the land as an integral part of agricultural waste management is a concept that must be recognized in applying available regulations. Flexible interpretation of legislation is required to control pollution from agricultural sources because of the above differences.

To assist the control of agricultural wastes while the most practical and economical waste management approaches are being developed, many states have suggested or prepared "Codes of Practice" or "Good Practice Guidelines" for the disposal of agricultural wastes and the expansion of existing or expansion of new livestock facilities. Examples of such codes or guidelines can be found in New York, Wisconsin, and Ontario, Canada, among others. These codes are intended to serve as a basis of reasonable operation for agricultural operations, including waste management, without being too specific in design requirements.

Key items in the codes or guidelines include suggestions of land area to dispose of wastes, waste storage capacity, distance to human dwellings, criteria for satisfactory waste handling and treatment facilities, odor control, solid waste disposal, and runoff control.

Where guidelines are available, an agricultural producer would be in a vulnerable position in respect to pollution problems if he did not adhere to them. Such guidelines do not insure that pollution or nuisance problems will not result. The advantage of codes or guidelines is that they represent best available practice, yet can be altered more simply than regulations when better procedures become known and proved. The following of existing guidelines may help eliminate punitive damages in a lawsuit since the operator was following good practice and using the best available knowledge rather than intentionally committing a wrongful act.

Violation of air pollution codes can be of concern for concentrated agricultural operations. The definition of air pollution in these cases is not always clear; however, one used in many states is

> "Air pollution" means the presence in the outdoor atmosphere of one or more air contaminants in quantities, of characteristics and of a duration which are injurious to human, plant or animal life or to property or which unreasonably interfere with the comfortable enjoyment of life and property throughout the state or throughout such areas of the state as shall be affected thereby.

The terms "unreasonable" and "comfortable enjoyment" require individual interpretation; however, the definition provides justification for complaints due to odors and dusts.

In several states, the odors generated by the confined agricultural production facilities have been a source of complaints and an important consideration in legal action against these operations. Enforcement of air pollution codes offers another avenue to assure maintenance of environmental quality but represents an additional constraint for agriculture.

Local

Nuisances caused by agricultural operations can be controlled by city or county governments or by individual action where the injured party seeks compensation for damages. Experience has shown that local legislation rarely is effective. Agricultural operations are adjacent to many small communities that hesitate to enforce legislation that may affect an enterprise which plays a large economic role in the success of the community.

Zoning can keep protective complaints to a minimum by requiring that nonfarm residents be located some distance from agricultural operations. Zoning laws do not, however, provide absolute protection against the application of nuisance laws. Automatic protection also is not provided because the agricultural operation existed prior to residential or recreational uses of adjacent land. An important aspect is whether changes in the operation have caused nuisances to arise.

Private

Individual action can be an effective legal restraint on agricultural operations. The basis for such action includes trespass, nuisance, negligence, and strict liability (8). Private regulation of environmental quality problems can occur by civil lawsuits based on nuisance laws. All persons have the basic right to enjoy their property. Unreasonable interference with such enjoyment is legally a nuisance. A nuisance may involve air, water, solid waste, and/or noise pollution. Nuisance law is of common law rather than statutory origin (9).

Both public and private types of nuisances exist. When the rights of a substantial number of people are interfered with, a public nuisance results. If only the rights of a few are interfered with, a private nuisance results. Cases have been brought against agricultural operations for odors, water pollution, air pollution, and aesthetic reasons and in a number of cases, injunctions were issued and/or damages awarded the plaintiff (9, 10). The potential of private legal action can exert a significant constraint when individual operations consider management decisions.

Foreign

Federal governmental concern with water quality standards is not confined to the United States. The governments of at least six European nations, Belgium, Bulgaria, France, Poland, U.S.S.R., and Yugoslavia, have adopted or are considering requirements for classifications of at least their major streams (11). Many countries have established standards for sewage and industrial effluents.

A variety of legislation relevant to river pollution in Great Britain has been enacted over the centuries (12). The Public Health Acts of 1937, 1951, and 1961 are the most pertinent to the problem of agricultural wastes. As a result of the 1961 Act, farm effluents are now regarded as trade effluents and their reception into a sewer is subject to conditions imposed by the local authorities.

June 1, 1963 was the day on which the discharge of farm effluent into any watercourse became illegal in England. The effluent from farms must either be retained on the farm or cleaned to a standard considered satisfactory by a river board. The legislation has had a significant effect in reducing pollution from agricultural effluents. Some farmers have ceased to discharge effluents and have made arrangements to dispose of their waste on the land. Others are separating uncontaminated surface and roof water and are only spreading waste water on land. A few operations have installed treatment plants.

The 1970 Surface Waters Pollution Act of the Netherlands forbids the discharge of waste into surface waters without governmental consent. The consent can set conditions on the degree of treatment and the limits of pollutional matter in the effluent. Every discharge of pollutional matter will be taxed on a population equivalent basis according to the quantity of chemical and nitrogen oxygen demand in the discharge. Inorganic discharges also can be taxed as appropriate. The Act is being applied to point source food and food product processing operations as well as to all other industries.

Social Constraints

Suburban development and the increase in farm technology have increased the interest in agricultural wastes. The number of urban oriented rural residents and open country recreational activities is increasing and the agricultural population of the United States continues to be a lesser proportion of the total population. Individuals can live and die without growing their own food or knowing what is involved in producing a gallon of milk, a dozen eggs, or a pound of meat. They are not accustomed to and may not tolerate the liquid and gaseous effluents associated with agricultural production. Concentrated agricultural production has intensified these effluents in the same period that the public has joined the quest for a pollution free environment.

It will be necessary to supplement and parallel any regulatory and enforcement activity with a planned educational program and a productive research program. The educational program is needed to alert the public and the agricultural community to the need for and cost of adequate treatment and disposal of agricultural wastes. The research program is needed to develop and demonstrate economically feasible treatment and disposal methods that can be used with these wastes.

When public concern is voiced about the manner in which agricultural operators dispose of their wastes, apply fertilizers to crops, or manage their waste systems, what is implicitly inferred is that the social costs of producing these products may have equaled or exceeded the benefits. The national search for a better quality of the environment will require agriculture to give a high priority to wastes in decisions concerning production, i.e., to continually "think waste management." The heightened public interest in the environment and the increase in the urban and suburban population will be an effective reminder to those operations that do not place a high priority on pollution control activities.

References

1. Brady, N. C. ed., "Agriculture and the Quality of Our Environment," Publ. No. 85. Amer. Ass. Advan. Sci., Washington, D.C., 1967.
2. Loehr, R. C., "Pollution Implications of Animal Wastes—A Forward Oriented Review." U.S. Dept. of Interior, Federal Water Pollution Control Administration, Ada, Oklahoma, 1968.

3. Wadleigh, C. H., Waste in relation to agriculture and forestry. *U.S., Dep. Agr., Misc. Publ.* **1065** (1968).
4. Anon., "A National Program of Research for Environmental Quality—Pollution in Relation to Agriculture and Forestry." Joint Task Force of the U.S. Dept. of Agriculture and the State Universities and Land Grant Colleges, Washington, D.C., 1968.
5. Freeman, O. F., and Bennett, I. L., "A Report to the President—Control of Agriculture—Related Pollution." U.S. Department of Agriculture and the Office of Science and Technology, Washington, D.C., 1969.
6. Committee on Public Works, U.S. House of Representatives, "Laws of the United States Relating to Water Pollution Control and Environmental Quality." GPO, Washington, D.C., 1970.
7. Gray, M. W., One state's approach to the problem of pollution from feedlot wastes. *42nd Annu. Conf., Water Pollut. Contr. Fed.* (1969).
8. Walker, W. R., "Legal Restraints on Agricultural Pollution." Proc. Agr. Waste Manage. Conf., pp. 233–241. Cornell University, Ithaca, New York, 1970.
9. Levi, D. R., "A Review of Public and Private Livestock Waste Regulations." Proc. Agr. Waste Manage. Conf., pp. 61–70. Cornell University, Ithaca, New York, 1972.
10. Willrich, T. L., and Miner, J. R., Litigation experiences of five livestock and poultry producers. *In* "Livestock Waste Management and Pollution Abatement," Publ. PROC-271, pp. 99–101. Amer. Soc. Agr. Eng., 1971.
11. Lyon, W. A., European practice in water quality control. *J. Sanit. Eng. Div., Amer. Soc. Civil Eng.* **93,** SA3, 37 (1967).
12. Wisdom, A. S., "The Law on the Pollution of Waters," 2nd ed. Shaw, London, 1966.

2

Changing Practices in Agriculture

Introduction

Agriculture has flourished in areas and regions with favorable conditions for plant growth. The importance of proximity to population centers has declined as transportation services and marketing techniques have improved. Environmental considerations, such as soil, water, climate, and pollution control, are of continuing importance in the location of agricultural production and related economic activity.

The quantity and quality of the world's food supply will be among man's most critical future problems. In recent years there have been large increases in the efficiency of agricultural production. Farm size and productivity per farm worker have increased. Both crop and livestock production per acre and total crop production are increasing with less land used for production. It is unlikely that the ultimate in agricultural productivity has been reached, and further increases in agricultural production and productivity will be achieved. Scarcity of labor and the need to replace it with mechanization has led to technical advances in crop production, harvesting, animal housing, feed handling, and waste management.

One of the benefits of these efficiencies has been reasonable food prices. A notable example has been broiler chickens. In 1934, United States production of commercial broilers was about 100 million pounds liveweight and the cost was about 19 cents per pound liveweight. In 1969, over 10 billion pounds were produced and the cost was 15.2 cents per pound liveweight. Other agricultural production items also illustrate these efficiencies. In the 1960–1969 period, the price of eggs received by the producer increased only 11% and the price received by farmers for beef cattle increased only 18%. These changes took place at a time when the personal disposable income increased markedly and when the cost of living was continually rising. Comparative changes in other food costs and in the food cost as a percent of disposable income are noted in Table 2.1 (1). Research, development, and better production management have made these efficiencies possible.

These efficiencies, however, have generated or been associated with a variety of problems related to the quality of our environment. The relationship between the efficiency of agricultural production and potential environmental quality problems is more pronounced as methods of agricultural production and processing become more intensive.

Many of these potential environmental quality problems are associated with the excesses from agricultural production operations. These excesses can be in the form of animal waste at animal production facilities, runoff and leachate from fertilized and manured fields, and liquid and solid wastes generated at food processing operations. Excesses of this nature always have been associated with agriculture but have become more noticeable because of ready availability of inexpensive commercial fertilizers, large concentrated production units, reduced availability of labor, narrow profit margins, and because land for disposal is becoming less available and more expensive.

Table 2.1

Changes in Retail Food Costs[a] and per Capita Consumption 1950–1970[b]

Item	Costs—% of 1950	Change in per capita consumption (%)
Cereal and bakery products	160	−11
Processed fruits and vegetables	145	+50
Fresh fruits and vegetables	180	−22
Sugar and sweets	150	+10
Dairy products	155	−10
Beef and veal	137	+50
Pork	140	− 6
Eggs	103	−18
Poultry meat	78	+100

[a] Food cost as percent of income: 1950, 22%; 1970, 16.5%.
[b] From reference 1.

Table 2.2

Shifts in Land Use Patterns in New York State[a]

	Actual (%)		Projected (%)		Index of change (1965 = 100)
Use category	1955	1965	1975	1985	1965 to 1985
Urban	5.9	7.5	8.5	9.8	143
Woodland	49.3	51.0	53.6	57.5	113
Open land	21.6	20.6	19.6	17.3	84
Cropland available	23.2	20.9	18.3	15.4	73
	100.0	100.0	100.0	100.0	
Nonurban	94.1	92.5	91.5	90.2	98
In farms	49.5	39.5	31.0	25.1	64
Cropland harvested	17.3	15.0	13.4	11.8	78

[a] From reference 2.

The expanded urbanization of the United States has lead to an increasing encroachment on agricultural lands. Data from New York illustrates typical current and predicted changes (Table 2.2) (2).

The uses and demands for agricultural lands are accompanied by an increase in land value. Agricultural enterprises located in these areas must intensify their operations if they are to survive. The intensification of agricultural industries results in concentration of wastes. The increase in the volume of waste in a smaller area and the putrescible nature of agricultural wastes require acceptable methods of treatment and disposal. The problem can be aggravated by marginal economies of some agricultural operations.

Agricultural Productivity

The productivity of American agriculture in recent decades has been remarkable. Fifty years ago, 37 million more acres were required to feed less than half as many people. In 1940, a farm worker produced enough farm products for nearly 11 persons. In 1971, one farm worker produced enough for 48 persons—over a 300% increase in three decades (Fig. 2.1).

In 1971, only 5% of the national work force was on farms compared to about 40% at the turn of the century. The farm workers are not the only contributors to agricultural production. In contrast to earlier periods, farm power units are produced in factories, repaired by service agencies,

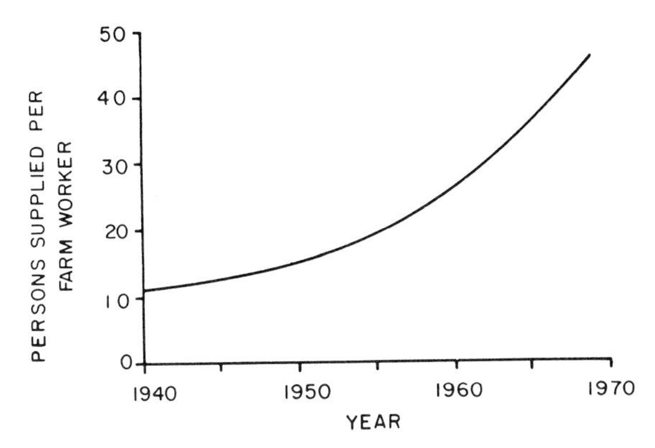

Fig. 2.1. Persons supplied with farm products by one farm worker (1).

and supported by fuels produced off the farm. The individuals contributing to agricultural production include both the on-farm workers and the nonfarm support personnel. The combined work force supporting agriculture is about 20% of the national work force. The growth of the farm service support area has been both a cause and result of the changes in agricultural productivity.

For the period 1950–1969 total farm input increased by 11%, farm labor declined by 78%, machinery inputs increased by 29%, fertilizer and lime inputs increased by 156%, and miscellaneous inputs which include pesticides increased by 48% (1). During the same period farm output increased by 35%. This increase in productivity has been achieved by many factors. The application of research results by farmers and industries serving farmers along with the application of modern business methods to farm management and to the investment of capital have been key elements that have made this record possible (Fig. 2.2).

American farmers are increasing output per man-hour at an average rate of 6% per year. That annual increase in productivity is more than the 2.5% annual increase for nonfarm private industry. In England the increase in labor productivity between 1961 and 1968 averaged 6.5% per year and was made possible by better management practices and increased mechanization. The agricultural industry in England produces about 45% of the food requirements of the United Kingdom with less than 2% of the total manpower of the United Kingdom directly employed and with about 5.5% of the manpower including that of the associated manufacturing and processing industries.

Both crop production per acre and total crop production are increasing even though less land is being used for production. Over one million acres have been released from crop production in recent decades. However, for

Fig. 2.2. Wheat harvesting equipment illustrates the mechanization and labor saving equipment in American agriculture (courtesy U.S. Dept. of Agriculture).

every two acres released, about one acre of new crop land was developed. Much of the new land is more productive than that abandoned. This differential land productivity coupled with improved crop varieties, higher use of fertilizer and pesticides, and land improvement practices such as drainage and irrigation have resulted in noticeable increases in per acre yields.

In 1969, about 3 million farms, half the number 30 years ago, supplied the nation's agricultural products. This number will decrease to about 2 million farms by 1980. At that time, perhaps only one-quarter to one-third of then existing farms will supply 90% of the products reaching the market. As a result, the average size of a farm will continue to increase in a manner similar to the increase during the last decade in which the size of the average United States farm increased 80 acres. The number of fárms, farm supply firms, and similar establishments in sectors of the food industry have and will continue to decline. Because business in all areas of agriculture is projected to increase, this means the concentration of more economic activity in the hands of fewer people.

To economically survive, the farmer has had to increase in size and obtain his profit on volume and efficiency. The decrease in people in agricultural occupations will release additional people from agriculture for urban employment. Rural residents have decreased to about 26% of the total population. The farm population will decrease further in the next decades.

Food production means employment for many individuals other than the initial producer or farmer (Table 2.3) (2). Due to the factors noted above, the numbers and proportion of farmers in the food and agricultural industry will decrease. The greatest increase in the agricultural industry will be in food retailing. Greater capital investment will be required in the food industry with the greatest increase occurring on farms (Table 2.4). Using the data in Tables 2.3 and 2.4 (2), the capital investment per employee on New York farms was $31,300 in 1963–1967 and is estimated to be $135,000 in 1985. Using similar data, the capital investment per farm acre in New York was $312 in 1964 and is estimated to be $815 in 1985. Comparable changes will occur throughout the nation.

Labor

There is a growing shortage and expense of labor for American farms. Factors contributing to the decrease include: (a) greater income for non-farm work, (b) greater mechanization, and (c) less motivation for specific farm tasks such as manure handling. The 1970 after tax farm disposable

Table 2.3

Employment in the Food and Agricultural Industry of New York State[a]

	Actual[b]	Projected[b]	Percent of total	
Activity	1965	1985	1965	1985
Farm supplies and services (last handler)	21,000	21,000	4.2	4.7
Farming	122,000	47,000	24.5	10.5
Food manufacturing	139,800	113,800	28.0	25.3
Food marketing (first handler)	23,400	24,000	4.7	5.3
Food wholesaling	33,000	39,000	6.6	8.7
Food retailing	156,000	201,000	31.3	44.7
Government and education	3,400	3,400	0.7	0.8
	498,600	449,200	100.0	100.0

[a] From reference 2.

[b] Numbers of people.

Table 2.4

Capital Investment in the Food Industry of New York State[a]

	Actual[b]	Projected[b]	Proportion (%)	
Activity	1963–67	1985	1963–67	1985
Farm supplies and services (last handler)	350.0	375.0	3.8	2.8
Farming	3,816.0	6,350.0	41.0	46.9
Food manufacture	2,500.0	3,100.0	26.8	22.9
Food marketing (first handler)	285.0	300.0	3.1	2.2
Food wholesaling	1,270.0	1,720.0	13.7	12.7
Food retailing	1,080.0	1,700.0	11.6	12.5
	9,301.0	13,545.0	100.0	100.0

[a] From reference 2.
[b] Data is given in millions of dollars.

income from all sources averaged about 75% of that for nonfarm people. This percentage may increase in the future. To obtain greater efficiency, agricultural production requires more sophistication and equipment. The demand for individuals to operate the equipment competes with the need in other industries. Even among individuals who are interested in working in agriculture there is a lack of interest in handling of wastes at concentrated production facilities.

Data on distribution of labor on a commercial hog farm (Table 2.5) (3) indicates that the largest portion of time was spent cleaning facilities. Time studies at other livestock production facilities also indicate that waste handling consumes a large fraction of the time. The result is the development of less time-consuming waste management methods.

Slatted floors have been a laborsaving approach for animals in large confinement buildings. The animals are housed on these floors and their wastes drop through to a collection pit. Treatment systems, such as the oxidation ditch, can be used in these pits. The slatted floor eliminates bedding and the requirement for daily cleaning. A comparison of the time spent in cleaning animal wastes (Table 2.6) (4.5) indicates the value of liquid manure systems and slatted floors. The use of this combination has made better use of human time in animal production but also has resulted in increased environmental problems for livestock operations that have adopted their use.

The number of individuals producing agricultural products will decrease (Table 2.3), although total farm production will stay the same or slightly increase. Farmers are able to increase their output per man because they purchase many services which they once performed themselves. At one time farmers produced their power and the fuel for it by raising horses and hay. They now purchase tractors and repairs as well as fuel to operate them. Progressive farmers hire services such as milk hauling, fertilizer spreading, and weed control that were done with farm labor in past years.

Table 2.5

Basic Tasks on a Commercial Hog Farm in the United Kingdom[a]

Task	Percent of working day
Cleaning	35–40
Feeding	15–20
Hog handling	4–7
Records	1
Feed mixing and handling	0–17
Travel about site	20

[a] From reference 3.

Table 2.6

Time Studies of Cleaning of Manure from Confinement Buildings

Dairy cattle[a]	
Method	**Time (average man minutes/cow per day)**
Manual	3.2
Semimechanical	2.6
Mechanical gutter cleaner	2.1
Liquid manure systems	1.3
Hogs[b]	
Type of floor	**Labor (man minutes/day)**
Solid	17
Solid	60
Slatted—total	2.2
Slatted—part	15.2
Slatted—part	12.2
Slatted—part	31.2

[a] From reference 4.
[b] From reference 5.

Agricultural production will become even more dependent on services purchased from other segments of the economy.

The number of seasonal workers used in agriculture has declined substantially as crops are harvested mechanically. Mechanical harvesters have been developed for many crops such as cherries, grapes, cabbage, onions, apples, potatoes, tomatoes, and lettuce. The harvesting of many other crops will be mechanized in the future.

In the United States the agricultural work force has decreased from 8.4 million to 4.6 million in the period from 1955 to 1969. Similar changes have been observed in other countries. In the 20 years from 1947 to 1967 the male agricultural work force in the United Kingdom decreased from 611,000 to 315,000 (3). In spite of this change, agricultural productivity increased.

In the developed countries, the agricultural labor force will be about 30 million people in 1985 as contrasted to about 70 million in 1950. By 1985, only about 2.3% of the North American labor force will be engaged in agriculture. In western Europe, developed areas of Oceania, and in other developed countries the proportion will be 9.6, 5.2, and 11.1% of the total labor force, respectively (5a). By contrast, the developing countries are expected to have considerable increases in the agricultural labor force. Estimates indicate that it will rise from 344 million people in 1950 to 500 million in 1985. While farming will still be the principal source of employment in the developing countries, the relative importance of farming will decrease.

The decline in the agricultural labor force will continue as countries develop but the rate in the developed countries will decrease. There is a limited movement of new labor into agricultural industries which results in a higher average age. Greater productivity will be required to maintain economic success in agriculture. The future will see fewer but larger production units managed by skilled staff earning higher wages. The time spent on waste handling will be reduced to a minimum. At the same time, these skilled individuals will be better able to understand and operate the waste management systems that will be necessary. In the future two types of agricultural production units will result, one farm based, extensive, and smaller in scale and the other nonfarm based, intensive, and larger in scale.

Livestock and Crop Production

Population growth and consumer desire are important factors in determining future demand for agricultural products, trends in the growth or decline of components of the industry, and in providing an estimate of future concerns. This section outlines changes that are occurring in agricultural production as well as the comparative size of specific segments of the industry. A pattern of the waste management challenges will emerge as the segments are described and as they are combined with information in Chapter 4 on the quantity and quality of the resultant wastes.

National crop and livestock requirements will continue to increase (Table 2.7) (6) due both to population growth and to increase in per capita consumption of certain products. Changes in consumer income, age distribution of the population, and developments of new uses for farm products will influence the actual future requirements.

Food Consumption

As the standard of living within a country improves, there is a greater demand for livestock and livestock products and a decline in the consumption of starchy foods. In the United States the per capita consumption of beef, broilers, and processed fruits and vegetables has increased while that of dairy products, eggs, cereal products, pork, and fresh fruits and vegetables has decreased (Table 2.1). Since 1950, United States per capita consumption of all red meat, beef, and broilers has increased at approximately 1.8, 2.0, and 1.0 lb per year, respectively. The increase in red meat consumption is due to the strong upward trend in beef consumption which has more than offset the decline in consumption of pork, lamb, and veal. In 1970, about 180% more chicken meat and about 240% more turkey meat were produced than in 1950. Chicken and turkey consumption per person increased about 110% during this period. Due to the growing population, total egg production increased about 14% during the 20-year

Table 2.7

Projection of Agricultural and Forestry Output by Major Products Groups[a]

Product group	Base[b]	Projections (index numbers)		
		1980	2000	2020
Feed crops	100	140	182	234
Food crops	100	150	196	261
Oil and fiber crops	100	144	182	235
Livestock and livestock products	100	142	196	273
Industrial timber products	100	124	150	165
Pulpwood	100	188	308	341

[a] From reference 6.
[b] For agricultural products, 1959–1961 = 100, and for forest products 1962 = 100.

period even though the per capita egg consumption decreased about 14%. Total U.S. per capita food consumption in 1970 was 6% greater than that in 1960.

If these trends continue, significant increases in beef cattle, broiler, and processed fruit and vegetable production will be needed to meet both the population growth and the increased per capita consumption. The production of other agricultural products will increase at a rate close to or slightly less than that of the population growth.

Livestock Production

Total animal production in the United States has shown a continued increase because of demands for meat noted above. The most dramatic increase has occurred in the broiler industry where the numbers of broilers raised in the United States increased from 1.8 billion in 1960 to almost 3 billion in 1970. During the same period, the number of beef cattle increased about 25 million head, an increase of about 38%.

To be economically competitive, livestock production has become concentrated in larger operations, become specialized in certain geographical areas, and used long distance transportation of inputs and outputs. The result has been increased confinement feeding of livestock and increased numbers of animals per livestock operation. Mechanization, improved production methods, and better nutrition and disease control have made it possible for the livestock producers to handle more animals with a minimum increase in help. The scarcity of inexpensive farm help has influenced the trend.

Increased efficiency of feed conversion and increased production per animal has occurred during the same time period as the trend toward confinement animal production operations. In the poultry industry, the feed consumption per dozen eggs decreased 20%, and the feed consumption per pound of broiler produced decreased 37% between 1945 and 1965. In Michigan, the feed consumption per dozen eggs during the 1959–1963 period was 5.5 lb while it was estimated to 4.0 in 1980 (7).

There are approximately 400,000 dairy farms in the United States and in 1968, 117 billion pounds of milk were produced. Total milk production has remained reasonably constant, decreasing slightly from about 123 billion pounds in 1955, with the per capita consumption decrease in milk consumption being about balanced by the population growth. The number of dairy cows has decreased from 21 million head in 1955 to 13 million head in 1968. In the same period, average milk production per cow increased from 5840 to 9000 lb/year to keep total production reasonably constant. Similar relationships are likely to exist in the future, i.e., decreased numbers of dairy animals with little change in total milk production. Fewer but larger dairy farms will exist in the future as dairymen improve techniques to produce livestock products more efficiently and profitably (Fig. 2.3).

Over two-thirds of the U.S. milk production is in the North Atlantic (19%), East North Central (28%), and West North Central (21%) states. The states making up the regions are indicated in Table 2.8. The three top milk producing states are Wisconsin, Minnesota, and New York. There are differences in milk production/cow throughout the regions. Milk production/cow varied from a low of 5230 lb/cow in Mississippi to 11,400 lb/cow in California. The average production was greatest in the western region, 10,420 lb/cow. Herds in California, Arizona, Washington, and New Jersey averaged more than 10,500 lb/cow. Herds averaging 14,000 lb/cow are not uncommon and some exceptional herds reach 20,000 lb/cow.

The change to confined livestock production has altered the complimentary relationship between crop and livestock production in which the grain and roughage produced on the land went into livestock production and the manure from the livestock went back on the land. Increasingly, many large confinement operations do not produce enough feed on their own land and import feed from adjacent farmers and states. One of the largest environmental quality problems associated with the confinement production of livestock involves waste handling and disposal. The size of livestock operations will increase as producers improve their techniques to produce livestock products efficiently and profitably.

Fig. 2.3. Large-scale commercial dairy farm.

Table 2.8

Agricultural Sections of the United States

Section	States
North Atlantic	Maine, New Hampshire, Vermont, Massachusetts, Rhode Island, Connecticut, New York, New Jersey, Pennsylvania
South Atlantic	Maryland, Delaware, West Virginia, Virginia, North Carolina, South Carolina, Georgia, Florida
East North Central	Ohio, Indiana, Illinois, Michigan, Wisconsin
West North Central	Minnesota, Iowa, Missouri, North Dakota, South Dakota, Nebraska, Kansas
South Central	Kentucky, Tennessee, Alabama, Mississippi, Arkansas, Louisiana, Oklahoma, Texas
Western	Montana, Idaho, Wyoming, Colorado, New Mexico, Arizona, Utah, Nevada, Washington, Oregon, California

Changes in sizes of livestock operations can be observed in many parts of the world. In Ontario, Canada (8), while the numbers of animals have increased for swine and poultry and slightly decreased for dairy operations, the number of farms decreased by 57, 65, and 50% for swine, poultry, and dairy operations respectively between 1950 and 1966. In England and Wales, about 123,000 farms produced milk in 1960 while in 1970 there were only about 80,000 dairy farms with approximately the same size of the national dairy herd.

A similar pattern has emerged in the United States. The demise of small, less-than-30-cow dairy farms in Michigan has been predicted by 1980. Dairy farms with 30–50 cows are expected to remain constant in number at 3200 while the farms having greater than 50 cows are expected to increase from 685 in 1959 to 2400 in 1980 and to contribute over 55% of the milk production in that state (9). In New York, it has been estimated that there will be 10,000 dairy farms in 1985, a decrease of 14,000 farms from those that existed in 1968. The average size of the herds in 1985 will be 80 cows, approximately double the herd size in 1968. The number of herds with 100 or more cows is estimated to triple by 1985 (10). Between 1960 and 1970, New York dairy farms with less than 30 cows declined by more than 70%. Farms with 50 or more cows increased by over 30% and farms with over 100 cows increased by about 110% during this period.

The changes in both the numbers and sizes of beef cattle feedlot operations have been dramatic. Specialization has removed cattle from pasture and grass land and has resulted in confinement of large numbers in small areas with an average density of one animal per 50 to 150 ft^2. The number of cattle on feed for slaughter was over 13 million head in 1970, an increase of 73% in the previous decade. During this period, the number of cattle on feed in the United States increased at the rate of over one-half million head per year.

Since midcentury, commercial feedlot operations in which feed and water are brought to the animals in confined areas have been expanding rapidly. Beef cattle enter such feedlots weighing 600–800 lb, are fed a ration high in grain and protein concentrate, and gain about 2–3 lb per day. The increase in commercial confined feedlot operations has occurred as a result of the availability of relatively inexpensive feed grains, proximity to an adequate supply of feeder cattle, and the strong demand for beef. The advantages of confinement include less space per animal, less labor, and economies of scale (Fig. 2.4).

Two types of cattle feeding operations exist, the farmer-feeder who has a feedlot with less than 1000 head capacity and the commercial feedlot which generally has greater than 1000 head and whose business is primarily cattle production. Neither the number of cattle fed or the capacity of the feedlot is the most important criterion to indicate the pollution potential of these two operations. From the pollution standpoint, the important factor is

Fig. 2.4. A typical large-scale commercial beef feedlot (courtesy U.S. Dept. of Agriculture).

whether land is avialable for waste disposal. Generally the smaller cattle feeders are farmers and have adequate land available to integrate waste disposal with crop production. The farmer-feeder operations are abundant in the midwestern cornbelt and the large commercial feedlots are abundant in the Texas, Oklahoma, Kansas, and Colorado high plains area.

There are four areas of concentrated cattle feeding in the United States. The area with the most dramatic increase is centered in Texas and Oklahoma where more than 5 million head are handled annually. A second area in the central corn belt feeds about 8 million head and a third area in Nebraska and eastern Colorado feeds about 6 million head annually. The fourth area is in southern California and Arizona which feeds about 3 million head per year. All but the last area have had large increases in fed cattle during the 1960–1970 period.

The Texas–Oklahoma feedlots are expected to increase in size and number. In this area, feedlots with 10,000 head or more capacity enjoyed a cost advantage over smaller lots and accounted for about 55% of the fed cattle marketed in this region during 1966–1967. Feedlots with less than 5000 head capacity generally were at a disadvantage in this region when competing with larger feedlots on the basis of annual fixed costs per pound of gain. The level of feedlot utilization was a major contribution to this economic difference. Feedlots with over 10,000 head capacity generally exhibited utilization rates of above 75% compared to utilization rates of 50% and lower for feedlots with less than 1000 head capacity (11).

Data on the number and size of beef feedlot operations are becoming available. In 1956, there were 182,000 head of cattle in feedlots in Kansas with only 30,000 head being fed in the five commercial feedlots having a capacity in excess of 1000 head. In 1969 there were 766,000 head of cattle on feed, of which 486,000 were fed in commercial feedlots having over 1000 head capacity (12). Statistics from other states indicate the same growth pattern. As of 1967 there were 1127 cattle feedlots in Colorado, 850 having a capacity of less than 500 head and 28 having greater than 5000 head capacity including reportedly the largest feedlot in the United States with over 100,000 head capacity. Of all the cattle on feed in Colorado, 625,000 in 1967, two-thirds were reported to be in the 28 larger feedlots (12). In 1955, there were 250,000 fed cattle marketed in Texas. In 1960 and 1970 approximately one million and four million head respectively were marketed reflecting the growth that took place in that state (13). In 1969, the six major cattle feeding states reported a 16% increase in cattle on feed. Texas, Arizona, and Colorado reported 50%, 41%, and 29% increases in cattle on feed for the year ending August 1969 (14).

The large beef feedlots will continue to market a greater proportion of the beef. Feedlots with 1000 head capacity or more account for slightly more than 1% of all cattle feedlots in the United States although they market over 50% of all the beef. The changes in feedlot capacity and marketing are noted in Table 2.9 (15). The total number of feedlots is decreasing as the smaller, inefficient lots go out of business.

The raising of dairy steers in feedlots for meat production is an emerging agricultural operation. A dairy beef enterprise offers a supplemental income for dairymen who have extra labor, feed, and facilities available. With the demand for beef, dairy beef feedlots will increase in number.

The poultry industry is another example of confined, intensive livestock production. In the major poultry producing regions, most laying hens and broilers are raised in confinement. Large poultry operations are highly mechanized and are able to handle over 100,000 birds per operation. Commercial egg production is almost 100% from confinement poultry houses. Present poultry management permits the concentration of egg-laying hens in buildings housing several hundred thousand birds on a site consisting of small acreage (Fig. 2.5).

Over 70 million eggs were produced in 1970 by about 320 million birds. Egg production is distributed throughout the United States with the greatest increase in production during 1960–1970 occurring in the Southern and Eastern Seaboard states. The top five egg producing states in 1970 were

Table 2.9

Beef Cattle Feedlot Numbers, Capacities, and Marketing

	Number of feedlots[a]			
Feedlot	1962	1964	1968	1970
Capacity (head)				
1,000– 1,999	752	808	967	1004
2,000– 3,999	373	421	522	549
4,000– 7,999	179	242	316	333
8,000–15,999	105	120	176	210
16,000–31,999	26	34	80	105
32,000+	5	10	19	41
Total number of (thousands)	236	224	199	182
Feedlot with capacity of 1000 or more cattle				
Percent of total feedlots	0.64	0.75	0.99	1.1
Percent beef cattle marketed	—	39	47	55

[a]From reference 15.

Fig. 2.5. Commercial egg production operation; manure held beneath cages until disposal (courtesy U.S. Dept. of Agriculture).

California, Georgia, North Carolina, Arkansas, and Pennsylvania. The per capita consumption of shell and processed eggs decreased from 370 in 1955 to about 317 in 1970. Total egg production increased during the same period. Over 90% of United States egg production is disposed of as shell eggs. Between 8–10% of the eggs are broken for commercial use as frozen eggs.

The marketing of poultry and eggs has undergone rapid and substantial changes which have been facilitated by advanced production and marketing technology. Per unit profits on these items have remained stable or declined even though wage rates and other prices have increased. Improvements in plant operating efficiency, larger production units, and higher density of production have been responsible for reduction of operational costs.

The swine industry represents some 10% of the total cash farm receipts, ranking only behind beef and dairy cattle among all animal commodities. Swine production is moving toward confined feeding although less than 10% of the hogs were produced in total confinement in 1970. Separate farrowing and finishing herds are common in operations having 1000 to 10,000 animals. A few 10,000- to 50,000-hog capacity operations exist. More large, specialized operations will develop incorporating complete feeding, waste treatment, and disposal in a single operation.

About 80% of the U.S. hogs are produced in the North Central states,

10% in the South Central states, 8% in the South Atlantic states, and the rest throughout the country. Six North Central states produce about 60% of the total U.S. hog production. Swine production contains two basic operations, the feeder-pig and growing-finishing operations. These can be under separate or integrated ownership and management. The swine industry has developed close to the corn production areas of the United States. Feed supplements are also fed to the animals. Because maximum productivity is the objective, swine are fed large amounts of high energy rations with small amounts of roughage.

The increased production per animal and the trend to confinement animal production operations produce situations that are analogous to modern industries. Inputs, the feed, and frequently raw material, the young stock, may be purchased elsewhere and brought to the production operation where under controlled conditions optimum production of meat, eggs, or milk occurs. Although the producer may be more interested in the quality and quantity of the product, the waste material that results must be disposed of without adversely affecting the environment.

Dairy Product Production

Fewer but larger dairy farms will exist in the future. A similar change will take place in the number of fluid milk processing plants. It has been predicted that the number of these plants will decrease to about 2000 in 1980, a decrease of 3500 from the number that existed in 1960. Production per plant will almost triple during this period.

Americans have experienced an increased taste for cheese. Per capita consumption increased from about 8.0 lb in 1955 to about 11.0 lb in 1970. Cheese consumption has increased from about 1.3 billion to 2.1 billion pounds in this period. It is expected that the number of cheese processing plants will decrease to less than 800 in 1980 or about one-half of the number that existed in 1960. Production per plant also will almost triple during this period. The per capita consumption of other dairy products such as butter and condensed and evaporated milk has decreased while that of ice cream, dry whole milk, and nonfat dry milk has remained relatively constant. These trends will continue. Plants processing these products also will have greater production per plant in the future.

The result of these changes will be to concentrate the wastes from dairy product production at larger operations. The changes emphasize the need to satisfactorily manage the increased wastes from milking parlors, whey from cheese production, and wastes from other enlarged dairy product operations.

Summary

Animal and dairy product production illustrate the type of changes that have occurred in agriculture in recent years. Meat, milk, dairy products, and eggs increasingly will be produced in confined large industrial type facilities. The trend toward controlled and enclosed facilities, which is virtually complete for smaller animals such as hens and broilers, is increasing even for the larger animals such as beef and dairy cattle.

The regional nature of livestock production in the future is unlikely to vary. The North Central states will continue to feed the majority of the hogs; the North Central, South Central, and Western states will continue to feed the majority of beef cattle; the South Central and Atlantic regions will continue to raise the majority of the broilers; and other poultry production as well as dairy production will be near the urban markets. The decreased numbers and increased size of production operations is expected to continue because of the enlargement of existing units and to the closing of uneconomic operations, of operations less specialized, and of those unable to adjust their size. The treatment and disposal of the wastes generated at these concentrated sites are the source of environmental problems that are likely to be magnified in the future.

Livestock Processing

While livestock production is becoming more centralized, marketing and slaughtering of animals, especially beef and hogs, has become more decentralized. Beef and hog processing and packaging facilities are being established near the sources of supply. The livestock slaughtering industry has changed in a functional manner. More slaughtering plants are specializing in only one kind of livestock. Processing plants show a trend to specialize in only one activity, i.e., slaughter or processing only rather than both. The slaughtering firms are decreasing in size as they move from the major terminal markets to the smaller communities near the supply.

The meat packing industry has had three major phases. Initially, it was a small industry with widely scattered local plants throughout the western United States. In the early part of this century, it became centralized in major midwestern cities which served as major terminal markets. The third phase is the decentralization phase now underway. As a part of the third phase, new regionally based concerns are capturing an important share of the market from the established packers.

Decentralization has been accelerated by improvements in truck transportation, rail freight rate adjustments, and refrigerated trucks. Weight loss in transportation of live animals lends a comparative advantage to the transportation of carcass meat. Another factor has been the age of the facilities of major packing concerns and the efficiency and automation obtainable in newer facilities.

An implication of this decentralization is that livestock processing wastes are being concentrated near the source of livestock supply. Wastes from livestock production and from slaughtering and processing operations are increasing around the smaller urban areas and towns of the nations. These industries are an economic advantage to the communities. However, their wastes represent a challenge for satisfactory management since they frequently are many times the magnitude of the wastes from the community.

Fruit and Vegetable Production

As with other agricultural products, Americans have changed their desires for fruits and vegetables (Table 2.10). Increased per capita consumption of canned and frozen vegetables has caused the per capita consumption of total vegetables to increase while that of fresh vegetables has declined. Processed fruit and vegetables have shown the greatest increase in total production and per capita consumption. A further increase in processed food consumption, especially frozen foods, will occur in the future as

Table 2.10

Fruit and Vegetable Production and Consumption, 1969[a]

	Production		Per capita consumption	
	Quantity (10^6 tons)	% of 1955	Pounds	% of 1955
Total vegetables	22.5	123	212	107
Fresh vegetables	13.1	109	97.6	93
Processed vegetables	9.4	152		
Canned[b]	—	—	52.9	122
Frozen[b]	—	—	18.9	275
Total fruits	22.6	130		
Fresh	7.9	98	79.1	79
Canned[b]	2.5	135	24.8	110
Dried[b]	0.26	90	2.6	72
Frozen[b]	0.9	130	9.3	107
Canned juices[b]	1.8	164	18.8	125

[a] From reference 1.
[b] Processed weight basis.

Americans increase their preference for convenience foods. Consumption and production of dried fruits (Fig. 2.6) had the greatest decrease in the period noted.

The fruit and vegetable canning and freezing industry includes operations in over 1800 plants. The industry utilized about 99 billion gallons of intake water, recirculated about 64%, and discharged about 96 billion gallons (16). Tomatoes, corn, and white potatoes (excluding dehydrated potatoes) are the largest tonnage of vegetables processed while citrus, peaches, apples, and pineapple represent the largest tonnage of fruits processed (Table 2.11). When all of the raw products of the food processing industry are considered, the distribution of the input has been estimated as 56% vegetables, 36% fruit, 10% specialties, and 3% seafood (17).

During the past decades, there has been a constant consolidation of smaller operations into larger, more centralized ones. The larger plants offer a greater opportunity for better control of waste discharges and for in-plant changes to reduce wastes. The processing plant wastes frequently

Fig. 2.6. Assembly line processing of apple slices (courtesy U.S. Dept. of Agriculture).

Table 2.11

Tonnage of Fresh Fruits and Vegetables Processed to Canned and Frozen Products (1968)[a]

Item	Million tons
Citrus	7.8
Tomato	5.0
Corn	2.5
White potato	2.4
Peach	1.1
Apple	1.0
Pineapple	1.0

[a]From reference 16.

are generated during a relatively small harvest period. Waste management systems must be geared to fluctuations in waste quantity and quality.

Potatoes represent one of the larger quantities of vegetables processed in the United States and processing patterns are changing. Annual production of potatoes in 1969 was over 300 million-hundred weight (Fig. 2.7), an increase of 135% over 1955 production. The increase in production reflected the increased processing of french fries and similar convenience foods. From 1955 through 1966, the quantity of frozen potato products increased from 129 to 1460 million pounds. In 1976, it has been estimated that about 4000 million pounds of frozen potato products will be processed (18).

Seafood Production

The world fish, mammal, and shellfish harvest in 1970 was about 70 million metric tons. This has increased from about 20 million tons in 1948 and 40 million tons in 1960. Ninety percent of this catch is fish with whales, crustacea, and mollusks making up the remaining 10%. In the 100-year period ending 1950, the world harvest of seafood increased at an average rate of 2.5% per year. Since 1950, the average rate has increased to about 5% per year. Future yields may increase by factors of 2 to 5. Although the United States fish harvests have declined about 20% since 1962, harvests in other countries such as Peru, Japan, and the Soviet Union have increased markedly. The seafood industry is not expected to increase rapidly in the United States.

Of the U.S. catch about 35% is rendered, 30% is marketed fresh, 20% is frozen, 1% is dried, and 4% is handled by miscellaneous methods (19).

Fig. 2.7. Mechanical harvesting of potatoes (courtesy U.S. Dept. of Agriculture).

Frozen fish products are increasing in popularity and a 150% increase has been predicted in the next 15 years. The annual United States per capita consumption of seafood has been about 11 lb per year for the past 20 years.

The major geographic seafood production regions in waters adjacent to the United States are those of the Gulf states, South Atlantic and Chesapeake Bay states, North and Middle Atlantic states, California, Alaska, and Oregon and Washington. The catches in these waters in 1967, in terms of millions of pounds, were 1036, 662, 552, 453, 360, and 242, respectively.

The distribution of the general types of fish and species that resulted in the largest U.S. catches is noted in Table 2.12. The menhaden, a fish primarily used for fish meal, fish solubles, and oil, represents the largest catch followed by seafood commonly used for public consumption. Public consumption of shrimp and tuna is likely to increase at more rapid rates than that of other seafoods.

The catfish industry has the potential to develop along lines analogous to that of livestock production. Farm pond grown catfish are grown in southern and south central states. Intensive catfish farming began in a significant manner in the 1960's. Commercial catfish pond acreage increased from about 400 acres in 1960 to 45,000 in 1970 and with about 75,000 acres forecast in 1975 (20). Yields per acre increased from about 800 to 1200 lb during that period and is expected to further increase in the future. The pounds of undressed catfish harvested in 1960 was about 300,000 lb while in 1970 it was 54 million pounds with over 115 million pounds expected in 1975. In pond culture of catfish, either fry or fingerling fish may be stocked. The fish are fed a commercial feed ration until maturity at which time they are processed into fillets or steaks which are marketed fresh or frozen.

Fertilizer Production

The sizable increases in crop and food production would not have been possible without the greater use of fertilizers. The increased inorganic fertilizer use in the United States has accounted for over half of the increase in crop production per acre since 1940 (21). Fertilizer has been substituted for land, labor, and other production inputs to increase agricultural production.

No single reason can be pinpointed for the high rate of fertilizer usage which averaged about a 10% increase per year between 1962 and 1969 and about 8.4% over the period of 1948 to 1969 in the United States. The significant factors for the increases include availability of fertilizer, competitive prices, interaction of other technology with fertilizer use, labor shortages, and the need to maintain economic viability. Future fertilizer consumption may not continue at this rate due to environmental concern and possible higher prices.

World fertilizer consumption reached about 60×10^{12} tons of plant

Table 2.12

Major United States Seafood Catches[a]

Fish	Average catch 1962–66 (10^6 tons)
Type	
Small, oily fish[b]	2094
Bottom fish[c]	628
Salmon	335
Tuna	305
Crab	289
Species	
Menhaden	1753
Shrimp	225
Blue crab	160
Sea herring	143
Pink salmon	141

[a] From reference 19.
[b] Menhaden, mackerel, anchovies, herring, alewives.
[c] Haddock, cod, ocean perch, whiting, flounder, hake, pollock.

nutrients in 1970, approximately eight times the consumption in 1946 and over double that of 1961. In the United States fertilizer use also has increased noticeably (Fig. 2.8). Nitrogen consumption has shown the greatest increase, about a tenfold increase between 1945 and 1969. The nitrogen content of commercial fertilizers doubled to slightly over 18% in the period from 1955 to 1968. In the same period, the phosphoric oxide content and potash content of fertilizers increased from 10.5 to 12.4% and from 8.8 to 10.3%, respectively. The use of liquid nitrogen as a fertilizer in the form of anhydrous ammonia and nitrogen solutions, increased about 5–6 times in the 1960–1970 period.

The use of fertilizers varies among nations and regions (Table 2.13) (21,22). Europe, excluding the U.S.S.R., was the heaviest user of fertilizer applying an average of 158 kg of nitrogen, phosphate, and potash per hectare of arable land under permanent crops. The world fertilizer consumption increased from 7.5 to 17 kg/capita over the period of 1955–1969. The most developed nations were the largest users of fertilizers. The largest consumers, in terms of quantity per person (kg/capita) in 1969, were North America, 66; Western Europe, 45; and Eastern Europe, 40. The consumption per hectare varies widely depending upon crop, farming intensity, and crop management. The changes in fertilizer use reflect the substitution of fertilizer for land, producing higher yields on fewer acres.

In the United States, fertilizer use varies among states, regions, crops

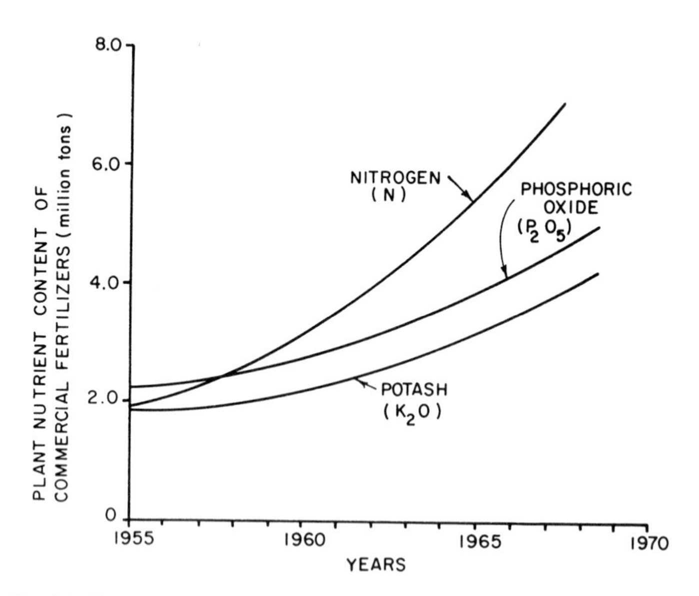

Fig. 2.8. Plant nutrients in commercial fertilizers used in the United States (1).

Table 2.13

National and Regional Fertilizer Use (kg/ha)

Region	Use[a]	Region or nation	Use
Europe	158	Netherlands	616[b]
North and Central America	66	Japan	400[b]
USSR	35	Latin America	22[c]
Oceania	33	India	10[b]
Mainland China	30	Western Europe	161[c]
Asia	22	Eastern Europe	49[c]
South America	14		
Africa	7		
Worldwide	42		

[a] 1969–1970; source, FAO.
[b] From reference 21.
[c] From reference 22.

and farming areas. In 1968, the South Atlantic states had the highest fertilizer consumption, 277 lb/harvested acre (23). The national average that year was 100 lb/acre. Eleven states used over 200 lb/acre with the top five states in terms of fertilizer consumption being Florida, 640 lb/acre; Rhode Island, 390 lb/acre; Georgia, 310 lb/acre; New Jersey, 273 lb/acre; and Alabama, 246 lb/acre. About 40% of the plant nutrients used in the United States were applied to corn. Hay and pasture accounted for about 15% of the total nutrients used because of the large acreage involved. Cotton was the third highest user accounting for about 8% of the plant nutrients consumed (21). Crops having the highest value per unit land utilize the greatest amount of fertilizer per unit of land area. Potatoes, fruits and nuts, vegetables, and sugar beets are among the crops on which the fertilizer use has been the highest.

Fertilizers have been profitable for farmers and have played an important part in providing consumers with an abundant and low cost food supply. On soils of low natural fertility, initial response in yields is large for a relatively small amount of fertilizer. Additional amounts of fertilizer produce successively smaller increases in yields. Ultimately, there is some point where additional fertilizer usage provides no additional production. Amounts applied beyond this point can result in decreased yield. In the United States increased fertilizer utilization will occur from increased fertilizer application on crops and in areas now using low amounts of fertilizer. Many of the crops grown with higher fertilizer use may be approaching maximum profitable application rates. Even before the maximum point is reached, there is the possibility that a portion of the fertilizer can be released to the environment. In the absence of sound crop management practices, this portion is increased. The low cost of fertilizers has caused many farmers to be less concerned with fertilizer nutrients that do not become part of the growing crop. These nutrient losses may reach groundwaters and surface waters.

Pesticides

The use of pesticides in agriculture has permitted the agricultural producer to produce food at a lower cost. As with fertilizers, pesticides have been substituted for labor and equipment. A marked increase in the use of pesticides started in the 1940's and lasted until the late 1960's. United States production peaked at about 1.2 billion pounds in 1968. Production decreased in 1969 and 1970, an average decrease of about 6.5% over those two years (24).

Pesticides have been categorized according to their use as insecticides, herbicides, fungicides, rodenticides, and fumigants. Pesticide use patterns have changed over the years and may be expected to change further as environmental constraints are imposed on their application. In 1970, although production of fungicides and insecticides decreased, that of herbicides increased.

Pesticides are primarily applied to corn and cotton (over 60%) with other major crops on which they are used being vegetables, field crops, and apples (about 20% combined). Future pesticide use will decrease and be selective. Concern of the environmental effect of pesticides has increased and has caused greater scrutiny of need, use, and dosages of pesticides. The regulatory powers of the federal government over the manufacture, sale, and use of pesticide products will become greater to provide more effective protection for the public and the environment while permitting the continued use of selected pesticides for beneficial uses.

Livestock, Fruit, and Vegetable Income

One of the compelling reasons that agricultural production operations have decreased in numbers and increased in size has been the income that the producer has been able to obtain. During a period when inflation was decreasing the value of a dollar, the income a livestock, fruit, or vegetable producer obtained remained constant or in some years decreased. Only within recent years has the per unit income from certain vegetables, fruits, and milk increased. Between 1950 and 1970, the output per man hour on

American farms increased 323%, retail food prices 154%, wholesale food prices 133%, and the price received by farmers increased by 118%.

The decreased or constant income from a unit of agricultural production has required the greater efficiency, increased size, greater mechanization, and automation described earlier. This in turn requires greater attention to the wastes generated at such production units.

Agriculture and the National Economy

Gross farm income has been increasing, reaching about 50 billion dollars in 1970, an increase of over 40% in the previous decade. The livestock sector contributed about 29 billion dollars of the income, up about 50%. The crop sector contributed about 21 billion dollars, up 37% from a decade earlier. If the contributions of processing, transporting, and marketing were added to the value of the farm products produced, the agribusiness sector of the economy contributed roughly 100 billion dollars to the domestic economy in 1970, as measured by consumer expenditures for domestic farm food products. Agribusiness also produced approximately 8 billion dollars worth of farm products for export.

Production expenses for farmers reached 40 billion dollars in 1970, an increase of about 150% from 1960. About one-third of total farm production expenses are produced in the farm sector, i.e., seeds, feeder cattle, and higher farm prices. Farm wage rates have increased almost 60% in the 1960's in response to minimum wage legislation and a smaller labor pool. The total wage cost has remained relatively stable due to a decline in the number of farm workers.

The gross farm product accounted for only 3% of the total private gross national product in 1970 compared to about 4.5% in 1960. Per farm net income has increased from $4050 in 1960 to $5410 in 1969, reflecting increased government payments, the demise of about one million small farms in this decade, and the rising price level.

Farms with annual sales of more than $20,000 increased by over 50% while those with sales of less than $20,000 decreased by one-third between 1960 and 1970. The number of farms with sales of less than $10,000 is expected to decrease by about one million before 1980. The top 200,000 farms, 6% of the 1970 total, accounted for over one-half of the marketing receipts. The top one-third of the farms accounted for almost 90% of the commercial output of U.S. agriculture.

Income of the farm population from nonfarm sources has increased substantially and in 1968 was almost one-half of their income from all sources. As a group, farm operator families with sales of $2500 to $5000 received about two-thirds of their total income from off-farm earnings while farms with farm sales of less than $2500 received over 85% of their income from off-farm sources. Only farms with sales of $20,000 or over exceeded or equaled the average income of the nonfarm population.

Consumer demand for certain farm products will increase but consumers may continue to spend a declining portion of their income on food. In 1970, about 16% of the disposable income was spent on food, a decline from the 20% spent in 1960.

The data presented in this chapter documented trends indicating that agricultural workers are a decreasing proportion of the national labor force, that agricultural income has not kept pace with national income, and that enlarged and mechanized food production operations will result. These changes represent a benefit for the American people and a challenge for agriculture.

Future policies for commercial agriculture must deal effectively with price mechanisms for commodities produced by fewer and larger producers as well as the effects of environmental interactions with agriculture, especially that of adequate and integrated waste management. An increasing need is information about the effects of agricultural operations on human health and the environment. These problems will not be solved easily and will require the efforts of public and private as well as agricultural and nonagricultural interests.

References

1. U.S. Department of Agriculture, "Agricultural Statistics—1970." U.S. Govt. Printing Office, Washington, D.C., 1970.
2. Forker, O. D., and Casler, G. L., "Toward the Year 1985—Summary Report: Implications, Issues, and Challenges for the People of New York State," Special Cornell Ser. No. 14. Cornell University, Ithaca, New York, 1970.
3. Scottish Farm Buildings Investigation Unit, *Farm Building Progr.* **18** (1969).
4. Oygard, G., "A Review of Recent Studies of Liquid Manure Handling and the Use of Slatted Floors," Bull. Agr. Econ. Res. No. 197. Dept. of Agricultural Economics, Cornell University, Ithaca, New York, 1966.
5. Robertson, A. M., Pig housing—future requirements and developments. Scottish Farm Buildings Investigation Unit. *Farm Building Progr.* **22,** 21–30 (1970).

5a. Anonymous, *CERES-FAO Rev.* **5,** 13 (1972).

6. Water Resources Council, "The First National Assessment of the Nation's Water Resources." Water Res. Counc., Washington, D.C., 1969.
7. Sheppard, C. C., and Larzelere, H. E., "Eggs and Poultry—Now and in 1980," Res. Rep. No. 51. Michigan State University, East Lansing, 1969.
8. Townshend, A. R., Reichert, K. A., and Nodwell, J. H., "Water Pollution Control Facilities for Farm Animal Wastes in the Province of Ontario," Proc. Conf. Anim. Waste Manage., pp. 131–144. Cornell University, Ithaca, New York, 1969.

9. Murray, D. L., Hoglund, C. R., and Rippen, A. L., "The Dairy Farm Enterprise," Res. Rep. No. 45. Michigan State University, East Lansing, 1969.
10. Conneman, G. J., "Toward the Year 1985—Milk Production and Consumption," Spec. Cornell Ser. No. 1. Cornell University, Ithaca, New York, 1970.
11. Dietrich, R. A., Costs and economics of size in Texas-Oklahoma cattle feedlot operations. *Tex., Agr. Exp. Sta., Rep.* **B-1083** (1969).
12. Rademacher, J. M., ed., "Proceedings of Animal Waste Management Conference." USDI-FWPCA, Missouri Basin Region, Kansas City, Missouri, 1969.
13. Wells, D. M., Grub, W., Albin, R. L., Meenaghan, G. F., and Coleman, E., Control of water pollution from southwestern cattle feedlots. Paper presented at the *Int. Water Pollut. Res. Conf., 5th, 1970* (1970).
14. Uhrig, J. W., Long range outlook bright for cattle. *Nat. Live Stock Prod.* pp. 16–17 (1969).
15. U.S. Department of Agriculture, "Number of Feedlots by Size Groups and Numbers of Feed Cattle Marketed," Statistical Reporting Service, Periodic Reports from 1966 to 1970. U.S. Dep. Agr., Washington, D.C.
16. Rose, W. W., Mercer, W. A., Katsuyama, A., Sternberg, R. W., Brauner, G. V., Olson, N. A., and Weckel, K. G., "Production and Disposal Practices for Liquid Wastes from Cannery and Freezing Fruits and Vegetables," Proc. 2nd Nat. Symp. on Food Process. Wastes, pp. 109–128. Pacific Northwest Water Lab., Environmental Protection Agency, 1971.
17. Hudson, H. T., "Solid Waste Management in the Food Processing Industry," Proc. 2nd Nat. Symp. on Food Process. Wastes, pp. 637–654. Pacific Northwest Water Lab., Environmental Protection Agency, 1971.
18. Dostal, K. A., "Secondary Treatment of Potato Processing Wastes," Rep. Proj. No. 12060. Water Quality Office, Environmental Protection Agency, 1969.
19. Soderquist, M. R., Williamson, J. J., Blanton, G. I., Phillips, D. C., Law, D. K., and Crawford, D. L., "Current Practice in Seafoods Processing Waste Treatment," Rep. Proj. No. 12060 ECF. Water Quality Office, Environmental Protection Agency, 1970.
20. Anonymous, "Producing and Marketing Catfish in the Tennessee Valley." Tennessee Valley Authority, 1971.
21. Nelson, L. B., Agricultural chemicals in relation to environmental quality: Chemical fertilizers, present and future. *J. Environ. Qual.* **1,** 2–6 (1972).
22. Harre, E. A., Kennedy, F. M., Hignett, T. P., and McCune, D. L., "Estimated World Fertilizer Production Capacity as Related to Future Needs—1970 to 1975." Tennessee Valley Authority, 1970.
23. Hargett, N., "1970 Fertilizer Summary Data." Tennessee Valley Authority, 1970.
24. Fowler, D. L., and Mahan, J. N., "The Pesticide Review—1971." USDA, Agricultural Stabilization and Conservation Service, Washington.

3

Environmental Impact

Introduction

Historically, national efforts to maintain and enhance the quality of the environment have been directed toward problems caused by urban centers and industrial operations. Had agricultural production practices remained static, environmental problems caused by agriculture might have remained minimal. However, real and potential environmental quality problems have accompanied the changes in agricultural productivity in recent decades.

Difficulties in obtaining precise relationships between agricultural practices and environmental problems occur because of the diffuse source of residuals from agriculture and the many factors that are involved, some of which are not controllable by man. The factors include topography, precipitation, cover crop, timing and location of chemical or fertilizer application, and cultivation practices. The variability of waste discharges complicates assessment of the environmental impact of agricultural production. Agricultural wastes frequently are not discharged on a regular basis; food processing wastes tend to be seasonal with production a function of crop maturity, feedlot runoff is a function of rainfall frequency and intensity, fertilizer applications are timed for ease of distribution and to maximize crop production, and land disposal of wastes are related to need for disposal and ability to travel on the land.

Available information suggests that potential environmental quality problems due to agricultural operations may be more dependent upon the production practices and waste management techniques utilized by farmers and processors than the size of the operation, the number of animals fed, or the amount of waste involved. This chapter will attempt to summarize the environmental problems that can or have been associated with agricultural production, to put these problems in perspective with those caused by other segments of society, and to indicate procedures to minimize these problems. Detailed waste management methods are described in subsequent chapters.

Water Quality

General

The amount of animal wastes that is retained on the land or reach ground waters and surface waters is not well documented. It is incorrect to use the amount of waste defecated by an animal to indicate the actual water pollution that may result. Only a small proportion of livestock wastes enter surface waters and groundwaters or the atmosphere. Data describing the total wastes from animals serves only to indicate that the problem of animal waste management is of considerable magnitude.

One of the most challenging problems of water quality management is the problem of excessive amounts of nutrients and the conditions they can cause. Of particular concern are nitrogen and phosphorus compounds. Although these elements are needed in small amounts for all living matter, excessive amounts in surface waters can result in over fertilization and can accelerate the process of eutrophication. Other concerns include excessive amounts of nitrates in groundwaters and surface waters, ammonia toxicity to fish, altered effectiveness of chlorination by ammonia, and the nitrogenous oxygen demand of reduced nitrogen compounds in surface waters.

All natural waters contain dissolved materials, including plant nutrients derived from natural processes. The quantities of eroded solutes are highest in areas of abundant precipitation and runoff while the concentrations of the dissolved matter are highest in areas of low precipitation. Agricultural drainage, as well as urban drainage and municipal and industrial waste discharges, can contribute significant quantities of nutrients to surface waters.

Until recently the major reason for investigating the nutrient content of the soil water below the crop root depth was to evaluate the loss of fertilizer nutrients and to estimate the loss of production or the need for more fertilizer to adjust for the losses. The increased concern on environmental quality has changed this emphasis to the study of nutrients and other contaminants in soil water because of potential water pollution. The quantity of rainfall or drainage water and the permeability of the soil are key factors in the leaching of contaminants from the soil.

An understanding of the forms of nitrogen and phosphorus in land may be helpful prior to discussing the concentrations of nutrients in agricultural drainage. The largest quantity of nitrogen in soil occurs as organic nitrogen. Microbial decomposition of the organic matter results in the production of ammonia nitrogen which with good aeration and suitable temperatures can be oxidized to nitrites and then to nitrates. Unless inhibition of the nitrate-forming bacteria occurs, practically no nitrite accumulation will occur. Nitrates can be reduced to gaseous nitrogen compounds and lost to the atmosphere if the nitrates are held under conditions of poor aeration and if suitable carbon sources are available for denitrification. Ammonium ions are held to the cation exchange sites in soils and the ammonium ion concentration in soil solution and leaching to the groundwater is low. Nitrate ions are soluble in the soil water. The form of nitrogen of greatest environmental quality concern in the soil is nitrate nitrogen since it is subject to leaching and will move with the soil water. Losses of organic nitrogen are related to surface runoff.

Both organic and inorganic forms of phosphorus exist in soil. The inorganic forms mainly are iron and aluminum phosphates in acid soils and calcium phosphates in alkaline soils. Any phosphorus added as fertilizer or released in the decomposition of organic matter rapidly is converted to one of these compounds. All inorganic forms of phosphate in soils are extremely insoluble and concentrations of phosphorus in soil water solutions are low, generally less than 0.2 mg/l. The phosphorus of environmental quality concern is associated with the soil itself and with the interchange of phosphorus from bottom deposits in bodies of water with the upper waters.

Soil management practices can reduce the losses of phosphorus caused by erosion. Losses of soil nitrogen are more difficult to control since soluble nitrate nitrogen can be transported through the soil. Soil conservation can reduce the quantity of organic nitrogen reaching surface waters by erosion.

Nutrient budget estimates of agricultural operations have indicated that in the humid East and parts of the West nutrients added to cropland are generally taken off in crops. Nitrogen applied in excess of the crop requirements can end up in surface waters or groundwaters or lost to the atmosphere by denitrification. In the prairie and plains area, residual fertility remains high and more nutrients are removed by the crops than are added as fertilizer. These high fertility soils can contribute nitrates to surface waters and groundwaters even in the absence of added fertilizers. In western states where irrigation is practiced, salts including nitrates tend to build up in the groundwater. Finally in the semi-arid and western states where intensive agriculture is not practiced, nutrients added to land may be about equal to the amount in the crop or brush (1).

Actual pollution caused by agriculture is difficult to document. The effect of agricultural contaminants can be observed in the number of fish kills that have occurred in the United States (Table 3.1) (2). Comparable data from other countries is less available. Information from Bavaria indicates that in one year, 10 out of 27 cases of fish kills originated from farm waste discharges (3).

The data in Table 3.1 represents an assessment of the source of pollution by the investigating organizations and as such are based on oral investigations and studies conducted after fish kills were noted. In the official reports, the term agriculture was used to describe fish kills that resulted from farming operations, i.e., use of insecticides and herbicides, fertilizers, and manure-silage drainage. Problems due to food production were included in the industrial waste category. In keeping with the terminology used in this book agricultural sources noted in Table 3.1 are more broadly defined to include both farming and food processing sources.

In 1967 and 1969, pollution from farming practices was the second largest source of pollution as recorded by the number of fish killed. Fertilizers were a minor cause of problems. Insecticides, food products, and manure-silage drainage caused greater problems than did fertilizers with insecticides and food products causing the greater problems in most years. The large number of fish killed by agricultural sources in 1969 and 1970 was due to food production waste discharges.

Natural differences in watershed characteristics such as rainfall, temper-

Table 3.1

Source of Fish Kills in the United States[a]

Source	1964	1965	1968	1969	1970
Total fish kill reports, all sources	385	446	438	465	634
Reports attributable to agricultural sources					
Farming					
Insecticides, etc.	93	74	51	80	63
Fertilizers	5	4	5	5	6
Manure, silage drainage	29	29	21	29	38
Food products	41	60	35	37	32
Percent of fish killed by farming sources	8	12	2.5	15	20
Primary source of fish kills in noted years					
Industrial	1	2	2	1	1
Municipal	2	1	1	4	2
Agricultural (farming[b])	3	3	4	2	4

[a] From reference 2.

[b] Wastes from food processing operations are classified as industrial wastes.

ature, drainage pattern, degree of erosion, soil permeability, and surface cover are important in determining the water quality of streams draining agricultural basins. The land is a natural treatment system for agricultural wastes. When proper management methods are employed, the contribution of agriculture to water pollution will be minimal.

Groundwater

Agricultural practices have the capacity to produce wastes that can enter the groundwater and alter its quality. The magnitude of agricultural contamination of groundwater is difficult to ascertain. There is limited information on the amount of agricultural contaminants that enter subsurface waters under given cropping and disposal practices and on the changes in groundwater quality caused by various agricultural contaminants. In addition, there is a scarcity of adequate groundwater monitoring programs to investigate these factors.

Of the rural water supply samples examined in Missouri, 50–75% contained sufficient nitrogen to be of concern in livestock production (4). The main contaminating source both in distribution and concentration was indicated to be waste matter at sites of animal habitation. Soil containing 200 to 4000 lb of nitrate nitrogen per acre was found below certain feedlots. Standard surface soils contained about 50 to 150 lb of nitrate nitrogen per acre. The increased nitrates remained after an area was abandoned from animal use.

Comparative studies in Colorado (5) and Wisconsin (6) have indicated that land use affected the nitrogen content of soils (Table 3.2). The highest soil nitrate nitrogen occurred under feedlots. The nitrogen content in virgin soil and under native grassland represents natural decomposition of organic matter as well as natural deposits in the soil rather than man's activities. The lower quantities below alfalfa are because of the ability of the crop to remove nitrates deep in the soil and the fact that this crop rarely is fertilized with nitrogen. Data of this type need not reflect groundwater nitrate concentrations since that concentration is a function of the water moving through the soil and the hydraulic characteristics of the aquifier. The nitrate nitrogen concentrations in the groundwaters under the lands studied in Colorado (Table 3.2) ranged from zero to more than 10 mg/l.

Water from shallow wells is more likely to contain contaminants than water from deeper wells, particularly shallow wells near barnyards, feeding lots, and manure piles. Occasionally, deep drilled wells contain considerable nitrate, either entering the well by surface leakage through poor well seals, or from nitrogen-rich deposits within lower soil zones.

Table 3.2

Soil Nitrogen Concentration in Agricultural Land

Land use	Average nitrate nitrogen (lb/acre)[a]	
Irrigated land, alfalfa	80	
Native grassland	90	
Nonirrigated cropland	260	
Irrigated cropland not alfalfa	506	
Feedlots	1436	
Land use	**Nitrate nitrogen (lb/acre)[b, c]**	**Ammonium nitrogen (lb/acre)[b, c]**
Virgin soil—Jack Pine	56	21
Cultivated	142	31
Barnyard	407	2200

[a] From reference 5. [b] 20-ft cores. [c] From reference 6.

At present, groundwater contamination from agricultural sources is, for the most part, below levels that have been demonstrated to cause disease, excessive removal costs, or aesthetic nuisance. However, the number of documented occurrences of groundwater contamination by agriculture continues to grow. In establishing enlarged agricultural operations and in choosing sites for the land disposal of wastes, the possibility of groundwater contamination should receive thorough investigation.

Sediment

About four billion tons of sediment wash into the streams and rivers of the United States each year (7). Approximately 75% of this material originates from forested and agricultural lands. This erosion loss is not distributed evenly throughout the United States. Losses are relatively small in areas of high density natural vegetation and larger in areas where agriculture is more intense.

The rates of erosion and resultant sediment have been accelerated by human use and management of lands, vegetation, and streams. Iowa data indicate that the soil loss in one county averaged 10 ton/acre/year since the land has been used agriculturally 125 years ago, whereas, before agricultural use the soil loss was about 1 ton/acre/year (8). In regions of erosive soils and well-defined drainage systems, 10–30 tons/acre/year are eroded if vegetative cover is poor. Generally, cultivated soils in warmer and more humid regions are more susceptible to erosion than soils in cooler, less humid areas. Erosion rates can increase due to agricultural activities because of removal of the natural vegetative cover of the land, plowing and cultivation which decrease the erosion resistance of the soil, and the trend to large planting and harvesting equipment which may not be compatible with soil conservation practices on small acreage and sloping land.

Sediment can reduce the recreational value of water, is a carrier of plant nutrients, crop chemicals, and plant and animal bacteria, and increases water treatment costs. Proper erosion control is a positive solution to sediment problems.

Soil erosion and sediment yield is a function of rainfall, soil properties, land slope length, steepness, and cropping practices. While little can be done to change the amount, distribution, and intensity of rainfall, measures can be taken to reduce the ability of rainfall to cause erosion. Examples include decreasing the amount and velocity of overland flow and decreasing the impact of the rain on the soil. Both flow and impact can be reduced by maintaining vegetative cover on the land through proper crops or mulch. The incorporation of crop residue and animal wastes in the soil increases soil porosity and aggregation and decreases runoff. Slowing runoff by vegetative cover and land modification (terraces, ponds, waterways) offers opportunities to provide greater infiltration time, conserve moisture, and decrease erosion. Wise land use planning and careful use and management of land can reduce erosion and control certain water quality problems arising from agricultural activities. Practices and structures to conserve soil and water are identical to those producing a reduction of pollutants found in agricultural runoff.

Agricultural Contributions

In the following sections, the characteristics of agricultural and other nonpoint sources are reported in both concentration units (mg/l) and potential area yield rates (kg/ha/year) whenever available data permits. The characteristics represent the reported data of many investigators and should not be utilized or extended beyond the conditions of the original study. Each situation will have its own runoff or leachate characteristics as well as potential control approaches.

The area yield rate information should be interpreted as a potential loading rate. It would be incorrect to multiply these yield rates by the area of a particular land use category and infer that the result is the actual amount of material entering a specific surface water body. The actual amount that enters a specific water body is a function of interrelated hydrologic, geochemical, agricultural, and human activities on the land.

The characteristics are presented only to indicate relative concentrations and rates and to compare the contributions from various sources. Application of the reported average characteristics is not likely to indicate the effect of an actual situation.

Forest Runoff. Forested areas represent areas that have not been grossly contaminated by man's activities. The runoff from these areas can indicate characteristics that result from natural conditions. However, the increased world demand for wood fiber has accelerated cultural production practices in managed forested areas. Practices such as forest fertilization, block cutting of mature trees, and other forest management practices will alter the characteristics of runoff from forested areas.

Data from the Hubbard Brook project (9) illustrate the effect of deforestation. All trees, saplings, and shrubs were cut in one area and dropped in place. No products were removed and care was taken to minimize surface erosion. Regrowth of vegetation was chemically inhibited. The annual runoff of water from the clear cut area exceeded the expected value of the undisturbed watershed by 39% during the first water year and by about 28% during the second water year after deforestation. Significant changes of stream water quality occurred as a result of deforestation. Ammonium, sulfate, and bicarbonate concentrations did not increase. The greatest increase in stream water ionic concentration was that of nitrate which increased by 41 times in the first year and by 56 times in the second year following deforestation. This type of deforestation is not typical in a managed forest but it does indicate that management of forest ecosystems can change the characteristics of streams in the area. In this experiment, nitrate concentrations have exceeded 10 mg/l for a continuing period and algal blooms have occurred. Other timbering or land development operations offer opportunities for water quality changes of this type.

The effect of forest fertilization on surface water quality has been reviewed (10). The leaching of nitrates into groundwater following fertilization was insignificant. Major loss of fertilizers to surface waters occurred by inadvertent application to water courses and compacted areas such as roads and trails. Nutrient problems caused by forest fertilization do not appear to be a problem under current conditions and practices.

Range Land. Where rainfall is sparse or the land not very fertile, intensive crop production is not practiced and few animals can be supported per acre of land. Some domestic animals are permitted to range over the land in search of available food. Range land, with its low level of animals and fertilized acreage, represents a natural situation. Runoff from these lands can represent background or "natural" contamination.

Runoff from range land that had low intensity agriculture and no evidence of chemical fertilizers contained 0.65 kg NO_3-N, 0.76 kg total P, and

0.024 kg soluble P per hectare per year (11). The characteristics of the runoff were due to the erosion and leaching of the soil in the area.

Rural and Crop Land. Rural land refers to land that is or could be under agricultural production. Examples include grassland and idle farm land. Crop land refers to land under cultivation.

Constituents of the runoff from rural and crop lands originate in rainfall, wastes from wildlife, leaf and plant residue decay, applied nutrients, nutrients and organic matter initially in the soil, and wastes from animals on pasture. It is complex to separate the natural from the other contaminant sources. To do so requires that water volumes be accurately measured, storm characteristics described, and representative sampling initiated. Constituents are released at varying rates from all soils and exposed geological formations. The natural weathering of rocks and minerals and the oxidation and leaching of organic matter will contribute to runoff and leachate characteristics even in the absence ot man's activities.

Topography plays a vital role in the quantity of nutrients discharged in land runoff. Vegetation on the land is a major factor in controlling the rate of runoff, subsequent erosion, and nutrient loss. Surface runoff waters from agricultural lands will have a higher load of soil particles especially clay and organic matter, a higher load of adsorbed phosphorus, and a lower load of soluble salts than will tile drainage water. In an area drained by tile, surface waters may be expected to receive both adsorbed and soluble phosphorus in times of high rainfall and runoff. The variation in crop land runoff can be wide. Data from a three-year study (12) demonstrated that the runoff from a cultivated wheat field contained 5–2070 mg/l of suspended solids, 30–160 mg/l of COD, 0.5–23 mg/l of BOD, 2.2–12.7 mg/l of total N, and 0.2–3.3 mg/l of total P.

Exposed plant residues and animal wastes will undergo weathering and leaching. The exposure of the residues prior to runoff will have an effect on the soluble material in the runoff. Snowmelt runoff occurs in the spring, frequently on frozen ground, and can carry a higher contaminant load than rainfall runoff that has an opportunity to infiltrate into the ground.

Base flow from agricultural land will contain only a small fraction of the potential nutrient loss from crop lands. In a rural area in which farms occupied 90% of the land, the average base flow nitrogen contribution was 1.25 kg/ha/year or about 3% of the nitrogen applied and the average base flow phosphorus contribution was 0.11 kg/ha/year or about 2% of that applied (13). Surface runoff and interflow during runoff can contribute greater amounts of nutrients to surface waters than can base flows.

Some nutrients leave crop lands. It does not necessarily follow that the same amount reaches surface waters and lakes. Any intermediate vegetation may remove most of the material in suspension and part of the material in solution.

The time distribution of nitrates in a number of rivers suggests that the nitrate concentration and load is the greatest when the flow is high, i.e., when rainfall has the greatest opportunity to leach the soluble nitrates from the soil. A seasonal variation in nitrate concentrations was found with the highest levels in surface waters found in spring and late summer. At many sampling stations, the nitrate concentrations closely paralleled stream flow (14). The relatively direct correlation between the two parameters indicated that the nitrate reached the stream seasonally by way of surface runoff, drainage, or interflow and originated from diffuse sources.

Similar correlations have been observed in an Iowa stream in which during low flow periods domestic and industrial waste waters were the principal sources of nitrogen to the stream (15). During high runoff periods large quantities of nitrogen entered the streams over extended periods. During the wet periods, phosphorus levels rose to as high as 19 times the level expected from municipal waste water alone. At low river flows, the phosphorus levels in the river were below that in the discharges of domestic and industrial waste waters. The phosphorus was bound to soil sediment and removed.

Data from a small agricultural watershed in Kansas indicated a relationship between total phosphate and nitrate concentrations and stream flow (16). Total phosphate concentrations were from 0.2 to 0.4 mg/l during low flows but increased to maximums of 2.4 to 4.0 mg/l during storm runoff. The phosphate load followed the turbidity pattern. The nitrate concentrations ranged from 0.2 to 4.2 mg/l as nitrate in low flows but increased to maximums of 40 to 45 mg/l during storm runoff.

Many factors affect the release of nitrate from the soil organic matter and the existing nitrogen pool in the soil. Conditions that favor decomposition of organic matter and oxidation of nitrogen are neutral to slightly basic conditions, warm temperatures, and aeration. When drainage losses increase so does the rate of loss of soluble constituents, such as nitrate, from the soil. A study of nitrate concentrations in English rivers (17) was unable to correlate nitrogen fertilizer usage in adjacent areas with the nitrate concentrations. A high proportion of the total nitrogen was carried by the rivers in a few periods of high flow.

In many cases, the nutrient concentration in a stream will vary with the flow in a predictable manner. A detailed study in an agricultural watershed noted that an equation of the type $L = AQ^{(B+1)}$ could be used to relate the load (L) carried by a stream in terms of kilograms/day to the stream flow (Q, ft^3/sec) (18). A and B were regression coefficients for the data. This equation was useful to describe the variation of magnesium, chlorides, total and soluble phosphorus, ammonia nitrogen, and soluble and particulate organic nitrogen. The variation of nitrate and nitrite nitrogen in the stream flow was related by an equation of the form $L = (A + BQ)\cdot(Q)$.

The watershed contained agricultural land including dairy farms (50% of the area), forest (35%), and the remaining 15% was water, residential, commercial, and recreational lands. The total basin population of 3500 included a small town of 1500.

Fertilized Lands. One of the requirements for satisfactory crop production is the availability of nutrients when the crop is growing. Intensive farming operations are designed to satisfy this requirement either by the addition of fertilizers or by the incorporation of readily decomposable organic material such as manures in the soil.

Crop fertilizers have been blamed as major contributors to nutrients in surface waters. Generally only 10–30% of the fertilizer phosphorus added to a soil is taken up by the following crop. The remaining applied phosphorus is converted to insoluble forms and may become a source of available phosphorus for crops in subsequent years. Rarely does phosphorus from commercial fertilizers or spread manures occur in groundwater except with very porous soils in areas with heavy rainfall or irrigation. Phosphorus added to soils as a fertilizer or released in the decomposition of organic matter will be rapidly converted to iron and aluminum phosphates in acid soils and to calcium phosphates in alkaline soils. The rate at which the phosphorus is converted to the insoluble form is regulated by the time taken for the phosphorus to come in contact with the soil. This is affected by the ground cover, rainfall rate, slope of the ground, and permeability of the soil. With proper erosion control measures, little of the applied phosphorus should reach surface waters. During low stream flows, the phosphorus inputs from agricultural runoff probably are limited to groundwater from porous soils.

The efficiency of applied nitrogen in terms of the increase in crop uptake per unit nitrogen applied is always less than one because of: (a) nitrogen uptake in the nonharvested plant parts, (b) denitrification in the soil, (c) available nutrients remaining in the soil, (d) loss by volatilization of applied ammonia, and (e) leaching into deeper soil horizons. Denitrification in soils is perhaps 10–20% of the total mineral nitrogen in the soil. Ammonia volatilization can be 5–10% of any ammonia applied as a fertilizer. Items (a) through (d) will contribute little to the nitrogen content in adjacent streams. On the average, no more than 50% of the applied nitrogen in fertilizer is recovered by grain crops. Grasses may recover 80% or more of the applied nitrogen. The recovery of the nitrogen varies with the season and the age of the crop.

Efforts have been made to attempt a direct relationship between fertilizer nitrogen applied and the concentration of nitrogen in stream and drainage waters. A direct relationship is not likely because of a large preexisting reservoir of nitrogen in the soil and the rate of biological transformation the reservoir can undergo. In the upper Rio Grande River, the nitrate concentration in the river had not increased although fertilizer use increased considerably over a 30-year period (19). Crop usage and denitrification were probable explanations for the disappearance of the applied nitrogen. In other areas where organic and inorganic nitrogen has accumulated for centuries, the nitrate concentration of the drainage waters are more a result of natural conditions than of fertilizer applications.

A study was made of the extent to which the increase in fertilizer usage in the Netherlands has affected the quality of groundwater over the years. It was not possible to determine whether observed increases in chlorides and nitrates were caused by the increased use of fertilizers or by the increase in well water demand. In the period from 1920 to 1966, the average chloride and nitrate concentration of unpurified groundwater in fertilized sandy soils increased by about 3 mg/l and 2.5 mg/l respectively. Over two-thirds of wells had no nitrates in their water during this period (20). During the same period, nitrogen fertilizer usage increased from 28 to 178 kg/ha/year. Denitrification was suggested as a principal reason for the low nitrate content of the groundwater. The change in nitrate and chloride content of the deeper groundwater in the Netherlands during a period of increased fertilizer usage was not large enough to reduce the quality of potable water prepared from the groundwater.

Conditions required to leach nitrates from soil depend upon the water percolating through the soil and upon the presence of excessive nitrate in the soil when percolation occurs. Situations conducive to nitrate leaching include conditions where infiltration exceeds water lost by evaporation and transpiration, where soils have low water-holding capacities, where irrigation is practiced, and where soils have high infiltration rates.

The best way to add fertilizer is as it is needed for crop growth. Attention has been called to the fact that fall application of fertilizers may not be used as effectively as that applied at crop planting time. Greater amounts of nitrogen may be needed in the fall than are required if the nitrogen is applied at planting and sidedressed (21). The nitrogen not incorporated into the crop will be lost in some manner and it is reasonable to expect that a fraction of the nitrogen will end up in the surface water or groundwater. The greater the excess of nitrogen above plant and soil bacteria needs, the greater the amount that may be in percolate water. A study on sandy soils indicated that about 88% of the fall added fertilizer nitrogen was lost to the succeeding spring crop (22). Nitrification and leaching were noted as the mechanisms for the loss even though the soil temperatures were below the optimum for nitrification.

Supplying fertilizer to increase crop yields is necessary for the economic production of low cost food. However, each succeeding increment of

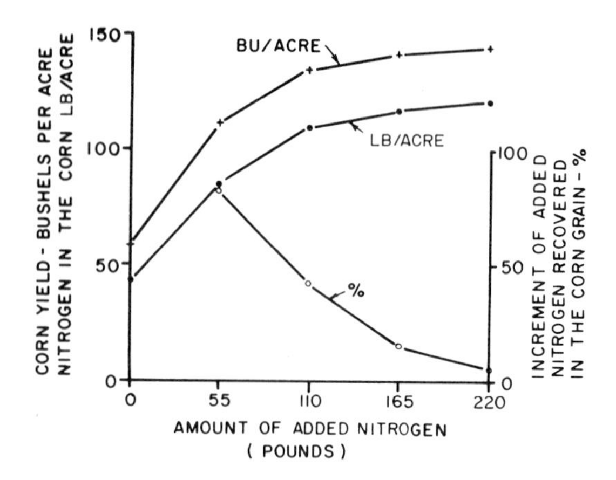

Fig. 3.1. Crop yield and nitrogen utilization; corn fertilization (23).

fertilizer is less effective for increasing yield. Smaller amounts of nutrients are recovered in the crop for each additional increment (Fig. 3.1). The cost of nitrogen relative to the value of additional crop produced is such that farmers are inclined to apply nitrogen nearly to the top of the yield response curve. At the point of maximum yield, the recovery of additional nitrogen is small.

It is evident that factors such as natural nitrogen content of the soil, soil type and topography, crop type and rate of fertilizer application, and climatic conditions complicate the relationship between fertilizer use and water quality changes. Aldrich (23) has provided a succinct summary of the influence of these factors on nitrates in surface waters and groundwaters. It is inevitable that there is a risk of nitrate loss during runoff and leaching. It is difficult to conceive of any approach that will reduce that risk to zero. It is possible to decrease the risk by proper application of inorganic fertilizers in periods when the crops are growing coupled with the use of slow release organic fertilizers.

Cropland Subsurface Drainage. The soils of many farms have poor drainage characteristics. Subsurface tile drains can be installed to permit better water movement and crop yields. The drainage from the lands will contain soluble constituents from the soil and materials added to the soils. The drainage can enter surface streams at many places and essentially is a nonpoint source of contaminants for the streams. Examples of the characteristics of such drainage are presented in Table 3.3 (24–26). In the Netherlands, the annual loss of nutrients on sandy soils averaged 32 kg N/ha/year and 0.6 kg P_2O_5/ha/year. These losses included leaching through tile drains, runoff, and erosion (24).

Table 3.3

Characteristics of Effluent from Drains under Cropland

Soil type[a]	P mg/l	P kg/ha/100 mm drainage water	N mg/l	N kg/ha/100 mm drainage water
Marine clay soil	0.06[b] (0–0.55)[c]	0.06	17 (2–53)	16.6
River clay soil	0.035 (0.01–0.06)	0.04	13 (1–23)	13
Sandy soil	0.02 (0.01–0.022)	0.02	25 (14–43)	25
Old cutover soil[d]	0.022 (0.01–0.035)	0.02	14 (5–18)	14
Young cutover soil	0.7 (0.035–1.1)	0.72	9.5 (2–17)	9.5

	No fertilizer (kg/ha/year)		Fertilizer added (kg/ha/year)	
Crop[e]	N	P	N	P
Corn	5.0	0.12	13.5	0.21
Oats–alfalfa	3.8	0.12	5.1	0.12
Alfalfa, first year	4.3	0.12	3.5	0.13
Alfalfa, second year	4.2	0.07	7.7	0.20
Continuous corn	5.9	0.23	12.5	0.26
Bluegrass sod	0.3	0.01	0.6	0.11

	River discharges from agricultural soils (kg/ha/year)[f]		
	England	Germany/Switzerland	
Land	N	N	P
Unused	4	3	0.9
Grassland	8	8	0.4
Arable	—	33	0.1
Clay	13	—	—

[a] In the Netherlands. From Sept. 1970–Mar. 1971. From reference 25.

[b] Average value.

[c] Range of values.

[d] Cutover soils are soils that have had the top organic peat layer removed for fuel; the remainder is an organic sandy soil.

[e] In the United States. From reference 26. Average of 1961–1967; 300 lb fertilizer per acre added to all crops; additional 100 lb per acre added to corn.

[f] From reference 24.

Pasture and Manured Lands. These lands contain either animals for production or are used for disposal of animal wastes. In either case, the runoff of these lands can contain contaminants from animal wastes. Where animals have direct access to streams, animal urine and feces can be discharged directly to these waters.

The data obtained by Howells *et al.* (27) illustrated the relative pollutional effects of manure disposal and pastured animals. Their major conclusions included: (a) Pollution indices for land drainage from manure spreading paralleled stream hydrographs with extended dragout on cessation of surface runoff; (b) the extent of water pollution from farm animal production units is more dependent on waste management methods than on the volume of the waste involved; (c) direct access of animal waste to surface waters should be prohibited; (d) points of animal concentration should be located away from streams and away from hillsides leading directly to streams; and (e) vegetation should be provided between areas of animal concentration and drainage paths or surface waters to intercept contaminants.

The access of animals and animal wastes to surface waters represents a possible controllable nonpoint pollution source. Animal manures should not be disposed of where rainfall or snow melt will result in their direct discharge to watercourses. The manure should be integrated with crop or disposal land shortly after disposal.

Land for Waste Disposal. The land will remain an acceptable point of disposal for most agricultural wastes. At many agricultural operations, the value of the nutrients contained in the wastes does not offset the investment and labor required to handle the wastes. Land disposal represents a least expensive method of disposal for most agricultural wastes.

Additional information is needed to define the proper crop and land management practice that can be used for waste disposal without causing additional environmental problems. Criteria need to be developed to provide a better understanding of the limitations of a soil to accept wastes. The maximum waste application rates will depend upon the type of soil, possible buildup of toxic materials in the soil, potential groundwater pollution, and available crop growth that can remove the added nutrients. Conservatively, the amount of nutrients in wastes disposed of on land probably should be no more than the amount that can be removed by the crop, weeds, or forest grown on the land.

The continued application of manure to the soil at rates which supply nitrogen in excess of the crop requirements may result in a loss of nitrate nitrogen from the root zone of the crop. This represents a potential groundwater problem although the magnitude of the problem will vary with water movement and type of soil. The annual application of 700 lb of nitrogen per acre in dairy manure slurry provided more nitrogen than the crop of orchard grass could use (28) and lesser rates were suggested. The manure was added with a subsoil injector about four inches below the soil surface.

Large applications of cattle wastes have depressed the yield of crops. The depressed yields were caused by accumulations of soluble salts, especially sodium, potassium, and ammonium (29). Significant accumulations of nitrate and total nitrogen occurred in the soil. The detrimental effects of excessive manure application may be reversed by continued cropping and adequate water penetration of the soil.

Manure Storage. Where manure cannot be disposed on land routinely, such as in the winter, it may be stacked or stored until conditions permit land disposal and integration with crop production. During storage, seepage can result which can contaminate surface waters. Table 3.4 (30) indicates the magnitude of contaminants from stacked dairy manure. Although the volume of seepage is small, the quantity of contaminants is not insignificant.

The release of stored manure seepage to surface waters should not be permitted. The seepage can be controlled by retention ponds and distribution on crop land in a nonpolluting manner. Even though manure seepage can occur in a large number of locations throughout the country, and as such approximates a nonpoint source of pollution, it can be considered as a controllable source of pollution.

Animal Feedlots. In many areas of the United States, animals remain an integral part of cropping systems and their wastes are returned to the fields

Table 3.4

Characteristics of Seepage from Stacked Dairy Cattle Manure and Bedding[a]

	Winter		Summer	
Parameter	Average	Range	Average	Range
Total solids (%)	2.8	1.8–4.3	2.3	1.7–2.9
Volatile solids (% TS)	55	52–59	53	50–58
Suspended solids (%)	0.35	0.2–0.8	0.24	0.2–0.3
BOD (mg/l)	13,800	4,200–31,000	10,300	4,400–21,700
COD (mg/l)	31,500	21,000–41,000	25,900	16,400–33,300
Total N (mg/l as N)	2,350	1,500–2,900	1,800	1,200–2,770
NH_4—N (mg/l)	1,600	980–1,980	1,330	780–2,200
Total P (mg/l as P)	280	64–560	190	90–340
Potassium (mg/l as K)	4,700	3,000–7,200	3,900	3,000–4,900
Total precipitation (inches)	15.0		9.4	
Seepage volume (gal/cow/day)	3.0		1.2	

[a] From reference 30.

that produce feed for the animals. In the Midwest and West, large feedlots now produce animals for slaughter and the manure tends to accumulate in the feeding areas. Until these wastes enter groundwaters or surface waters, they do not represent a serious water pollution problem. The characteristics of the livestock wastes are such that they will usually stay within a confinement area until the area is cleaned or until runoff washes them away. The water pollution problem associated with livestock wastes is a drainage problem. Runoff from uncovered confinement areas and from land used for disposal of waste occurs during and following rainfall.

Surface drainage from cattle feedlots and other areas where confined animals are housed in the open has caused considerable concern. In the early development of open confined animal operations, feedlots were situated in locations where the natural drainage facilitated the removal of the wastes. The water pollution potential was given little if any consideration. The intensification of the feedlot industry coupled with runoff pollution problems have caused interest in methods for minimizing runoff and preventing the runoff that does occur from reaching surface waters.

The quantity and quality of feedlot runoff will depend upon previous weather conditions, the number of cattle per feedlot area, the method of feedlot operation, soil characteristics, topography of the area, and intensity of rainfall. Actual documentation of the characteristics of feedlot runoff is rare because of the variable nature of rainfall–runoff relationships and the logistics of being at the proper sampling location when runoff occurs. In the past few years, studies that have included natural and simulated rainfall events have provided data on the characteristics of feedlot runoff under a variety of environmental conditions.

An early investigation illustrated the water quality changes that can occur when feedlot runoff reaches surface waters (31). Considerable length of streams were devoid of oxygen due to the runoff. BOD and ammonia concentrations were as high as 90 and 12 mg/l, respectively (Table 3.5). Ammonia concentrations from the runoff were detectable before other parameters. Such runoff provides little warning to downstream users and can trap game fish in the polluted waters.

Detailed studies on the environmental factors affecting the quantity and quality of feedlot runoff have been conducted. The greatest pollutant concentrations occurred during warm weather, during periods of low rainfall intensity, and when the manure had been wet (32). Ammonia nitrogen concentrations ranged from 16 to 140 mg/l, suspended solids concentrations from 1500 to 12,000 mg/l, and COD concentrations from 3000 to 11,000 mg/l in the runoff from these studies.

Part of the organic matter in animal manure is stabilized by bacterial action after it is deposited in the feedlot. Some nitrogen contained in the manure and urine can be lost to the atmosphere. About 50% of the organic

Table 3.5

Water Quality Changes Caused by Feedlot Runoff, Fox Creek, Kansas (mg/l)[a]

Time	DO[b]	BOD_5	COD	Cl	NH_3
Dry weather average	8.4	2	29	11	0.06
After rainfall (hr)					
13	7.2	8	37	19	12.0
20	0.8	90	283	50	5.3
26	5.9	22	63	35	—
46	6.8	5	40	31	0.44
69	4.2	7	43	26	0.02
117	6.2	3	22	25	0.08

[a] From reference 31.
[b] Dissolved oxygen.

matter can be decomposed into carbon dioxide and water. The amount of decomposition that takes place is dependent upon temperature and moisture conditions. The greater the bacterial action and the moisture content of the wastes, the greater the degree of solubilization and the greater the amount of soluble constituents in subsequent runoff. Because of a decrease in bacterial action in the winter, there will be a greater accumulation of wastes in an open feedlot during this period than during the rest of the year.

The loss of carbon and nitrogen from the waste by bacterial action increases the mineral content, reduces the organic content, and results in an accumulation of undegradable material. This residual material is not readily amenable to treatment by biological treatment methods and will result in a high COD/BOD ratio. While the manure is decomposing on the lot, it is constantly being mixed with fresh wastes and with soil if the feedlot is unsurfaced. The characteristics of the actual waste in a feedlot will be a mixture of these physical and biochemical changes.

The runoff from a surfaced (concrete) feedlot contained higher organic matter and nitrogen concentrations than did the runoff from an unsurfaced lot (32). It was postulated that the manure on the unsurfaced lot became mixed with the soil to form a structure which held the manure in position.

The highest concentration of pollutants in the feedlot runoff can occur in the initial runoff and decreases to "equilibrium" conditions as the runoff continues. Figure 3.2 illustrates the pattern for a number of rainfall rates and antecedent feedlot conditions. The "equilibrium" conditions were caused by the runoff continually dissolving the surface layer of manure on the lot. Once the ground surface was covered with manure, the depth of manure was not an important parameter of runoff water quality. The runoff was affected only by the surface of the manure on the lot.

The characteristics of runoff from other beef cattle feedlots (Fig. 3.3) (33)

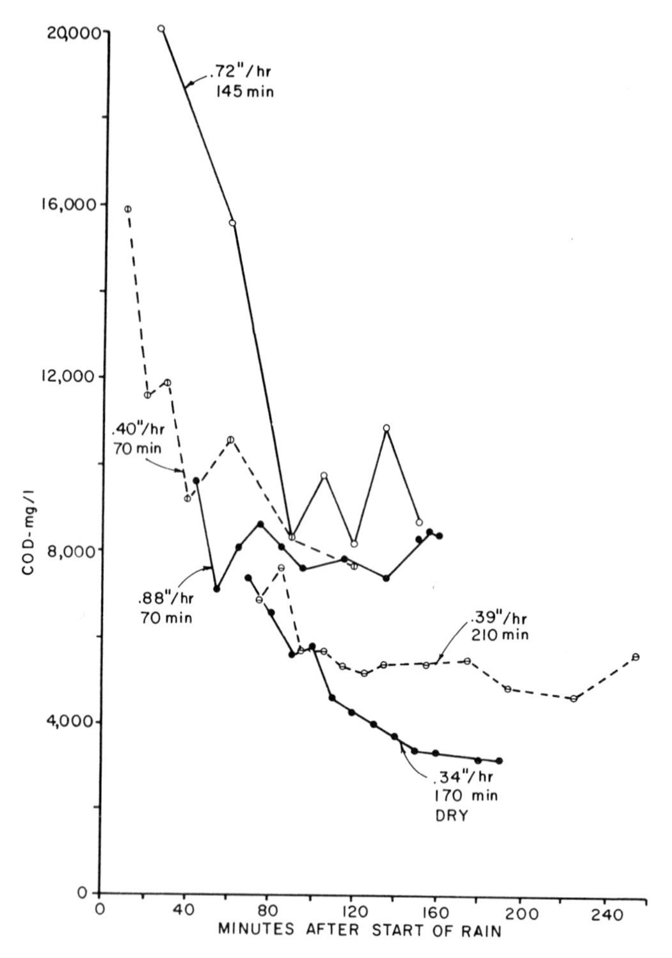

FIG. 3.2. The variation of feedlot runoff quality is affected by the rate and length of precipitation. Except as noted, the waste on the feedlot was slightly moist before the rainfall event. The first sample was collected as runoff began (prepared from data in Ref. 32).

showed even larger contaminant concentrations. Data from additional studies corroborate the strength of feedlot runoff. Reported characteristics are presented in Tables 3.6 and 3.7. An acre foot of feedlot runoff may contain 5–60 lb of NH_4—N, 2–23 lb of NO_3—N, and about 100 lb of P (40).

In addition to the beef cattle feedlots, there are a large number of cattle herds of less than 100 head of beef cattle. These herds are confined in barn lots in the winter and pastured in the summer. The barn lots have intermixed mud and manure as a surface in the wet, cool seasons and a hard

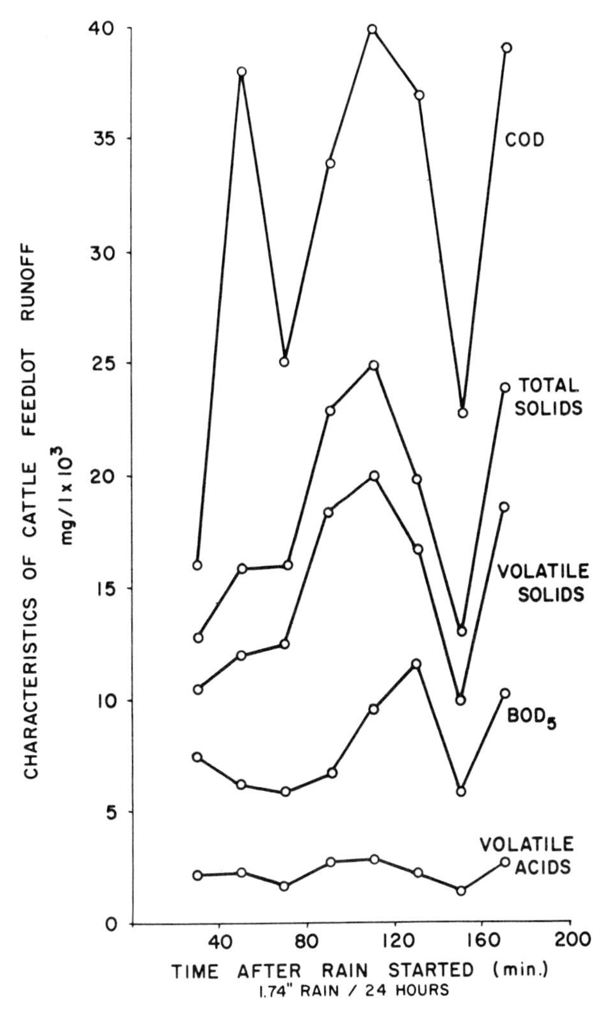

Fig. 3.3. Characteristics of natural runoff from a beef cattle feedlot. The feedlot surface was moist before the rain (33).

packed surface in the warm, dry periods. The runoff from barn lots also can be a source of surface water contamination. The runoff from a small barnlot housing 60 beef steers had the following range of characteristics: total solids, 0.1–1.2% on a wet basis; volatile solids, 17–38% of total solids; COD, 350–5650 mg/l; and BOD, 9–560 mg/l (41).

Runoff from feedlots can have a detrimental effect on surface waters. Runoff collection ponds can reduce the sediment load in such runoff and can produce an effluent with less variable characteristics. The quality of the

Table 3.6

Beef Cattle Feedlot Runoff Characteristics

Location and condition	Total solids	Volatile solids	COD	BOD	TKN	NH_4—N	NO_3—N	Total N	Total P	Alkalinity	Ref.
					Concentration (mg/l)						
Nebraska											
Snowmelt runoff	—	—	41,000[a] (14,100–77,000)[b]	—	—	780 (6–2,020)	17 (0–280)	2,100 (190–6,530)	290 (5–920)	—	34
Rainfall runoff	—	—	3,100 (1,300–8,200)	—	—	140 (2–1,240)	10 (0–220)	920 (11–8,590)	360 (4–5,200)	—	34
Texas											
Dirt lots	—	—	9,500 (2,900–28,000)	1,460 (1,010–2,200)	128 (9–280)	56 (2–85)	23 (0–103)	—	—	730 (100–1,400)	35 (Aug.–Nov. 1969)
Concrete lots	—	—	21,500 (8,400–32,800)	8,000 (3,300–12,700)	550 (70–1,070)	350 (33–775)	220 (0–880)	—	—	1,250 (85–2,400)	35
Colorado	—	100–7,000	—	300–6,000	—	—	—	—	—	—	36
Kansas	10,000–25,000	—	4,000–40,000	1,000–11,000	200–450	—	—	—	—	—	37
Texas	3,100–28,880	—	1,440–16,320	1,075–3,450	—	4–173	0–2.3	—	21–223	—	38
				Areal yield rate (kg/ha/year)							
Nebraska											
Snowmelt runoff											
100 ft²/head											
1969	60,000	31,000	—	—	—	—	—	1,600	620	—	34
1970	1,100	770	—	—	—	—	—	100	10	—	
200 ft²/head											
1969	14,000	7,200	—	—	—	—	—	450	200	—	34
1970	640	450	—	—	—	—	—	360	60	—	
Rainfall runoff											
100 ft²/head											
1969	7,500	3,400	—	—	—	—	—	330	22	—	34
1970	27,000	16,000	—	—	—	—	—	900	130	—	
200 ft²/head											
1969	7,300	3,400	—	—	—	—	—	220	33	—	34
1970	18,000	10,000	—	—	—	—	—	350	70	—	
South Dakota[c]	10,500 (1,600–23,200)	4,900 (800–10,000)	7,200 (720–16,000)	1,560 (135–3,800)	510 (46–3,110)	—	—	—	130 (9–470)	—	39

[a] Average. [b] Range of values. [c] Runoff from beef feeding, dairy confinement, and lamb finishing operations.

Table 3.7

Major Ions in Cattle Feedlot Runoff

Location	Ion	Mean (mg/l)	Range (mg/l)
Nebraska[a]	Na	840	40–2750
	K	2520	50–8250
	Ca	790	75–3460
	Mg	490	30–2350
	Zn	110	1–415
	Cu	7.6	0.6–28
	Fe	765	24–4170
	Mn	27	0.5–146
Texas[b]	Na	408	130–655
	K	760	226–1350
	Ca	700	194–1620
	Mg	69	28–89

[a] From reference 34.
[b] From reference 38.

effluent from runoff retention ponds is inadequate for release to streams or reservoirs without further treatment. Discharge of large quantities of runoff collected in retention ponds to a 45-acre flood control reservoir killed essentially all the game fish in the reservoir (42). The cause of the kill was zero dissolved oxygen concentrations and high free ammonia concentrations.

Although climatic conditions and size of beef feedlots are different in Canada, studies on the quality of feedlot runoff in Ontario have produced data similar to that noted in the United States. Runoff from the Canadian feedlots contained from 800 to 7500 mg/l BOD_5, 265 to 3400 mg/l total Kjeldahl nitrogen, and 93 to 2180 mg/l phosphorus as P_2O_5. The events causing the runoff were 0.3- to 1-inch rains with the higher concentrations occurring shortly after the larger rains (43).

Runoff from winter thawing conditions has contained greater waste concentrations than that contained in the runoff caused by rainfall under warmer conditions (44). Runoff from high density lots (100 ft^2/head) contained 130–170% greater runoff quantities and 4–5 times more total solids than did runoff from lower (200 ft^2/head) density lots.

The solids content of feedlot runoff is a function of the rainfall intensity and the degree of erosion that occurs on the feedlot surface. Phosphorus content is closely related to solids removal and therefore is directly affected by rainfall intensity. Ammonia and nitrate content is related to prior bacterial action in the manure pack. The concentration of ammonia and nitrate in feedlot runoff will decrease with continuing precipitation since these compounds are rapidly leached from the feedlot surface. BOD and COD content of feedlot runoff is related to the solids content which is affected by the rainfall intensity and to the prior bacterial action which affects the type of soluble components.

The concentrations of pollutants in feedlot runoff are relatively independent of the type of ration fed the animals. In the semiarid regions of the southwestern United States, cattle wastes are dehydrated in a short time and remain that way until wetted by precipitation. The wetting may reconstitute the wastes to almost original composition. In more humid climates, the wastes may remain moist for longer periods of time before natural drying occurs. The longer the wastes remain moist, the greater the opportunity for bacterial action and solubilization of the solid matter.

Pollution from an uncovered livestock area, such as a feedlot, is related to the fraction of precipitation that becomes runoff and reaches surface streams. Only after a portion of the rainfall soaks into the manure does runoff occur. This fraction will depend upon previous weather and precipitation conditions. The relationship between precipitation and runoff at feedlots has not been well defined but is needed to estimate the quantity of feedlot runoff that will occur for a given storm and therefore to design control systems for the runoff.

The general relationship relating precipitation (inches) and runoff (inches) from cattle feedlot surfaces is

$$\text{Runoff} = A \times \text{rainfall} - B \tag{3.1}$$

Results obtained from small surfaced and unsurfaced cattle feedlots are shown in Fig. 3.4. The data represent both natural and simulated rainfall events occurring over separate two- to three-hour periods from June to November in Kansas.

The coefficients of Eq. 3.1 will be affected by the moisture content of the manure pack prior to the rain. Values of the coefficients from different investigations are reported in Table 3.8.

The rainfall–runoff relationships at beef feedlots indicate that, after a minimum amount of rainfall, most of the rainfall ends up running off the feedlot. The surface of the feedlot acts much the same as a paved surface in this regard. These relationships were corroborated by a study which observed a 1:1 relationship between rainfall and runoff after runoff became constant (36). A dry hard crust frequently is observed below the manure surface. The silt, clay, and manure mixtures of feedlots result in the clogging of the porous soil voids and in the development of a less permeable layer. Below this developed hardpan, soil densities are less and permeability values larger.

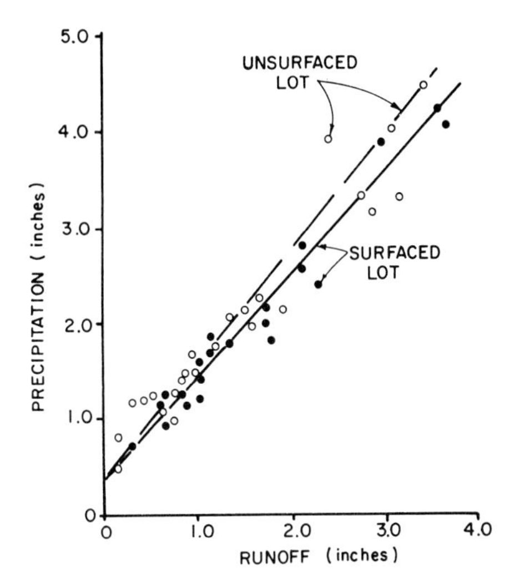

Fig. 3.4. Precipitation–runoff relationships for beef cattle feedlots (developed from data in Ref. 32).

These conditions suggest that infiltration to groundwaters can be neglected in determining the total runoff from operating feedlots. When feedlots are abandoned, the accumulated wastes may remain, the hard crust break up, and greater downward motion of pollutants may occur. The quantity of runoff from a wet manure pack is independent of its depth. The runoff quantity from dry manure is dependent upon manure depth, feedlot slope, and rainfall intensity.

A relatively dry manure pack may hold about 0.5 inches of rainfall per inch of manure depth (47). When runoff occurred from a relatively dry manure pack during high intensity storms, channelling was produced and erosion of the feedlot surface occurred. If the manure pack had about 60% moisture prior to runoff, runoff flowed primarily from the upper surface without appreciable penetration or erosion.

A difference was noted in the moisture retention capacity of the manure from animals fed different rations. The manure from feed containing 12% roughage had greater water-holding capacity than manure that resulted from an all concentrate ration feed (35). Data of this nature will vary depending on previous weather conditions and lot management operations.

Based upon detailed studies, the annual average BOD discharged per acre of feedlot was estimated to be 2500 lb for concrete feedlots and 1200 lb for nonsurfaced feedlots (32). On this basis the annual average pollutional population equivalent per acre of feedlot was estimated to be 40 for flat surfaced lots and 20 for similar nonsurfaced feedlots. Data from South

Table 3.8

Runoff and Rainfall Relationships on Beef Cattle Feedlots

A^a	B^a	Minimum rainfall to produce runoff (inches)	Conditions	Ref.
0.945	0.34	0.4	Surfaced lot	32
0.882	0.37	0.4	Unsurfaced lot	25
0.53	0.14	0.4–0.5	3 to 9% slopes	44
0.69	0.20	0.3	Data from 1.5-yr study	45
—	—	0.38	August to October	35
—	—	0.5	Barn lot	41
0.93	0.41	0.45	1968 runoff	40
0.45	0.05	0.5	1969 runoff	40
0.49	0.06	0.5	1970 runoff	40
0.57	0.14	0.25	—	46
0.59	0.13	0.22	—	46
0.50	0.12	0.2–0.32	1969–70	38

[a] Coefficients in the equation: runoff (inches) = $A \times$ rainfall (inches) $- B$.

Dakota (39) indicated that less than 5% of the oxygen demanding materials produced in a beef feedlot were removed by runoff. With this data, it was estimated that the population equivalent of the runoff was about 25. The cattle stocking rate in this study was about 250 head of cattle per acre. The pollutional effect of feedlot runoff occurs on an intermittent high rate basis with a potentially greater effect on surface waters than if it were released at a continuous lower rate.

In groundwater, high concentrations of dissolved solids originating from cattle feedlots normally are restricted to the vicinity of the feedlot especially in the drier parts of the country. This is due to the low permeability rates in the feedlots, to low rainfall, and to high runoff and evaporation rates. The quality of the groundwater near feedlots is closely related to the surface material of the feedlot and local soil characteristics. It was not found to be correlated with the number of cattle per feedlot area, feedlot slopes, or drainage basin surface area (48).

A study of nitrate nitrogen in soil under feedlots in Kansas noted accumulations from almost zero to 4540 lb/acre in a 4-m soil profile (49). The amount of accumulation was directly related to the age of the feedlot with the older feedlots having more nitrate nitrogen in the soil profile than did the younger lots. In areas with higher rainfall, nitrate nitrogen accumulated to greater depths than in areas with low rainfall. On soils of fairly constant composition, soil texture did not effect nitrate-nitrogen distribution patterns beneath feedlots markedly. In soil with varied profiles, as the clay content increased, moisture and nitrate-nitrogen content increased. Soil phosphorus analysis indicated little if any movement of phosphorus except on soils with an extremely low exchange capacity. Concentrations of up to 78 mg/l nitrate nitrogen were found in the groundwater below certain feedlots.

Other studies have indicated the quality of the groundwater in and around animal feeding operations. Under and adjacent to cattle feedlots, 10–18 mg/l of nitrate nitrogen was found in the groundwater (50). In areas removed from the feedlot but which were under farming, the groundwater contained 1–7 mg/l of nitrate nitrogen. Samples of groundwater under feedlots in the South Platte River Valley, an area containing most of the cattle in Colorado, has been observed to contain ammonium nitrogen up to 38 mg/l, organic carbon up to 300 mg/l, and to have had an offensive odor (51).

In general, there is little uniform or continuing movement of organic compounds from a feedlot manure pack into the subsoil profile or into groundwater. Wells in one feedlot area contained less organic material than did water from a local river (52). Although free and polymerized phenolics were identified in manure, none could be isolated from the organic polymers found in trace amounts in groundwater. At depths of 20 cm, the organic carbon content of the feedlot soil was indistinguishable from subsoil away from the feedlots.

Runoff and percolation are not the only ways that feedlot contaminants can affect water quality. Investigations have shown that ammonia volatilized from cattle feedlots can contribute to the nitrogen enrichment of surface water in the vicinity of the feedlots (53). Similar results can be expected from other accumulations of ammonia-containing wastes exposed to the atmosphere. The exact impact of increased levels of ammonia in the air around feedlots and other confined animal operations on surrounding crops and bodies of water is unknown at this time. In establishing enlarged agricultural operations and in choosing sites for the land disposal of wastes, the possibility of pollution from all sources should receive thorough investigation.

The water pollution potential of livestock feedlots is related to the waste production per animal, the number of animals in the confinement unit, days confined, frequency of cleaning, climate, waste characteristics, and waste degradation in the lots. The contribution of feedlot runoff to surface water pollution will be a function of the temperature, magnitude of rainfall, slope of the confinement area, surface area of the feedlot, type of lot surface, and management practices. Range-fattened cattle represent a smaller runoff pollution problem than feedlot cattle since they are more widely distributed on the land. As the density of animals per acre decreases, the wastes are less concentrated and the nature can absorb more of the wastes. The trend, however, is toward confinement livestock feeding.

Irrigation Return Flows. Irrigation return flow is any water that finds its way back to a water supply source after it has been diverted for irrigation purposes. The term includes: (a) tailwater runoff, the portion of the applied water that runs off the land surface; (b) seepage, water seeping from the distribution system; (c) deep percolation, applied water that contributes to groundwater recharge or to a subsurface drainage system; and (d) bypass water, water that is diverted but which is returned directly to the source of supply without being used for irrigation. The soluble salt concentration in the irrigation return flows is the major contaminant of concern. The environmental irrigation return flow problems primarily are located in the arid areas. However, both the salt and nutrient content of irrigation return flows can contribute to the contaminant load of water sources wherever irrigation is practiced.

An increase in soluble salts is an unavoidable result of the use of water for irrigation. As water is transpired by the crop and evaporated from the soil, the salts in the applied water are concentrated. Most of the salts in the applied irrigation water remain soluble in the soil solution. The amount of mineral uptake by crops rarely is significant in the overall salt balance of irrigated agriculture. Almost the entire amount of salts present in the applied irrigation water is contained in the soil solution. A favorable salt balance is obtained when the output of salts in the drainage water equals or exceeds the input to the soil. To maintain a favorable salt balance in the root zone of the irrigated crops, i.e., to prevent an accumulation of soluble salts in the soil, a more saline water must leave the land by percolating below the root zone to the groundwater or to a natural or artificial drainage system. The quantity of water applied for irrigation includes the amount required by the crop and a sufficient amount to leach the soluble salts from the soil.

The characteristics of surface and subsurface irrigation return flows will be different. Surface runoff from irrigated land generally will have the following characteristics: total dissolved salts only slightly greater than in the applied water; presence of precipitated or adsorbed contaminants will be a function of the amount of erosion; colloidal and sediment load will be greater than in subsurface flow; and characteristics will be variable for any given area.

The general characteristics of subsurface return flows will include: a greater concentration of dissolved solids than in the applied water; little colloidal or particulate matter; a different distribution of cations and anions than those in the applied water; a nitrate content that will depend upon the concentration in the applied water, the nitrogen content of the soil, and the nitrogen uptake by the crop; and low phosphorus concentrations. Examples of characteristics of irrigation drainage are noted in Table 3.9.

Table 3.9

Major Ions in Irrigation Return Flows[a]

	Irrigation season return flow[b]				Nonirrigation season return flow	
	Surface		Subsurface		Subsurface	
Ion	mg/l	kg/ha	mg/l	kg/ha	mg/l	kg/ha
HCO_3	56	110	266	1000	303	1250
CO_3	0.5	2.0	0	0	0	0
Cl	1	4.0	12	49	19	79
NO_3	1.1	4.3	11	46	11.3	47
PO_4, soluble	0.2	0.8	0.66	2.7	0.7	3.0
SO_4	5.4	21	39	160	50	206
Ca	10	40	44	182	51	212
Mg	5	20	20	81	24	99
Na	4.1	16	38	156	46	190
K	1.4	5.5	4.7	19	5.3	22

[a]From reference 54. [b]Irrigation season: April through December, Yakima River.

Fruit and Vegetable Processing. The food processing industry creates considerable quantities of liquid and solid wastes for disposal. Estimates vary on the quantity and quality of the food processing waste that is produced. Table 3.10 indicates reasonable characteristics of these wastes.

Food processing wastes not disposed of as solid waste are disposed of as part of the plant waste water with or without treatment. On a national basis, about 10% of the solid waste from food processing operations is disposed of in plant waste water. Of the total amount of food waste disposed of in this manner, vegetables generated 50%, seafood 30%, fruit 12%, and specialty items 8% (56). The distribution of these residuals among the major product categories is shown in Fig. 3.5.

Table 3.10

Characteristics of Canned and Frozen Fruit and Vegetable Processing (National Estimates)[a]

	Year			
Waste	1963	1968	1972	1977
Total flow (10^9 gal)	66–87	—	94	—
BOD (10^6 lb)	660	785	845	905
Suspended solids (10^6 lb)	750	890	960	1035
Total dissolved solids (10^6 lb)	710	845	910	980

[a]From reference 55.

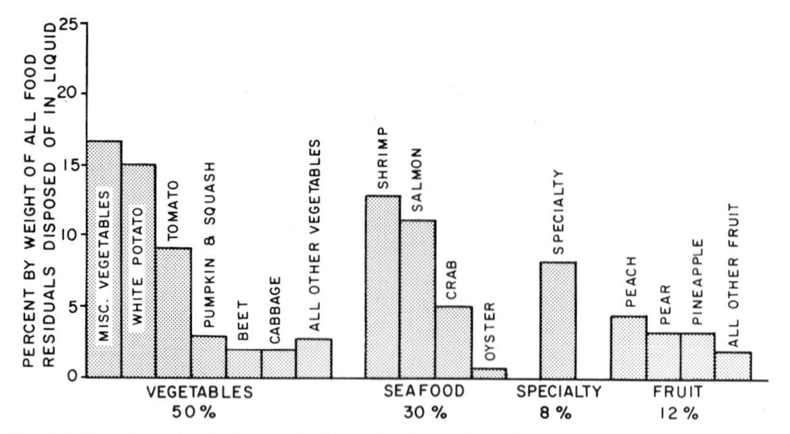

Fig. 3.5. Food residuals disposed of in a liquid medium for the major product lines (56).

These liquid wastes are discharged in the following manner; about 35–38% to public treatment systems, about 11–17% to land or ground disposal systems, and about 42–51% to surface waters (55). Discharges to the land are primarily by spray irrigation. Waste water discharged to surface waters usually received some degree of treatment before discharge. Plants in urban areas send almost all of their wastes to city systems, whereas plants in nonurban areas primarily use spray irrigation or treatment units such as oxidation ponds or aerated lagoons.

Food processing wastes are organic and can be treated by processes generally used at municipal treatment plants. If, however, these plants are not designed to treat the variable and high strength food processing wastes, the municipal plants may be overloaded and unable to attain the efficiency for which they were designed. Excessive food processing wastes in municipal plants can tax the solids handling and disposal facilities of a municipal waste treatment plant as well as the oxygen capacity of aerobic units.

The seasonal nature of the food processing industry compounds the problem of designing treatment facilities for the wastes or of joint treatment of the wastes in municipal treatment plants. The preponderance of processing activity takes place during the summer and fall months (Fig. 3.6) (57).

In food processing, the raw product becomes either primary product yield or food residues in a solid or liquid form. Knowledge about the residuals from the processing of a raw product permits an estimate of the wastes that can be expected from a given plant. Each type of product will have a different quantity and percent residual. Percent solid residuals vary from 5 to 10% for tomato processing, 30–35% for white potato processing, and 60–65% for corn processing (56). About 40% of the total weight of solid wastes produced by vegetable processing results from tomato processing. White potato processing produces about 25% of the total weight of these solid wastes, and corn about 15%.

The processing of citrus fruits dominates the fruit industry. About 35–40% of the citrus fruit ends up as solids residual in citrus processing. Of all the raw fruits processed, citrus residuals account for about 65% of the total weight of residuals produced in fruit processing. About 95% of total residuals from fruit processing originates from only four products, citrus, pineapple, peach, and apples.

In the seafood processing industry, percent residuals for different products range from 10 to 90%. Ranges of percent residuals obtained from processing specific products include tuna (15–20%), salmon (30–35%), and shrimp (50–55%). Tuna solid residuals account for almost 60% of the total weight of solid residuals produced by the processing of seafood.

About 9–10 million tons of total food residuals are generated each year.

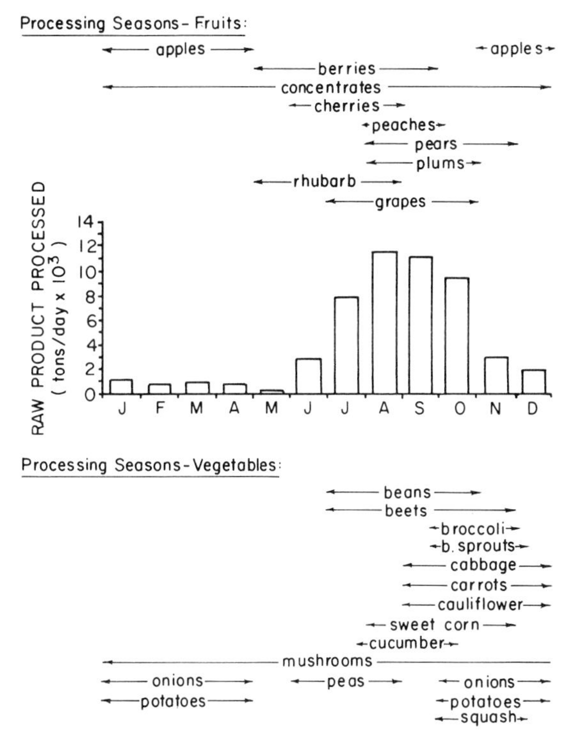

FIG. 3.6. Seasonal fruit and vegetable processing operations in Oregon (57). [Reprinted from *J. Environ. Qual.* **1** (1), 82 (1972), Fig. 1.]

Fruit processing accounts for 46%, vegetables processing 47%, specialty items 4%, and seafood 3% of this total. About 80% of these residuals are used for by-products of one type or another. The remaining 20% represents food processing wastes for disposal (56).

The distribution of these residuals disposed of as solid wastes in the major product categories is noted in Fig. 3.7. Almost all of the food processing solid wastes are disposed of by dumping or spreading on the land. Only about 1% of the solid wastes are burned on-site. The magnitude of these solid residuals represent a challenge to assure that they are disposed of in a manner that will not result in environmental quality problems.

Dairy Product Processing. The dairy food industry represents an important part of the food industry and contributes significant liquid wastes for treatment and disposal. There has been a decrease in the number of plants processing dairy foods with resultant increased waste loads per plant. About 11% of the total dairy plants in the country process about 65% of the total milk supply. Waste loads per plant range from 2000 to 10,000 lb BOD/day (58).

The per capita consumption of dairy products has changed. Consumption of low fat fluid milk, ice milk, sherberts, and cheese has increased at the expense of fluid whole milk, butter, fluid cream, and evaporated milk. These changes indicate the segments of the industry that will have increased waste problems.

Wastes from dairy plants consist primarily of varying quantities of milk solids, sanitizers, detergents, and human wastes. The wastes produced in urban plants are sent to municipal waste treatment facilities with a minimum of prior treatment. Dairy plant wastes produced in nonurban areas are

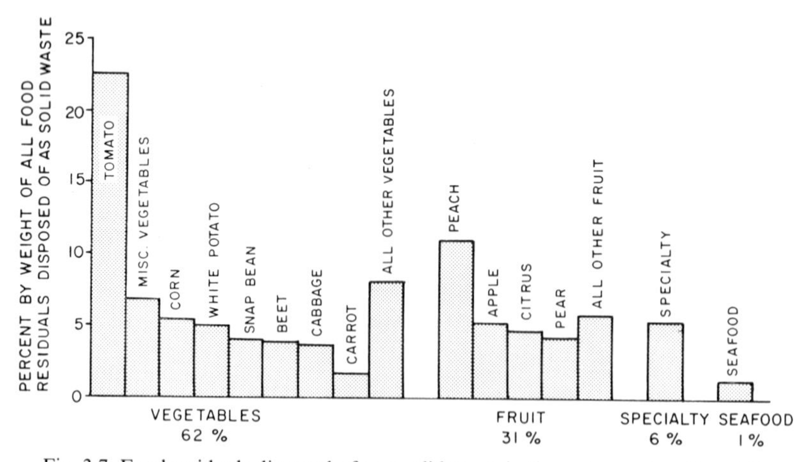

Fig. 3.7. Food residuals disposed of as a solid waste in the major project lines (56).

treated in aerated treatment facilities, by irrigation on the land, or by uncontrolled disposal to land and surface waters.

Approximately 32 billion gallons of wash water, 3.5 billion gallons of cheese whey, and 66 billion gallons of condenser water are discharged from the nation's dairy plants each year. Whey wastes contain about 4% solids and have a high oxygen demand (30,000–45,000 mg/l BOD) which makes disposal to streams undesirable. Only about one-third of the whey produced each year is used to any extent. The remainder is discharged from the plants. The nutritive content of whey offers interesting possibilities for by-product recovery and utilization if suitable equipment and markets are available.

Nonagricultural Sources

Agriculture can be both a point and a nonpoint source of contaminants to surface waters and groundwaters. Food processing plants and similar operations represent point sources of agricultural wastes. General agricultural runoff is a diffuse, nonpoint source that is difficult to control. Agriculture, however, is not the only source of contaminants in a watershed. Effluent from sewage treatment systems, septic tank drainage, urban land runoff, precipitation, the effluent from industries, and the runoff from agricultural land all contribute to the contaminants in the watershed. The relative contribution from each source will be a function of the type and mix of activities in a specific drainage basin. This section will attempt to present the characteristics of other nonpoint sources so that agricultural nonpoint sources can be placed in perspective.

Precipitation. The constituents of precipitation are influenced by both man-made and natural events. Fuel burning, automobiles, manufacturing operations, forest and other fires, volcanic eruptions, and wind erosion are examples of activities that contribute to the constituents in dry, dust fallout and in rain or snow that are deposited on the sample collecting surfaces. The distribution of contaminants between dry fallout and rainfall is rarely indicated by various investigations. Data are presented as reported in the literature and are assumed to be typical of bulk precipitation.

Available data indicate variable characteristics. Tables 3.11 and 3.12 summarize reported data in both concentration (mg/l) and area yield (kg/ha/year) units. Area yield rates are not constant but are a function of rainfall which is variable in both time and space. The most appropriate area yield rate would be in terms of weight of contaminant per surface area per unit time per unit of precipitation such as kg/ha/year/100 mm precipitation. Few investigators provide sufficient data to report yield rates in this manner. It would be useful if future data on the characteristics

Table 3.11

Precipitation Characteristics (mg/l)[a]

Constituent (mg/l)	References							
	12	12	59	24	9	60	61	62
Nitrogen								
NH_4—N	—	—	—	—	0.16	0.17–1.5	1.1	—
NO_3—N	—	—	0.14	—	0.33	0.56	1.15	—
Inorganic N[b]	0.7	0.9	—	—	—	—	—	—
Total N[b]	1.27	1.17	—	—	—	—	—	0.73
Phosphorus								
Total PO_4—P	—	—	—	0.39	—	—	0.02	0.04
Hydrolyzable PO_4—P	0.24	0.08	—	—	—	—	—	—
Suspended solids	13	11.7	—	—	—	—	—	—
COD	16	9	—	—	—	—	—	—
Major ions								
Ca	—	—	0.65	—	0.21	—	—	—
Cl	—	—	0.57	—	—	—	—	—
Na	—	—	0.56	—	0.13	—	—	—
K	—	—	0.11	—	0.05	—	—	—
Mg	—	—	0.14	—	0.04	—	—	—
SO_4	—	—	2.18	—	3.2	—	—	—
HCO_3	—	—	—	—	0	—	—	—
Year	1963–1964	1963–1964	—	—	1965–1968	—	4 years data	—
Environment or location	Urban	Rural	—	Mixed	Forest	—	Ohio	Florida

[a] Data are primarily yearly averages.
[b] Inorganic N = NH_4, NO_2, and NO_3—N; total N = all four forms.

Table 3.12

Precipitation Characteristics—Average Nitrogen and Phosphorus Area Yield Rates (kg/ha/year)[a]

Location	N		P	Ref.
	Total	$NO_3 + NH_4$—N		
World mean	6.2 (0.8–7.0)[c]	—	—	63 (before 1954)
World mean	8.7 (1.8–22.2)	—	—	63 (1905)
Temperate zone	—	6.8	—	63 (1938)
Humid temperature zone	5.6	—	—	63 (1960)
Europe and United States	—	0.8–2.1	—	63 (before 1952)
United Kingdom				
Upland	8.2	—	0.27	64
Northern	8.7–19	—	0.2–1.0	64
Netherlands				
Rural	8.5	—	—	24
Industrial	16–100	—	—	24
Canada				
Hamilton, Ontario	—	6.0	—	65 (1948)
Ottawa	7.7 (4.8–12.9)	—	—	65 (1924–1925)
Ceylon	—	12.9	—	65 (1941)
New York	10.0	—	—	63 (before 1948)
Ithaca, New York	—	7.4	0.05	66
Aurora, New York	—	7.6	0.06	66
Geneva, New York	—	8.3	0.05	66
Hubbard Brook, New Hampshire	—	5.8	0.10	67
Cincinnati, Ohio[b]	9.6	5.2	0.6	68

[a] 1.12 lb/acre = 1.0 kg/ha.
[b] Average U.S. rainfall of 30 inches/year assumed.
[c] Data in parentheses indicate range of data.

of precipitation would include total rainfall, area covered by the precipitation, and characteristics of the contaminants.

Soil and road dust are among the principal sources of metal ions in precipitation. Where there is little disruption of ground surface, as in forests, lower concentrations of these ions occur in the precipitation. Sulfate concentrations in precipitation have increased probably because of pollution of the air masses passing over an area (66, 69). The sulfate in such precipitation can result from sulfur dioxide and other sulfur products added by combustion. The sulfate and other ions placed in the atmosphere by man's activities at one location will return at other locations. Air movement patterns can assist in evaluating where these contaminants will be deposited by precipitation.

The sulfur in precipitation collected close to industrial locations was 10 to 15 times greater than that collected from urban locations. Sulfur in precipitation followed seasonal variations being lower in the summer and increasing in the winter (70). The amount of ammonia and organic nitrogen in precipitation was highest during the spring and lowest during the winter months and was higher in areas adjacent to barnyards than in areas removed from the barnyards. The distribution of nitrogen and sulfur in rural and urban situations are shown in Table 3.13. The average total sulfur and total nitrogen contributions caused by precipitation were estimated as 30 kg S/ha/year and 20 kg N/ha/year, respectively.

Both in terms of concentration and area yield rates, the characteristics of precipitation may not be insignificant. In terms of eutrophication potential the nitrogen and phosphorus contribution of precipitation should not be neglected. Precipitation is a variable and intermittent source of constituents in surface waters. Once contaminants are in precipitation, they are uncontrollable. Man, however, can exert some control over the contaminants that are released to the atmosphere through management practices that minimize particulates and gases from combustion processes, particulates from disturbing the land, and volatile gases from industrial operations.

Urban Land Drainage. Street litter, gas combustion products, ice control chemicals, rubber and metals lost from vehicles, decaying vegetation, domestic pet wastes, fallout of industrial and residential combustion products, and chemicals applied to lawns and parks can be sources of contaminants in urban runoff. A portion of the urban runoff can drain to sewerage systems while the remainder may reach surface waters by natural drainage channels without receiving treatment. The contaminant load to

Table 3.13

Average Sulfur and Nitrogen Loadings Due to Precipitation (kg/ha/year)[a]

	Rural			
	Adjacent to barnyard	Removed from barnyard	Urban	Industrial
Sulfur, as S	16		42	108
Nitrogen				
Organic N	14.4	7.5	6.2	—
NH_4—N	12.2	2.9	3.6	—
NO_3—N	3.5	2.7	3.7	—
Total N	30.2	13.1	13.5	—

[a] From reference 70. [Hoeft, R. G., Keeney, D. R., and Walsh, L. M. *J. Environ. Qual.* **1**, 203–208 (1972).]

streams or sewage treatment plants results in a slug load effect that the receptors may be unable to handle.

The characteristics and control of urban runoff have been studied and the fundamental factors delineated. Characteristics of urban runoff are presented in Table 3.14. Besides the conventional water pollution parameters, a number of other contaminants such as fecal coliform bacteria, chlorinated hydrocarbon and organic phosphate compounds, a number of heavy metals, and polychlorinated biphenyls have been found in urban runoff (78).

The major constituent of street surface contaminants was found to be inorganic, minerallike matter. The greatest portion of the pollutional potential was associated with the fine solids fraction of street surface contaminants. The quantity of contaminants per unit length of street increased as the time since the last street cleaning increased (78). Runoff from residential streets contained the highest concentrations of total phosphorus while runoff from arterial streets contained the highest concentrations of soluble phosphorus and runoff from arterial highways contained the highest concentration of nitrogen (71).

Comparison of Sources

It is difficult to compare data from various studies because of the variations in sampling methods, analytical methods, and the fact that all studies did not measure comparable parameters. The characteristics of nonpoint sources are the result of complex interactions in and on the soil making definitive comparisons of these nonpoint sources difficult. For a comparison of these sources, the order of magnitude of the characteristics and the differences between sources are more significant than the values themselves. The range of values reported for many of the sources has been summarized in Table 3.15 for different parameters and in Figs. 3.8 and 3.9 for total nitrogen and total phosphorus. All of the comparisons are presented in terms of concentration or yield rate potentially contributed *to* a surface water and are not in terms of the constituents *in* a surface water.

Comparing the data in terms of concentration units, precipitation, forest land runoff, and surface irrigation return flow have comparable values. The wide range in crop land phosphorus is due to the effect of erosion. With erosion control, the total phosphorus of cropland runoff should be near the lower end of the range. Subsurface irrigation return flow nitrogen concentrations are greater than those from surface irrigation return flows, undoubtedly because of the leaching of nitrogen compounds from the soil. Cropland tile drainage has nitrogen and phosphorus concentrations comparable to subsurface irrigation return flows.

The total phosphorus concentration of cropland runoff, irrigation re-

Table 3.14

Urban Land Drainage and Stormwater Overflow Characteristics[a]

Location	Total solids	Suspended solids	COD	BOD	TKN	NO_3—N	Total N	Soluble P	Total P	Cl	Ref.
				Concentration (mg/l)							
Cincinnati, Ohio	—	227 (5–1,200)	111 (20–610)	17 (1–173)	—	—	3.1 (0.3–75)	—	1.1 (0–7.3)	—	12 (1962–1964)
Seattle, Wash.	—	—	—	—	2.0	0.53	—	0.08	0.21	—	71
Tulsa, Okla.	545[e] (200–2.240)[f]	367 (84–2,050)	85 (42–138)	12 (8–18)	0.9[b] (0.4–1.5)	—	—	0.38 (0.18–1.2)	—	12 (2–46)	72
East Bay Sanitary District San Francisco, Calif.	1,400	1,400	—	87 (3–7,700)	—	—	—	—	—	5,100	73
Los Angeles County, Calif.	2,910	—	—	160	—	—	—	—	—	200	73
Washington, D.C.	—	2,100 (26–36,200)	—	126 (6–625)	—	—	—	—	—	42 (11–160)	73
Detroit, Mich.	(310–910)	(102–210)	—	(96–234)	—	—	—	—	—	—	73
Madison, Wisc.	—	(20–340)	—	—	—	(0.4–2.0)	—	(0.2–1.8)	(0.5–4.0)	—	74
Russia											
Rainwater runoff	—	(450–5,000)	—	(12–145)	—	—	—	—	—	(6–32)	75
Street washing water	—	(31–14,500)	—	(6–220)	—	—	—	—	—	(11–17)	75
Melting snow	—	(570–4,950)	—	(5–105)	—	—	—	—	—	(6–58)	75
Moscow	(1,000–3,500)	—	—	(186–285)	—	—	—	—	—	—	73
Leningrad	14,500	—	—	36	—	—	—	—	—	—	73
Stockholm	(30–8,000)	—	—	(17–80)	—	—	—	—	—	—	73
				Area yield rate (kg/ha/year)							
Cincinnati, Ohio											
Storm water runoff	—	640	310	47	—	—	8.8	—	1.1	—	68
Combined sewer overflow	—	280	—	—	—	—	6.8	—	5.6	—	68
Tulsa, Okla.	1,400 (550–5,700)	—	220 (67–530)	30 (13–54)	2.1 (1.2–4.0)	—	—	1.0 (0.4–3.0)	—	—	72
Detroit, Mich., (kg/ha)	—	220	—	100	8.8	0.2	—	2.1	4.0	—	76 (June Aug., 1965)
Ann Arbor, Mich. (kg/ha)	—	1230	—	35	1.2	0.9	—	0.3	1.0	—	76
Rock Creek, Md. (Potomac)[c]	—	—	—	—	2.9	12[d]	—	—	1.6	—	77

[a] Average characteristics, range of values shown in parentheses.
[b] Organic N.
[c] Urban, 60%; farm, 30%.
[d] $NO_2 + NO_3$—N.
[e] Average.
[f] Range of values.

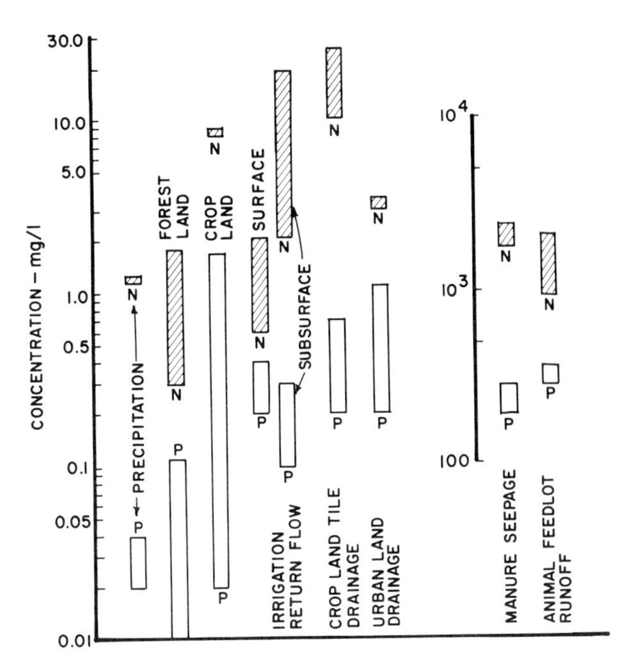

Fig. 3.8. Comparison of sources; range of total nitrogen and total phosphorus concentrations (mg/l).

turn flows, cropland tile drainage, and urban land drainage span comparable ranges. Because of the difficulties of comparing different studies, it is impossible to state that the concentrations of total phosphorus in these five nonpoint sources are likely to be grossly different. There is, however, a wide range of values for all nonpoint sources.

Animal feedlot runoff and manure seepage have characteristics that are many orders of magnitude greater than those of the other nonpoint sources.

The water pollution effect on a surface water can be evaluated better when the data are in terms of a yield per unit area since although constituent concentrations may be large, the volume of flow may be low. The data in Fig. 3.9 again illustrates the range of reported data but permits a reasonable comparison of the relative contribution from the various nonpoint sources. The total nitrogen yield from precipitation, forest land, crop land, land receiving manure, surface irrigation return flows, and urban land drainage span a comparable range. The total phosphorus yield of precipitation and range land are comparable and the total phosphorus yield of land receiving manure, surface irrigation return flows, and urban land drainage are comparable. The total phosphorus yield from forest land and cropland span a wide but comparable range, again undoubtedly due to the effect of erosion.

The nitrogen and phosphorus yield of animal feedlot runoff also is many orders of magnitude greater than that of the other nonpoint sources.

Although these comparisons cover wide ranges and may be considered gross, they do permit an assessment of the nonpoint sources that may require control. Actual decisions on the control of nonpoint sources should be made on the basis of the relative importance of the respective sources in specific locations and on what is technically and logically controllable.

Nonpoint sources also can be compared to other sources to assess their relative pollution contribution. A comparison of the nitrogen and phosphorus contribution from nonpoint sources and point sources such as municipal and industrial waste waters is presented in Table 3.16. Contributions of this type frequently are determined by difference, where only the characteristics of a portion of the inputs are known, and by inference from intermittent stream water quality information. Detailed mass and water balances are necessary to obtain contributions of greater accuracy. The comparison noted in Tables 3.16 and 3.17 should be used only to indicate relative relationships. The comparisons also indicate the problems inherent in applying pollution control technology to specific portions of the economy.

A study of the Great Ouse River in England (64) reported constituents

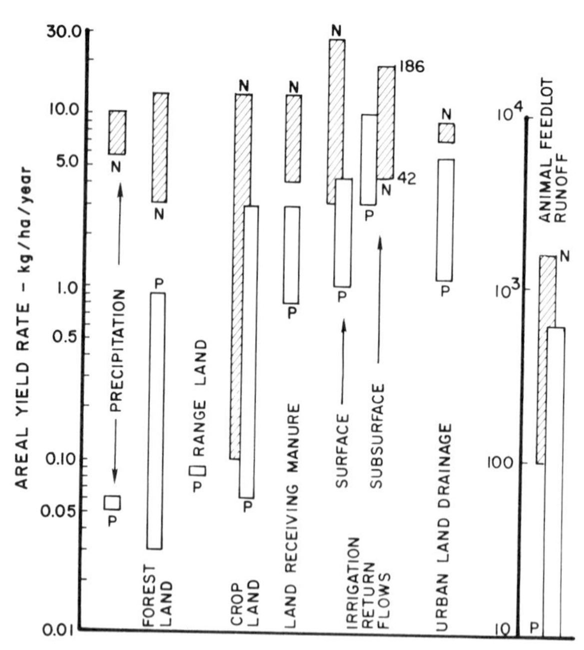

FIG. 3.9. Range of potential yield rates of total nitrogen and total phosphorus from various sources (kg/ha/year).

Table 3.15

Summary of Nonpoint Source Characteristics[a]

Source	Concentration (mg/l)					Area yield rate (kg/ha/year)					Surface area of interest
	COD	BOD	NO_3—N	Total N	Total P	COD	BOD	NO_3—N	Total N	Total P	
Precipitation	9–16	12–13	0.14–1.1	1.2–1.3	0.02–0.04	124	—	1.5–4.1	5.6–10	0.05–0.06	Total land area
Forested land	—	—	0.1–1.3	0.3–1.8	0.01–0.11	—	—	0.7–8.8	3–13	0.03–0.9	Forest area
Range land	—	—	—	—	—	—	—	0.7	—	0.08	Range land
Agricultural crop land	80	7	0.4	9	0.02–1.7	—	—	—	0.1–13	0.06–2.9	Active crop land
Land receiving manure	—	—	—	—	—	—	—	—	4–13	0.8–2.9	Crop or unused land used for manure disposal
Irrigation tile drainage, western United States											
Surface flow	—	—	0.4–1.5	0.6–2.2	0.2–0.4	—	—	—	3–27	1.0–4.4	Irrigated western soils
Subsurface drainage	—	—	1.8–19	2.1–19	0.1–0.3	—	—	83	42–186	3–10	Irrigated western soils
Crop land tile drainage	—	—	—	10–25	0.02–0.7	—	—	—	0.3–13	0.01–0.3	Active crop land requiring drainage
Urban land drainage	85–110	12–160	—	3	0.2–1.1	220–310	30–50	—	7–9	1.1–5.6	Urban land areas
Seepage from stacked manure	25,900–31,500	10,300–13,800	—	1,800–2,350	190–280	—	—	—	—	—	Manure holding area
Feedlot runoff	3,100–41,000	1,000–11,000	10–23	920–2,100	290–360	7,200	1,560	—	100–1,600	10–620	Confined, unenclosed animal holding areas

[a] Data do not reflect the extreme ranges caused by improper waste management or extreme storm conditions; the data represent the range of average values reported in previous tables.

Table 3.16

Comparison of Sources—Percent Contribution of Nitrogen and Phosphorus

Location	Item	Inputs						Ref.
			Land runoff			Waste water		
		Direct precipitation	Urban	Agricultural or rural	Forest	Municipal	Industrial (direct)	
Lake Erie[a]	Phosphate P	5.1	4.3	14.7	—	73	4.3	68
Lake Mendota, Wisconsin[b]	N	17	5	9	—	8 (Municipal and Industrial combined)		79
	P	2	17	42	—	36 (Municipal and Industrial combined)		
Wisconsin surface waters[d]	Total P	—	10	27	—	59[c]	—	80
Potomac River	Total P	—	1	8	4	87 (Municipal and Industrial combined)		77
Lake Canadarago, New York	Total P	2	52 (Urban, Agricultural and Forest combined)			46	0	18
	Soluble P	—	28 (Urban, Agricultural and Forest combined)			72	0	
	Total N	3	91 (Urban, Agricultural and Forest combined)			6	0	
Lake Ontario[e]	P	—	9 (Urban, Agricultural and Forest combined)			86	5	26

Lough Neagh, Ireland	P	—	7	56	—	37	—	81
Florida Lakes[f]								
Santa Rose	N	40	0	0	47	13	—	82
	P	59	0	0	30	11	—	
Santa Fe	N	37	7	13	41	2	—	82
	P	54	15	4	25	2	—	
Orange	N	17	1	25	57	0	—	82
	P	35	5	11	49	0	—	
Newman	N	15	8	20	56	1	—	
	P	26	22	10	41	1	—	
Hawthorne	N	13	36	5	14	32	—	82
	P	12	57	2	6	23	—	
Dora	N	6	4	74	1	15	—	82
	P	12	12	14	1	61	—	

[a] Excludes input from Lake Huron which was 12.6% of the total input.
[b] Input from nitrogen fixation was estimated as 14%, from groundwater as 45% for N and 2% for P.
[c] Includes input from private systems.
[d] Groundwater, precipitation, and other inputs estimated as 4%.
[e] Excludes input from Lake Erie.
[f] Except for Lake Dora, the municipal input was due to septic tank contributions.

Table 3.17

Estimated Quantity of Inputs from Various Sources

Source[a]	Nitrogen Tons/year	Nitrogen %
Industrial smoke	44,000	33
Auto exhaust	38,000	29
Domestic waste	16,500	12
Animal wastes	13,500	11
Dairy and poultry feed	13,000	10
Agricultural fertilizers	4,600	3
Nonagricultural fertilizer	2,800	2

Source[b]	N[c]	P[c]
Domestic waste	500–800	100–250
Industrial waste	>500	?
Rural runoff		
Agricultural land	750–7,500	60–600
Nonagricultural land	200–850	75– 85
Farm animal waste	>500	?
Urban runoff	55–550	6– 85
Rainfall	15–300	1.5–4.5

[a] In Connecticut. From reference 1. (Reprinted from *J. Amer. Water Works Ass.*, vol. 59 by permission of the Association. Copyrighted 1967 by the Amer. Water Works Ass.)
[b] In the United States. From reference 83.
[c] Data in thousand tons/year.

comparable to those in Table 3.15. Sewage contributed the bulk of the phosphorus but only a small fraction of the nitrogen, potassium, silicon, chlorides, and sulfates to the river. Domestic and industrial waste waters are major contributors of phosphorus in many lake and river basins.

Studies in the Potomac River Basin (77) demonstrated that 13% of the total phosphorus and 47% of the total nitrogen entering the surface waters of the Potomac came from land runoff. Nutrient loadings from land runoff were affected by land use. About 65% of the total nitrogen in the land runoff came from agricultural areas. In the Hudson River Basin, land runoff contributed 27% of the total phosphorus as PO_4 and 37% of the total nitrogen, as N, to the river above the New York City line (77). In both the Potomac and Hudson River basins, the major source of nutrients to the aquatic ecosystem was from waste water discharges.

Other sources of environmental constituents are compared in Table 3.17. These and previous comparisons indicate that the relative contribution in a watershed will be related to the type and magnitude of the activities controlled by man.

Summary

The review of available data illustrates the difficulty of quantifying the contribution of agricultural operations to characteristics of surface waters. The differences between the types of agricultural situations are apparent. Forest land runoff and range land runoff represent baseline, natural conditions that do not need to be controlled. The concentrations and yield of constituents from these sources are comparable to that of precipitation. Precipitation, forest runoff, and range land runoff can be considered as background nonpoint contaminant sources. Erosion control should be practiced in forests and on range land as with all other lands to avoid silt and associated contaminants from entering surface waters.

Nonpoint agricultural sources that may require control include cropland runoff, land receiving manure, cropland tile drainage, and irrigation return flow. Routine control of cropland runoff and runoff from land receiving manure technically is difficult and may not be needed in all cases. Each situation requires individual evaluation before deciding upon the need for control and the type of control if it is needed. In most cases, agricultural nonpoint sources can be controlled through proper management of the land rather than through the application of treatment techniques. When erosion of crop land is controlled, the contaminants in crop land runoff should be comparable to that of background nonpoint sources. If manure is disposed of on land under good crop and land management conditions, contaminants from these lands should be comparable to that of cropland.

Cropland tile drainage and irrigation return flows are difficult to intercept and control. The nitrogen content of this drainage is the major item of concern. Where the drainage will have a significant adverse effect on water quality, control of these sources will be necessary.

Nonpoint agricultural sources that should be controlled include manure seepage, feedlot runoff, and the direct discharge of animal wastes to surface waters. Manure seepage and feedlot runoff can contribute considerable pollutional loads. Nonsophisticated control procedures are available for the control of these sources and should be included in water pollution control policies.

Direct discharge of animal urine and manure can occur when animals have direct access to surface waters. Although control of wild animals is not possible, it is possible to fence confined animals, deny them access from streams, and supply them with water by other means. Where large numbers of confined animals have access to streams, it may be necessary to restrict their access as part of local or regional water pollution control policy.

Bacteria

General

The list of infectious disease organisms common to man and livestock is lengthy, including a number that can be water borne. When drainage or runoff from animal production units reach a watercourse, a potential chain for the spread of disease has been initiated. Salmonella organisms have been isolated from animal fecal specimens, runoff from animal confinement operations, carcasses of dead animals, and from waterholes from which the animals drank (84–86). Two organisms, *S. dublin* and *S. typhimurium* were the salmonella organisms most commonly found in the cattle and contaminated water investigated. *S. dublin* is essentially a pathogen of cattle but can cause meningitis and septicemia in humans. Children are more susceptible than adults. *S. typhimurium* can infect practically all species of birds, animals, and man. Animal wastes contain other organisms, such as *Staphylococcus aureus,* which also can be human pathogens. Evidence indicates, however, that the animal waste environment or the soil environment is not very hospitable to the survival of these pathogens. The disease potential inherent in the disposal of animal wastes on land is unknown but is considered not to be critical. Organism survival in soil is determined by the viability of organisms in the soil environment, their interaction with the soil, and the nature of their transport. Bacteria and viruses in soil have moved from inches to several hundred feet from the point of disposal on or in the land. Salmonella species have survived over 40 days in soils with high organic content. Survival was less in soils of lower organic content (87).

A study of the effectiveness of soil to remove bacteria in animal wastes indicated that coliform and enterococcus bacteria were removed by adsorption during soil percolation and by die-off because of the inability to compete against the established soil microflora (88). Greater than 98% removal occurred in 14 inches of soil. It was concluded that there was little concern that the bacteria would move any great distance from the point of application of the manures or manure water on undisturbed soil.

Coliform organisms are used as an index of water-borne disease hazard. The per capita coliform output for cows, hogs, sheep, and ducks ranged from 2.9 to 9.3 times that of man (89). Fecal coliform organisms are more indicative of pollution from human and animals. Fecal streptococci also have been suggested as a reliable and definitive measure of human or animal pollution. Enterococci counts of bovine manure have ranged from 3.5 to 17 million per gram while coliform counts have ranged from 0.3 to 0.6 million per gram (90). The total coliform, fecal coliform, and fecal streptococcus counts in cattle feedlot runoff averaged 105, 72, and 324 million organisms per 100 ml, respectively, for concrete surfaced lots (32).

The ratio of fecal coliform to fecal streptococci has been used to delineate between human and animal pollution. The ratio in human and domestic wastes was shown to be over 4 while for fecal material from animals, the ratio was less than 0.6. A ratio of 0.2 has been noted in cattle feedlot runoff (32). The bacterial types found in a lagoon treating bovine wastes have led one group of investigators to conclude that the pollutional potential for true *E. coli* and *S. faecalis* in lagoon waters is minor (90).

In the Snake River Basin, which contained at least 300,000 cattle over about 5300 square miles, the wastes from cattle feedlots were considered to be the principal source of the high bacterial counts in the Snake River (91). The coliform concentrations in the river averaged about 1000 MPN/100 ml and at times have been above 10,000 MPN/100 ml.

Summary

Agricultural operations such as livestock and crop production are natural contributors to bacteria in surface waters. Runoff from cropland can increase the bacteria in surface waters which must be removed in water treatment plants to produce potable water. This runoff rarely contains human pathogenic bacteria. The possible presence of human and animal pathogens in runoff and effluent from livestock waste treatment systems suggests that caution should be exercised in reusing such water without bacterial control.

Air Quality

General

In the broad sense, air pollution can be defined as the presence of air contaminants under conditions that are injurious to human, plant, or animal life or property or which unreasonably interfere with the enjoyment of life and property. Agriculture can be a recipient of the air contaminants of municipalities and industries and sustain the associated injuries. It is rare that agriculture is the producer of air contaminants that are injurious to plants, animals, humans, or property. Agriculture can, however, produce odors and dusts that interfere with the enjoyment of life and property and thus be a source of localized air pollution. The use of air pollution codes to stop the release of odors and dusts frequently hinges on the definition of the word "unreasonably." An agricultural industry that offends its neighbors with the production of air pollutants can be forced to cease if the nuisances are not corrected.

Odors

The subjective nature of odors is a difficulty in evaluating their offensiveness and effects. Individuals have varying personal response to odors based upon age, sex, smoking habits, occupation, mental attitudes, and respiratory differences. Each can have varying thresholds of detection. Regular and prolonged exposure to an odor can raise the detection threshold. However for any group of people, as the concentration of odor increases, the odor eventually will be detected and become obnoxious or discomforting.

In spite of the sophisticated technical equipment that has been developed, the human odor sensing system remains the most suitable detector available. Most odors are a complex mixture of chemical compounds. In some odors a specific compound such as ammonia or hydrogen sulfide can be predominant and detected instrumentally to obtain a measure of its strength. Because most odors are caused by small amounts of a number of compounds, a human odor panel remains one of the best odor detection and evaluation systems. Methods to develop odor evaluation panels have been standardized (92) and such panels are helpful for comparable studies.

Odors from animal wastes are a persistent problem arising from the confinement of large numbers of animals. The problem is especially prevalent near feedlots, in and around confined and enclosed animal production operations, and where field spreading of unstabilized waste from confinement operations is practiced. There is an acute need for effective methods of odor control if agricultural industries are to coexist with their neighbors. Effective odor control must be based on an understanding of the fundamentals of odor generation and control and on a knowledge of the odor-causing compounds.

The conditions under which the odorous compounds are produced offer suggestions as to possible control measures. With current practice, most agricultural wastes are held in piles or in storage tanks or pits prior to disposal. The storage period varies from weeks to years depending on the flexibility desired by the operator. The environmental conditions during such storage are uncontrolled and anaerobic microbial waste decomposition products such as ammonia, sulfides, mercaptans, amines, organic acids, and methane will result. Many of the unoxidized compounds have their own unique and, even in low concentrations, objectionable odors.

Some of the odorous compounds are tenacious, can cling to clothing and other articles, persist for long periods, and carry great distances.

The fact that the odor-causing materials are reduced organic compounds suggests the possibility of an oxidative microbial process for odor control. If the stored wastes are aerated, adequate oxygen can be added so that anaerobic end products are not produced. Under these conditions the primary odorous compound will be ammonia since during the aeration of animal wastes, ammonia can be released from the mixture. If sufficient oxygen is added to the aerated mixture the ammonia can be microbially oxidized to nitrites and nitrates eliminating ammonia as an air pollution problem in these circumstances. The key in reducing anaerobically caused odors is to be sure that the oxygen supplied is equal to or greater than the oxygen demand. If the oxygen supply is less than that of the demand, objectionable odors may still result.

The greater part of the odor associated with confined animal housing originates from the manure. Liquid manure handling and disposal have generated many of the odor nuisance conditions that have resulted in complaints. Anaerobic liquid systems, in which the waste is diluted with water for ease of handling, provide a better environment for the production of the odorous compounds than does a "dry" or an adequately aerated system. Many people have observed and research has documented that odors arising from diluted, anaerobic "liquid" manure are more offensive than odor from the undiluted, fresh manure.

Compounds identified as part of the odors from anaerobic poultry manure included ammonia, hydrogen sulfate, two- to five-carbon organic acids, indole, skatole, diketones, mercaptans, and sulfides (93). An odor panel indicated that the organic acids, mercaptans, and sulfides were the major malodorous components of the odor. Ammonia, methylamine, ethylamine, and triethylamine were components of hog house odors (94).

Odor causing compounds exist not only as gases but also as volatile material associated with the particulate matter in the animal confinement operations. Analyses of the volatiles carried by the particulates in a poultry house revealed a number of odor causing compounds. The concentration of particulate matter in a poultry house averaged 0.093 mg/ft^3, consisted of fecal matter, feed particles, and feather and epidermal fragments, and when collected had a "chicken house" odor (95). Filters can remove the particulate material and reduce the intensity of the odor.

Gases

The gases generated by microbial degradation of the wastes are of concern to the health of animals and humans in confined livestock environments. The gases of greatest concern are carbon dioxide, ammonia, hydrogen sulfide, and methane. In addition, aliphatic aldehydes and sulfur dioxide were measured in the atmosphere of a commercial poultry house (93). Under conditions with adequate ventilation, there has been no evidence that these gases reached harmful concentrations for animals or humans. Under abnormal conditions, gas concentrations have exceeded threshold values and have had adverse effects on animals (96, 97).

The conditions leading to accidents have been animals in close proximity to anaerobic wastes, inadequate ventilation, and the release of anaerobic gases as the anaerobic wastes were agitated for movement or removal. Toxic gas concentrations can exist in a waste storage unit and the unit should be adequately ventilated prior to human entry after mixing anaerobic wastes.

Other Effects

Flies can be a nuisance in and around agricultural production and waste treatment facilities. Flies are attracted to freshly excreted animal waste if the moisture content ranges from 55 to 85%. Fly production can be controlled by covering manure pits.

Effluents from animal production waste treatment facilities have a yellow to brown color. An effluent with this characteristic may be aesthetically undesirable and, depending upon the amount of dilution in the receiving stream, may cause color problems for downstream users.

Dusts result from feed, bedding, manures, and the livestock in animal production. In a poultry house, the quantity of dust varies with the type of feed used and type of litter used for broilers. Dust production is inversely proportional to the relative humidity, is less in cages than on litter for poultry, and is greater during periods of illumination than of darkness in animal confinement operations. In cattle feedlots, dust production is related to the lot cleaning operations and the dryness of the manure. In food processing operations it is a function of the movement of the crop and the type of processing that is done. Dusts can be an irritant to animals housed and humans working in enclosed facilities and a nuisance to neighbors.

All of the environmental impacts described in this chapter can be minimized or eliminated by the use of proper production and waste management methods. Common sense and the elevation of the concept of environmental impact to a higher level of consciousness in agricultural operations are important first steps in the control of real or potential environmental quality problems caused by agriculture.

References

1. Frink, C. R., Plant nutrients and water quality. *Agr. Sci. Rev.* pp. 11–25, Sec. Quarter (1971).
2. Anonymous, "Pollution Caused Fish Kills." Federal Water Pollution Control Administration, Federal Water Quality Administration, Dept. of Interior, and the Environmental Protection Agency, Washington, D.C., 1964–1970.
3. Sheltinga, H., Farm wastes. *J. Inst. Water Pollut. Contr.* **68,** 403–413 (1969).
4. Smith, G. E., Fertilizer nutrients as contaminants in water supplies. *In* "Agriculture and the Quality of Our Environment," Publ. No. 85, pp. 173–186. Amer. Ass. Advan. Sci., Washington, D.C., 1967.
5. Stewart, B. A., Viets, F. G., Jr., and Hutchinson, G. L., Agriculture's effect on nitrate pollution of ground water. *J. Soil Water Conserv.* **23,** 13–15 (1968).
6. Beatty, M. T., Kerrigan, J. E., and Porter, W. K., "What and Where are the Critical Situations with Farm Animal Wastes and By-Products in Wisconsin?," Proc. Farm Anim. Wastes and By-Prod. Manage. Conf., pp. 36–57. University of Wisconsin, Madison, 1969.
7. Wadleigh, C. H., Wastes in relation to agriculture and forestry. *U.S., Dep. Agr., Misc. Publ.* **1065** (1968).
8. Johnson, H. P., and Moldenhaver, W. C., Pollution by sediment: Sources and the detachment and transport process. *In* "Agricultural Practices and Water Quality," DAST 26, 13040 EYX, pp. 3–20 Federal Water Pollution Control Administration, Dept. of Interior, 1968.
9. Likens, G. E., Bormann, F. H., Johnson, N. M., Fisher, D. W., and Pierce, R. S., Effects of forest cutting and herbicide treatment on nutrient budgets in the Hubbard Brook watershed. *Ecol. Monogr.* **40,** 23–47 (1970).
10. Groman, W. A., "Forest Fertilization: A State of the Art Review and Description of Environmental Effects," Environ. Protect. Technol. Ser., EPA-R2-72-016. National Environmental Research Center, Environmental Protection Agency, Corvallis, Oregon, 1972.
11. Campbell, F. R., and Webber, L. R., Contribution of range land runoff to Lake Eutrophication. *Int. Water Pollut. Res. Conf., 5th,* San Francisco (1970).
12. Weibel, S. R., Weidner, R. B., Cohen, J. M., and Christianson, A. G., Pesticides and other contaminants from rainfall and runoff as observed in Ohio. *J. Amer. Water Works Ass.,* **58,** 1075–1084 (1966).
13. Minshall, N., Nichols, M. S., and Witzel, S. A., Plant nutrients in base flows of streams. *Water Resour. Res.* **25,** 706–713 (1969).
14. Harmeson, R. H., Sollo, F. W., and Larson, T. E., The nitrate situation in Illinois. *J. Amer. Water Works Ass.,* **63,** 303–310 (1971).
15. Baumann, E. R., and Kelman, S., Effects of agricultural pollutants on municipal uses of surface waters. *In* "Agricultural Practices and Water Quality," DAST 26, 13040 EYX. pp. 344–364. Federal Water Pollution Control Administration, 1969.
16. Stoltenburg, G. A., "Water Quality in an Agricultural Watershed," Trans. 20th Annu. Conf. Sanit. Eng. University of Kansas, Lawrence, 1970.
17. Tomlinson, T. E., Trends in nitrate concentrations in English rivers in relation to fertilizer use. *Water Treat. Exam.* **19,** 277–293 (1970).
18. Hetling, L. J., and Sykes, R. M., Sources of nutrients in Canadarago Lake. *J. Water Pollut. Contr. Fed.* **45,** 145–156 (1973).

19. Bower, C. A., and Wilcox, L. V., Nitrate content of the Upper Rio Grande as influenced by nitrogen fertilization of adjacent irrigated lands. *Soil Sci. Soc. Amer., Proc.* **33,** 971–973 (1969).
20. Kolenbrander, G. J., Does leaching of fertilizers affect the quality of ground water at the waterworks? *Stikstof* **15,** 8–15 (1972).
21. Lathwell, D. J., Bouldin, D. R., and Reid, W. S., "Effects of Nitrogen Fertilizer Applications in Agriculture," Proc. Agr. Waste Manage. Conf., pp. 192–206. Cornell University, Ithaca, New York, 1970.
22. Krause, H. H., and Batsch, W., Movement of fall applied nitrogen in sandy soil. *Can. J. Soil. Sci.* **48,** 313–365 (1968).
23. Aldrich, S. A., "The Influence of Cropping Patterns, Soil Management, and Fertilizers on Nitrates," Proc. 12th Sanit. Eng. Conf., pp. 153–176. University of Illinois, Urbana, 1970.
24. Kolenbrander, G. J., The eutrophication of surface waters by agriculture and the urban population. *Stikstof* **15,** 56–67 (1972).
25. Henkens, H., Fertilizer and the quality of surface waters. *Stikstof* **15,** 28–40 (1972).
26. Cywin, A., and Ward, D., "Agricultural Pollution of the Great Lakes Basin," Rep. No. 13020. Environmental Protection Agency, 1971.
27. Howells, D. H., Kriz, G. J., and Robbins, J. W. D., "Role of Animal Wastes in Agricultural Land Runoff," Final Rep., Proj. 13020 DGX. Environmental Protection Agency, 1971.
28. Marriott, L. F., and Bartlett, H. D., "Contribution of Animal Waste to Nitrate Nitrogen in the Soil," Proc. Agr. Waste Manage. Conf., pp. 435–440. Cornell University, Ithaca, New York, 1972.
29. Murphy, L. S., Wallingford, G. W., Powers, W. L., and Manges, H. L., "Effect of Solid Beef Feedlot Wastes on Soil Conditions and Plant Growth," Proc. Agr. Waste Manage. Conf., pp. 449–464. Cornell University, Ithaca, New York, 1972.
30. Cramer, C. O., Converse, J. C., Tenpas, G. H., and Schlough, D. A., The design of solid manure storages for dairy herds. *Winter Meet., Amer. Soc. Agr. Eng., 1971* Paper 71-910 (1971).
31. Smith, S. M., and Miner, J. R., "Stream Pollution from Feedlot Runoff," Trans. 14th Annu. Conf. Sanit. Eng., pp. 18–25. University of Kansas, Lawrence, 1964.
32. Miner, J. R., Water pollution potential of cattle feedlot runoff. Ph.D. Thesis, Kansas State University, Manhattan, Kansas, 1967.
33. Loehr, R. C., "Treatment of Wastes from Beef Cattle Feedlots—Field Results," Proc. Agr. Waste Manage. Conf., pp. 225–241. Cornell University, Ithaca, New York, 1969.
34. McCalla, T. M., Ellis, J. R., Gilbertson, C. B., and Woods, W. R., "Chemical Studies of Solids, Runoff, Soil Profile and Groundwater from Beef Cattle Feedlots at Mead, Nebraska," Proc. Agr. Waste Manage. Conf., pp. 211–223. Cornell University, Ithaca, New York, 1972.
35. Wells, D. M., Grub, W., Albin, R. C., Meenaghan, G. F., and Coleman, E., Control of water pollution from southwestern cattle feedlots. *Int. Water Pollut. Conf., 5th,* San Francisco (1970).
36. Norton, T. E., and Hansen, R. W., "Cattle Feedlot Water Quality Hydrology," Proc. Agr. Waste Manage. Conf., pp. 203–216. Cornell University, Ithaca, New York, 1969.
37. Loehr, R. C., Drainage and Pollution from Beef Cattle Feedlots. *J. Sanit. Eng. Div., Amer. Soc. Civil Eng.* **96,** SA6, 1295–1309 (1970).
38. Kreis, R. D., Scalf, M. R., and McNabb, J. F., "Characteristics of Rainfall Runoff from a Beef Cattle Feedlot," Final Rep., Proj. 13040 FHP, Environ. Protect. Tech. Ser. EPA-R2-72-061. Environmental Protection Agency, 1972.

39. Madden, J. M., and Dornbush, J. N., Measurement of runoff and runoff carried waste from commercial feedlots. *In* "Livestock Waste Management and Pollution Abatement," Publ. PROC.-271, pp. 44–47. Amer. Soc. Agr. Eng., 1971.
40. Swanson, N. P., Mielke, L. N., Lorimore, J. C., McCalla, T. M., and Ellis, S. R., Transport of pollutants from sloping cattle feedlots as affected by rainfall intensity duration, and recurrence. *In* "Livestock Waste Management and Pollution Abatement," Publ. PROC-271, pp. 51–55. Amer. Soc. Agr. Eng., 1971.
41. White, R. K., and Edwards, W. M., "Beef Barnlot Runoff and Stream Water Quality," Proc. Agr. Waste Manage. Conf., pp. 225–235. Cornell University, Ithaca, New York, 1972.
42. Scalf, M. R., Duffer, W. R., and Kreis, R. D., Characteristics and effects of cattle feedlot runoff. *Proc. Purdue Ind. Waste Conf.* **25,** 855–864 (1970).
43. Townshend, A. R., Janse, J. F., and Black, S. A., Beef feedlot operations in Ontario. *J. Water Pollut. Contr. Fed.* **42,** 195–208 (1970).
44. Gilbertson, C. B., McCalla, T. M., Ellis, J. R., Cross, O. E., and Woods, W. L., The effect of animal density and slope on characteristics of runoff, solid wastes, and nitrate movement on unpaved feedlots. *J. Water Pollut. Contr. Fed.* **43,** 483–493 (1971).
45. Gilbertson, C. B., McCalla, T. M., Ellis, J. R., and Woods, W. R., Methods of removing settleable solids from outdoor beef cattle feedlot runoff. *Ann. Meet., Amer. Soc. Agr. Eng., 1970* Paper 70-420 (1970).
46. Manges, H. L., Schmid, L. A., and Murphy, L. S., Land disposal of cattle feedlot wastes. *In* "Livestock Waste Management and Pollution Abatement," Publ. PROC-271, pp. 62–65. Amer. Soc. Agr. Eng., 1971.
47. Keeton, L. L., Grub, W., Wells, D. M., Meenaghan, G. F., and Albin, R. C., Effects of manure depth on runoff from southwestern cattle feedlots. *Annu. Meet., Amer. Soc. Agr. Eng., 1970* Paper 70-910 (1970).
48. Miller, W. D., "Infiltration Rates and Groundwater Quality Beneath Cattle Feedlots, Texas High Plains," Rep. Proj. 16060 EGS. Water Quality Office, Environmental Protection Agency, 1971.
49. Murphy, L. S., and Gosch, J. W., "Nitrate Accumulation in Kansas Ground Water," Proj. Completion Rep., OWRR Proj. A-016-Kan. Kansas State Univ., Manhattan, 1970.
50. White, N. K., and Sunada, D. K., "Groundwater Quality of Severence Basin, Weld County, Colorado." Civil Eng. Dept., Colorado State University, Fort Collins, 1966.
51. Stewart, B. A., Viets, F. G., Hutchinson, G. L., and Kemper, W. D., Nitrate and other water pollutants under fields and feedlots. *Environ. Sci. Technol.* **1,** 736–739 (1967).
52. Mosier, A. R., Haider, K., and Clark, F. E., Water soluble organic substances leachable from feedlot manure. *J. Environ. Qual.* **1,** 320–323 (1972).
53. Hutchinson, G. L., and Viets, F. G., Nitrogen enrichment of surface water by absorption of ammonia volatilized from cattle feedlots. *Science* **166,** 514–515 (1969).
54. Sylvester, R. O., and Seabloom, R. W., "A Study on the Character and Significance of Irrigation Return Flows in the Yakima River Basin," Final Report to the Public Health Service. U.S. Dept. of Health, Education and Welfare, University of Washington, Seattle, 1962.
55. Rose, W. W., Mercer, W. A., Katsuyama, A., Sternberg, R. W., Brauner, G. V., Olan, N. A., and Weckel, K. G., "Production and Disposal Practices for Liquid Wastes from Cannery and Freezing Fruits and Vegetables," Proc. 2nd Nat. Symp. Food Process. Wastes, pp. 109–127. Pacific Northwest Water Lab., Environmental Protection Agency, 1971.
56. Hudson, H. T., "Solid Waste Management in the Food Processing Industry," Proc. 2nd Nat. Symp. Food Process. Wastes, pp. 637–654. Pacific Northwest Water Lab., Environmental Protection Agency, 1971.

57. Soderquist, M. R., Waste management in the food processing industry. *J. Environ. Qual.* **1,** 81–86 (1972).
58. Harper, W. J., and Blaisdell, J. L., "State of the Art of Dairy Food Plant Wastes and Waste Treatment," Proc. 2nd Nat. Symp. Food Process. Wastes, pp. 509–545. Pacific Northwest Water Lab., Environmental Protection Agency, Washington, D.C., 1971.
59. Cooper, C. F., Nutrient output from managed forests. *In* "Eutrophication: Causes, Consequences, Correctives," pp. 404–445. Nat. Acad. Sci., Washington, D.C., 1969.
60. Feth, J. H., Nitrogen compounds in natural water—a review. *Water Resour. Res.* **2,** 41–58 (1966).
61. Taylor, A. W., Edwards, W. M., and Simpson, E. C., Nutrients in streams draining woodland and farmland near Coschocton, Ohio. *Water Resour. Res.* **7,** 81–89 (1971).
62. Joyner, B. F., Appraisal of chemical and biological condition of Lake Okeechobee. *U.S., Geol. Surv., Open File Rep.* **71006** (1970).
63. Johnson, J. D., and Straub, C. P., "Development of a Mathematical Model to Predict the Role of Surface Runoff and Groundwater Flow in Overfertilization of Surface Waters," Bull. No. 35. Water Resour. Res. Cent., University of Minnesota, Minneapolis, 1971.
64. Owens, M., Nutrient balances in rivers. *Water Treat. Exam.* **19,** 239–252 (1970).
65. Matheson, D. H., Inorganic nitrogen in precipitation and atmospheric sediments. *Can. J. Technol.* **19,** 406–412 (1951).
66. Johnson, N. M., Reynolds, R. C., and Likens, G. E., Atmospheric sulfur: Its effect on the chemical weathering of New England. *Science* **177,** 514–516 (1972).
67. Hobbie, J. E., and Likens, G. E., The output of phosphorus, dissolved organic carbon, and fine particulate carbon from Hubbard Brook watershed. *Limnol. Oceanogr.* (1973) (in press).
68. Weibel, S. R., Urban drainage as a factor in eutrophication. *In* "Eutrophication: Causes, Consequences, and Correctives," pp. 384–403. Nat. Acad. Sci., Washington, D.C., 1969.
69. Fisher, D. W., Gambell, A. W., Likens, G. E., and Bormann, F. H., Atmospheric contributions to water quality of streams in the Hubbard Brook Experimental Forest, New Hampshire. *Water Resour. Res.*, **4,** 1115–1126 (1968).
70. Hoeft, R. G., Keeney, D. R., and Walsh, L. M., Nitrogen and sulfur in precipitation and sulfur dioxide in the atmosphere in Wisconsin. *J. Environ. Qual.* **1,** 203–208 (1972).
71. Sylvester, R. O., Nutrient content of drainage water from forested, urban, and agricultural areas. *In* "Algae and Metropolitan Wastes," SEC-TR-W61-3, pp. 80–87. U.S. Dept. of Health, Education and Welfare, Washington, D.C., 1961.
72. Avco Economic Systems Corp., "Storm Water Pollution from Urban Land Activity," Final Rep., Contract 14-12-187. Federal Water Quality Adminstration, Dept. of Interior, Washington, D.C., 1970.
73. American Public Works Association, "Water Pollution Aspects of Urban Runoff," Final Rep., Contract WA 66-23. Federal Water Pollution Control Administration, Department of Interior, Washington, D.C., 1969.
74. Kluesener, J. W., and Lee, G. F., Nutrient loading from a separate storm sewer in Madison, Wisconsin. *Water Pollut. Contr. Fed. Conf., 45th, 1972.*
75. Pracoshinsky, N. A., and Gatello, P. D., Calculations of water pollution by surface runoff. *Water Res.* **2,** 24–26 (1968).
76. Burm, R. J., Krawezyk, D. F., and Harlow, G. L., Chemical and physical comparison of combined and separate sewer discharge. *J. Water Pollut. Contr. Fed.* **40,** 112–126 (1968).
77. Jaworski, N. A., and Hetling, L. J., "Relative Contribution of Nutrients to the Potomac River Basin from Various Sources," Proc. Agr. Waste Manage. Conf., pp. 134–146. Cornell University, Ithaca, New York, 1970.
78. Sartor, J. D., Boyd, G. B., and Agardy, F. J., Water pollution aspects of street surface contaminants. *Water Pollut. Contr. Fed. Conf., 45th, 1972.*

79. Anonymous, The Lake Mendota pilot project. *In* "Farm Animal Wastes, Nitrates, and Phosphates in Rural Wisconsin Ecosystems" (T. S. Brevik and M. T. Beatty, eds.), Pp. 280–291. Univ. Extension, University of Wisconsin, Madison, 1971.
80. Powell, R., and Dunsmore, J., Phosphorus in the rural ecosystem: Runoff from agricultural land. *In* "Farm Animal Wastes: Nitrates, and Phosphate in Rural Wisconsin Ecosystems" (T. S. Brevik and M. T. Beatty, eds.), pp. 156–166. Univ. Extension, University of Wisconsin, Madison, 1971.
81. Graham, T. R., Pollution prevention in Northern Ireland. *Eff. Water Treat. J.* **7,** 87–89 (1967).
82. Shannon, E. E., and Brezonik, P. L., Relationships between lake trophic state and nitrogen and phosphorus loading rates. *Environ. Sci. Technol.* **6,** 719–725 (1972).
83. Task Group 2610P, Sources of nitrogen and phosphorus in water supplies. *J. Amer. Water Works Ass.* **59,** 344–366 (1967).
84. Oglesby, W. C., Bovine salmonellosis in a feedlot operation. *Vet. Med. & Small Anim. Clin.* **59,** 172–174 (1964).
85. Gibson, E. A., Disposal of farm effluent—animal health. *Agriculture* **74,** 183–192 (1967).
86. Hibbs, C. M., and Foltz, V. D., Bovine salmonellosis associated with contaminated creek water and human infection. *Vet. Med. & Small Anim. Clin.* **59,** 1153–1155 (1964).
87. Mallman, W. L., and Mack, W. N., "Biological Contamination of Groundwater," Tech. Rep. WP 61-5, pp. 35–43. Robert A. Taft Sanit. Eng. Center, Cincinnati, Ohio, 1961.
88. McCoy, E., Removal of pollution bacteria from animal wastes by soil percolation. *Annu. Meet., Amer. Soc. Agr. Eng., 1969,* Paper 69-430 (1969).
89. Geldrich, E. E., Bordner, R. H., Huff, C. B., Clark, H. F., and Kabler, P. W., Type distribution of coliform bacteria in the feces of warm blooded animals. *J. Water Pollut. Contr. Fed.* **34,** 295–301 (1962).
90. Witzel, S. A., McCoy, E., Polkowski, L. B., Attoe, O. J., and Nichols, M. S., "Physical, Chemical and Bacteriological Properties of Farm Wastes (Bovine Animals)," Proc. Nat. Symp. Anim. Wastes Manage., Publ. SP-0366, pp. 10–14. Amer. Soc. Agr. Eng., 1966.
91. Anonymous, "Water Quality Control and Management in Snake River Basin." Fed. Water Pollut. Contr. Adm., Portland, Oregon, 1968.
92. American Society for Testing Materials, Basic principles of sensory evaluation. *Amer. Soc. Test. Mater., Spec. Tech. Publ.* **433** (1968); Manual on sensory testing methods. *ibid.* **434**; Correlation of subjective-objective methods in the study of odors and taste. *ibid.* **440.**
93. Burnett, W. E., and Sobel, A. T., "Odors, Gases, and Particulate Matter from High Density Poultry Management Systems as They Relate to Air Pollution," Proj. Rep. No. 2, N.Y.S. Contract No. 1101. Cornell University, Ithaca, New York, 1968.
94. Miner, J. R., and Hazen, T. W., Ammonia and amines: Compounds of swine building odor. *Annu. Meet., Amer. Soc. Agr. Eng.,* Paper 68-910 (1968).
95. Burnett, W. E., Odor transport by particulate matter in high density poultry houses. *Poultry Sci.* **48,** 182–184 (1969).
96. Miniats, O. P., Willoughby, R. A., and Norrish, J. G., Intoxication of swine with noxious gases. *Can. Vet. J.* **10,** 51–53 (1969).
97. Charles, D. R., and Payne, C. G., The influence of graded levels of atmospheric ammonia on chickens. *Brit. Poultry Sci.* **7,** 177–198 (1966).

4

Waste Characteristics

Introduction

Knowledge of agricultural waste characteristics is fundamental to the development of feasible waste management systems. Treatment and disposal methods that have been successful with other industrial wastes may be less successful with agricultural wastes unless the methods are modified to accommodate the characteristics of specific agricultural wastes. The wastes produced by agriculture vary in quantity and quality. Wastes from food processing are low strength, high volume liquid wastes while those from livestock operations tend to be high strength, low volume wastes. Both liquids and solids result from food processing and livestock production requiring that both liquid and solid waste management possibilities be considered.

An understanding of the characteristics of a waste permits judgments on the type of treatment and/or disposal methods that may be effective. With a liquid waste containing dissolved organic solids, biological treatment is appropriate. Solid wastes with a high organic content are amenable to incineration or composting. Other alternatives are noted in Table 4.1. In general the bulk of the oxygen-demanding material is in the dissolved state for food processing waste waters whereas with livestock wastes most of the oxygen demanding material is in form of particulate matter.

Information on the frequency and quality of waste discharges allows design of facilities to handle constant as well as intermittent discharges such as those due to the seasonal nature of fruit and vegetable processing and to the variable nature of livestock waste runoff. Identification of the waste

Table 4.1

Feasible Treatment and Disposal Methods with Wastes of Different Characteristics

Wastes	Treatment and disposal methods
Liquid	
Dissolved organic wastes	Biological treatment, land disposal
Dissolved inorganic matter	Land disposal, physical or chemical treatment
Suspended organic wastes	Sedimentation, biological treatment, chemical precipitation, land disposal
Suspended inorganic matter	Sedimentation, land disposal, chemical treatment
Solid	
Organic wastes	Incineration, composting, land disposal, dehydration, soil conditioner, animal feed
Inorganic wastes	Land disposal

sources within a processing plant provides information for in-plant separation of the waste streams in the plant, for reuse of the less contaminated waters, and for changes in processes that produce large quantities and/or concentrated wastes. Knowledge of the waste characteristics permits consideration of reuse of waste liquids for the transport of raw products, recovery of specific waste components, irrigation, by-product development, and fertilizer value. Similar reuse possibilities exist with solid wastes, such as for animal feed and as a soil conditioner. The prevention of waste is of economic importance in all processing operations. The quantity and quality of the waste affects the complexity and size of waste treatment systems as well as recovery opportunities. Waste treatment and disposal are only part of agricultural waste management systems.

Agricultural wastes consist of food processing wastes, liquid and solid animal wastes, waste packaging materials, agricultural chemical losses, crop and field residues, greenhouse and nursery wastes, dead livestock and obsolescent vehicles, equipment, and buildings. The characteristics of food processing wastes and animal wastes receive major emphasis in this chapter.

Food Processing

General

The function of the food processing industry is to serve the farm operations which produce perishable products on a seasonable basis and consumers who desire a variety of nutritional foods throughout the year. Changes in food marketing have resulted in alterations of food processing methods to accommodate consumer desires. Requirements of supermarkets have encouraged producers to wash, prepare, and prepack fruits, vegetables, and meats at the source. Greater water volumes may be used and more wastes are left at the processing plant and in the fields. The food processing industry is a diversified enterprise processing items such as fruit and fruit products, vegetables and vegetable products, meat and meat food products, milk and milk products, and seafood and seafood products.

Even small agricultural production items can be a source of waste problems. The mushroom industry, a small part of total agriculture, has a yearly waste product that exceeds one million tons. The spent compost is dumped wherever convenient causing fly and other problems. Table 4.2 indicates the estimated pollutional loads from other food processing industries.

Solid agricultural wastes also can be of considerable magnitude. The solid wastes from fruit harvesting have ranged from 1.4–3.0 tons/acre/year. Greenhouse and nurseries produce about 25 tons of solid wastes/acre/year (2).

Each food processing plant has wastes of different quantity and quality. No two plants are the same and predictions of loadings expected from proposed plants must be regarded as approximate. Few plants have adequate knowledge of the volumes, characteristics, and fluctuations of their wastes. At each processing plant, material and water balances on the usage in the plant can provide reasonable estimates of the total waste flow and contaminant load. These balances, coupled with in-plant surveys of individual waste flows, can pinpoint opportunities for in-plant changes to reduce wastes and help identify unnoticed waste discharges.

Much of the available information on food processing waste character-

Table 4.2

Estimated Pollution Loadings from Agricultural Processing Industries in the United States (1968)[a]

Source	Potential daily BOD discharges (million pounds)
Meat slaughtering and packing	2.17
Dairy products	1.99
Canning	1.3
Sugar refining	0.8
Potato products	0.35
Poultry	0.22

[a] From reference 1.

istics is of limited use and not comprehensive. Average waste loads reported in the literature should be regarded as guidelines rather than exact values that can be extrapolated to different plants. Because of the differences that exist between processing plants, only characteristics that are indicative of the ranges likely to be encountered are presented in subsequent sections.

The effluent from fruit and vegetable processing plants consists mainly of carbohydrates such as starches and sugars, pectins, vitamins, and other components of the product cell walls which have been leached during processing. Of the total organic matter, 70–85% is present in the dissolved form. These dissolved solids are not removed by mechanical or physical separation methods although they can be stabilized and/or removed by biological or chemical oxidation and adsorption.

The characteristics of food processing wastes are such that they generally are low in nitrogenous compounds, have a high BOD and suspended solids concentration, and undergo rapid decomposition. Some wastes, such as those from beet processing, are highly colored. Fresh wastes have a pH close to neutral. During storage the pH decreases. Storage of the effluent and neutralization of the wastes may not be effective. In addition to the organic content, fruit and vegetable processing wastes can contain other pollutants such as soil, lye, heat, and insecticides. Food processing wastes result from the washing, trimming, blanching, pasteurizing, juicing of raw materials, the cleaning of processing equipment, and the cooling of the finished product. In most plants, cooling waters have low contamination and may be reused for washing and transport of the raw product.

Fruit and vegetable processing is a seasonal operation occurring during summer and fall. The seasonable nature of the operations requires that the cost of waste treatment equipment be kept at a minimum. The main problems for a fruit or vegetable processing waste treatment system are the fluctuations in flow rate and organic load that may vary from hour to hour and day to day.

The wastes from food processing plants are largely organic and can be treated in municipal waste treatment plants. The flow and organic load fluctuations must be adequately evaluated when these wastes are permitted to enter municipal treatment facilities. Fruit and vegetable processing can cause overloading in these facilities because of the variable characteristics and the short processing season. Meat, poultry, and milk processing is less seasonal.

Fruit and Vegetables

In a cannery, the largest water usage is needed for cooling, plant cleanup, and product washing. Water used for peeling, sorting, slicing, processing, and product washing will contain the highest waste concentration but will have lower flows.

Ranges of characteristics for fruit and vegetable processing wastes are presented in the Appendix, Tables A-1 to A-12. Waste flows and pollutional characteristics can be reduced by in-plant changes and close control over processing operations. The variations in waste characteristics are a result of the type of product, method of picking the crop, processing methods and conditions, and plant activities at the time of sampling. The higher pH values result from the use of caustics such as lye in peeling. These caustic solutions can have a pH about 12–13 and are discharged intermittently as they lose their strength resulting in slug loads to the waste treatment facilities. Pickle and sauerkraut wastes are acidic and contain large chloride concentrations as well as organic matter.

The peeling of apples, apricots, and peaches produced 0.47, 0.16, and 0.42 lb of BOD per case of finished product, respectively (14). High percentages of the total dissolved organic solids in fruit and vegetable processing waste water originate in peeling and blanching operations. For peaches, peeling and the following rinse contributed about 40% of the BOD discharged from a cannery. The use of a dry caustic peeling method can reduce the organic waste load from peeling operations. An experimental unit for peaches reduced the waste water volume by 93%, the COD by 70%, and the suspended solids by 70% (14). More extensive studies with the dry caustic peeling of potatoes have shown that there is about four times more water used in the wet peeling process than in the dry caustic process. The pounds of dry solids per ton of potatoes resulting from the wet caustic peeling system was about 3.75 times that for the dry caustic system. Similarly the pounds of BOD per ton of potatoes was about 4.5 times larger with the wet system than with the dry caustic system (15). These systems are illustrative of the in-plant changes that can be made in the processing operation to reduce the waste load.

Analyses of the edible portion of the common vegetables show that apart from water, which is present to the extent of 74–94% of the total weight, the vegetables consist chiefly of carbohydrate (3.2–19.1%) with relatively small amounts of protein (1.1–6.7%) and fat (0.1–0.5%). The carbohydrate consists partly of sugars and partly of starch. The remainder (10–20%) is mostly cellulose fiber (12).

When the vegetable is brought into contact with water, the soluble substances, mainly the sugars, are leached out. The rate of leaching is initially high and decreases exponentially with time and with the number of washes. Particles of vegetable matter are lost by abrasion. The ratios of the amount of sugar to nitrogen and phosphorus leached from peas during repeated washing are close to the optimum for aerobic biological treatment (17).

The ratios were similar to those to be expected from the composition of the vegetable. The rate of wash water required depends on the type of machine, the quantity of vegetables processed per day, and water management in the processing plant.

The processing of grapes into wine produces a variety of wastes which must be adequately treated prior to disposal. Winery waste waters include continuous year-round wastes and seasonal wastes. Production and operational activities increase during the fall grape pressing season and create additional wastes which are grape-washing waters and spent solutions from the pressing operations. Wine-making processes include grape crushing and pressing, grape juice fermentation, racking, blending, filtering, and bottling the wine. Most of these operations are batch operations that result in batch waste water discharges. The volume and strength of winery waste waters will vary during the day and between operating days. Continual waste waters are those from filter cleanup, floor washing, cooling water, stillage from the wine process, and bottle wash. Typical characteristics of these wastes for both the grape pressing and nonpressing seasons are illustrated in the Appendix, Tables A–1 to A–12.

Meat and Poultry Processing

The meat slaughtering and meat-packing industry consists of a number of small plants. The trend is toward decentralization and establishment of new plants near the source of animal production. Meat-packing plants range in size from plants with annual kills of from less than one million pounds to those with greater than eight hundred million pounds. The main wastes originate from killing, hide removal or dehairing, paunch handling, rendering, trimming, processing, and clean-up operations. The wastes contain blood, grease, inorganic and organic solids, and salts and chemicals added during processing operations. The BOD and solids concentrations in the plant effluent will depend on in-plant control of water use, by-product recovery, waste separation at the source, and plant management.

At a beef cattle slaughtering plant, as high as 50 lb of blood, 50 lb of paunch manure, and 40 lb of animal manure per animal can result. Blood from beef cattle had a BOD_5 of 156,500 mg/l, a COD of 218,300 mg/l, a moisture content of 82%, and a pH of 7.3 (80). The average weight of wet blood produced per 1000 lb of beef animal was 32.5 lb. Recovery of the blood is an important aspect of pollution control and should be a part of all animal processing plants.

In ruminants, the first stomach or paunch contains undigested material or paunch manure. The method by which the paunch manure is removed and disposed of affects the waste load of a plant. The paunch material will have a moisture content of about 88%, with an average COD of 177,300 mg/l and an average BOD_5 of 50,200 mg/l. The solid portion of the paunch material contains the greatest pollutional load, about 73% of the COD and 40% of the BOD. The paunch material resulted in about 8.8 lb COD and 2.5 lb BOD per 1000 lb liveweight killed (80). Separation of the paunch material at the source combined with nonliquid handling and offsite disposal will decrease the total liquid waste load at a slaughtering plant.

Rendering is a process for recovery of grease from meat scraps. Dry rendering, i.e., cooking under vacuum and low temperature at 118°F with no water added, results in a minimum of pollutants added to the plant waste stream. Wet rendering results in tank water containing dissolved organic matter and having a high BOD, about 30,000 mg/l.

The nature of meat processing results in variable waste flows and concentrations. To determine the composition of a plant effluent at least daily composite samples, preferably over a full working week, are required. Grab samples seldom provide accurate information because of the variation in composition that can take place. Not every plant will contain all of the above processes or practice maximum by-product recovery. The processes and waste reductions that can be expected are outlined in the Industrial Waste Profile series of the Department of Interior (17). Characteristics of meatpacking and meat processing wastes are presented in the Appendix, Tables A-13 to A-16.

At poultry processing plants, wastes originate from killing, scalding, defeathering, evisceration, washing, chilling, and clean-up operations (Fig. 4.1). Waste quantity and quality depend on the manner in which the blood, feathers, and offal are handled, the type of processing equipment used, and the attitude of the plant management concerning pollution control. In most modern plants wastes discharged from evisceration and the feathers, dirt, and blood from the defeathering machines are carried in the waste water streams. These streams normally pass through screens which remove the larger solids. Usually the chilling and packing waters and clean-up waste also pass through the screens. About 75% of the daily waste volume, BOD, and suspended solids are discharged during processing with the remainder discharged during clean-up periods.

The growth rate of the broiler processing industry has caused changes in processes, processing equipment, and procedures. Water use per bird has increased from 7 gal/bird to 12–15 gal/bird in different plants. The BOD production has increased from about 27 lb/1000 birds to better than 60 lb/1000 birds while suspended solids production has increased from 13–14 lb/1000 birds to about 47 lb/1000 birds. These increases in the past decade have occurred because of more stringent federal standards for cleanliness and

Fig. 4.1. Broiler processing assembly line (courtesy U.S. Dept. of Agriculture).

sanitation along with additional processing that was not done earlier. The hand wash facilities are one of the largest water using areas in a poultry processing plant.

In the processing of broilers, about 70% of the original weight of the bird represents the finished product. The remaining 30% includes feathers, intestines, feet, head, and blood which require liquid and solid disposal at the processing plant. The waste of greatest pollutional significance is the blood from the killing operation. About 8% of the body weight of chickens is blood and about 70% is drainable. The drainable blood has a pollutional load of about 17 lb BOD/1000 chickens processed (24). With reasonable blood recovery, BOD and suspended solids loads can be reduced by 15 lb and 10 lb/1000 birds processed, respectively (25). Most plants attempt to recover some of the blood. Characteristics of wastes from poultry processing plants is presented in the Appendix, Tables A-17. Washing compounds used in cleanup increase the pH and alkalinity of the wastes.

With the exception of the turkey processing industry, seasonal variations are not of importance in the poultry processing industry. The poultry slaughter and waste load were estimated to be 3.3 billion birds, 150 million pounds of BOD, and 25 billion gallons of waste water per year in 1970 (17).

Milk Processing

Milk processing wastes result from manufacturing and transfer operations after the milk from a dairy farm reaches a central receiving station. The wastes consist of whole and processed milk, whey from cheese production, and wash water. Fresh milk processing wastes are high in dissolved organic matter and very low in suspended matter. The BOD of whole milk is about 100,000 mg/l and can exert a significant oxygen demand even in small quantities. The BOD and COD of milk plant waste water will be a function of the type of product manufactured. Various dairy products differ widely in their relative organic constituents. These differences are reflected in the oxygen demand of the resultant wastes. The constituents of interest are noted in the Appendix, Tables A-18 to A-22. The three major constituents contributing to an oxygen demand are lactose, milk fat, and milk proteins. These items have BOD/pound values of 0.65, 0.89, and 1.03, respectively. The high oxygen demand of these products indicates that wastes from their processing will ferment rapidly if stored and require aeration to minimize odors. Concentrated wastes such as from butter, buttermilk, or dried milk operations should be separated from other waste waters and used for animal feed or by-product recovery.

A survey over 50 plants revealed the range of unit flow and BOD values reported in the Appendix, Tables A-18 to A-22. These plants used either advanced technology or a mixture of advanced and typical processing technology. Over half of the plants produced more than one type of dairy product. The controlling factor in waste volume and BOD production appeared to be management. Under extremely good management, about 0.5 lb of waste flow per pound of milk processed and 0.5 lb of BOD per 1000 lb of milk processed were obtainable. A realistic average of 1.5 lb of waste flow per pound of milk processed and 2.0 lb of BOD per 1000 lb of milk processed appeared to be achievable under good management. Factors above 3.0 in both categories were considered excessive and an indication of poor waste management. Plants producing whey will have BOD factors considerably in excess of the above.

Data on the wastes from milk and cheese processing plants are presented in the Appendix, Tables A-18 to A-22. The pH and alkalinity variations in these wastes are caused by fermentation and acid production. The major source of nitrogen, phosphorus, and potassium in milk plant wastes is milk spilled or lost during processing. The phosphorus also results from cleaning compounds used in the plants. Such compounds also contribute to the sodium and magnesium in milk plant wastes.

The cleaning compounds contribute to the oxygen demand of these wastes. Wetting agents and surfactants vary widely with BOD values, pound per pound of product, ranging from 0.05 to 1.2 (28). The most commonly used surfactant and acid detergent used in cleaning food plant equipment had BOD values of about 0.65 lb per pound of product. Alkalies, such as sodium hydroxide, also are used for dairy plant cleaning. Under average conditions in a modern milk plant, the amount of BOD contributed by surfactant and acid detergent would be about 0.1 lb/1000 lb of milk processed.

Milk processing plants will have numerous short term fluctuations in characteristics and flow during the day. There are seasonal peaks due to milk production variability throughout the year. The wastes are amenable to biological treatment and can be treated in municipal waste treatment facilities provided that the plant has capacity for the fluctuation and increased oxygen demand.

The largest pollutants in dairy food plant waste waters are whey from cheese production operations followed by wash water and pasteurization water in that order. The manufacture of cheese from either whole or skim milk produces cheese and a greenish yellow fluid known as whey. Whole milk is used to produce natural and processed cheeses such as cheddar and the resultant fluid is called sweet whey with a pH in the range of 5 to 7. Skim milk is used to produce cottage cheese and the by-product fluid is called acid whey with a pH in the 4 to 5 range. The lower pH is the result of the acid developed during or employed for coagulation. Each pound of cheese produced results in 5–10 lb of fluid whey. About 70% of the nutrients in skim milk are part of the acid whey.

The BOD of whey ranges from 32,000 to 60,000 mg/l depending on the specific cheese-making process used. Whey contains about 5% lactose, 1% protein, 0.3% fat, and 0.6% ash. Because the protein has been precipitated as cheese, evaluation of the nitrogen content of whey should be made to assure adequate nitrogen if biological treatment is contemplated.

The production of cheese removes about one-half the total solids of whole milk. The remainder is discharged in the about 20 billion pounds of whey produced annually. The protein and lactose in whey could be used in many by-products if excess salts were removed. Investigations on the use of whey in food products and on the use of electrodialysis and reverse osmosis to remove the salts have been successful (81). The high bulk and low values of whey make it impractical to transport long distances and handling is difficult and costly.

Seafood Processing

The degree of waste in seafood processing varies widely. Fish which are rendered whole, such as menhaden, to produce fish meal result in no solid waste. Crab processing results in up to 85% solid waste. Each fish processing operation will produce significant liquid flows from the cutting, washing, and processing of the product. These flows contain blood and small pieces of fish and skins, viscera, condensate from cooking operations, and cooling water from condensers. The latter two flows have large volumes but contain small organic loads.

Variations in a freshwater fish processing plant are caused by the type of fish processing techniques, plant size, water usage, and the time the solid wastes were in contact with the waste water. The pollutional strength of perch and smelt waste water increase considerably with increased contact time between the solid waste material and the liquid waste (77). Samples of mixed smelt and perch wastes in the processing plant effluent exhibit greater pollution characteristics than are obtained by simple addition of the characteristics of the separate perch and smelt waste waters. The greater strength is due to longer contact time of the solid fish wastes and the transporting water. Fish processing waste solids should be kept separate from the waste water whenever possible. Considerable day to day variability exists in fish processing characteristics.

Few comprehensive, detailed studies have been made to determine fish processing waste strengths. General ranges of reported values for various types of fish processing operations are presented in the Appendix, Tables A-23 and A-24.

Livestock Production

General

Large numbers of livestock are raised in the United States with the result that over 2 billion tons of manure are produced each year. About 3 lb of wet manure is defecated for each quart of milk produced and from 6 to 25 lb of wet manure is produced per pound of weight gain of livestock (31). Part of the total livestock waste production remains in the pasture and rangeland, but large volumes accumulate in feedlots and buildings and must be collected, transported, and disposed of in an economical and inoffensive manner.

The term "livestock wastes" may mean any one of a number of things: (a) fresh excrement including both the solid and liquid portions, (b) total excrement but with bedding added to absorb the liquid portion, (c) the material after liquid drainage, evaporation of water, or leaching of soluble nutrients, (d) only the liquid which has been allowed to drain from the total excrement, or (e) material following aerated or anaerobic storage.

The characteristics of these items are different. The moisture content of fresh waste is a function of the type of feed and environmental temperature and humidity. Evaporation of water may occur under certain conditions. Addition of water occurs from rainwater, wash water, or water added to increase the flow and pumping characteristics of the wastes. Differences in animal waste characteristics also can be a result of changes in the environment and the level of productivity among animals.

The best approach to obtain characteristics of animal wastes is to obtain representative samples of the waste and conduct appropriate analyses. To determine the relationship between animal ration and waste characteristics, nutritional trials can be conducted and the quantity and quality of the wastes determined. While accurate, this process is time consuming and costly.

It is possible to estimate the quantity and quality of a waste if the digestion coefficients for the feed components are known. Such data have been obtained as feed manufacturers improve their feeds. Data are available on the quantity of protein, fat, fiber, nonfiber extract, and mineral matter in the feed. Other nutritional data of interest are frequently available.

Digestion coefficients are not constants for a given feed or animal species and are influenced by factors such as the characteristics and relationships of the nutrients fed. The digestibility of a mixture is not necessarily the average of the values for its constituents determined separately or indirectly. Each feed component may exert an influence on the digestibility of another. In addition, individual lots of a particular feed may differ from the average.

Even though these differences and variations are recognized, digestion coefficients can be used to estimate the quantity and quality of waste from a given feed. The volatile content of the waste can be estimated by the sum of the protein, fat, fiber, and nonfiber extract percentages of the feed after digestion. The mineral content can be used as an estimate of the ash content. The composition of the fiber and nonfiber extract can be estimated as combinations of lignin, cellulose, and hemicellulose. Materials and water balances on livestock production operations can provide reasonable estimates of the waste generated at a specific facility.

With both swine and poultry, the diets are highly digestible. The waste from ruminants, such as cattle and horses, has a different composition than waste from simple stomached species. The diet consumed by ruminants is more resistant to digestion. The bacteria that inhabit the stomach of the ruminants enable these animals to utilize cellulosic feeds. These feeds include compounds such as lignin which accompany cellulose in plants and which are difficult to digest in the rumen. Urine from ruminants tends to be more alkaline because their diets are higher in compounds such as potassium, calcium, and magnesium.

Wastes from grass fed animals, mature stock, and milk animals will be digested to a greater degree than that from growing livestock. As a result these wastes will be less biodegradable than that from animals being fattened or from animals liberally fed on concentrates. Animals in confinement are fed feed of a composition to cause the greatest weight gain in the shortest period of time. Highly efficient consumption of the feed by the animal is subordinate to continuous and rapid weight gain or to egg production. Animals of the same kind that are fed more concentrates excrete more of the nutritive material because the food contains more. As the level of protein feeding is raised beyond a certain point, the protein is less effectively digested and more passes into the feces.

The characteristics of livestock wastes are a function of the digestibility and the composition of the feed ration. The feces of livestock consist chiefly of undigested food, mostly cellulose fiber, which has escaped bacterial action. A portion of the other nutrients also escape digestion. Undigested proteins are excreted in the feces and the excess nitrogen from the digested protein is excreted in the urine as uric acid for poultry and urea for animals. Potassium is absorbed during digestion but eventually almost all is excreted. Feces also contain residue from the digestive fluids, waste mineral matter, worn-out cells from the intestinal linings, mucus, bacteria, and foreign matter such as dirt consumed along with the food. Calcium, magnesium, iron, and phosphorus are voided chiefly in the feces. Livestock wastes can contain feed spilled in the animal pens.

There are obvious variations in the characteristics of wastes from livestock feeding operations. One must be cautious in assuming that results of waste management studies using human wastes or wastes from one species of animal will be applicable to other animal species. These variations occur because of the items described above, the kind of surface upon which the manure accumulates, and the frequency with which the operation is cleaned. Data accumulated on the quantity and quality of animal wastes that is more than ten years old may not represent the characteristics of current wastes because of the changes in feeding, housing, and environment that have taken place. The characteristics noted in this section represent data from investigations reported since 1960.

Animal feces frequently contain inorganic feed additives to increase the weight gain of the animals. Some of these additives are inhibitory to microorganisms and thus may affect the performance of biological treatment units for animal wastes. Large amounts of copper salts in commercial pig feed supplements, between 11 and 300 mg/l (73), offer an example of such additives. Copper concentrations from 80 to 1420 mg/l have been found in pig feces and concentrations up to 750 mg/l in slurries of animals fed the copper-supplemented diets. It is not known whether the copper in these slurries will inhibit biological treatment. Organic compounds can complex metals such as copper so that it will not be inhibitory to microorganisms. When pig waste containing 50 to 500 mg/l of added copper was aerated,

COD reductions were inhibited at all copper concentrations (73). The possible inhibition of biological treatment by feed additives in animal feeds raises questions about the need and appropriate concentrations. There is no reason for unwarranted or random additions of additives to feeds.

Municipal wastes, food processing wastes, and other industrial wastes are water borne and result in liquid waste handling, treatment, and disposal systems. Livestock wastes are solid, semisolid, or liquid depending upon how the production operation is designed and operated. The characteristics of livestock wastes are affected by decisions on how the wastes are to be handled. A broader set of waste management alternatives exist for livestock wastes than for other industrial wastes since opportunities exist for both liquid and solid waste handling, treatment, and disposal. Livestock wastes are generated as a semisolid and there is logic in handling and disposing of the wastes in this condition. The labor requirements and the lack of suitable solids handling equipment have caused many operations to consider slurry or liquid waste systems where the waste can be removed as liquid and transported by pumps, spreaders, or irrigation equipment.

Mechanical handling equipment for livestock wastes is needed in their removal from the livestock area, transfer to storage and/or treatment, and transfer from storage and/or treatment site. Figure 4.2 indicates ranges of the physical characteristics of livestock waste and handling requirements.

Individuals concerned with pollution control activities are interested in characteristics of these wastes in terms of BOD, COD, solids, and nutrient content. Suspended solids have little relevance to most livestock wastes since they usually exist as a solid or slurry rather than a dilute liquid waste. A number of ways have been used to report the characteristics of livestock wastes.

The pollutional characteristics can be reported in terms of milligrams/liter of the liquid slurry that results. Since the water content of the waste slurry will vary depending on the quantity of water used in cleaning, spilled

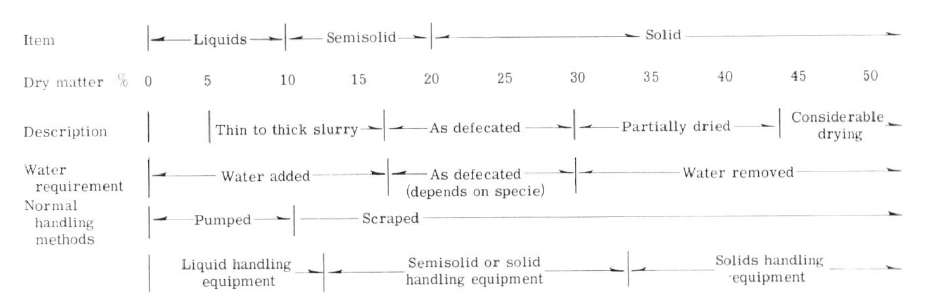

Fig. 4.2. General characteristics and handling methods for livestock wastes.

by the animals in drinking, excreted by the animals, and by any evaporation or rainfall that occurs between cleanings, values based on milligrams/liter can be expected to vary widely.

Livestock wastes differ from municipal and industrial wastes in that livestock wastes are solid matter which contains some water rather than liquid waste containing some solids. Although the water content will vary, the solid content is a function of the ration fed and the specie of animal and should be relatively constant per animal. Data can be presented as milligrams of a pollutional characteristic per milligram of total or volatile solids.

Parameters such as BOD:COD ratios and BOD:total or suspended solids ratios are used to estimate the feasibility of biological waste treatment processes with specific wastes. Small ratios indicate a considerable nonbiodegradable fraction. Livestock wastes contain sizeable amounts of solids which may be resistant to further biodegradation. Examples are salts, residual feed, and microbial cells. In hog waste water, the resistant or only slowly biodegradable fraction can amount to about 60–70% of the total solids. Similar fractions occur in dairy cattle, beef cattle, and poultry wastes. Removal of these solids prior to biological treatment processes would benefit subsequent processes and improve treatment kinetics.

Other data relate the quantity of waste to the animal in terms of pounds per head of animal. This approach is the most realistic and valuable in estimating the gross wastes generated at a particular production unit. A detailed waste sampling and analytical program may be justified before treatment and disposal methods are chosen and designed. The traditional analytical techniques for characteristics such as BOD, COD, and suspended solids were developed for use with liquid wastes. Livestock wastes and waste slurries are concentrated and many dilutions must be made before the traditional methods are used. Livestock wastes and waste slurries can contain concentrations of contaminants that interfere with some of the traditional analytical techniques.

For estimation purposes, population equivalents have been used to indicate the relative quantity or quality of livestock wastes. Population equivalents (PE) are a ratio of the amount of a given pollutant in a waste to the amount present in municipal sewage. Population equivalent values can be based upon any parameter of interest and a specific waste will have different PE values for different parameters. PE values are usually based on an average contribution of 0.17 lb BOD, 0.55 lb total solids, and 0.08 lb total nitrogen per person per day in municipal sewage. Although PE values can be based upon liquid flow, such values have questionable meaning for livestock waste because of the variability of liquid volumes from livestock production units.

Municipal wastes are liquid and are disposed of in surface waters after suitable treatment. When applying PE values to livestock wastes, there is the hazard that the values may be used as indications of water pollution problems. Population equivalent values are useful for relative comparisons of the magnitude of the wastes contributed from municipal sources and from livestock waste sources. However, the use of PE values to estimate actual stream pollution by defecated livestock wastes is erroneous since these wastes will cause surface water pollution only when they enter a stream. At present most animal wastes are disposed of on land either directly from the animal or by spreading and do not represent a direct water pollution source. Even from beef cattle feedlots, it is estimated that less than 5–10% of the wastes may reach a stream when runoff is large. On a national basis, the production of wastes defecated by beef cattle, hogs, and poultry in the United States exceeds the wastes produced by the human population by at least factors of 5:1 on a BOD basis, 10:1 on a total dry solids basis, and 7:1 on a total nitrogen basis.

Representative sampling is difficult with wastes as concentrated and heterogenous as animal wastes. Until more experience is gained with analytical methods on animal wastes and slurries, it is best to consider the available information as guidelines or reasonable estimates rather than accurate and precise data. Characteristics indicated in this book and in other literature should be used with an understanding of the variations in the data and their relative value. Although it is difficult to apply average livestock waste production values to a specific location and problem, knowledge of average values is useful to develop order of magnitude information concerning current and potential livestock production units. Comparative data on animal wastes are in the Appendix.

Poultry

Studies on the physical composition of fresh poultry manure have shown that it contains 75–80% moisture, 15–18% volatile solids, and 5–7% ash with an average particle density of 1.8 and a bulk density of about 65 lb/ft^3. About 50% of the solids were finer than 200 mesh. Based upon 75% moisture, the manure contained 1375 BTU/lb of wet manure (33). Manure excreted from a chicken per day represents about 5% of the body weight of the bird. Waste volume ranges from 0.05–0.06 gal per bird per day. The gross energy value of chicken feces ranges from 3.2 to 4.5 cal per gram of dry matter and the nitrogen content ranges from 0.03 to 0.07 gm of nitrogen per gram of dry matter depending upon the feed ration (34). Other data indicated that chicken manure had an energy value of 1.37 cal/bird/day (2).

Essentially all of the broilers produced in the United States are grown on litter which absorbs moisture and provides a nesting material for the birds. The litter may be sawdust, peanut hulls, wood shavings, or other suitable material. About 1.3 billion cubic feet of used broiler litter must be disposed of every year. The frequency of removal and disposal of broiler manure and litter depends on the number of growing cycles that can be completed on one batch of litter without disease problems. Practices vary among producers but broiler houses are cleaned about 2–3 times/year. The wastes are spread on land and used for fertilization of pasture for beef cattle.

Combined broiler manure and litter contained an average of 25% moisture, 1.7% nitrogen, 0.81% phosphorus, and 1.25% potassium. Hen manure averaged 40% moisture, 1.3% nitrogen, 1.2% phosphorus, and 1.1% potassium (35). Other estimates of the characteristics of broiler and poultry wastes are presented in the Appendix, Tables A-25 to A-27.

Pollutional characteristics are of value when estimating the effect of poultry wastes on land or water resources but available data vary. European data indicate poultry manure production of 0.047 gal/day/bird and a BOD concentration of 100 g/l for fresh manure (44). Data from England indicated manure and BOD production from layers to be about 0.14 gal/day and 71,000 mg/kg, respectively (36). BOD concentration of undiluted chicken wastes have ranged from 8500 to 40,000 mg/l (40) and from 13,000 to 24,000 mg/l (45). The data presented represent average values obtained in specific

Fig. 4.3. Production of ducks showing exterior duck runs and available flowing water.

investigations. Poultry manure from different farms, geographic areas, and different age birds may be the cause of these different characteristics.

Turkey production experiences a seasonal variation with the major producers active during the summer and fall. Waste accumulation in both confinement houses and in open feeding areas must be considered. Young turkeys are kept in houses on litter for the first eight weeks and then placed in open range areas at densities of from 400–2500 birds/acre for the remainder (12–16 weeks) of their growing cycle. The wastes in turkey houses usually are removed twice a year and are handled identically to broiler litter.

Duck production operations have greater water usage than do other poultry operations. The volume of wastes from the duck farms on Long Island ranged from 10–34 gal per duck with an average of 18 gal per duck (42). More recent water usage is near the lower end of the range as duck producers have sought to minimize their waste flows. The characteristics of duck waste water are variable due to variable water flow rates. Sand, undigested food, and manure particles make up the wastes (Fig. 4.3). Duck production is seasonal with the peak period being from April to October. The characteristics of waste water from duck farms are noted in the Appendix, Tables A-25 to A-27.

Swine

The daily waste production from a hog is a function of the type and size of animal, the feed, the temperature and humidity within the building, and the amount of water added in washing and leakage. The quantity of feces and urine produced increases with the weight and food intake of the animal. The feeding regime will affect the properties of the pig manure. Approximately 30% of the consumed feed is converted to body tissue and the remainder is excreted as urine and manure. Feed conversion efficiencies averaged 3.2 lb commercial meal consumed per pound of live weight gained. The data ranged from 2.6 to 3.8 lb feed/lb weight gain. The manure production ranges from 6–8% of the body weight of a hog per day. The total amount of waste water to be handled in a swine operation can be affected more by the amount of wastage from the pressure waterers than any other factor. Volume of manure production can average about 1 gal per 100 lb animal per day. Wet manure can contain 5–9% total solids. Of these 83% may be volatile.

Studies in Scotland have noted the following quantities of manure production per week per animal: meal fed bacon pigs—1.5 ft^3 of feces and urine; skim milk or whey fed pigs—up to 4 ft^3; and systems producing pigs from weaning to sale weight—1 ft^3 (46). Other data indicated manure volumes of 1.75 and 0.72 gal/100 lb/animal/day for pigs and sows, respectively (47).

The manure averaged 80–85% moisture. Swine wastes which have been stored anaerobically for any period have a smell which could be considered as a nuisance when exhausted from a building or spread on the land.

Additional studies noted feed efficiencies of hogs ranging from 2.6 to 5.5 lb feed/lb gain. In this study (79), there was an average of 3.4 part urine to 1 part feces. The average elimination rate of the hogs was 5.0 lb feces plus urine/day/100 lb live weight of animal. There was a greater elimination rate for the smaller animals, i.e., 7.0 lb waste/day/100 lb live weight for 160–200 lb hogs.

Because the land is the ultimate acceptor of swine wastes, nutrients in the wastes are of interest. Comparative data on nutrients and pollutional characteristics are presented in the Appendix, Tables A-28 and A-29. Urine contains about 50% of the fertilizer value of swine manure.

Dairy Cattle

Two types of wastes result from a dairy farm, the solid waste from the animals and the liquid wastes from the cleaning of the milking parlor. At some farms, these wastes are kept separate for treatment and disposal. At other farms, the liquid wastes are added to the manure in a storage pit to make the manure more amenable to transport and disposal as a liquid. Because dairy cattle manures normally are handled as a solid or slurry, concentrations in terms of mg/l have little meaning. Manure production from dairy cattle ranged from 73 to 143 lb with an average of 86 lb/animal/day (33) and represented 7–8% of the body weight of the animal per day. Moisture content averages 80–88% in fresh mixed manure and urine. Urine makes up about 30% of the weight of the manure. Characteristics of dairy cattle manure are compared in the Appendix, Tables A-30 and A-31. Fresh manure contains from 80–85% volatile matter.

Dairy farms using stanchion barns will use bedding such as sawdust, wood shavings, straw, or chopped corn stalks for the animals. The mixture of the manure and bedding is removed daily and transported by solid manure handling equipment. Free stall dairy barns use no bedding and have a more liquid material of combined feces and urine.

An average dairy cow will produce between 14 and 18 tons of feces and urine/year. Manure has about 300 to 400 lb of dry matter/ton. There are about 10 lb of nitrogen, 4 lb P_2O_5, and 9 lb K_2O in each ton of wet manure defecated by the cow. The combined feces and urine collected over a 7-day period from dairy cows fed alfalfa hay and corn silage in equal portions indicated the following waste characteristics: wet feces—average 60 lb/day/1000 lb body weight, range 47–69 lb/day; urine—average 16 lb/day/1000 lb body weight, range 11–27 lb/day; and percent dry matter—average 17%, range 15–18% (72). The average volume of stanchion barn wastes was 1.8 ft^3/day/1000 lb body weight with an average density of 61.7 lb ft^3. The value of dairy manure as a fertilizer can be reduced because of runoff, leaching, volatilization of nitrogen, and the presence of nutrients in unavailable forms.

In areas where dairy cattle manure cannot be spread on the land during the winter, stanchion barn manure and bedding can be stacked in storage areas until it can be hauled to available fields in the spring and summer. The manure and bedding will drain while stacked. The seepage liquid has a high pollutional potential. The quantity and quality of seepage differs in the winter and summer periods. Detailed characteristics of fresh and stacked stanchion barn manure and bedding and of the stack seepage are presented in the Appendix, Tables A-30 and A-31. The in-place volume of stacked manure was about 1.9 ft^3/cow/day with an average density of 53 lb/ft^3. Control of the seepage is necessary to avoid contamination of streams and creeks. Seepage collection tanks or ponds followed by irrigation of the collected seepage on crop land is a positive control method for this liquid.

Each dairy farm has a milking area or parlor where the animals are milked twice a day generally on a 12-hr cycle. On small dairy farms, the milking areas may be used only a fraction of the day while on the very large farms, the milking parlors are in almost continuous use. Milk sanitation regulations require that the milking parlor and milk transfer lines be cleaned after every milking session. A chlorine cleaning solution is used for the pipelines and fresh water is used for the milking parlor itself. The volume of cleaning water is related to the size of the milking parlor and the total length of milk transfer lines rather than to the number of animals milked. The milking parlor wastes will contain the milk residue in the pipelines plus any manure and debris flushed from the parlor. Where the milking parlor wastes are treated separately from the cattle manure, the manure and other solids in the parlor should be swept out prior to cleaning the parlor and should be handled with the other solid wastes. The characteristics of milking parlor wastes are related to the waste management at the dairy farm. Comparative data on these wastes is presented in the Appendix, Tables A-32 and A-33. Other data on milking parlor wastes indicated from 70–85% volatile suspended solids with an average of 78%, soluble BOD of 260–350 mg/l with an average of 290 mg/l (0.2 lb/day/animal) (61).

Beef Cattle

Most of the cattle raised for meat production are produced in the feedlots of the Midwest and West, particularly in the grain producing states of the Great Plains. Few beef cattle are raised in confinement housing, although there is an increase in this method especially in the colder climates. The amount and kind of beef cattle feedlot wastes depend on the ration.

The wastes of beef cattle feedlots will be somewhat different from that of cattle on general farms or on pasture. Since the early 1960's, more efficient rations and feeding programs have helped maintain the economic position of the cattle feedlot operation. Increasingly, feeder cattle are fed high concentrate rations in the feedlots. These rations lend themselves to mechanical processing and handling. A concentrate:roughage ratio of 4:1 has been suggested as satisfactory with respect to rate of gain, feed efficiency, and carcass merit for commercial feeders.

Cattle in feedlots are started on high roughage rations and shifted to high concentrate rations. A high concentrate ration has about 75–85% digestible material and 5–7% minerals, resulting in about 60 lb of wet manure per day which includes about 17 lb of urine from a 900-lb steer. Urine contains about 6% dry matter and is the major source of moisture in the manure. Urine serves as the major carrier of mineral wastes from the animal. Fresh manure has a moisture content of about 85%.

A large quantity of cattle are still fed on farms and ranches that produce and use large quantities of harvested roughage or pasture for feed. Because of the higher quantity of nondigestive matter in roughage, cattle on pasture or roughages may be expected to produce a greater quantity of waste per day than cattle fed a large quantity of concentrates.

Fattening cattle will consume an amount of dry feed equal to about 2.5–3.0% of their body weight. The type of roughage utilized in a feed ration will be a function of the operator's choice and relative feed costs.

Animals on a high roughage diet can produce as much as 10 lb of dry matter in their feces. The waste produced from beef cattle separately fed on all concentrate ration, cottonseed hulls, and sweet sorghum silage was 2.4, 6.6, and 1.6 lb dry solids/day/animal, respectively (63). Lignins and hemicellulose are relatively stable materials that are only slowly degraded by microorganisms. These materials form a large fraction of high roughage feeds and resultant wastes. Cattle feces also contain ligno–protein complexes which are produced in the digestive tract of the animal and which are similar to the humus found in soil. The total dry weight of cattle manure may contain up to 20–25% of these humuslike compounds.

Waste removed from beef cattle feedlots will have characteristics different from fresh manure. The characteristics change as the manure undergoes drying, microbial action, wetting by precipitation, and mixing and compaction by animal movement. The characteristics of beef cattle waste are compared in the Appendix, Tables A-34 and A-35. Beef wastes average 80–85% volatile matter. Analysis of a manure slurry from beef cattle manure indicated that each ton of wet slurry contained 5.8, 3.6, and 6.6 lb of nitrogen (N), phosphorus (P_2O_5), and potassium (K_2O), respectively (66). The COD concentration was 121,000 mg/l.

Veal Cattle

Production of animals for veal requires the confinement and growth of young cattle to the 200–300-lb weight level. The calves are purchased when approximately one week old weighing about 100 lb and sold when 13 weeks old weighing about 300 lb. The animals are kept in stalls with the wastes flushed using a gutter system. The animals are fed high concentrate diets and milk and milk products and produce a more liquid waste than do mature animals. Waste production has been estimated at 2 gal/day/head and 0.16 BOD/day/head with a BOD concentration of about 10,000 mg/l (44). Grab samples of wastes generated at a veal operation indicated a flow of 0.9 gal/day/animal with a BOD of 28,000 mg/l when the animals were maximum age. The BOD production was 0.22 lb BOD/animal/day. Veal production is not increasing because of decreased per capita consumption.

Dust and Gases

Dusts in livestock facilities occur from feed, attrition of building materials, drying of waste products, and the removal of hair and skin tissues from the animals. The dust content of the atmosphere will be a function of the species of animal, the stocking rate and activity of the animals, the temperature, humidity and air movement, the methods and materials involved in feeding, and the absence or presence of bedding.

Dusts usually have the same chemical composition as the substances from which they were derived. Their particulate size will determine the duration of their suspension in the air. The atmosphere of swine facilities has contained 80–95 dust particles/cm^3 of air of which 75–80% were between 2–5 μm in diameter. Similar particle sizes (0.3–10 μm) have been found in dairy and other swine facilities (67). Dusts in the atmospheres of caged laying hen facilities have ranged in diameter from 1–450 μm. Much of this material was identified as skin debris from the birds and feed particles (67). An analysis of the dust in the exhaust air from a poultry house indicated that it contained 90% dry matter which contained 60% crude protein, 11% ash, 9% fat, 3% cellulose, and 17% other carbohydrates. Quantities of dust produced ranged up to 50 mg/bird/day (68).

Gases in a livestock building occur from the respiration by the animals and fermentation in ruminants. The two primary gases from these sources are carbon dioxide and methane. When manure is held under anaerobic conditions in the same building as the livestock, ammonia, hydrogen sulfide, and volatile compounds such as mercaptans, indole, and skatole also can be in the atmosphere of the livestock buildings. Normal ventilation can keep the concentration of potentially toxic gases at levels that do not affect the animals or human workers. When agitation of the stored anaerobic wastes occur, the level of certain gases can increase to toxic levels.

At summer ventilation rates with no agitation of the stored manure in a confined hog house, concentrations of hydrogen sulfide, ammonia, and carbon dioxide were 7.6, 9.1, and 600 ppm, respectively. At maximum winter ventilation rates, the concentrations of H_2S, NH_3, and CO_2 were 24, 12, and 700 ppm, respectively. As the stored manure was permitted to flow under the animals, hydrogen sulfide concentrations reached over 70 ppm (74). At this time, the ventilating fans were off. The concentrations of the above gases were inversely related to the ventilation rate.

Other Animals and Agricultural Products

The above data provide information on the characteristics on livestock for food production. Knowledge of the characteristics of wastes from other animals may be of interest such as those from pet stores, stables, veterinary facilities, animal farms, and pharmaceutical testing laboratories. Data of the pollutional characteristics of pets and small animals are meager. Some available information is presented in the Appendix, Table A-37.

A complete compilation of data on the composition of organic manures and agricultural waste products has been prepared by the National Agricultural Advisory Service of England (70). The data summarizes results of analysis performed from 1947 to 1968 in the Soil Chemistry Department of NAAS and is grouped into three sections: poultry manures, animal manures, and industrial waste products and miscellaneous materials. The compilation has been reproduced in the Appendix, Tables A-38 to A-47.

References

1. Hoover, S. R., and Jasewicz, L. B., Agricultural processing wastes—magnitude of the problem. *In* "Agriculture and the Environment," Publ. No. 85, pp. 187–203. Amer. Ass. Advan. Sci., Washington, D.C., 1967.
2. Golueke, C. G., "Comprehensive Studies of Solid Waste Management—Third Annual Report," SERL Rep. 90–2. University of California, Berkeley, 1970.
3. Splittstoesser, D. F., and Downing, D. L., Analysis of effluents from fruit and vegetable processing factories. *N.Y., Agr. Exp. Sta., Geneva, N.Y. Res. Circ.* 17 (1969).

4. Eckenfelder, W. W., Woodward, C., Lawler, J. P., and Spinna, R. J., "Study of Fruit and Vegetable Processing Wastes Disposal Methods in the Eastern Region," Final Rep., USDA Contract 12-14-100-482 (73). Research and Marketing, U.S. Dept. of Agriculture, 1958.
5. Esvelt, L. A., "Aerobic Treatment of Liquid Fruit Processing Waste," Proc. 1st Nat. Symp. Food Process. Wastes, pp. 119–143. Pacific Northwest Water Lab., Federal Water Quality Administration, 1970.
6. Liner, G. H., and Stepp, J. M., Summary of selected literature on lagoon and spray systems of treating fruit and vegetable processing wastes. *S.C., Agr. Exp. Sta., Rep.,* Clemson, S.C., AE**316** (1968).
7. Eckenfelder, W. W., and O'Connor, D. J., "Treatment of a Cannery Waste in an Aerated Lagoon," Report prepared for Duffy-Mott Corp., 1958.
8. Barnes, G. E., and Weinberger, L. W., Internal housekeeping cuts waste treatment at pickle packing plants. *Wastes Eng.* Jan. (1958).
9. Dostal, K. A., "Aerated Lagoon Treatment of Food Processing Wastes," Rep. 12060. Water Quality Office, Environmental Protection Agency, 1968.
10. Dostal, K. A., "Secondary Treatment of Potato Processing Wastes," Rep. 12060. Water Quality Office, Environmental Protection Agency, 1969.
11. Anonymous, "Pilot Plant Installation for Fungal Treatment of Vegetable Canning Wastes," Final Rep., Proj. 12060 EDZ. Environmental Protection Agency, 1971.
12. Wheatland, A. B., and Borne, B. J., Treatment, use, and disposal of wastes from modern agriculture. *Water Pollut. Contr.* **69,** 195–208 (1970).
13. Tofflemire, T. J., Smith, S. E., Taylor, C. W., Rice, A. C., and Hartsig, A. L., Unique dual lagoon system solves difficult wine waste treatment problem. *Water Wastes Eng.* **7,** #11 F1–F5 (1970).
14. Ralls, J. W., Mercer, W. A., Graham, R. P., Hart, M. R., and Maagdenberg, H. J., "Dry Caustic Peeling of Tree Fruit to Reduce Liquid Waste Volume and Strength," Proc. 2nd Nat. Symp. Food Process. Wastes, pp. 137–167. Pacific Northwest Water Lab. Environmental Protection Agency, 1971.
15. Cyr, J. W., "Progress Report: Study of Dry Caustic vs. Conventional Caustic Peeling and the Effect on Waste Disposal," Proc. 2nd Nat. Symp. Food Process. Wastes, pp. 129–136. Pacific Northwest Water Lab., Environmental Protection Agency, 1971.
16. Anonymous, "Canning Waste Treatment, Utilization and Disposal," Publ. No. 39, p. 55. State Water Resources Control Board, California, 1968.
17. Anonymous, "The Cost of Clean Water," Vol. III, No. 8, Publ. IWP-8. Federal Water Pollution Control Administration, Dept. of Interior, 1967.
18. Anonymous, "Study of Omaha, Nebraska Meat-Packing Wastes." Robert A. Taft Sanit. Eng. Cent. U.S. Dept. of Health, Education and Welfare, 1965.
19. Prohoska, J., and Szeflel, A., Studies on sewage from slaughter houses. *Chem. Abstr.* **5994C** (1960).
20. Paulson, W. L., Kneck, D. R., and Kramlich, W. E., "Oxidation Ditch Treatment of Meat Packing Wastes," Proc. 2nd Nat. Symp. Food Process. Wastes, pp. 617–635. Pacific Northwest Water Lab., Environmental Protection Agency, 1971.
21. Dart, M. C., The treatment of meat trade effluents. *Eff. Water Treat. J.* **7,** 29–33 (1967).
22. Anonymous, An industrial waste guide to the meat industry. *U.S., Pub. Health Serv., Publ.* **386** (1958).
23. Crandall, C. S., Kerrigan, J. E., and Rohlich, G. A., Nutrient problems in meat industry wastewaters. *Proc. Purdue Ind. Waste Conf.,* **26,** (1971).
24. Porges, R., Wastes from poultry dressing establishments. *Sewage Ind. Wastes* **22,** 531–539 (1950).

25. Porges, R., and Struzeski, E. J., Characteristics and treatment of poultry processing wastes. *Proc. Purdue Ind. Waste Conf.* **17,** 583–601 (1962).
26. Nemerow, N. L., Baffled biological basins for treating poultry plant wastes. *J. Water Pollut. Centr. Fed.* **41,** 1602–1612 (1969).
27. Teletzke, G. H., Chickens for the barbeque—wastes for aerobic digestion. *Wastes Eng.* 134–138, March (1961).
28. Harper, W. J., and Blaisdell, J. L., "State of the Art of Dairy Food Plant Wastes and Waste Treatment," Proc. 2nd Nat. Symp. Food Process. Wastes, pp. 509–545. Pacific Northwest Water Lab., Environmental Protection Agency, 1971.
29. Lawton, G. W., Engelbert, L. E., Rohlich, G. A., and Porges, H., Effectiveness of spray irrigation as a method for disposal of dairy plant wastes. *Wisc., Eng. Exp. Sta., Res. Rep.* 15 (1960).
30. Soderquist, M. R., Williamson, K. J., Blanton, G. I., Phillips, D. C., Law, D. K., and Crawford, D. L., "Current Practice in Seafoods Processing Waste Treatment," Proj. 12060 ECF. Water Quality Office, Environmental Protection Agency, 1970.
31. Hart, S. A., and McGauhey, P. H., Wastes management in the food producing and processing industries. *Pac. Northwest Ind. Waste Conf., 11th, 1960*
32. Morrison, F. B., "Feeds and Feeding," 22nd ed. Morrison Publ. Co., Ithaca, New York (1956).
33. Sobel, A. T., "Some Physical Properties of Animal Manures Associated with Handling," Proc. Nat. Symp. Anim. Waste Manage., Publ. SP-0366, pp. 27–32. Amer. Soc. Agr. Eng., 1966.
34. Pryor, W. J., and Connor, J. K., A note on the utilization by chickens of energy from faeces. *Poultry Sci.* **43,** 833–834 (1964).
35. Perkins, H. F., Parker, M. B., and Walker, M. L., Chicken manure—its production, composition, and use as a fertilizer. *Ga., Agr., Exp. Sta., Bull.* [N. S.] **123** (1964).
36. Taiganides, E. P., and Hazen, T. E., Properties of farm animal excreta. *Trans. ASAE (Amer. Soc. Agr. Eng.)* **9,** 375–376 (1966).
37. Dornbush, J. N., and Anderson, J. R., Lagooning of livestock wastes in South Dakota. *Proc. Purdue Ind. Waste Conf.* **19,** 317–325 (1964).
38. Hart, S. A., and Turner, M. E., Lagoons for livestock manure. *J. Water Pollut. Contr. Fed.* **37,** 1578–1596 (1965).
39. Baines, S., Some aspects of the disposal and utilization of farm wastes. *Inst. Sewage Purif., J. Proc.* **63,** 578–588 (1964).
40. Little, J. F., Agriculture and the prevention of River pollution as experienced in the west of Scotland. *Inst. Sewage Purif., J. Proc.* **65,** 452–454 (1966).
41. Stewart, T. A., and McIlwain, R., Aerobic treatment of poultry manure using an oxidation ditch. *In* "Livestock Waste Management and Pollution Abatement," Publ. PROC-271, pp. 261–263. Amer. Soc. Agr. Eng., 1971.
42. Sanderson, W., Studies of the character and treatment of wastes from duck farms. *Proc. Purdue Ind. Waste Conf.* **8,** 170–176 (1953).
43. Loehr, R. C., and Johanson, K. J., Removal of phosphates from duck wastewaters—1970 laboratory study, Rep. 71-2. Agr. Waste Manage. Program, Cornell University, Ithaca, New York, 1971.
44. Scheltinga, H. M. J., and Poelma, H. R., "Treatment of Farm Wastes," Proc. Symp. Farm Wastes, pp. 138–142. Inst. Water Pollut. Contr., Newcastle-upon-Tyne, 1970.
45. Niles, C. F., Egg laying house wastes. *Proc. Purdue Ind. Waste Conf.* **22,** 334–341 (1967).
46. Soutar, D. S., and Baxter, S. H., Disposal of effluent from the piggery. *Agriculture* **67,** 165–170 (1960).

47. Pontin, R. A., and Baxter, S. H., "Wastes from Pig Production Units," *Inst. Sewage Purif. J. Proc.* **67,** 632–643 (1968).
48. Taiganides, E. P., Hazen, T. E., Baumann, E. R., and Johnson, D., Properties and pumping characteristics of hog wastes. *Trans. ASAE* (*Amer. Soc. Agr. Eng.*) **7,** 123 (1964).
49. Poelma, H. R., "The Biological Breakdown of Pig Urine," Bull. No. 18. Inst. Farm Buildings, Wageningen, Netherlands, 1966.
50. Townshend, A. R., Reichert, K. A., and Nodwell, J. H., "Status Report on Water Pollution Control Facilities in the Province of Ontario," Proc. Agr. Waste Manage. Conf., pp. 131–149, Cornell University, Ithaca, New York, 1969.
51. Clark, C. E., Hog waste disposal by lagooning. *J. Sanit. Eng. Div., Amer. Soc. Civil Eng.* **91,** SA6, 27–42 (1965).
52. Schmid, L. A., and Lipper, R. T., "Swine Waste Characterization and Anaerobic Digestion," Proc. Agr. Waste Manage. Conf., pp. 50–57. Cornell University, Ithaca, New York. 1969.
53. Weller, J. B., "Building Design," Proc. Symp. Farm Wastes, pp. 84–93. Inst. Water Pollut. Contr., Newcastle-upon-Tyne, 1970.
54. Muehling, A. J., "Swine Housing and Waste Management," Agr. Eng. Dept., University of Illinois, Urbana, 1969.
55. Scheltinga, H. M. J., Aerobic purification of farm waste. *Inst. Sewage Purif., J. Proc.* **65,** 585–588 (1966).
56. Irgens, R. L., and Day, D. L., Laboratory studies of aerobic stabilization of swine wastes. *J. Agr. Eng. Res.* **11,** 1–10 (1966).
57. Jeffrey, E. A., Blackman, W. C., and Ricketts, R. L., "Aerobic and Anaerobic Digestion Characteristics of Livestock Wastes," Eng. Ser. Bull. No. 57. University of Missouri, Columbia, 1963.
58. Witzel, S. A., McCoy, E., Polkowski, L. B., Attoe, O. J., and Nichols, M. S., "Physical, Chemical and Bacteriological Properties of Farm Wastes (Bovine Animals)," Proc. Nat. Symp. Anim. Waste Manage., Publ. SP-0366, pp. 10–14. Amer. Soc. Agr. Eng., 1966.
59. Anonymous, "Water Pollution Research—1963," pp. 73–77. Dept. Sci. Ind. Res., HM Stationary Office, London, 1964.
60. Wheatland, A. B., and Borne, B. J., Treatment of farm effluents. *Chem. Ind.* (*London*) pp. 357–362 (1964).
61. Ruf, J. A., and Loehr, R. C., Anaerobic treatment of milking parlor wastes. *J. Water Pollut. Contr. Fed.* **40,** 83–94 (1968).
62. Zall, R. R., Characteristics of milking center waste effluents from New York State dairy farms. *J. Milk Food Technol.* **35,** 53–55 (1972).
63. Grub, W., Albin, R. C., Wells, D. M., and Wheaten, R. Z., "The Effect of Feed, Design and Management on the Control of pollution from Beef Cattle Feedlots," Proc. Agr. Waste Manage. Conf., pp. 217–222, Cornell University, Ithaca, New York, 1969.
64. Loehr, R. C., and Agnew, R. W. Cattle wastes—pollution and potential treatment. *J. Sanit. Eng. Div., Amer. Soc. Civil Eng.* **93,** SA4, 72–91 (1967).
65. McCalla, T. M., Frederich, L. R., and Palmer, G. L., Manure decomposition and fate of breakdown products in soil. *In* "Agricultural Practices and Water Quality," DAST 26, 13040 EYX, pp. 241–255. Federal Water Pollution Control Administration, 1969.
66. McCalla, T. M., and Viets, F. G., "Chemical and Microbial Studies of Wastes from Beef Cattle Feedlots," Semin. Manage. Beef Cattle Feedlot Wastes. University of Nebraska, Lincoln, 1969.
67. Anonymous, "The Environmental Complex in Livestock Housing," Farm Building Rep. Scottish Farm Buildings Investigation Unit, Aberdeen, Scotland, 1969.

68. Eby, H. J., and Wilson, G. B., "Poultry House Dust, Odor, and Their Mechanical Removal," Proc. Agr. Waste Manage. Conf., pp. 303–309. Cornell University, Ithaca, New York, 1969.
69. Howe, R. H. L., Research and practice in animal wastes treatment. *Water Wastes Eng.* **6,** A14–A18 (1969).
70. Berryman, C., "Composition of Organic Manures and Waste Products Used in Agriculture," N.A.A.S. Advisory Pap. No. 2. Ministry of Agriculture Fisheries and Food, London, 1970.
71. Anonymous, "The Economics of Clean Water," Vol. III. Federal Water Pollution Control Administration, 1970.
72. Cramer, C. O., Converse, J. C., Tenpas, G. H., and Schlough, D. A., The design of solid manure storages for dairy herds. *Winter Meet., Amer. Soc. Agr. Eng.* Paper 71–910 (1971).
73. Robinson, K., Draper, S. R., and Gelman, A. L., Biodegradation of pig waste: Breakdown of soluble nitrogen compounds and the effect of copper. *Environ. Pollut.* **2,** 49–56 (1971).
74. Robertson, A. M., and Galbraith, H., "Effect of Ventilation on the Gas Concentration in a Part-Slatted Piggery," Farm Buildings R & D Studies. Scottish Farm Buildings Investigation Unit, Aberdeen, Scotland, 1971.
75. O'Callaghan, J. R., Dodd, V. A., O'Donoghue, P. A. J., and Pollock, K. A., Characterization of waste treatment properties of pig manure. *J. Agr. Eng. Res.* **16,** 399–419 (1971).
76. Layton, R. F., An industrial waste survey of a poultry processing plant for broilers. *Proc. Purdue Ind. Waste Conf.* **27,** (1972).
77. Riddle, M. J., and Murphy, K. L., An effluent study of a fresh water fish processing plant. *Proc. Purdue Ind. Waste Conf.* **27,** (1972).
78. LaBella, S. A., Thaker, I. H., and Tehan, J. E., Treatment of winery wastes by aerated lagoon, activated sludge process, and rotating biological contactors or 'RBC'. *Proc. Purdue Ind. Waste Conf.* **27,** (1972).
79. Ngoddy, P. O., Harper, J. P., Collins, R. K., Wells, G. D., and Heidor, F. A., "Closed System Waste Management for Livestock," Final Rep., Proj. 13040 DKP, Environmental Protection Agency, 1971.
80. Beefland International Inc., "Elimination of Water Pollution by Packinghouse Animal Paunch and Blood," Final Rep., Proj. 12060 FDS. Environmental Protection Agency, 1971.
81. Crowley Milk Co., "Membrane Processing of Cottage Cheese Whey for Pollution Abatement," Final Rep., Proj. 12060 DXF. Environmental Protection Agency, 1971.
82. Battelle Memorial Inst., "Inorganic Fertilizer and Phosphate Mining Industries—Water Pollution and Control," Final Rep., Proj. 12020 FPD. Environmental Protection Agency, 1971.
83. Longhouse, A. D., "Reduction in Moisture and Daily Removal of Wastes from Caged Laying Hens," Proc. Agr. Waste Manage. Conf., pp. 173–185. Cornell University, Ithaca, New York, 1972.
84. McCalla, T. M., Ellis, J. R., Gilbertson, C. B., and Woods, W. R., "Chemical Studies of Runoff, Soil Profile and Groundwater from Beef Cattle Feedlots at Mead, Nebraska," Proc. Agr. Waste Manage. Conf., pp. 211–223. Cornell University, Ithaca, New York, 1972.
85. Ludington, D. C., Sobel, A. T., Loehr, R. C., and Hashimoto, A. G., "Pilot Plant Comparison of Liquid and Dry Waste Management Systems for Poultry Manure." Proc. Agr. Waste Manage. Conf., pp. 569–580. Cornell University, Ithaca, New York, 1972.

FUNDAMENTALS AND PROCESSES

5

Biological Processes

Introduction

Biological degradation of waste is a natural process that has occurred since the beginning of time. Controlled and uncontrolled biological systems are the major systems used to treat organic wastes. An understanding of the factors affecting the biological processes occurring in these systems is essential to the design and operation of treatment facilities for agricultural wastes. The systems may treat liquid or solid wastes, may be aerobic, anaerobic, or facultative, and may be within controlled structures or unconfined on the land. Examples of biological treatment processes include oxidation ponds, aerated lagoons, oxidation ditches, anaerobic lagoons, anaerobic digesters, composting, and land disposal.

Because the processes are biological, an understanding of the processes must be based upon the fundamentals of microbiology and on the transformations in biological waste treatment units. If this understanding can be achieved, rational predictions of performance become possible and the capabilities of a process can be better utilized. Without an understanding of the fundamentals, the processes can be treated only as "black boxes" in which the performance is subject to parameters seemingly beyond our control. Lack of proper understanding means that successful design and operation of biological processes must be based only on prior performance which may be difficult to translate to different wastes and environmental conditions.

Biochemical Reactions

In the biological systems, microorganisms utilize the wastes to synthesize new cellular material and to furnish energy for synthesis. The organisms also can use previously accumulated internal or endogenous food supplies for their respiration and do so especially in the absence of external or exogenous food sources. Synthesis and endogenous respiration occur simultaneously in biological systems with synthesis predominating when there is an excess of exogenous food and endogenous respiration dominating when the exogenous food supply is small or nonexistent.

Regardless of the biological system utilized, the principles of energy, synthesis, and endogenous cellular respiration are basic. The rates at which these reactions occur are a function of the environmental conditions imposed by and/or on a given biological treatment process.

The general reactions that occur can be illustrated in Eq. (5.1).

$$\text{Energy containing metabolizable wastes} + \text{microorganisms} \rightarrow \text{end products} + \text{more microorganisms}. \tag{5.1}$$

Equation (5.1) represents energy–synthesis reactions in which the wastes are metabolized for energy and for the synthesis of new cells. The energy utilized in Eq. (5.1) is obtained during the metabolism of the wastes. Synthesis or growth is affected by the ability of the microorganisms to metabolize and assimilate the food, the presence of toxic materials, the temperature and pH of the system, and the presence of adequate accessory nutrients and trace elements.

The wastes must contain sufficient carbon, nitrogen, phosphorus, and trace minerals to satisfy nutritional requirements. In a biological system, the indispensable nutrient that is present in the smallest quantity needed for microbial growth will become the limiting factor. With most organic wastes, adequate nutrients are available and the biological reactions proceed at a rate constrained only by environmental factors such as temperature, pH, and inhibitory compounds. In batch biological waste treatment units containing adequate nutrients at nontoxic concentrations initially, available carbon becomes a limiting nutrient as the carbonaceous material is metabolized and lost from the systems as carbon dioxide in aerobic systems or as carbon dioxide and methane in anaerobic systems.

When growth becomes limited, the microorganisms die and lyse releasing the nutrients of their protoplasm for utilization by still living cells in an autoxidative or endogenous cellular respiration process:

$$\text{Microorganisms} \rightarrow \text{end products} + \text{fewer microorganisms} \tag{5.2}$$

Endogenous respiration proceeds in the presence as well as in the absence of an external food source. The rate of cellular oxidation and endogenous respiration is related to the mean time the cells have undergone treatment, i.e., solids retention time.

In the presence of waste material (food), microbial metabolism will occur to produce new cells and energy and the microbial solids will increase. In the absence of food, endogenous respiration will predominate and a reduction of the net microbial solids will occur. The microbial mass will not be reduced to zero even with a long endogenous respiration period, however. A residue of about 20 to 25% of synthesized microbial mass will remain (1, 2). Even in a long term biological treatment unit, there will be a minimum rate of solids accumulation. Any inert solids in the raw waste will increase the rate of solids buildup in the unit. Eventually these solids must be removed from the units.

When the organic matter is metabolized and converted into microbial cells, the waste is only partially stabilized. As indicated in Eq. (5.2), the microbial cells are capable of further degradation. Only when the microbial cells are oxidized or removed does a stabilized effluent result.

It is possible to design and operate a biological treatment unit to function in any portion of a synthesis-endogenous respiration relationship. The specific design will depend on the characteristics of the desired effluent. If the treated wastes are to be discharged to surface waters, a high quality effluent will be required. This can be attained by operating well into the endogenous region and removing residual solids from the effluent before it is discharged. If the land is the ultimate disposal point, a high quality effluent may not be necessary. In this case the treatment unit can be operated in synthesis phase and without separation of the solids since further degradation will take place on the soil.

Basic Biological Processes

Biological processes can be defined by the presence or absence of dissolved oxygen, i.e., aerobic or anaerobic, by their photosynthetic ability, or by the mobility of the organisms, i.e., suspended or adherent growth. Common examples of processes that are utilized for waste treatment are shown in Table 5.1. Since the terms are not mutually exclusive, some processes can be defined in more than one manner.

Aerobic

As used with biological waste treatment processes, the term refers to processes in which dissolved oxygen is present. The oxidation of organic

Table 5.1

Common Biological Treatment Processes

Aerobic	Anaerobic
Activated sludge units	Anaerobic lagoons
Trickling filters	Digesters
Oxidation ponds	Anaerobic filters
Aerated lagoons	
Oxidation ditch	Photosynthetic
	Oxidation ponds
Suspended growth	
Activated sludge	Adherent growth
Aerated lagoons	Trickling filters
Mixed digesters	Rotating biological contactors
Oxidation ditch	Anaerobic filters
	Denitrification columns

matter using molecular oxygen as the ultimate electron acceptor is the primary process yielding useful chemical energy to most microorganisms in these processes. Microbes that use oxygen as the ultimate electron acceptor are aerobic microorganisms.

Anaerobic

Some microorganisms are able to function without dissolved oxygen in the system. Such microorganisms can be called anaerobic organisms or anaerobes. Certain anaerobes cannot exist in the presence of dissolved oxygen and are obligate anaerobes. Examples of these are the methane bacteria commonly found in anaerobic digesters, anaerobic lagoons, and swamps. Anaerobes obtain their energy from the oxidation of complex organic matter but utilize compounds other than dissolved oxygen as oxidizing agents. Oxidizing agents are defined broadly as electron acceptors. Oxygen is not necessary to have an oxidation reaction. Oxidizing agents other than oxygen that can be used by microorganisms include carbon dioxide, partially oxidized organic compounds, sulfate, and nitrate. The process by which organic matter is degraded in the absence of oxygen frequently is called fermentation.

Facultative

Only a few species of organisms are obligate anaerobes or aerobes. A large number of organisms can live either in the absence or presence of oxygen. Organisms that function under either anaerobic or aerobic conditions are facultative organisms. When oxygen is absent from their environment, they are able to obtain energy from degradation of organic matter by nonaerobic mechanisms but, if dissolved oxygen is present, they metabolize the organic matter more completely. Organisms can obtain more energy by aerobic oxidation than by anaerobic oxidation.

Biological waste treatment units may be designed to be either aerobic or anaerobic. There are occasions when anaerobic conditions occur in units that are designed to be aerobic. Examples of these conditions are organic matter that has settled to the bottom of oxidation ponds and streams, i.e., benthic deposits, when aerobic systems are overloaded because of an increase in the strength of the raw waste, and in the interior of activated sludge floc particles and trickling filter growths. The majority of the organisms in biological waste treatment processes are facultative organisms.

Photosynthetic

Photosynthesis is the utilization of solar energy by the chlorophyll of green plants for the incorporation of carbon dioxide and other inorganic constituents in the production of cellular material. In this process molecular oxygen is formed. The photosynthetic organisms of interest in biological treatment systems are algae and rooted or floating plants. Examples of such biological treatment systems include oxidation ponds, streams, reservoirs, lakes, and high rate algal production systems to recover the nutrients in wastes.

Suspended Growth

This term refers to mixtures of microorganisms and the organic wastes. The microorganisms are able to aggregate into flocculant masses and are able to move with the liquid flow. Agitation of the liquid keeps microbial solids in suspension. Suspended growth processes may be either aerobic or anaerobic. Anaerobic suspended growth units can be agitated by mechanical mixing and gas diffusion. Activated sludge units, aerated lagoons, oxidation ditches, and well-mixed anaerobic digesters are suspended growth processes.

Adherent Growth

Microbial growth is adherent when the microorganisms grow on a solid support medium and the wastes flow over or come in contact with the organisms. The support media can be large stones, rocks, slag, corrugated plastic sheets, or rotating disks. Commonly the organic wastes flow over or through the openings of the supporting media. Although the vast majority of adherent growth systems currently used for waste treatment are aerobic, a few are anaerobic. Examples of adherent growth units are trickling filters, rotating biological disks, and anaerobic filters.

Summary

The above terminology describes the functional aspects of the common biological waste treatment processes. Their operational forms are described in detail in subsequent sections.

Energy Relationships

Knowledge of the energy relationships of microbial cells permits an understanding of energy available for synthesis and respiration, of production of microbial cells in biological waste treatment units, and of the nature of the expected end products under certain conditions. All cells, whether animal, plant, or microbial, use similar fundamental mechanisms for their energy transforming activities. These activities involve transferring chemical energy from food to the processes which utilize energy for the functions and survival of living cells. In both aerobic and anaerobic cells, the energy of the food material is conserved chemically in the compound adenosine triphosphate (ATP). ATP is the carrier of chemical energy from the oxidation of foods, either aerobic or anaerobic, to those processes of the cells which do not occur spontaneously and can proceed only if chemical energy is supplied. These processes are involved in the performance of osmotic, mechanical, or chemical work. In the context of this book, the food for the cells would be organic wastes of agriculture.

During the oxidation in the cells, ATP is formed from adenosine diphosphate (ADP). ATP is the high energy form of the energy transporting system and ADP is the lower energy form. A portion of the energy of the oxidation thus is conserved as the energy of the ATP. This process operates in a continuous dynamic cycle, receiving energy during the oxidation of foods and releasing energy during the performance of cellular work. A molecule of inorganic phosphate (P_i) is released when ADP is formed and incorporated in ATP when the ATP is formed. The principle of the cellular energy cycle is shown in Fig. 5.1. Although ATP is not the only energy carrying compound in every cellular reaction, it is the common intermediate in the energy transformation in the cells.

The purpose of biological waste treatment is to stabilize or oxidize the organic wastes of man, industry, and agriculture. Oxidation is the process in which a molecule or compound loses electrons. Reduction is the process in which a molecule or compound gains electrons. Examples of these pro-

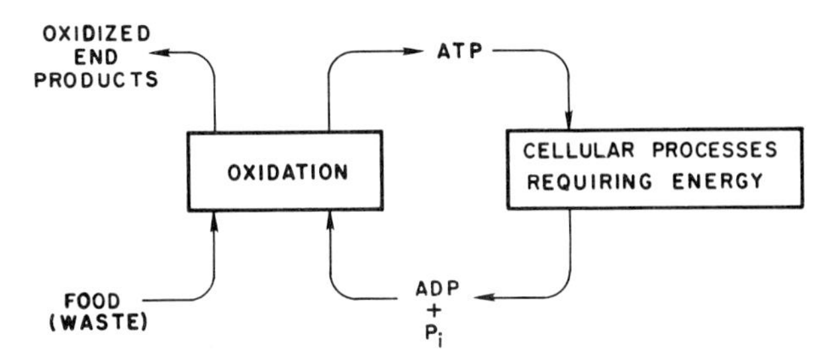

Fig. 5.1. Schematic pattern of energy transfer in the ATP–ADP system.

cesses are noted in Fig. 5.2. In the first case ferrous iron is oxidized to ferric iron with the release of an electron. In the second case carbon dioxide is reduced to methane. The carbon gains electrons and is reduced.

The general oxidation–reduction relationship can be indicated by:

$$A \times H + B \rightarrow A + B \times H \tag{5.3}$$

where B is the electron (hydrogen) acceptor and is being reduced and A is the compound being oxidized. Although all the reactions involve oxidation, they sometimes are referred to in terms of the type of hydrogen acceptor. Transformations in which oxygen is the hydrogen acceptor are called oxidation; when nitrate is the hydrogen acceptor, denitrification; when sulfate is the hydrogen acceptor, sulfate reduction; and with carbon dioxide as the hydrogen acceptor, the transformation is methane fermentation. Oxidation and reduction reactions do not occur independently but as coupled reactions. When a compound is oxidized, another compound must be reduced.

The reducing agent is an electron donor and an oxidizing agent is an electron acceptor. Each electron donor has a characteristic electron pressure and each electron acceptor has a characteristic electron affinity (3). Electron donors may be arranged in a series of decreasing electron pressures. The tendency will be for electrons to flow from compounds having the highest electron pressure to compounds lower in the series.

A schematic diagram of the electron flow in the aerobic oxidation of an

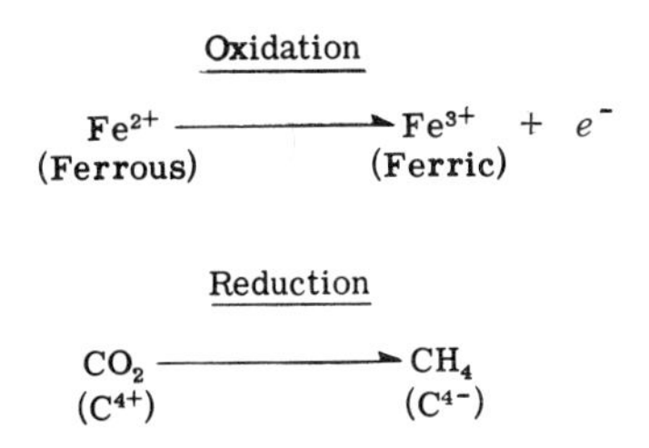

Fig. 5.2. Examples of oxidation and reduction.

organic compound illustrates the oxidation–reduction sequence (Fig. 5.3). In each step, two electrons are passed along. The electrons flow from compounds having the highest electron pressure with the oxidized forms of the electron carriers serving as electron acceptors. The electron carriers noted in Fig. 5.3 are NAD, nicotinamide adenine dinucleotide; FAD, flavin adenine dinucleotide; CYT b, c, a, cytochromes; CYT a_3, cytochrome oxidase. The cytochromes are iron-containing enzymes. The iron is reduced and oxidized as the oxidation–reduction reactions occur. The respiratory sequence involving the cytochromes is the final common metabolic pathway by which all electrons derived from the oxidation of different organics flow to oxygen, the final oxidant, or acceptor of electrons in aerobic cells.

In anaerobic cells, the cytochrome pathway does not exist. The energy-conserving steps of the cytochrome system are not available to organisms that do not use oxygen as the terminal electron or hydrogen acceptor. An example of the difference in relative energy conversions by aerobic and by anaerobic cells can be illustrated by the metabolism of glucose (Fig. 5.4). Anaerobic cells obtain energy from the conversion of glucose to lactate which then leaves the cell as a metabolic waste. The energy available to anaerobic cells from this conversion is only about 7% of the amount that would be available if glucose were oxidized aerobically. Aerobic organisms can conserve for themselves a greater portion of the available energy from the metabolism of organic matter than can anaerobic organisms. Thus anaerobic organisms must process a greater quantity of food to obtain the same amount of energy.

This information is useful in predicting the products and the efficiency of aerobic and anaerobic biological treatment processes. The synthesis of organisms per molecule of ATP should be the same for most bacteria since once the substrate energy is converted to ATP or biological energy, the growth of organisms follows generally similar biochemical pathways. Both

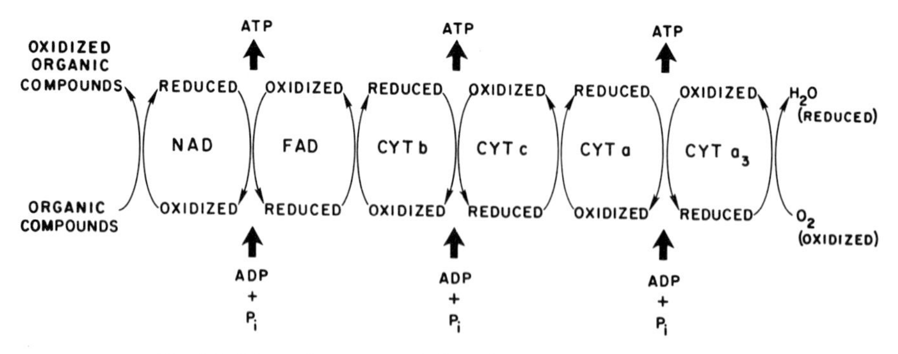

Fig. 5.3. Schematic diagram of the oxidation of organic compounds.

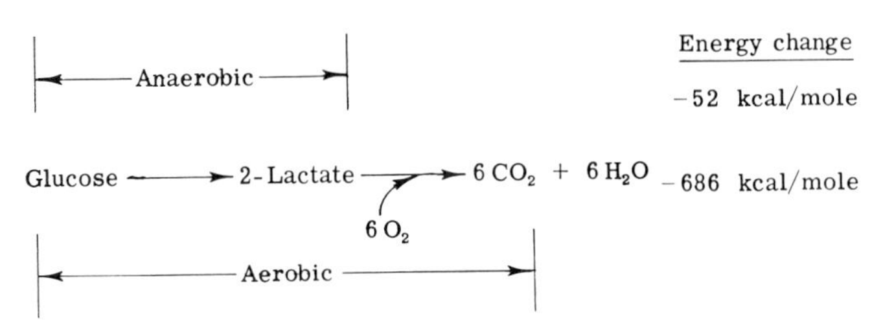

Fig. 5.4. Aerobic and anaerobic metabolism of glucose.

aerobic and anaerobic bacteria have essentially the same composition and both contain ATP. The concept of ATP as an energy resource permits the formulation of a general relationship between substrate and growth (4). Because the energy recovery per unit of food is so small for anaerobic organisms, it follows that the amount of microbial cells synthesized per unit of food metabolized will be significantly less than for aerobic organisms.

Figure 5.4 shows the end products of aerobic metabolism in the most oxidized state, carbon dioxide and water. The end products of anaerobic metabolism are in a partially oxidized state and, if oxygen were available, would have an oxygen demand.

For a given organic loading, aerobic conditions will produce a more oxidized end product or effluent than will anaerobic conditions and will permit synthesis of a greater quantity of microbial cells. These additional cells are an asset because it is thus possible to have a larger amount of active microbial solids to increase the removal of organic wastes. The greater synthesized microbial cells in an aerobic unit will increase the sludge disposal problem however. Anaerobic conditions will produce smaller quantities of microbial cells for ultimate disposal but because of the decreased rates of synthesis, there may be problems in maintaining adequate microbial solids in anaerobic units. The end products of anaerobic units are in the partially oxidized state. Some of the end products, i.e., methane, carbon dioxide, hydrogen, and nitrogen gases, can be exhausted to the atmosphere without problems. Others can produce disagreeable odors, i.e., mercaptans, amines, or volatile acids, and will exert an oxygen demand when released to the environment.

Microorganisms

Biological waste treatment processes contain a mixture of microorganisms capable of metabolizing organic wastes. Within limits, they can adjust to varying organic loads and environmental influences, such as temperature and pH, that may be imposed. Extreme temperature, high concentrations of metal ions, or toxic chemicals can decrease or eliminate the activity of the microorganisms. The microorganisms in various biological treatment systems include bacteria, fungi, algae, protozoa, rotifers, crustacea, bacteriophage, worms, and insect larvae depending upon environmental conditions.

Bacteria

Bacteria are the most important group of microorganisms in waste treatment systems. The diverse biochemical activities of bacteria, as a group, enable them to metabolize most organic compounds found in municipal, industrial, or agricultural wastes. Aerobes and facultative bacteria are active in all aerobic treatment units. Facultative and obligate anaerobic bacteria are active in anaerobic treatment units. Bacteria are single cell microorganisms that metabolize soluble food. Insoluble foods are converted into soluble food by microbial enzymes. The bacteria exist in a variety of forms, usually some modification of a cylinder or ovoid, with dimensions on the order of a few micrometers. They exist in waste treatment processes in agglomerations of varying arrangements and species.

A microbe is a complex organized system. A representative formula for bacterial cells is $C_5H_7O_2N$ or $C_{75}H_{105}O_{30}N_{15}P$. The composition of bacteria is not constant and varies according to the stage of growth and the particular substrate utilized. Storage of reserve materials can occur during the growth phase and will alter the representative composition of bacterial protoplasm. This empirical formula expresses only the average proportions of the principal constituents in a bacterial cell. The cell contains many other elements in small amounts.

Bacteria can be divided into groups depending upon their source of carbon used for synthesis of protoplasm. Organisms that use organic carbon as their carbon source are heterotrophic organisms while organisms that utilize carbon dioxide for cellular carbon are autotrophic organisms. Heterotrophs are the most numerous and important group of organisms in the common biological waste treatment processes.

The microbial species having the fastest growth rate and the ability to utilize most of the available organic matter will be the predominant species. Shifts in microbial predominance occur in waste treatment systems as environmental conditions such as temperature, pH, available dissolved oxygen, ultimate hydrogen acceptor, or available food vary in the system.

A useful characteristic of some bacteria is their ability to flocculate. Such flocculation permits the removal of microbial solids in a subsequent solids separation unit and assists in the production of a good quality effluent.

Fungi

Fungi are nonphotosynthetic, multicellular, aerobic, branching, filamentous microorganisms that metabolize soluble food. Both bacteria and fungi can metabolize the same kinds of organic material. The environmental conditions will determine which group of organisms will predominate. Fungi will predominate at low pH levels, low moisture content, in low nitrogen wastes, and when certain nutrients are missing. The composition of fungal cells can be represented empirically by $C_{10}H_{17}O_6N$.

Fungi are not active in anaerobic systems. Since fungal cells contain less nitrogen than bacterial cells, fungi may compete more favorably in wastes having a lower nitrogen content than required for bacterial synthesis. Many fungi grow well at pH levels of 4 to 5, levels at which it is difficult for bacteria to compete.

The filamentous nature of the fungi make them less desirable in biological waste treatment units because they do not settle well. Under the normal environmental conditions that exist in most waste treatment processes, the fungi will not predominate. Fungi will be of secondary importance in common, properly operating aerobic biological treatment units.

Algae

Algae are photosynthetic autotrophs. The composition of algal cells can be represented by $C_{106}H_{180}O_{45}N_{16}P_1$ (5). Since the nutritive requirements of algal species are different, this formula is an empirical average. Algae obtain their energy from sunlight and utilize inorganic materials such as carbon dioxide, ammonia or nitrate, and phosphate in the synthesis of additional cells. In photosynthesis molecular oxygen is formed. It is released to the environment and utilized by bacteria as the bacteria metabolize available organic matter. The design and management of oxidation ponds attempts to balance and exploit both groups of organisms.

Algae obtain carbon dioxide from the following sources in water or waste water: (a) absorption from the atmosphere, (b) respiration of aerobic and anaerobic heterotrophic organisms, and (c) bicarbonate alkalinity. As the carbon dioxide is removed from a waste water by growing algae, the pH will increase. pH values as high as 10 are not uncommon in active algal systems such as oxidation ponds and similar units. Although algal growth can be controlled by carbon limitation, carbon from alkalinity and bacterial carbon dioxide production provides an ample amount of carbon for algal growth. Carbon in natural systems rarely limits algal growth.

The type of nitrogen assimilation by the algae affects the alkalinity of the liquid. Nitrate utilization increases total alkalinity by hydroxide ion production. Ammonia utilization produces hydrogen ions and decreases the total alkalinity.

Algae are of consequence only where sufficient sunlight can penetrate the liquid. Algae will not predominate where there is high turbidity as in activated sludge units and aerated lagoons, where sunlight is excluded, or where the liquid is dark in color.

In the absence of sunlight, photosynthesis ceases and the endogenous respiration of the algae continues in the same manner as it does with bacteria. The algae thus present an additional oxygen demand on the unit in which they exist. The decomposition of the biodegradable organic fraction of algae under anaerobic conditions was described by first-order kinetics with an average decay constant of 0.022/day (6). The extent of algal decomposition will depend upon the algal age prior to decomposition.

After 0.5–1 year of aerobic decomposition, an average of 50% of the initial nitrogen and phosphorus remained in the undecomposed algal fraction while the other 50% was regenerated. Under anaerobic conditions 40% of the nitrogen and 60% of the phosphorus were regenerated (7). Nutrients from dead algal cells are released to surface waters for a long period of time. To accomplish a high degree of organic carbon, nitrogen, and phosphate removal in biological treatment units utilizing algae, the algal cells must be removed from the unit effluent before discharge.

Protozoa

Protozoa are single-celled organisms that can metabolize both soluble and insoluble foods. The protozoa found in aerobic treatment systems include flagellates, free-swimming ciliates, and stalked ciliates which are attached to solid particles by stems. Protozoa reduce the concentration of bacteria and nonmetabolized particulate organic matter in a treatment system and assist in producing a higher quality and clearer effluent.

Activated sludge units free of protozoa produced effluents of high turbidity. The turbidity was caused by the presence of large numbers of dispersed bacteria. As a result, effluent BOD and nonsettleable solids were high. The addition of ciliated protozoa to these units increased effluent quality and decreased bacterial numbers (8). The succession of protozoa types in aerobic systems has been delineated and related to the degree of treatment in the system (9).

Protozoa generally have more complex nutritional requirements than do bacteria or fungi. Because of their utilization of particulate organic matter and their need for dissolved oxygen, protozoa will exist in well-stabilized systems in which the soluble food has been converted to microbial cells and in which the oxygen supply exceeds the oxygen demand. Since they are sensitive to dissolved oxygen changes, they can serve as indicators of the status of aerobic biological waste treatment.

Protozoa have been observed in anaerobic treatment of sewage solids (10) and in systems treating animal wastes, especially ruminant wastes. The role of protozoa in these systems is unknown, but it is postulated as the same as in aerobic systems, i.e., metabolism of particulate material and bacteria, and clarification of the resultant effluent.

Rotifers

Multicellular organisms that can metabolize solid food, such as rotifers, are found in highly stabilized systems having dissolved oxygen at all times. Rotifers metabolize solid particles some of which the protozoa can not use and also assist in producing a nonturbid effluent.

Crustacea

Crustacea are multicellular organisms with hard shells. They grow in well-stabilized systems using smaller organisms as their major source of food. In doing so, they assist in producing a clarified effluent and are indicative of a high quality effluent from aerobic treatment systems.

Summary

The predominance of the various forms of microorganisms in biological systems may at times be indicative of the performance and environmental conditions in the systems. Microscopic examination of the biological system can be utilized as a tentative guide to the quality of the effluent, the degree of treatment that has been accomplished, and changes occurring in the systems. Knowledge of the environmental factors affecting the typical microorganisms can be useful in understanding and operating biological waste treatment units. Many biological treatment units will have their individual peculiarities and with experience the observer can learn to relate the microscopic pattern in a unit with trends in effluent quality and process performance.

Biochemical Transformations

A number of changes take place in biological units. Some of the transformations affect the constituents of the wastes undergoing treatment thus affecting the quality of the unit effluent. Others affect the properties and the quantity of microbial solids. Many of the important transformations in biological waste treatment systems will be discussed in subsequent sections. These are by no means the only ones but they are fundamental transformations in a variety of treatment systems.

Carbon

The oxidation of organic carbon-containing compounds represents the mechanism by which heterotrophic organisms obtain the energy for synthesis. The process is called respiration. The general relationships were noted in Eq. (5.1). In aerobic treatment systems organic carbon is transformed, via many steps, to synthesized microbial protoplasm, $C_5H_7O_2N$, and carbon dioxide.

$$\text{Organic carbon} + O_2 \rightarrow C_5H_7O_2N + CO_2 \tag{5.4}$$

The uptake of oxygen and formation of carbon dioxide represent the effects of respiration.

In anaerobic systems, molecular oxygen cannot be the terminal electron acceptor and all of the respired carbon will not be transformed to carbon dioxide. Under anaerobic conditions, organic carbon is converted to microbial solids, carbon dioxide, methane, and other reduced compounds. Anaerobic metabolism leading to the formation of methane occurs in a series of steps. For simplicity these can be summarized as the conversion of complex organics to simpler compounds:

$$\text{Organic carbon} \rightarrow \text{microbial cells} + \text{organic acids, aldehydes, alcohols, etc.} \tag{5.5}$$

and the conversion of the simpler compounds to gaseous end products:

$$\text{Organic acids} + \text{oxidized organic carbon} \rightarrow \text{microbial cells} + \text{methane} + \text{carbon dioxide} \tag{5.6}$$

Little stabilization of organic matter occurs in the first step [Eq. 5.5]. Stabilization of the organic matter occurs in the second step [Eq. 5.6] in which the carbon compounds, carbon dioxide (CO_2) and methane (CH_4), are released to the atmosphere and removed from the substrate. The oxygen demand of the waste is thus reduced. At standard conditions, the production of 5.6 ft^3 of methane results in the stabilization of 1 lb of ultimate oxygen demand.

Because of the smaller amount of energy available from incomplete oxidation in anaerobic systems, fewer microbial solids are generated per pound of organic waste processed in comparison to aerobic processes. The effluents from anaerobic units will contain organic compounds that are incompletely oxidized and will exert an oxygen demand when discharged to an aerobic environment. A number of these compounds can cause odor problems when the contents of anaerobic units are agitated or disposed of.

Nitrogen

Nitrogen is an important nutrient in biological systems. Nitrogen is about 12% of bacterial protoplasm and 5 to 6% of fungal protoplasm. In waste matter, nitrogen will be present as organic and ammonia nitrogen, the proportion of each depending upon the degradation of organic matter that has occurred. In biological systems, organic nitrogen compounds can be transformed to ammonium nitrogen and oxidized to nitrite and nitrate nitrogen.

$$\text{Organic N} \rightarrow \text{ammonium N} \rightarrow \text{nitrite N} \rightarrow \text{nitrate N} \tag{5.7}$$

The oxidation of ammonia to nitrite and nitrate is termed nitrification and occurs under aerobic conditions. A more basic definition of nitrification is the biological conversion of inorganic or organic nitrogen compounds from a reduced to a more oxidized state. In waste treatment the term usually is used to refer to the oxidation of ammonia. A residual dissolved oxygen concentration of about 2 mg/l has been found necessary to have optimum nitrification. Autotrophic bacteria, such as *Nitrosomonas,* which obtain energy from the oxidation of ammonia to nitrite, and *Nitrobacter,* which obtain energy from the oxidation of nitrite to nitrate, are organisms that in combination can accomplish the complete oxidation of nitrogen.

Ammonia nitrogen is the main soluble nitrogen end product in anaerobic units. The release of ammonia nitrogen to aerobic treatment units or to receiving streams creates an added oxygen demand to these systems. The oxidation of 1 lb of ammonia nitrogen to nitrate nitrogen will require 4.57 lb oxygen. The oxygen demand of ammonia is significant and requires consideration when evaluating the effect of discharging wastes to the environment and when evaluating the design of adequate biological treatment processes.

Dentrification is the process by which nitrate and nitrite nitrogen is reduced to nitrogen gas and gaseous nitrogen oxides under anoxic conditions. This process requires the availability of electron donors (reducing agents). The necessary donors can be organic material such as methanol, addition of untreated wastes, unmetabolized organic matter, or the endogenous respiration of microbial cells.

Dentrification offers the opportunity to reduce the nitrogen content of waste effluents by having a fraction of the nitrogen exhausted to the atmosphere as an inert gas. Because of the role nitrogen plants in the eutrophication and oxygen demand of surface waters, control of nitrogen by denitrification in biological waste treatment systems will play a larger role in the future.

Phosphorus

Phosphorus is an important nutrient in biological processes. The phosphorus content of bacterial cells is about 2%. Microbial synthesis rarely will be an important mechanism for the removal of phosphorus from waste waters. Removal of phosphorus from all waste waters is one of the key methods in minimizing eutrophication of surface waters and will receive increasing emphasis in the future.

The sources of phosphorus in waste waters include organic matter, phosphates originating in cleaning compounds used for process cleanup, and the urine of man and animals. The organic phosphorus is transformed to inorganic phosphorus during biological treatment. In human beings 50 to 65% of the phosphorus discharged is found in urine and the remainder in feces (11). The phosphorus in urine exists as orthophosphate.

Condensed phosphates constitute a substantial portion of the phosphorus in municipal sewage. The form of phosphates in waste waters is of interest since phosphate removal techniques generally are evaluated on their ability to remove orthophosphates. The hydrolysis of condensed phsophates to orthophosphate [Eq. (5.8)] is affected by environmental conditions such as temperature and microbial concentration.

$$\text{Tripolyphosphate } (P_2O_{10}{}^{-5}) + H_2O \rightarrow \text{orthophosphate } (PO_4{}^{-3}) + H^+ \tag{5.8}$$

The rate of hydrolysis of condensed phosphates in the following systems decreases in the order given: activated sludge, untreated waste water, algal cultures, and natural waters.

The orthophosphate concentrations of fresh municipal waste water range from 20 to 30% of the total phosphate concentration, with the influent to biological treatment units ranging from 20 to 85%. The effluent from aerobic treatment units will contain from 60 to 95% orthophosphates, the remainder being condensed phosphates (12). Phosphorus-removing processes should handle condensed phosphates as well as orthophosphates.

Aerobic biological treatment will convert condensed phosphates to orthophosphates. Anaerobic treatment will result in other changes. A primary step in anaerobic treatment is the liquefaction of organic matter and inorganic phosphorus compounds will be released from organic compounds. The effluent from an anaerobic unit can contain a greater concentration of soluble phosphorus compounds than the influent. The release of such effluents to other parts of a waste treatment facility or to the environment can complicate and/or negate phosphorus removal processes at the facility.

Sulfur

Microbial transformations of sulfur are similar to those of nitrogen. Both sulfide and ammonia are decomposition products of organic compounds. Both are oxidized by autotrophic bacteria, as are other incompletely oxidized inorganic sulfur and nitrogen compounds. Sulfate and nitrate are reduced by microorganisms under anaerobic conditions.

Inorganic unoxidized sulfur compounds and elemental sulfur are oxidized by photosynthetic and chemosynthetic bacteria as well as by certain heterotrophic microorganisms. Under anaerobic conditions, sulfide is the reduced endproduct and under aerobic conditions, sulfate is the oxidized end product.

All organisms contain sulfur and are involved in the transformations of sulfur to some degree. The assimilation of sulfur into cellular protoplasm is the primary reaction of heterotrophic organisms. With other organisms, sulfur transformations can provide the energy for metabolism and sulfur compounds can be hydrogen donors or acceptors. Because of these reactions, certain bacteria are designated as sulfur bacteria. These bacteria are autotrophic, can utilize sulfur or incompletely oxidized inorganic sulfur compounds as reducing agents, i.e., direct or indirect hydrogen donors, and can assimilate carbon dioxide as their sole source of carbon.

Food and Mass

The primary purpose of biological waste treatment is to oxidize the organic content of the waste, i.e., the food for the microorganisms. The waste concentration decreases as the microbial mass increases. In aerobic systems, approximately 0.7 lb of cell mass is synthesized for every 1.0 lb of food, as BOD, that is oxidized. Following extensive endogenous respiration, or aerobic digestion of the cells, the 0.7 lb of cells will be reduced to about 0.17 lb of residual cellular material that remains for ultimate disposal. The actual residual cellular solids in a system will be somewhere between the latter two values depending upon how the aerobic system is operated, i.e., the degree of endogenous respiration that takes place. Changes similar to these also occur in anaerobic systems.

Engineers generally use the volatile suspended solids concentration of a biological treatment unit as an estimate of the concentration of active microorganisms in the unit. While this parameter is an imperfect measure of the active mass, it has been a useful design and management parameter. Other parameters have been explored as better measures of both biomass and bioactivity in treatment units. These include dehydrogenase enzyme activity to measure overall rates of cellular oxidation reactions, specific enzymes involved in intermediary metabolism, and DNA concentration. ATP is a specific measure of microbial activity and can be used to estimate viable microorganism concentrations in a biological treatment unit. Measurements using the latter parameters have indicated that in general, less than 20% of volatile suspended solids in a typical activated sludge unit are active microorganisms.

Long-term biological treatment studies have indicated that if all the solids in the effluent from a biological treatment unit can be captured and returned to the system, it may be possible to operate the system without a great deal of sludge wasting and a total oxidation system can be approximated. Changes in bacterial population and in predator population such as protozoa and higher animals were suggested as a reason for the lack of buildup of net solids (13).

Oxygen

Oxygen plays a critical role in biological systems since when it serves as an ultimate hydrogen acceptor, the maximum energy is conserved for the microorganisms. Minimum dissolved oxygen concentrations of from 0.2 to 0.6 mg/l have been suggested as necessary to maintain active aerobic systems (14). The dissolved oxygen concentrations in aerobic treatment units should be kept above about 1.0 mg/l if oxygen limitations are to be avoided. The BOD in the effluent from most aerated units is due to microbial solids.

pH

Biological activity can alter the pH of a treatment unit. Photosynthesis, denitrification, organic nitrogen breakdown, and sulfate reduction are examples of biological reactions that can cause an increase in pH. Sulfate oxidation, nitrification, and organic carbon oxidation are examples of biological reactions that can cause a decrease in pH. The relative changes in pH will be affected by the buffer capacity of the liquid and amount of substrate utilized by the microorganisms.

Nutrient Needs

To achieve satisfactory biological treatment of wastes, the wastes must contain sufficient carbon, nitrogen, phosphorus, and trace minerals to sustain optimum rates of microbial synthesis. In most wastes, nutritional balance is not a problem since there usually is more than enough nitrogen, phosphorus, and trace minerals with respect to the carbon used in cell synthesis. These excess nutrients can be a cause of eutrophication in surface waters when the treated effluent is discharged. Methods to control or manage these excess nutrients are becoming required before discharge of the treated effluents.

Certain wastes, such as some food processing wastes, may have a deficiency of specific nutrients which need to be added in proper amounts to accomplish satisfactory biological waste treatment. Knowledge is required of the amount of nutrients that are needed both to assure that adequate nutrients are available and to avoid excess nutrients in the resultant effluent. Besides being uneconomical, added nutrients appearing in the effluent have the potential of causing environmental quality problems. Inadequate quantities of nutrients, such as nitrogen and phosphorus, tend to decrease the rate of microbial growth, decrease the rate of BOD removal, and impair the settling characteristics of the sludge.

The common approach of avoiding nitrogen or phosphorus limitations is to add nutrients to obtain a BOD:N:P ratio of 100:5:1. This ratio is satisfactory if one wishes to assure no nutrient deficiency but is of little use if the purpose is to have the nitrogen and phosphorus levels be low in the effluent. The above ratio was designed to assure adequate nutrients in high rate biological treatment. Studies with nutrient deficient wastes established that 3–4 lb N/100 lb BOD_5 removed and 0.5–0.7 lb P/100 lb BOD removed would avoid nutrient deficient conditions (15). This results in a BOD:N:P ratio of 100:3:0.6. Other studies with food processing wastes have noted that a BOD to nitrogen ratio of 100:2 or 100:1.5 was satisfactory in treating cannery wastes without a decrease in process efficiency (16). Release of nutrients during endogenous respiration helped lower the nitrogen that had to be added. The treatment of citrus wastes indicated that a BOD:N ratio of about 100:3 and a BOP:P ratio of about 100:2 was satisfactory (17).

The actual nutritional needs will be related to the manner in which the biological treatment process is operated. A high rate process will have a high rate of microbial synthesis and a higher nutrient requirement. However for a stationary or declining growth biological treatment system, such as most treatment systems, a lower rate of microbial synthesis and nutrient requirement will prevail. With the long solids retention time, a matter of days in the common treatment systems, endogenous respiration of the microbial cells will release nutrients to the system. These nutrients will be used in the synthesis of new microbial cells. Approximately 0.11 lb of nitrogen will be released from the oxidation of 1 lb microbial cells.

The required nutrients should be added in relation to the rate of cell synthesis. Practically, when wastes are nutrient deficient, the nutrients should be added to the system in proportion to the nutrients in the microbial solids that are lost in the effluent and/or wasted from the system.

Vegetable processing wastes have been treated using a system in which fungi predominate. In this system, a mycelium growth of 50% of the BOD in the influent was obtained (18). The nutrients were assimilated into fungal cells. The clarified effluent had an ammonia nitrogen concentration of 0.2 mg/l and a soluble phosphorus concentration of 0.01 mg/l. The data emphasize that even if nutrients are tied up in microbial cells, the cells must be removed from the effluent to achieve positive nutrient control. Degradation of these solids in the receiving stream will release a portion of the nutrients. The need for solids separation for nutrient control is apparent if an oxidation pond is used. These ponds are organic matter generators and high removals of BOD, solids, and nutrients will not be achieved unless the solids in the effluent are removed.

Nutrients will not need to be added to animal wastes or meat and poultry processing wastes since nutrients are in excess of the amounts required for synthesis. Wastes having excess nutrients such as domestic sewage or animal processing wastes can be combined with nutritionally deficient carbonaceous wastes for mutual benefit.

Oxygen Demand Measurements

Biochemical Oxygen Demand

The BOD test is one of the most widely applied analytical methods in waste treatment and water pollution control. The test attempts to determine the pollutional strength of a waste in terms of microbial oxygen demand and is an indirect measure of the organic matter in the waste. It evolved as an estimate of the oxygen demand that a treated or untreated waste will have on the oxygen resources of a stream. The acceptable BOD test is described in *Standard Methods* (19).

Experience with a number of organic wastes have indicated that the change in the oxygen demand of a waste (BOD) can be characterized by a first-order equation:

$$\frac{dC}{dt} = -kC \tag{5.9}$$

C is the waste concentration and k is a proportionality constant referred to as the BOD rate constant. Expressing the waste concentration in terms of the amount of oxygen required to biologically oxidize the waste, Eq. (5.9) can be written as:

$$Y = L(1 - e^{-kt}) \tag{5.10}$$

in which Y is the oxygen or BOD exerted in time t, and L is the ultimate amount of oxygen to biologically oxidize the carbonaceous waste or the ultimate first-stage BOD (Fig. 5.5a). The BOD test is conducted as a batch aerobic experiment in BOD bottles. Other oxygen demand determinations can be conducted in manometric respirometers and similar large containers.

Microorganisms can oxidize both carbon-containing compounds (carbonaceous demand) and nitrogen compounds (nitrogenous demand). The nitrogen-oxidizing bacteria are autotrophs, normally not in large concentrations in untreated waste waters. These organisms can be present in well-oxidized waste waters such as aerobically treated waste water effluents from activated sludge and trickling filter plants, and in streams. If a low concentration of nitrifying organisms is present in a BOD bottle, a lag period can exist before the nitrifiers are present in large enough numbers to exhibit a noticeable nitrogeneous demand. In waste waters containing a number of organic compounds, such as agricultural waste waters, a two-stage oxygen demand frequently can be observed if the oxygen demand is measured over a long enough period.

The BOD test is standardized at a 5-day period at 20°C using prescribed quality dilution water to permit comparison of results. The test does not

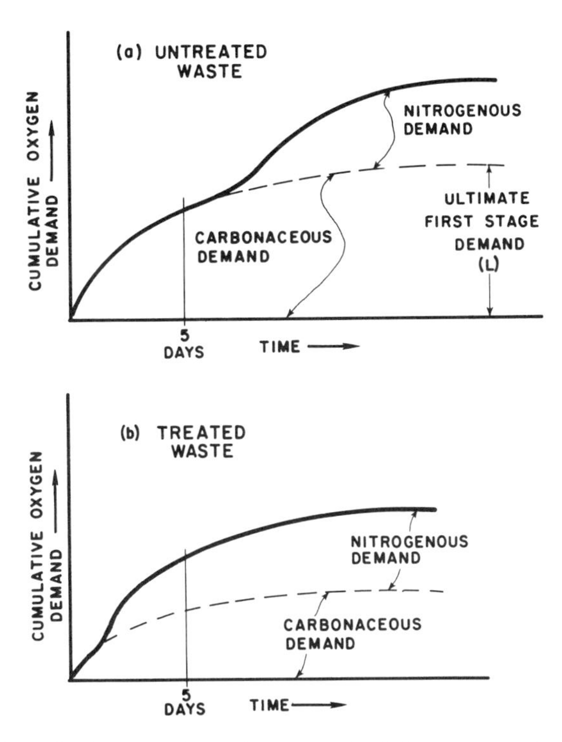

Fig. 5.5. Generalized oxygen demand patterns for heterogenous untreated and treated wastes.

provide an absolute measure of the oxygen-consuming organic matter in the waste. Unless otherwise noted, results reported as BOD indicate that the test has been conducted under standard conditions (19). BOD values are affected by time and temperature of incubation, presence of adequate numbers of microorganisms capable of metabolizing the waste, and toxic compounds. In practice a constant temperature of 20°C does not occur and water is not in a state of rest for five days in the dark. The bacteria used in the test are the real link with practice.

The k value in Eqs. (5.9) and (5.10) changes with temperature, increasing with increasing temperatures. Relationships are available to determine k at temperatures other than 20°C:

$$k_t = (k_{20}) \cdot \theta^{(t-20)} \tag{5.11}$$

Experiments have shown that θ can vary from 1.016 to 1.077. Theta (θ) is commonly used as 1.047. The maximum or total oxygen demand is not affected by temperature since it is a function only of the quantity of organic matter available.

With heterogenous wastes, the five-day period frequently occurs before the beginning of significant nitrification. With adequately treated wastes, the carbonaceous demand is caused primarily by microbial cells and is low. In addition, a significant population of nitrifiers could be present. Under these conditions, the nitrogeneous demand can occur early and before the five-day test period (Fig. 5.5b). Because of these conditions, a treated waste water may exhibit both a carbonaceous and nitrogenous demand while an untreated waste water may exhibit only a carbonaceous demand. When this occurs, evaluation of the BOD removal at a waste water treatment facility will be in error since the oxygen demand of the influent and effluent samples does not measure the same materials.

The maximum oxygen demand (OD_m) will occur if all of the unoxidized organic and inorganic matter is oxidized completely. In most biological treatment units, this means that the organic carbon in a waste is oxidized to carbon dioxide and all of the organic and ammonia nitrogen is oxidized to nitrates. Mathematically, this can be expressed as

$$OD_m = 2.67\,\mathrm{C} + 4.57\,\mathrm{N} \tag{5.12}$$

where C is the organic carbon concentration and N is the sum of the organic and ammonia nitrogen expressed as nitrogen. In practice not all of the organic carbon in a waste may be oxidized, not all of the organic nitrogen in the waste may be converted to ammonia, not all of the ammonia may be oxidized, and a portion of the synthesized microbial cells will not be completely oxidized. Thus Eq. (5.12) will not determine the actual oxygen demand of a waste in a treatment system or a stream. It can, however, be used to estimate the maximum oxygen demand that could occur.

The BOD rate constant, k, is a function of the oxidizability of the waste material. Wastes having a high soluble organic content, such as milk wastes, will exhibit a higher k than a waste having a high particulate organic content, such as a cellulosic waste. Microorganisms must first solubilize particulate organic material before it can be metabolized while they can metabolize soluble wastes more rapidly.

The first-order equations expressed in Eqs. (5.9) and (5.10) represent a best estimate of the rate of oxygen demand and have been the subject of considerable discussion over the years. Because of the heterogeneous nature of most waste mixtures, the different waste constituents can have varying reaction rates. For a specific waste, the oxygen demand may be better described by a second-order equation or a composite exponential equation. Heavy reliance should not be placed on the monomolecular or first-order oxygen demand relationship since it is only an estimate of the complex reactions that are taking place in the biological reactor and frequently reflects only the carbonaceous demand. A nitrogenous demand and in some wastes an initial chemical oxygen demand also can occur.

The small quantity of oxygen present in the BOD test bottle, 2–3 mg, means that high strength wastes, such as many food processing waste waters and animal wastes, must be diluted prior to analyzing the waste. Prior to BOD analysis, animal wastes may require dilutions of 1:100 to 1:1000 or more. The difficulty of diluting wastes that are neither physically or chemically uniform decreases the precision of the standard BOD test which is estimated to have a precision of $\pm 20\%$ (20).

An understanding of the limitations of the BOD test permit utilization of BOD results in a proper manner. The 5-day BOD cannot be considered a quantitative result without an approximation of the rate of oxidation and the ratio of the BOD_5 to the ultimate oxygen demand. In addition, biological treatment plant efficiency is not accuractely determined on the basis of the BOD_5 of the influent and effluent. Throughout a biological treatment plant, the characteristics of the initial organic compounds change as the easily degradable compounds are metabolized, as microbial solids are synthesized, and as more resistant organic compounds remain. Decreases in BOD_5 values through treatment systems provide no information of the changes in specific organic compounds. Nitrification will affect BOD_5 values in a well-oxidized effluent.

In spite of the problems associated with the BOD test, it remains an important analytical tool in water pollution control work since it is one of the few analyses that attempts to measure the effect of a waste under conditions approximating natural stream conditions.

Chemical Oxygen Demand

The lengthy analytical time of the BOD test as well as a desire to find a more precise measure of the oxygen demand of a waste led to the development of the COD test. The COD, chemical oxygen demand, test is a wet chemical combustion of the organic matter in a sample. An acid solution of potassium dichromate is used to oxidize the organic matter at high temperatures. Various COD procedures, having reaction times of from 5 min to 2 hr, can be used (19, 21).

The use of two catalysts, silver sulfate and mercuric sulfate, are necessary to overcome a chloride interference and to assure oxidation of hard to oxidize organic compounds, respectively. Animal wastes and certain food processing wastes such as those from sauerkraut, pickle, and olive processing can contain high chloride concentrations and will require the use of the mercuric sulfate in COD analyses or a chloride correction factor. Compounds such as benzene and ammonia are not measured by the test. The COD procedure does not oxidize ammonia although it does oxidize nitrite.

BOD and COD analyses of a waste will result in different values because the two tests measure different material. COD values are always larger than BOD values. The differences between the values are due to many factors such as chemicals resistant to biochemical oxidation but not to chemical oxidation, such as lignin; chemicals that can be chemically oxidized and are susceptible to biochemical oxidation but not in the five-day period of the BOD test such as cellulose, long chain fats, or microbial cells; and the presence of toxic material in a waste which will interfere with a BOD test but not with a COD test.

In spite of the inability of the COD method to measure the biological oxidizability of a waste, the COD method has value in practice. For a specific waste and at a specific waste treatment facility, it is possible to obtain reasonable correlation between COD and BOD values. Examples of the BOD and COD values of agricultural wastes are noted in Chapter 4. The method is rapid, more precise ($\pm 8\%$) (20), and in most circumstances provides useable estimates of the total oxygen demand of a waste.

Changes in both the BOD and COD values of a waste will occur during treatment. The biologically oxidizable material will decrease during treatment while the nonbiological but chemically oxidizable material will not. The nonbiologically oxidizable material will exist in the untreated wastes and will increase because of the residual cell mass resulting from endogenous respiration. The COD/BOD ratio will increase as the biologically oxidizable material becomes stabilized.

The COD/BOD ratio can be used to estimate the relative degradability or oxidizability of a waste. A low COD/BOD ratio would indicate a small nonbiodegradable fraction. A waste with a high COD/BOD ratio such as animal wastes have a large nonbiodegradable fraction that will remain for ultimate disposal after treatment. Wastes that have been treated, such as waste activated sludge or waste mixed liquor from oxidation ditches, have a high COD/BOD ratio indicating that most of the organic matter has been metabolized and that further treatment might not be economically rewarding. Examples of these relationships are noted in Table 5.2.

As with BOD rates, COD data must be used with caution and judgement. Both can be used, separately and together to estimate the oxidizability of a waste and its effect on a stream.

Total Organic Carbon

Total organic carbon (TOC) is measured by the catalytic conversion of organic carbon in a waste water to carbon dioxide. No organic chemicals have been found that will resist the oxidation performed by the equipment now in use. The time of analysis is short, from 5–10 min, permitting a rapid estimate of the organic carbon content of waste waters. The relationship of TOC values of wastes and effluents to pollution control results requires further understanding. A TOC value does not indicate the rate at which the carbon compounds degrade. Compounds analyzed in the TOC test,

Table 5.2

Oxygen Demand Parameters during Treatment

Waste water	COD/BOD Ratio[a]	Organic carbon/BOD ratio[a]
Raw sewage	2.27	
Settled sewage	1.92	—
Settleable solids	3.57	—
Unsettleable solids	1.82	—
Organics in solution	2.08	—
Activated sludge effluent	2.78	1.06
Activated sludge sedimentation tank effluent	3.85	—
Trickling filter effluent	5.55	2.0
Untreated citrus wastes	1.73[b]	—
Treated citrus wastes	3.34[b]	—
Primary tank effluent	—	0.7
Sand filter effluent	—	2.5

[a] From reference 22.
[b] From reference 17.

such as cellulose, degrade only slowly in a natural environment. Values of TOC will change as wastes are treated by various methods.

BOD and COD utilize an oxygen approach. TOC utilizes a carbon approach. There is no fundamental correlation of TOC to either BOD or COD. However, where wastes are relatively uniform, there will be a fairly constant correlation between TOC and BOD or COD. Once such a correlation is established, TOC can be used for routine process monitoring.

Total Oxygen Demand

The TOD of a substance is defined as the amount of oxygen required for the combustion of impurities in an aqueous sample at a high temperature (900°C) using a platinum catalyst. The oxygen demand of carbon, hydrogen, nitrogen, and sulfur in a waste water sample is measured by this method. The interpretation of TOD values to treatment plant efficiency or to stream quality requires further investigation but generally can be related to BOD and COD values. A portion of both TOC and TOD values will represent nonbiodegradable matter. TOD and TOC methods are rapid and can be incorporated in waste water and treatment plant control systems.

Temperature

General

Temperature is an important factor affecting waste treatment since it influences the physical properties of the liquid under treatment, the rates of the biological and chemical treatment processes, and the waste assimilation capacity of land or water. Most investigations have concluded that, within specific temperature ranges, an equation of the form

$$k = a \cdot e^{bt} \tag{5.13}$$

can be used to express the relationship of a reaction rate k and temperature t (°C). The values of a and b are a function of the specific process and reaction. Reaction rates or coefficients are determined or known at a specific temperature and can be estimated at other temperatures by:

$$k_{T_2} = k_{T_1} e^{c(T_2 - T_1)} = k_{T_1} \cdot \theta^{(T_2 - T_1)} \tag{5.14}$$

where T_1 is the temperature at which the reaction rate or coefficient is known (k_{T_1}) and T_2 is the temperature at which the rate or coefficient is desired.

Equation (5.14) is only an approximation since the temperature variation is theoretically exponential in nature. However over a limited range, the above relationship can be utilized for practical purposes. The temperature effect is the alteration of the rates of specific enzyme activity. In general, only the overall effect of temperature is of engineering and design interest and Eq. (5.14) has proven adequate to describe the temperature effects on biological systems. The value of θ will be different for different treatment systems and different physical, chemical, and biological conditions. Table 5.3 (17, 23, 24) summarizes the temperature effects on a number of biological processes.

For simplicity, it is assumed that θ is not a function of temperature. The error introduced in assuming θ to be independent of temperature is in the range of 10–15% which may be acceptable in most engineering work. Where more precise results are necessary, specific studies may be necessary.

Because temperature has an effect on many fundamental factors, i.e., viscosity, density, surface tension, gas solubility, diffusion, and enzyme activity, it is illogical to assume that θ should be constant over the entire temperature range affecting a treatment process. A number of investigations have demonstrated the interrelationship of θ with temperature. An example is shown in Fig. 5.6a. Other evaluations of θ at different temperatures for specific processes and rates are noted in Table 5.4. After the optimum temperatures range for the specific microorganisms, the reaction rate will decrease and θ will become negative (Fig. 5.6).

If $\theta = 1.072$, a reaction rate is doubled for an increase in temperature of 10°C. This value, 1.072, is in the same range of the coefficients for many processes and reactions (Tables 5.3 and 5.4) and has led to the common phrase that, in general, biological reaction rates are doubled for each 10°C rise in temperature. This estimation generally is useful for only a given temperature range, around 20°C. Its use is less applicable at extreme temperature conditions.

Table 5.3

Temperature Coefficients (θ) for Biological Waste Treatment Processes

	Temperature coefficients		
Treatment	Range	Average	Ref.
Activated sludge	1.0 –1.04	1.03	23
Aerated lagoon	1.085–1.1	—	23
Anaerobic lagoon	1.08 –1.09	1.085	23
Trickling filter	1.035–1.08	—	23
Aerated lagoon	—	1.035	24
Aerated lagoon	—	1.05	17

Table 5.4

Variations of θ with Temperature

Item	Temp. range (°C)	θ, Average	Ref.
BOD reaction rate, k	5–15	1.109	25
	15–30	1.042	25
	30–40	0.967	25
Carbonaceous BOD reaction rate	10–20	1.077	26
	20–30	1.048	26
Soluble COD removal in sewage	5–20	1.089	27
Organic and ammonium nitrogen removal	5–20	1.092	27
Nitrification	5–20	1.106	27
Nitrogenous BOD reaction rate	10–22	1.097	26
	22–30	0.877	26
Denitrification diffused growth	15–25	1.06	28
Ammonia loss-air stripping	10–35	1.063	29
Food processing, aerobic waste treatment			
Endogenous respiration rate	7–20	1.14	30
COD removal rate	7–20	1.16	30

In a biological treatment system, the reduction of the temperature of the biological unit by 10°C will require about double the active organisms in the unit to achieve equivalent process efficiencies. This can be accomplished by increasing the mixed liquor suspended solids concentration in the unit. Such an increase in MLSS may affect the viscosity of the liquid in certain cases and may reduce the solids settling rate. Although temperature variations of 10°C rarely occur abruptly, an uncovered unit can experience considerable temperature differences between winter and summer. Systems with solids recycle permit the operator to compensate for lower temperatures with increased microbial solids in the system. Systems without solids recycle do not have this flexibility. A knowledge of the effect of temperature on a biological system permits better design and operation of the system.

Kinetics

General

The proper utilization of biological treatment processes depends upon an understanding of the kinetics of the processes and the effects of environmental factors on the kinetics. If organic matter, i.e., waste is added at a constant rate to a continuous flow biological treatment unit, the unit

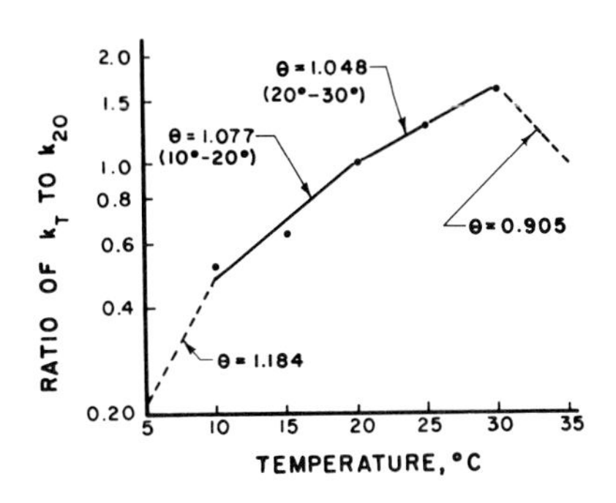

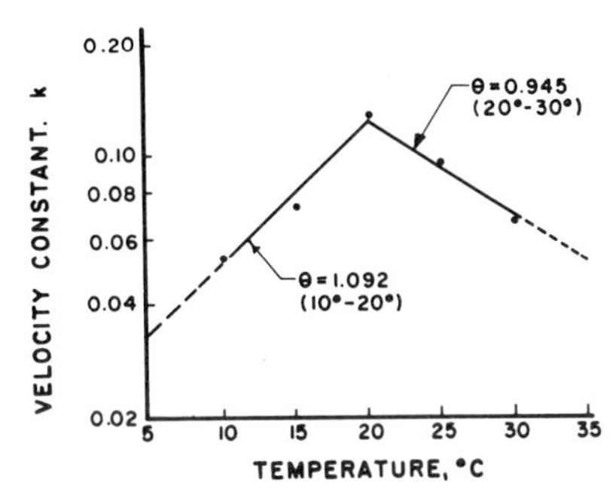

Fig. 5.6. Variation of carbonaceous and nitrogenous temperature coefficients with temperature (26).

eventually will reach equilibrium conditions. Until equilibrium conditions occur, the microorganisms will respond to the waste addition and synthesize new organisms until the microbial mass is in equilibrium with the available food supply, i.e., waste. At equilibrium, the net microbial concentration is related to the available substrate and to the decay rate, or endogenous respiration of the organisms. At equilibrium the unit is a food-limited unit.

In this explanation, the substrate, i.e., the carbonaceous material, BOD, or COD is assumed to be the limiting material. The same relationship would be observed for cases where some other nutrient, i.e., nitrogen, phosphorus, or a trace metal, would limit the maximum reaction rates. For agricultural wastes, nutrients other than carbon rarely are the limiting nutrient.

The terms used in this section are defined when they are first used. A summary of the notations and definitions is presented at the end of this chapter. Wherever possible the symbols and notations used by different investigators are used when the approach and equations of that investigator are discussed. In all of the equations, the various terms must have the proper units to make the equations dimensionally correct.

Continuous Growth

An equation for the growth of microorganisms may be expressed as

$$\frac{dX}{dt} = uX \tag{5.15}$$

where X is the microbial concentration and u is the microbial growth coefficient, the net specific growth rate. If growth is unlimited, the microbial population should grow at an exponential rate. However in a biological system, unlimited growth will not occur indefinitely and may not occur at all. Biological waste treatment systems generally have a single substrate limiting the growth of the microorganisms. Growth-limiting kinetics can be used in the analysis of such systems.

Under growth-limiting conditions the growth coefficient is not a constant but depends upon the limiting substrate concentration:

$$\mu = \frac{\bar{\mu} S}{K_s + S} \tag{5.16}$$

μ is the rate of growth corresponding to a concentration S of the limiting nutrient in the system; $\bar{\mu}$ is the maximum rate which prevails when S is large; and K_s is a constant, characteristic of the microorganisms and the given substrate. The units of the terms are quantity of cells produced/unit time/quantity of existing cells for μ and $\bar{\mu}$, and mass/volume for S and K_s. K_s is the concentration at which the value of μ would be half of $\bar{\mu}$, i.e., the substrate concentration at one-half the maximum growth rate (Fig. 5.7). One advantage of this relationship is that it is a continuous function describing the reaction kinetics throughout a biological system, i.e., describing both substrate excess and substrate-limited situations.

This type of kinetic equation was developed by Michaelis and Menton to explain enzymatic reactions and applied by Monod (31), Hinshelwood (32), and others to microbial growth systems. This equation is a useful relationship which was derived empirically and is simply an analytical expression which has been found to fit a large amount of experimental absorption, transport, and enzymatic data related to the microbial metabolism of organic matter.

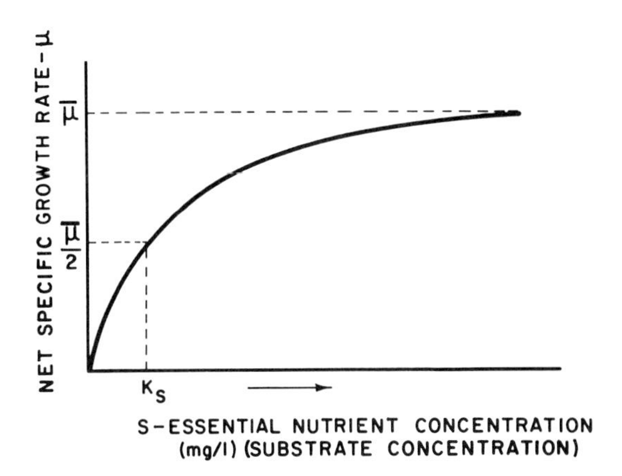

Fig. 5.7. Microbial growth rate pattern.

Utilization of kinetic models developed for simple enzyme systems to describe the response of biological systems treating complex wastes should be done with caution. However, the performance of biological systems does appear to fit such simple models. The models permit the performance of biological waste treatment systems to be estimated mathematically under varying substrate concentration and environmental factors. A number of sanitary engineers subsequently utilized these models for the design of aerobic and anaerobic waste treatment processes (33–37).

Equation (5.16) can be transformed to a form that plots as a straight line:

$$\frac{1}{\mu} = \frac{K_s}{\bar{\mu}} \cdot \frac{1}{S} + \frac{1}{\bar{\mu}} \tag{5.17}$$

Experimental or field data can be utilized to determine K_s and $\bar{\mu}$ graphically.

The cell growth rate, μ, is equal to a substrate removal rate, q (quantity of substrate removed/unit time/quantity of existing cells) times a yield coefficient, Y (quantity of cells produced/quantity of substrate removed):

$$\mu = q \cdot Y \tag{5.18}$$

The substrate removed per unit time, dS/dt, can be expressed as $q \cdot X$ where X is the quantity of cells in the system. Then

$$\frac{dS}{dt} = qX = \frac{\mu}{Y} \cdot X = \frac{\bar{\mu}SX}{Y(K_s + S)} \tag{5.19}$$

where $\bar{\mu}/Y$ is the maximum quantity of substrate removed/quantity of cells in the system/unit time, i.e., the maximum rate of substrate utilization that can occur in a system per unit weight of microorganisms. For a given system, this maximum rate is a constant, k. Therefore, the rate of substrate removal can be expressed as

$$\frac{dS}{dt} = \frac{kSX}{(K_s + S)} \quad (5.20)$$

While these general equations have been developed to express relationships throughout a microbial growth cycle, the majority of biological waste treatment systems, including those treating agricultural wastes, operate with relatively constant input under relatively constant environmental conditions and can be considered to be systems in equilibrium. Ranges of values for these kinetic constants are noted in Table 5.5. Comparative data for agricultural and other waste treatment systems are presented in Table 5.6.

Nonrecycle Systems

The application of these equations can be examined by considering a typical biological treatment system used with agricultural wastes (Fig. 5.8). It is assumed that this system is completely mixed with the incoming wastes of constant quantity and quality. Reasonable examples of such systems include aerated lagoons, oxidation ditches, and mixed anaerobic digestion units.

A balance on substrates would yield

$$\frac{dS}{dt} \cdot V = QS_o - QS_i - X_1 \cdot V \cdot q \quad (5.21)$$

(Change in substrate in the system) = (input) − (output) − (substrate removal by cells)

At equilibrium $dS/dt = 0$ and

$$S_o - S_i = \frac{X_1 \cdot V \cdot q}{Q} = \frac{X_1 \cdot V \cdot S_i \cdot \bar{\mu}}{Q \cdot Y \cdot (K_s + S_i)} \quad (5.22)$$

Table 5.5

Ranges of Kinetic Coefficients for Aerobic and Anaerobic Treatment Processes

		Treatment	
Coefficient	Units	Aerobic	Anaerobic
Y	lb volatile suspended solids/lb BOD_{ult}	0.3–0.4	0.03–0.15
K_D	day^{-1}	0.02–0.06	0.01–0.04
K_S	mg/l	1–50	20–300
q	lb BOD/day/lb volatile suspended solids	4–24	4–20
SRT_c	days	0.1–0.3	2–6

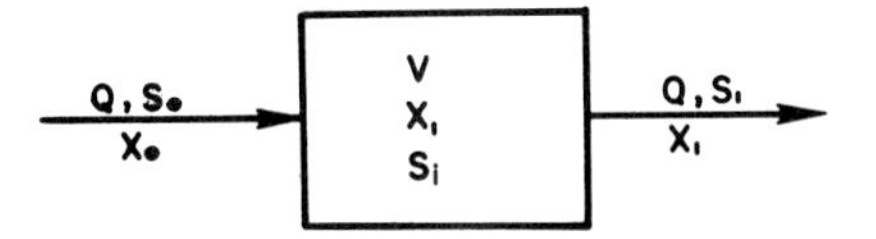

Q = FLOW
S = SUBSTRATE (WASTE) CONCENTRATION
X = MICROBIAL CONCENTRATION
V = REACTOR VOLUME

Fig. 5.8. Completely mixed, no recycle biological treatment system.

The theoretical liquid residence time in the unit, HRT, is V/Q and Eq. (5.22) can be simplified to

$$S_o - S_i = \frac{X_1 \cdot S_i \cdot k(\mathrm{HRT})}{(K_s + S_i)} \tag{5.23}$$

The equation indicates that for a given waste influent concentration, the amount of substrate removed, and hence substrate removal efficiency, is a function of the liquid detention time, the microbial cells in the system, and the constants of the basic equation. These constants are determined by the

Table 5.6

Kinetic Coefficients Determined in Specific Waste Treatment Systems

Waste	Y	K_D (day^{-1})	SRT$_c$ (days)	Temperature (°C)	Ref.
Packinghouse waste	0.76 lb VSS/lb BOD[a]	0.17	—	35	36
Synthetic milk waste	0.37 lb VSS/lb COD	0.07	—	20–25	36
Synthetic carbohydrate and protein wastes	—	—	0.52	10	37
	—	—	0.21	20	37
	—	—	0.14	30	37
Ammonia oxidation	0.05 lb VSS/lb N	—	1.5	19	38
	—	—	0.7	27	38
Nitrite oxidation	0.02 lb VSS/lb N	—	1.2	19	38
Pear processing	0.49 lb VSS/lb COD	—	0.115	20	30
Peach processing	0.46 lb VSS/lb COD	—	0.055	15	30
Apple processing	0.56 lb VSS/lb COD	—	0.03	10	30
Potato processing	0.6–0.8 lb VSS/lb BOD	—	—	—	39
Field extended aeration system	0.54 lb VSS/lb BOD	0.014	—	—	40
Municipal waste water Summary of data from 6 plants	0.336 lb VSS/lb COD	0.16	—	20–25	41

[a] VSS = volatile suspended solids.

microbial and environmental characteristics of the system. The constants can be estimated by data from laboratory or field experiments with a specific waste under conditions that approximate actual treatment conditions.

The microbial cell concentration is a function of the substrate in the system. With a given and constant substrate input and a desired substrate effluent or removal efficiency, a continuous, nonrecycle process will contain a specific quantity of microbial cells. Thus for this system the only factor that a design engineer can vary to produce different removal efficiencies and effluent quality is the hydraulic residence time (HRT). When a removal efficiency or an effluent substrate concentration is specified, there is only one set of conditions that will meet the specified removal efficiency.

The change in microbial concentration in a substrate limited unit can be expressed combining Eqs. (5.15) and (5.16)

$$\frac{dX}{dt} = \frac{\bar{\mu} S X}{(K_s + S)} \tag{5.24}$$

For an existing unit such as in Fig. 5.8, a materials balance can be written to relate the microbial cell changes:

$$\frac{dX_1}{dt} \cdot V = QX_0 + \mu \cdot X_1 \cdot V - QX_1 - K_D \cdot X_1 \cdot V \tag{5.25}$$

(Change in cell mass) = (input) + (growth) − (output) − (microbial respiration)

The term $K_D \cdot X_1 \cdot V$ represents the endogenous respiration of the cells, i.e., the decrease in cellular mass caused by a cell using its reserves and the decrease caused by cell death and lysis. K_D represents the organism decay rate, quantity of cells decreased/unit time/quantity of cells, in the system and is temperature dependent. The numerical value for this constant should be based upon the quantity of active microbial solids in the system rather than the suspended or volatile solids since these latter items will contain unknown quantities of nonmicrobial solids.

At equilibrium $dX_1/dt = 0$ and

$$(\mu - K_D)X_1 = \frac{Q}{V}(X_1 - X_0) \tag{5.26}$$

In many cases, the microbial mass entering the system, X_0, is small compared to the microbial mass in the system, X_1. This is true for many dilute wastes such as municipal sewage, milking parlor wastes, and liquid food processing wastes. In this case, X_0 can be considered negligible and Eq. (5.26) simplified to:

$$\mu - K_D = \frac{QX_1}{VX_1} \tag{5.27}$$

The term $\mu - K_D$ is the net growth rate of the microbial solids in the system which is important since it estimates the accumulation of microbial solids that ultimately must be removed from the system. Equation (5.27) also verifies an earlier statement that synthesis and respiration occur simultaneously in biological systems with synthesis predominating when food is in excess ($\mu > K_D$) and respiration predominating when food, or essential nutrient, is limited, ($K_D > \mu$).

The term $Q \cdot X_1/V \cdot X_1$ is of interest since it represents the quantity of cells leaving the system per day divided by the quantity of cells in the system. The inverse of this term represents the theoretical time the cells stay in the biological reactor and is the mean cell residence time or solids residence time (SRT). In the example used, Fig. 5.8, the theoretical liquid residence time (HRT), V/Q, is equal to the SRT. As will be shown, there are biological systems in which the SRT is different from the HRT thus giving a design engineer another variable to utilize when designing appropriate biological waste treatment systems.

Equation (5.27) can be written as:

$$\mu - K_D = \frac{1}{\text{SRT}} \tag{5.28}$$

and illustrates that numerically the reciprocal of solids retention time is the net growth rate of the system. For a biological process to function, the reciprocal of the design solids residence time of the process should be in excess of the minimum time it takes for the microorganisms to reproduce in the process. If this does not occur, the microbial cells will be removed from the system at a faster rate than they can multiply and failure of the system will result.

Equation (5.28) can be written as:

$$\frac{1}{\text{SRT}} = Y \cdot q - K_D \tag{5.29}$$

and the critical SRT or cell washout time can be written as:

$$\frac{1}{\text{SRT}_c} = Y \cdot q_{\text{max}} - K_D \tag{5.30}$$

SRT_c is the minimum microbial residence time at which a stable microbial cell population can be maintained. Because the yield coefficient, Y, is smaller for anaerobic metabolism than for aerobic metabolism, the value of SRT_c for anaerobic processes is larger than for aerobic processes.

The concept of a minimum SRT is important in all biological waste treatment systems, especially in anaerobic treatment systems. In aerobic treatment systems, the energy relationships are such that microorganisms can reproduce rapidly, a matter of hours or less, depending upon the specific system and its management. It is extremely rare to note an aerobic treatment system that failed because its SRT was less than the minimum time necessary for microbial reproduction.

The substrate removal rate, q, may be expressed as the mass of substrate removed from the waste per unit mass of active organisms per unit of time,

$$q = \frac{S_o - S_i}{X_1 \cdot \mathrm{HRT}} \tag{5.31}$$

This equation is comparable to Eq. (5.19) except it is in terms of common treatment process parameters. The maximum substrate removal rates or process loadings for aerobic and anaerobic systems are in the range of 4–24 lb COD/day/lb active organisms (42). Biological systems operating in this range are high rate systems requiring close control and in general the process efficiencies are lower than for low rate systems. Systems requiring high efficiencies are more conventional and are operated at process loadings of 0.5 lb COD/day/lb active organisms or less. Lower loadings also are used to have minimum solids production for disposal.

In anaerobic systems, the microorganisms reproduce less rapidly and a longer minimum SRT is required to accommodate the slower net growth rate. An example of the relationship between the SRT and the efficiency of substrate removal is shown in Fig. 5.9.

Combining Eqs. (5.22) and (5.27) will yield

$$X_1 = \frac{Y \cdot \mathrm{SRT} \cdot (S_o - S_i) \cdot Q}{[1 + K_D(\mathrm{SRT})] \cdot V} \tag{5.32}$$

but V/Q = HRT which for the example equals SRT, and Eq. (5.32) simplifies to

$$X_1 = \frac{(S_o - S_i)Y}{(1 + K_D\mathrm{SRT})} \tag{5.33}$$

The effluent substrate concentration can be expressed as

$$S_i = S_o - (1 + K_D \cdot \mathrm{SRT}) \cdot \left(\frac{X_1}{Y}\right) \tag{5.34}$$

For a given system such as Fig. 5.8, Y, K_D, S_o, and SRT are fixed and X_1 is a function only of the equilibrium waste concentration of the system, which in this case is equal to the effluent waste concentration. The cell mass in the system will adjust to the conditions imposed on the system and the effluent substrate concentration, S_i, is a function of the microbial solids, X_1, in the system if the indicated factors are fixed.

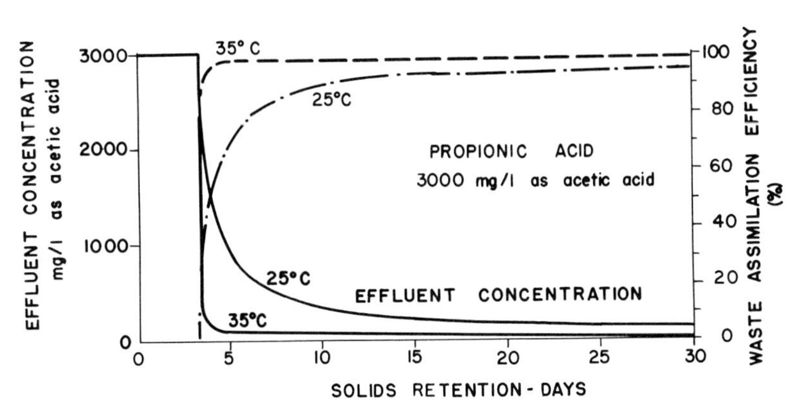

Fig. 5.9. Process efficiency as related to SRT in an anaerobic treatment unit (35).

The net growth of microorganisms in the system, dX'/dt, is equal to cell growth less cell degradation, $dX'/dt = \mu \cdot X - K_D \cdot X$ but $\mu = dS/dt \cdot Y/X$ and the net growth can be represented by

$$\frac{dX'}{dt} = Y \cdot \frac{dS}{dt} - K_D \cdot X \tag{5.35}$$

Eq. (5.35) can be used to estimate the quantity of excess microbial solids that will accumulate with time in a batch system or will need to be removed in a continuous system. The quantities of solids will vary with the type of biological treatment, the method of operation, and the environmental conditions.

The net microbial solids production is inversely related to SRT:

$$\frac{dX'}{dt} = \frac{X}{\text{SRT}} \tag{5.36}$$

Thus a biological unit with a long SRT will have less net microbial solids for wasting than will a unit with a shorter SRT. The amount of microbial solids to be wasted can be varied by controlling the SRT of the unit. A long SRT biological unit is less effective in removing inorganic nutrients because net microbial solids production and incorporation of the nutrients into microbial cells is less. Biological systems with high microbial solids production, i.e., high rate synthesis systems, should be utilized where high rates of inorganic nutrient removal by microorganisms are desired.

The total quantity of solids that will accumulate or require removal will include the excess microbial solids and the relatively stable organic and inorganic solids contained in the untreated waste and which are not altered in the treatment process. The total quantity of excess solids can be estimated by

$$\frac{dX_s}{dt} = \frac{dX'}{dt} + \frac{dX_i}{dt} \tag{5.37}$$

where dX_s/dt represents the rate of total sludge solids produced, dX'/dt represents the rate of net microbial solids produced, and dX_i/dt represents the rate of stable inorganic and organic solids accumulation.

The value for dX_i/dt varies with the type of waste. Agricultural wastes can have values of dX_i/dt ranging from close to zero for wastes such as milk processing wastes to large values with wastes having large quantities of nonbiodegradable material such as cattle wastes.

The rate at which solids will accumulate in a biological treatment system is dependent upon the waste loading rate, the rate at which solids are lost from the system, the rate at which inert solids enter the system, the rate at which volatile waste solids and microbial solids are oxidized in the system, and the rate at which the waste is being converted into microbial cells.

In aerobic systems, the above equations can be used to estimate the oxygen requirements. The quantity of oxygen required will be a summation of that required for substrate removal (synthesis) and that required for respiration. For a process such as in Fig. 5.8,

$$V \cdot \frac{dO}{dt} = a \cdot Q \cdot (S_o - S_i) + b \cdot K_D \cdot X_1 \cdot V \tag{5.38}$$

where dO/dt is in terms of mg/l/time, a is a constant used to convert substrate units to oxygen units utilized in synthesis (units of oxygen/unit of BOD or COD), and b is a constant, used to convert cell mass units to oxygen units (unit of oxygen/unit of cell mass). Cell mass generally is expressed as volatile suspended solids although this parameter is an inadequate measure of the active microorganisms. The rate of oxygen utilization is

$$\frac{dO}{dt} = \frac{(S_o - S_i) \cdot a}{\text{HRT}} + b \cdot K_D \cdot X_1 \tag{5.39}$$

Using Eq. (5.31), Eq. (5.39) simplifies to

$$\frac{dO}{dt} = \frac{(S_o - S_i)}{\text{HRT}} \left(a + \frac{b}{q} \cdot K_D \right) \tag{5.40}$$

for the special case where SRT = HRT (Fig. 5.8).

The oxygen requirement varies directly with the substrate to be removed and inversely with the hydraulic residence time of the system. The longer the detention time of the biological system, the lower the rate of oxygen demand in the system and the greater the total oxygen utilization. The constants a and b and Eqs. (5.39) and (5.40) normally do not include the oxygen required for nitrification. The oxygen requirement for nitrification is a function of the amount of ammonia that is nitrified and the equations can be modified by including the following term:

$$\frac{4.57 \cdot \Delta NH_3}{HRT} \tag{5.41}$$

where ΔNH_3 is the ammonia oxidized in the unit during aerobic treatment.

Solids Recycle

For a given substrate removal rate, a larger quantity of waste can be treated if the mass of organisms in the system is greater. Systems with solids recycle (Fig. 5.10) can increase the concentration of active organisms in a biological unit and can provide engineers and operators with additional alternatives to obtain satisfactory performance and to decrease the size of the biological unit. The solids separator, generally a sedimentation unit, plays an important role in such systems since the quantity of recycled solids and the effluent quality depends on the efficiency of the solids separator. The microbial solids must be separated easily otherwise the system will approximate a system without recycle (Fig. 5.8), the advantages of recycle will be lost, and the system may not produce an effluent of the desired quality.

A substrate, microbial mass, and oxygen requirement balance can be made for the recycle system in the same manner as was done for the non-recycle system. A microbial mass balance on the entire system operating at equilibrium and assuming the influent solids concentration to be negligible illustrates that

$$\mu - K_D = \frac{QX_2}{VX_1} = \frac{1}{HRT \cdot r} = \frac{1}{SRT} \tag{5.42}$$

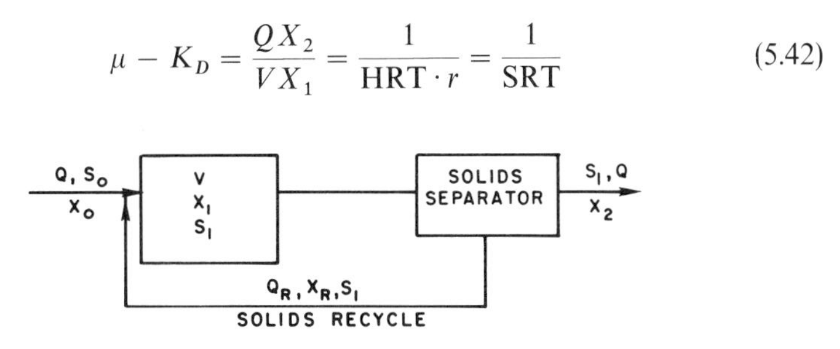

Fig. 5.10. Completely mixed biological treatment system with recycle.

where r is the mixed liquor to effluent cell ratio, X_1/X_2 and is numerically greater than 1.0. The relationship between the cell mass in the system, X_1, and the substrate removed, $S_o - S_i$, for the system in Fig. 5.12 will be as expressed by Eq. (5.32). Substituting HRT $\cdot\, r$ for SRT will yield

$$X_1 = \frac{Y \cdot r \cdot (S_o - S_i)}{(1 + K_D \cdot \mathrm{SRT})} \tag{5.43}$$

which indicates that the microbial mass in the system will be a function of both the substrate removal and the cell ratio. The microbial mass in a biological system with recycle is greater than one without recycle all other factors being equal. The effluent substrate concentration can be expressed as

$$S_i = S_o - \left[\frac{X_1(1 + K_D \cdot \mathrm{SRT})}{Y \cdot r}\right] \tag{5.44}$$

The oxygen requirements for a system with recycle also can be described by Eq. (5.39). The value of X_1 is now greater because of solids recycle. The effect of solids recycle on the oxygen requirement can be shown by using Eqs. (5.18), (5.28), and (5.43) to modify Eq. (5.39) to

$$\frac{d\mathrm{O}}{dt} = (S_o - S_i)\left(\frac{a}{\mathrm{HRT}} + \frac{b \cdot K_D \cdot r}{q \cdot \mathrm{SRT}}\right) \tag{5.45}$$

which demonstrates that the oxygen requirement for substrate removal for synthesis is a function of hydraulic residence time while the oxygen requirement for respiration is related to the microbial cell residence time.

The HRT of a system with recycle would be less than one without recycle to achieve a specific substrate removal efficiency. When design values are used in these equations, the oxygen demand rate for a system with recycle will be greater, because of a shorter HRT and a larger quantity of microbial solids in the system, than for a system without recycle used to treat the same wastes.

As indicated by Eq. (5.42), the SRT of the system is a function of both the HRT and the mixed liquor to effluent cell ratio of the system. This permits the system to have a long SRT to obtain a high treatment efficiency, low S_i with a short HRT, and a smaller size biological unit. Solids recycle also offers the opportunity to maintain a SRT greater than the minimum required for cell growth with an HRT which could be less than the minimum SRT.

A system with cell recycle offers greater flexibility for design and operation than a system without recycle. When a treatment process is subject to variable waste loads, the microbial solids in the biological unit can be varied by changing the recycle ratio to adjust to the waste loads and produce a consistent effluent. This requires a qualified operator, a consideration when deciding to use a system with or without recycle.

A system without recycle will require a longer HRT and a larger biological unit than will a system with recycle to obtain the same quality of waste effluent. A large biological unit can dampen the surges caused by variable waste load and may provide adequate time for the microbial solids to adjust to an increased waste concentration in the unit. For many agricultural wastes, the decision to use a system with or without solids recycle will hinge on the availability of an operator to manage a system with recycle. Because many facilities producing agricultural wastes are located where land costs are not excessive, treatment systems without recycle, such as oxidation ponds, aerated lagoons, and oxidation ditches are common.

Inhibition

The basic relationships have been shown to be valid for a wide variety of organisms and limiting substrates where process inhibition does not occur. The relationship must be modified for substrates which limit growth at low concentrations and are inhibitory to organisms at higher concentrations. This reasoning has been extended to a number of biological systems by modifying Eq. (5.16) to include an inhibition factor, K_i, (43)

$$\mu = \frac{\bar{\mu} S}{(K_s + S)\left(1 + \dfrac{S}{K_i}\right)} \quad \text{or} \quad \frac{\bar{\mu} S\left(1 + \dfrac{S}{K_i}\right)}{\left(K_s + S + \dfrac{S^2}{K_i}\right)} \tag{5.46}$$

In the usual continuous treatment systems operated near steady state, substrate concentrations are low and the term S/K_i is much less than 1.0. Under these conditions, such inhibition equations reduce to the usual basic relationship [Eq. (5.16)]. In batch treatment systems, the term S/K_i may be significant because of higher substrate concentrations present in the early stages of growth and after batch feeding.

Separate Growth Phases

The kinetics described above were based on the use of a single equation to describe the growth cycle of microbial cells. Other researchers have utilized separate equations for the different growth phases of a microbial system, i.e., log growth, declining growth, or stationary growth. Equations based on these growth phases have been utilized to develop suitable design criteria for biological waste treatment systems.

Both approaches are empirical. A choice between the two frequently is made by engineers on their familiarity and experience with one approach or the other.

The work of Eckenfelder (44, 45) exemplified the approach used in separating the microbial growth curve into different phases. Equations used to describe sludge growth as BOD removal in the log growth phase are

$$\frac{dS}{dt} = k_1 S \tag{5.47}$$

$$L_R = \frac{S_0}{a}(e^{k_1 t} - 1) \tag{5.48}$$

In these equations S = solids concentration in the system, S_0 = initial solids concentration in the system, L_R = substrate (BOD) removed, a = fraction of the substrate (BOD) synthesized to microbial (sludge) solids, and t = aeration time. The growth rate in this phase is limited only by the generation time of the organisms for the environmental conditions imposed on the biological unit. When a biological system is in log growth, a plot of log S versus aeration time will yield a straight line, the slope of which is k_1, the logarithmic growth rate. A decreasing slope of the line indicates that the system is in the declining growth phase rather than the log growth phase (Fig. 5.11).

In the declining growth phase, the substrate removal rate is limited by

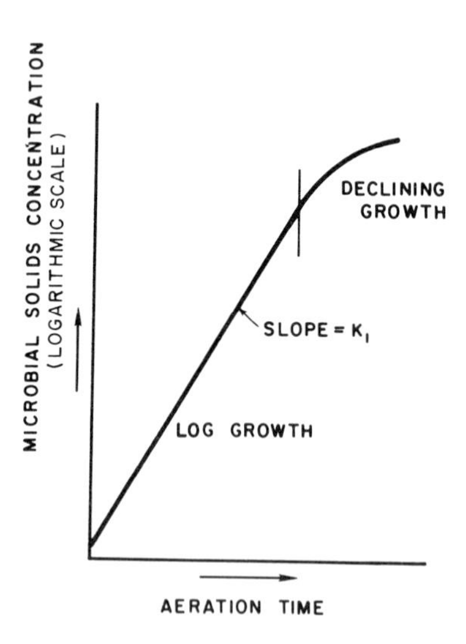

Fig. 5.11. Microbial growth phases.

the substrate available. A term for the average solids concentration in the system, S_a, is included:

$$\frac{dL}{dt} = -k_2 \cdot S_a \cdot L \tag{5.49}$$

For a particular waste, a relationship of mixed liquor suspended solids and time to BOD removal can be established (Fig. 5.12). Experiments with many wastes have indicated that a relationship of this type, where the BOD removal is a first-order reaction related to time at the average solids concentration, can be utilized to develop suitable design criteria.

In the autoxidation or endogenous phase, the microbial solids decrease is a function of the existing solids concentration:

$$\frac{dS}{dt} = -k_2 S \tag{5.50}$$

The endogenous constant, k_3, is considerably smaller than the log growth constant, k_1. The constant, k_3, is not a true constant but may vary with time.

Although the above equations can be used to define solids and removal relationships within any microbial growth phase, common biological treatment systems will function in a declining or stationary growth phase since they generally have a continuous input and output and are well mixed. Equation (5.49) describes the relationships in this phase.

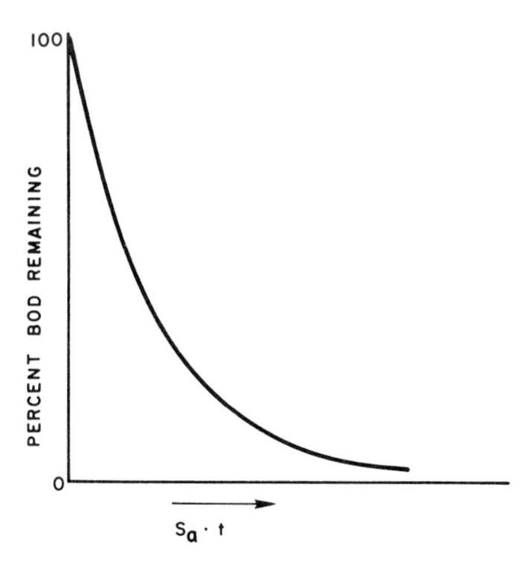

Fig. 5.12. BOD removal as a function of mixed liquor solids and time.

With some wastes, the BOD removal curve can be approximated by a succession of exponential terms (44), i.e.,

$$L = L_a e^{-K_a S_a t} + L_b e^{-K_b S_a t} + \cdots L_n e^{-K_n S_a t} \tag{5.51}$$

when the waste contains compounds having recognizably different reaction rates.

The sludge growth and oxygen requirements of aerobic systems are related to the substrate or BOD removed in the system. At any time t, the sludge concentration in the system is equal to the initial concentration plus any change. In a continuously mixed solids system at equilibrium, the mixed liquor solids concentration as the waste enters the unit is equal to the average solid concentration in the unit, S_a. During the aeration period, a solids increase occurs as microbial solids are synthesized and a solids decrease occurs as a portion of the microbial solids undergo endogenous respiration. The quantity of solids synthesized per unit time is related to the BOD removed per unit time and the solids decrease is related to the microbial solids in the system. The net change of solids in a continuous output system would be

$$\frac{\Delta S}{\Delta t} = \frac{aL_r}{\Delta t} - bS_a \tag{5.52}$$

which is comparable to Eq. (5.35).

The values of a and b for a specfic waste are estimated from experimental or actual units if the data is plotted as $S/S_a \cdot t$ versus $L_r/S_a \cdot t$ (Fig. 5.13).

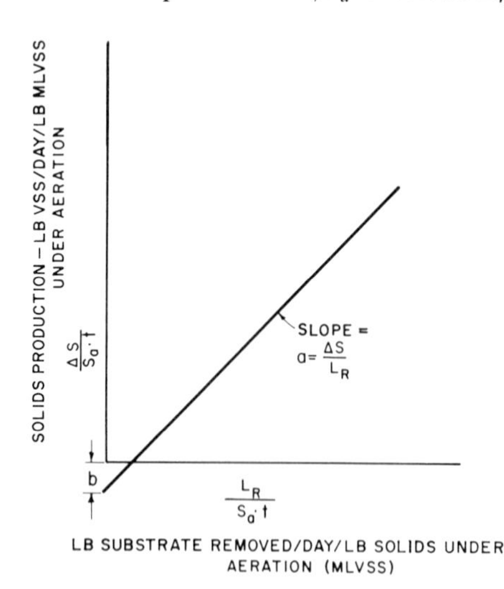

Fig. 5.13. Graphical method to determine synthesis and endogenous constants.

Relationships of this type were used to determine the values of a and b for a poultry processing waste. In a continuous system, the value of a was 0.20 and that of b was 0.04 while in a batch system treating the same wastes, the values were 0.5 and 0.07, respectively (46). The difference in the values, especially for b emphasizes the caution that should be used in using data data from one type of a system to design other systems. In a continuous process without separate sludge wasting or removal, the net increase in solids plus the inert and unmetabolized solids in the influent would be in the effluent. In a nonoverflow system, the net increase in microbial solids, the inert, and the unmetabolized solids will accumulate in the system until they are removed at periodic intervals.

The oxygen requirements in an aerobic system are related to synthesis and endogenous respiration,

$$\frac{\Delta O_2}{\Delta t} = a' \frac{L_R}{\Delta t} + b' S_a \tag{5.53}$$

Eq. (5.53) is comparable to Eq. (5.40).

In nonaerobic systems, the synthesis and respiration constants are related to the use of other electron acceptors, such as nitrites and nitrates in a denitrification system and partially oxidized organic matter in methane producing systems. The term b' is the endogenous respiration rate in terms of quantity of oxygen required per day per quantity of solids under aeration, and a' is the fraction of BOD removed which is used to provide energy for growth in terms of quantity of oxygen required per quantity of BOD removed. Numerical values of a' and b' will vary for different wastes since the

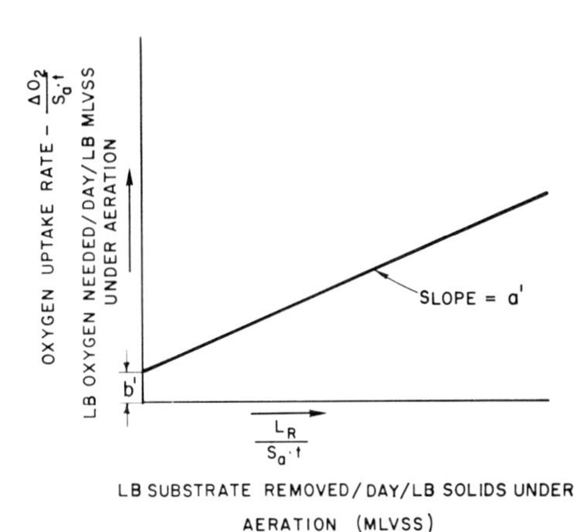

Fig. 5.14. Graphical approach to determine the synthesis and endogenous oxygen requirements.

constants usually are determined on the basis of mixed liquor volatile suspended solids (MLVSS) in the biological treatment unit rather than active mass. Oxygen utilization will be greater with nitrifying systems. In low rate systems, the endogenous demand, b', can be a significant fraction of the total demand.

The values of a' and b' generally are determined from experimental or field studies in which the data is plotted as $O_2/S_a \cdot t$ versus $L_R/S_a \cdot t$ (Fig. 5.14). The constants a and a' as well as b and b' can be related to each other. If a microbial solids composition of $C_5H_7O_2N$ is assumed, the oxygen demand per unit of solids is 1.42 for complete oxidation (endogenous respiration). This value is commonly used to relate solids concentration in oxygen units. Closely controlled experiments have shown (1, 2) that in common biological waste treatment processes all of the microbial solids are not oxidized—approximately 20–25% of the synthesized microbial solids are not oxidized and remain in system. A more reasonable value to relate microbial solids to oxygen demand is about 1.2. The terms b and b' are related by

$$b' \sim 1.2b \tag{5.54}$$

In a similar manner, a and a' are related. In terms of ultimate oxygen units, $a + a' = 1$ since a is the fraction of BOD used for synthesis and a' is the fraction used for energy. As defined above (Fig. 5.13), a has the units of solids and must be converted to oxygen units to be related to a'. When done, the two terms can be related.

$$1.2a \sim 1 - a' \tag{5.55}$$

The generalized Eqs. (5.52) and (5.53) do not take the solids in the raw waste into account in estimating sludge accumulation or oxygen requirements. The equations have been modified to include the effect of volatile suspended solids in the raw waste (47). The amount of influent volatile suspended solids degraded was a function of the solids retention time of the system. For municipal wastes, 14% of the influent volatile solids were degraded per day. Other percent influent solids degradation will occur with influent solids of different composition.

In the above equations, (5.47–55), the terms sludge, sludge solids, and microbial solids are used interchangeably. Theoretically, the values for S in these equations should represent viable microbial solids concentration since the equations describe the changes in microorganisms. No reliable, routine analytical method has been developed to determine the active mass or active viable organisms in a heterogenous biological waste treatment process. Because it is recognized that microbial solids are organic solids, changes in the volatile suspended solids concentration in a biological treatment unit are used as a reasonable estimate of the microbial solids change. While not theoretically satisfying, this approach has been used to obtain useful data for the design of biological waste treatment systems.

Realizing that most biological waste treatment systems operate as continuous flow systems at equilibrium, McKinney (48) developed a series of equations for a variety of systems operating under these conditions. These equations relate the change in the substrate concentration, F, to the active mass, M_a, in the system. The equations for a system such as in Fig. 5.8 are

$$F = \frac{F_i}{K_5 t + 1} \tag{5.56}$$

$$M_a = \frac{K_6 F}{1/t + K_7} \tag{5.57}$$

$$\frac{dO}{dt} = K_9 F + K_2 M_a \tag{5.58}$$

where F_i is the organic matter (BOD) in the raw wastes and the K values are rate constants for the respective equations. The constants are temperature dependent. In addition, the energy-synthesis constant, K_1, used as a basis for these equations, and constants dependent upon K_1, i.e., K_5 and K_6, are a function of the type of waste (49). Lower K_1 values were determined for carbohydrate wastes. Certain food processing wastes have a high carbohydrate content. Biological systems designed by Eqs. (5.56–58) for such wastes should utilize the appropriate constants.

Comparable equations were developed for a system with recycle, and with recycle and separate sludge wasting (48). In the equations used to determine the substrate concentration, F, in the system effluent, such as Eq. (5.56), it was recognized that the active mass in the system does affect the removal of the substrate. Because the system is at equilibrium and the active mass will be constant for the given waste influent, the role of the active mass was implied but not used.

Based upon the basic synthesis–oxidation relationships, a system with a food to mass (F/M) ratio of 2:1 or greater will function as an active mass limited system and is in the log growth phase. A system with an F/M ratio less than 2:1 is a food limited system. Since in an actual waste treatment system the F/M ratio is on the order of 0.1 to 0.2, conventional waste treatment systems are food or nutrient limited systems.

Equations that include the active mass term, M_a, permit an evaluation of the quantity of active mass in a treatment system and the value of the volatile suspended solids in the system to approximate the active mass. The active mass in a waste treatment system is a function of the suspended solids, the solids retention time, and the aeration time of the system (Fig. 5.15). The assumptions made in preparing Fig. 5.15 are that the characteristics are of average domestic sewage, that 1% of the mixed liquor suspended solids is lost in the effluent of a system such as in Fig. 5.10, and that additional mixed liquor solids will be wasted separately each day to permit the solids rentention times that are noted. The active mass is insensitive to SRT at higher aeration periods. These relationships are logical since at long SRT's and long aeration periods the waste concentration in the unit is low, microbial synthesis is low, and endogenous rates predominate. The value of systems with solids recycle is apparent since the increased solids retention time increases the level of active mass in the biological unit.

The active mass will be a small fraction of the suspended solids in a treatment unit. Low values are indicative of processes with long solids retention times and of processes treating wastes with a high nonbiodegradable solids content of the raw waste. Examples of these wastes would be wastes with a high cellulose or lignin content such as paper pulp wastes, animal wastes, and wastes from vegetable pulping operations.

The oxygen requirements are related to the solids and hydraulic retention time of an aerobic process. Figure 5.16 was developed with the same assumptions used for Fig. 5.15 and portrays the oxygen demand in a bio-

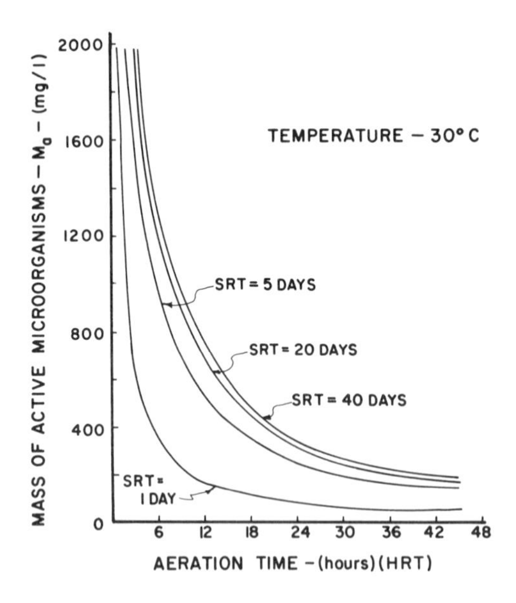

Fig. 5.15. Active mass of microorganisms in a biological waste treatment unit as a function of solids retention time and aeration time.

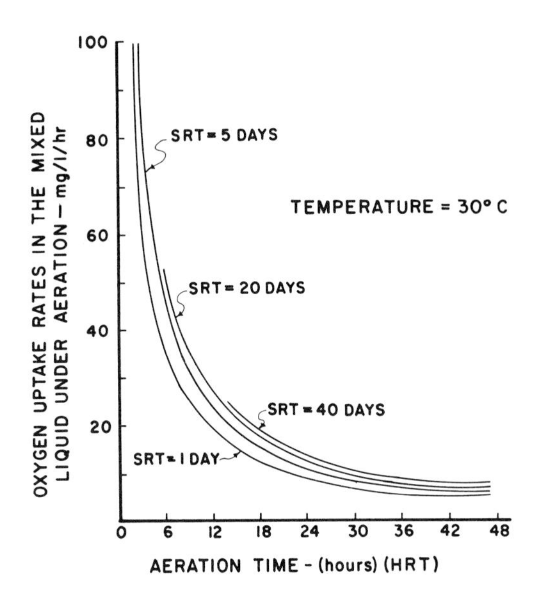

Fig. 5.16. The oxygen demand in a biological waste treatment unit as a function of solids retention time and aeration time.

logical system as related to SRT and aeration period. The oxygen uptake rate is less sensitive to SRT since the largest demand is caused by microbial synthesis. At short aeration periods synthesis predominates while at longer aeration periods, endogenous respiration predominates.

Comparison of Equations

With a biological treatment process that exhibits a combination of the log and declining growth phases, such as the start up of a treatment system, the use of a single equation to describe microbial growth and substrate removal relationships in these phases is desirable. Almost all actual biological waste treatment systems operate in the declining or stationary growth phases and equations describing equilibrium conditions in the systems can be used.

The choice of equations used to design a specific biological waste treatment system will be based upon the experience and success of a design engineer with a set of equations, and by the convenience in developing the constants necessary to apply the equations to a specific waste. The use of all the equations depends upon the availability and accuracy of the constants in the equations. The numerical value of the constants will be related to the parameters used to develop the constant, i.e., BOD, COD, TOC, or a specific substrate. Methods to determine these constants have been noted and original references have been provided.

The equations noted in the previous paragraphs are not the only ones that have been used successfully by engineers to design biological waste treatment systems. Because of difficulties in applying the previous equations to specific wastes, other investigators have developed additional empirical approaches and equations. All of the successful approaches utilize the fundamentals of synthesis and respiration described in this chapter.

Application to Agricultural Wastes

While fundamental to all biological systems, the mathematical relationships and equations noted above rarely have been applied to systems treating agricultural wastes. The use of these relationships offers an opportunity to those who wish to develop better biological processes for agricultural wastes and to better understand the processes currently handling these wastes. Because agricultural wastes are organic wastes, some type of biological treatment will be utilized prior to ultimate disposal on the land or to surface waters. The current biological treatment processes for agricultural wastes rest almost entirely on uncoordinated empirical approaches that offer little hope of success when utilized with wastes and under conditions different from those where success has been obtained.

Design equations, such as the above, have been developed based on studies with liquid wastes. Agricultural wastes can vary from highly dilute wastes to pumpable slurries. Agricultural waste management systems include a broad spectrum of possible biological systems ranging from continuous flow units in equilibrium to batch type units used primarily for waste storage prior to discharge. It is obvious that care must be used in applying generalized mathematical relationships to specific agricultural wastes.

The design engineer can utilize the mathematical relationships to obtain satisfactory performance. The actual design of a waste treatment facility will include an estimation of possible overloads to the system, temperature effects, operator ability and interest, differences in waste characteristics, and allowance for incalculable risks.

Some of the specific differences that are involved with agricultural wastes include the solids concentration in the raw waste, long detention time, nonhomogeneity in the biological system, and the oxygen requirements. With municipal wastes and certain industrial wastes, the quantity of solids in the untreated waste is small compared to the quantity of solids in the treatment system. Such an assumption cannot be made for agricultural wastes, especially food processing and animal wastes. These wastes contain a high concentration of solids which must be considered in all mass balance and mathematical relationships. These wastes frequently have a large non-biodegradable fraction which will not be removed in a biological treatment unit and will remain for ultimate disposal. In the previous mathematical relationships, the effect of an influent nonbiodegradable fraction was not considered. The total solids for disposal will include the net increase in microbial solids and the nonbiodegradable solids.

Because of the high concentration of BOD in agricultural wastes, long detention times are necessary to produce an effluent suitable for discharge to surface waters. Where discharge on land is used, long detention times in storage units prior to disposal are common. Detention times can range from days with dilute waste where discharge to surface water is contemplated to weeks and months where storage and land disposal is practiced. The long detention times will have an obvious effect on the quantity of active organisms in the system. As predicted from earlier equations, the active mass will be a minor fraction of the suspended solids in the biological unit. The long detention times permit accommodation of decreased reaction rates caused by lower temperatures or by microbial inhibition caused by materials in the wastes.

One of the basic assumptions made in developing the basic equations was that the biological treatment process was completely mixed. This is not always the case with processes treating agricultural wastes. The solids content and particle size of some of the wastes preclude complete mixing and solids sedimentation and accumulation in the quiescent areas of the biological unit can occur. In evaluating the performance of biological units treating agricultural wastes, this possibility should be closely checked.

The oxygen requirements, Eq. (5.40), (5.53), and (5.58), are based upon maintaining conditions where there is an excess of dissolved oxygen, generally at least 1–2 mg/l. When lesser amounts of dissolved oxygen exist in the system, the system is oxygen limited and the microbial reaction rates and treatment efficiency decrease. In long detention time units having a residual dissolved oxygen content, a population of nitrifying organisms can be established and an oxygen demand caused by nitrification can result. The nitrification demand will be in excess of that noted in the above three equations. The nitrification demand can be high since every unit of ammonia oxidized to nitrate requires 4.57 units of oxygen.

For most agricultural wastes, especially those disposed of on the land, a residual dissolved oxygen concentration is not required. The minimum input oxygen requirement would be to avoid the odors that are produced under highly reduced conditions. The minimum quantity of oxygen necessary to avoid odor production has not been established as yet. These minimum oxygen requirements will be less than those described by Eqs. (5.40), (5.53), and (5.58).

The mathematical relationships described in this chapter assume that the rate at which organic matter is utilized by microorganisms is a function of a limiting nutrient. In most waste treatment systems the organic waste itself is the limiting nutrient, i.e., substrate-limited. With certain food processing wastes, inorganic nutrients such as nitrogen and phosphorus can be limiting and may have to be added to achieve optimum rates of treatment and desired treatment efficiency.

These generalized equations should be applied to agricultural waste treatment processes with an understanding of how the characteristics of agricultural wastes differ from the idealized conditions used to develop the equations as well as with a knowledge of the requirements for ultimate disposal of the treated wastes. Not all agricultural wastes are released to surface waters. The characteristics of treated wastes that will be disposed of on land can be quite different from those discharged to receiving waters. This in turn will affect the design of the appropriate treatment system.

Notation

Equations (5.15) through (5.46)

- x microbial cell concentration in a biological treatment unit, mass/volume
- μ net specific growth rate, unit of cells produced per unit of existing cells per unit time, time^{-1}
- $\bar{\mu}$ maximum net specific growth rate at infinite substrate concentration, time^{-1}
- K_s a velocity constant equal to the substrate concentration at one-half the maximum net specific growth rate, mass/volume
- S substrate concentration, mass/volume
- q substrate removal rate, unit of substrate removed per unit of existing cells per unit time, time^{-1}
- Y microbial yield coefficient, unit of cells produced per unit of substrate removed, mass/mass
- k maximum substrate removal rate, i.e., maximum q, time^{-1}
- Q waste flow rate to a treatment process, volume/time
- S_o influent waste concentration, mass/volume
- S_i effluent waste concentration, mass/volume
- V reactor volume, volume
- X_1 mixed liquor microbial cell concentration, generally measured as suspended solids or volatile suspended solids, mass/volume
- HRT mean hydraulic residence time in a biological reactor, V/Q, time
- X_0 microbial cell concentration in the influent wastes, generally measured as suspended or volatile suspended solids, mass/volume

K_D microorganism decay rate, unit of cells decreased per unit of existing cells per unit time, time^{-1}
SRT mean microbial solids residence time in the biological unit, time
SRT_c minimum microbial solids residence time, occurs at maximum q and when $S_o = S_i$, time
X' net microbial solids concentration in the biological process, cell growth less degradation, mass/volume
X_s total sludge solids concentration in the process, inert plus microbial solids, mass/volume
X_i stable inorganic and organic solids in the biological process, mass/volume
a a constant to convert substrate units to oxygen units, unit of oxygen per unit of substrate
b a constant to convert cell mass units to oxygen units, unit of oxygen per unit of cell mass
O oxygen requirement, mass/volume
ΔNH_3 ammonia oxidized to nitrate during an aeration time of HRT under aerobic conditions, mass/volume
X_2 microbial cell concentration in the effluent of a secondary solids separator in a biological treatment process with solids recycle, mass/volume
Q_r recycle solids flow rate, volume/time
X_r microbial cell concentration in recycle flow, mass/volume
r mixed liquor to effluent microbial cell ratio, X_1/X_2
K_i inhibition factor, mass/volume

Equations (5.47) through (5.55)

S microbial solids concentration in a biological unit, mass/volume
k_1 microbial solids growth rate constant during the logarithmic microbial growth phase, time^{-1}
S_0 initial microbial solids concentration at the start of a logarithmic growth phase, mass/volume
L_R substrate removed during biological treatment, mass/volume
a constant which relates substrate removal to microbial solids synthesis, unit of microbial cell mass per unit of substrate
L substrate concentration, mass/volume
k_2 empirical substrate removal constant in the declining growth phase, mass/volume/time
S_a average microbial solids concentration in the biological unit in terms of mixed liquor suspended or volatile suspended solids, mass/volume
k_3 empirical microbial solids decrease constant during an autoxidation or endogenous growth phase, time^{-1}
L_a, L_b, L_n initial concentrations of different substrates in a mixed waste, mass/volume
K_a, K_b, K_n empirical removal rate constants for the removal of substrates L_a, L_b, and L_n, respectively, in a mixed waste, mass/volume/time
b microorganism decay rate, unit of cells decreased per unit of existing cells per unit time, time^{-1}
O_2 oxygen requirement, mass/volume
a' constant to convert substrate units to oxygen units, unit of oxygen per unit of substrate
b' constant to convert all mass units to oxygen units per unit time, unit of oxygen per unit of cell mass per unit time, time^{-1}

Equations (5.56) through (5.58)

F	substrate concentration in the effluent from a biological treatment unit, mass/volume
F_i	substrate concentration in the influent to a biological treatment unit, mass/volume
M_a	viable, active microbial mass concentration in the mixed liquor of a biological treatment unit, mass/volume
O	oxygen requirement, mass/volume
$K_1, K_2, K_5, K_6, K_7, K_9$	constants in McKinney's equations; units depend upon the type of recation or equation being described
F/M	food to mass ratio; used to relate the substrate to active microorganism relationships in a biological treatment unit; generally in terms of units of BOD or COD per unit of suspended or volatile suspended solids—substrate mass/volume per microbial cell mass/volume.

References

1. Kountz, R. R., and Forney, C., Metabolic energy balances in a total oxidation activated sludge system. *Sewage Ind. Wastes* **31,** 810–825.
2. Washington, D. R., and Symons, J. M., Volatile sludge accumulation in activated sludge systems. *J. Water Pollut. Contr. Fed.* **34,** 767–790 (1962).
3. Lehninger, A. L., "Bioenergetics." Benjamin, New York, 1965.
4. McCarty, P. L., Thermodynamics of biological synthesis and growth. *Advan. Water Pollut. Res.* **2,** 169–200 (1964).
5. Stumm, W., and Morgan, J. J., "Stream Pollution by Algal Nutrients," Trans. Sanit. Eng. Conf., pp. 16–26. University of Kansas, Lawrence, 1962.
6. Foree, E. G., and McCarty, P. L., Anaerobic decomposition of algae. *Environ. Sci. Technol.* **4,** 842–844 (1970).
7. Foree, E. G., Jewell, W. G., and McCarty, P. L., The extent of nitrogen and phosphorus regeneration from decomposing algae. *Int. Water Pollut. Res. Conf.*, San Francisco (1970).
8. Curds, C. R., Cockburn, A., and Vandyke, J. M., An experimental study of the role of ciliated protozoa in the activated sludge process. *J. Inst. Water Pollut. Contr.* (*Brit.*) **61,** 312–324 (1968).
9. McKinney, R. E., and Gram, A., Protozoa and activated sludge. *Sewage Ind. Wastes* **28,** 1219–1231 (1956).
10. Rudolfs, W., Effect of temperature on sewage sludge digestion. *Ind. Eng. Chem.* **19,** 241–243 (1927).
11. Rudolfs, W., Phosphates in sewage and sludge treatment. I. Quantities of phosphates. *Sewage Works J.* **19,** 43–47 (1947).
12. Finstein, M. S., and Hunter, J. V., Hydrolysis of condensed phosphate during aerobic biological sewage treatment. *Water Res.* **1,** 247–254 (1967).
13. Gaudy, A. F., Ramonathan, M., Yang, P. Y., and DeGeare, T. V., Studies on the operational stability of extended aeration processes. *J. Water Pollut. Contr. Fed.* **42,** 165–179 (1970).
14. Porges, N., Jasewicz, L., and Hoover, S. R., A microbiological process report—aerobic treatment of dairy wastes. *Appl. Microbiol.* **1,** 202–278 (1953).

15. Helmers, E. N., Frame, J. D., Greenberg, A. E., and Sawyer, C. N., Nutritional requirements in the biological stabilization of industrial wastes. III. Treatment with supplementary nutrients. *Sewage Ind. Wastes* **24,** 496–507 (1952).
16. Parker, C. D., and Skerry, G. P., "Cannery Waste Treatment in Lagoons and Oxidation Ditch at Shepparton, Victoria, Australia," Proc. 2nd Nat. Symp. Food Process. Wastes, pp. 251–270. Pacific Northwest Water Lab., Environmental Protection Agency, 1971.
17. Anonymous, "Treatment of Citrus Processing Wastes," Final Rep., WPRD 38-01-67. Water Quality Office, Environmental Protection Agency, 1970.
18. Church, B. D., Nash, H. A., Erickson, E. E., and Brosz, W., "Continuous Treatment of Corn and Pea Processing Waste Water with Fungi Imperfecti," Proc. 2nd Nat. Symp. Food Process. Wastes, pp. 203–226. Pacific Northwest Water Lab., Environmental Protection Agency, 1971.
19. "Standard Methods for Examination of Water and Wastewater," 12th ed. Amer. Pub. Health Ass., New York, 1965.
20. Analytical Reference Service Sample Type VII, Water Oxygen Demand. U.S. Dept. of Health, Education and Welfare, R. A. Taft Sanit. Eng. Center, Cincinnati, Ohio, 1960.
21. Jeris, J. S., A rapid COD test. *Water Wastes Eng.* **4,** 89–91 (1967).
22. Oldham, G. F., Oxygen demand parameters. *Eff. Water Treat. J.* **8,** 234–242 (1968).
23. "The Cost of Clean Water and Its Economic Impact," Vol. IV. Cyrus W. Rice and Co. Federal Water Pollution Control Administration, Dept. of Interior, Washington, D.C., 1969.
24. Timpany, P. L., Harris, L. E., and Murphy, K. L., Cold weather operation in aerated lagoons treating pulp and paper mill wastes. *Proc. Purdue Ind. Waste Conf.* (1971).
25. Gotaas, H. B., Effect of temperature on the BOD reaction rates. *Sewage Works J.* **20,** 441–456 (1948).
26. Zanoni, A. E., Secondary effluent deoxygenation at different temperatures. *J. Water Pollut. Contr. Fed.* **41,** 640–659 (1969).
27. Terashima, S., Koyama, K., and Mazara, Y., *In* "Biological Sewage Treatment in a Cold Climate Area," (R. S. Murphy and D. Nyquist, eds.), EPA Rep. 16100EXH, pp. 263–385. University of Alaska, College, Alaska, 1971.
28. Stensel, H. D., Biological kinetics of the suspended growth denitrification process. Ph.D. Thesis, Cornell University, Ithaca, New York, 1971.
29. Loehr, R. C., Prakasam, T. B. S., Srinath, E. G., and Joo, Y. D., "Development and Demonstration of Nutrient Removal from Animal Wastes," Rept. EPA-R2-73-095, 340 pages, Office of Research and Monitoring, Environmental Protection Agency, January 1973.
30. Esvelt, L. A., and Hart, H. H., Treatment of fruit processing waste by aeration. *J. Water Pollut. Contr. Fed.* **42,** 1305–1326 (1970).
31. Monod, J., La technique de culture continué théorie et applications. *Ann. Inst. Pasteur, Paris* **79,** 390–410 (1950).
32. Hinshelwood, C. N., "The Kinetics of the Bacterial Cell." Oxford Univ. Press, London and New York, 1946.
33. Stewart, M. J., and Pearson, E. A., "Reaction Kinetics and Operational Parameters of Continuous Flow Anaerobic Fermentation Processes," Rep. No. 4. Sanit. Eng. Res. Lab., University of California, Berkeley, 1958.
34. Agardy, F. J., Cole, R. D., and Pearson, E. A., "Kinetics and Activity Parameters of Anaerobic Fermentation Systems," SERL Rep. 63-2. Sanit. Eng. Res. Lab., University of California, Berkeley, 1963.
35. Lawrence, A. W., and McCarty, P. L., Kinetics of methane fermentation in anaerobic treatment. *J. Water Pollut. Contr. Fed.* **41,** R1–R17, 1969.
36. Gates, W. E., A rational model for the anaerobic contact process. *J. Water Pollut. Contr. Fed.* **39,** 1951–1970 (1967).

37. Garrett, M. T., and Sawyer, C. N., Kinetics of removal of soluble B.O.D. by activated sludge. *Proc. Ind. Waste Conf.* **7,** 51–77 (1952).
38. Knowles, G., Downing, A. L., and Barrett, M. J., Determination of kinetic constants for nitrifying bacteria in mixed culture, with the aid of an electronic computer. *J. Gen. Microbiol.* **38,** 263–278 (1965).
39. Guttormsen, K., and Carlson, D. A., "Status and Research Needs of Potato Processing Wastes," Proc. 1st Nat. Symp. Food Process. Wastes, pp. 27–38. Pacific Northwest Water Lab., Federal Water Quality Administration, 1970.
40. Middlebrooks, E. J., and Garland, C. F., Kinetics of model and field extended aeration wastewater treatment units. *J. Water Pollut. Contr. Fed.* **40,** 586–599 (1968).
41. Ecknoff, D. W., and Jenkins, D., "Activated Sludge Systems—Kinetics of the Steady and Transient States," SERL Rep. 67-12. College of Eng., University of California, Berkeley, 1967.
42. McCarty, P. L., "Biological Treatment of Food Processing Wastes," Proc. 1st Nat. Symp. Food Process. Wastes, pp. 327–346. Pacific Northwest Water Lab., Federal Water Quality Administration, 1970.
43. Edwards, V. H., Influence of high substrate concentrations on microbial kinetics. *Biotechnol. Bioeng.* **12,** 679–712 (1970).
44. Eckenfelder, W. W., and O'Connor, D. J., "Biological Waste Treatment." Pergamon, Oxford, 1961.
45. Eckenfelder, W. W., Comparative biological waste treatment design. *J. Sanit. Eng. Div., Amer. Soc. Civil Eng.* **93,** SA6, 157–170 (1967).
46. Benedek, P., Farkas, P., and Horvath, I., Discussion—comparative biological waste treatment design. *J. Sanit. Eng. Div., Amer. Soc. Civil Eng.* **94,** SA5, 1028–1034 (1968).
47. Schmidt, R. K., and Eckenfelder, W. W., Parameter responses to influent suspended solids in the activated sludge treatment process. Presented at the *Annu. Meet., Water Pollut. Contr. Fed.* (1970).
48. McKinney, R. E., Mathematics of complete-mixing activated sludge. *J. Sanit. Eng. Div., Amer. Soc. Civil Eng.* **88,** 87–113 (1962).
49. Burkhead, C. B., and McKinney, R. E., Application of complete mixing activated sludge design equations to industrial wastes. *J. Water Pollut. Contr. Fed.* **40,** 557–570 (1968).

6

Ponds and Lagoons

Introduction

Ponds and lagoons are among the simplest treatment systems in current use. They have found wide use with municipal and agricultural wastes. The major types of ponds and lagoons can be classified as facultative, aerobic, anaerobic, or aerated.

Facultative ponds are the most common ponds in use. The term facultative describes ponds which have aerobic conditions in the upper layers and anaerobic processes occurring in the bottom layers, especially in the settled solids. The ponds are frequently termed oxidation ponds or waste stabilization lagoons. Design loadings rarely exceed 50 lb BOD/acre/day and effluents may have BOD concentrations in the 20–40 mg/l range. Particulate matter is the main source of the BOD in the effluent since the influent wastes have been converted to bacterial or algal protoplasm.

In aerobic ponds, organic matter is decomposed solely through aerobic oxidation with the oxygen obtained by mixing and photosynthesis. Few natural ponds in practice are designed to operate in this manner. Aerobic ponds are designed with a large surface area to volume ratio. Oxygen is introduced by liquid recirculation, wind, or mechanical mixing and by photosynthesis. Continuous mechanical movement of the liquid must be employed if photosynthetic ponds are to be aerobic and not facultative. Loadings of 100–200 lb BOD/acre/day can be employed with effluents containing 10–20 mg/l BOD if the algal cells are removed.

Anaerobic ponds are systems in which the concentration of organic wastes applied per unit area is sufficient to bring about complete depletion of dissolved oxygen by limiting photosynthesis, by high bacterial oxygen demand, or both. In these ponds, up to 75% of the applied BOD can be accounted for as methane and carbon dioxide released to the atmosphere. Methane formation is the primary biological process for carbonaceous BOD removal in these units. Anerobic ponds are deep, have a small surface area to volume ratio, and may be loaded in excess of 400 lb BOD/acre/day. Effluents have a high BOD concentration.

Aerated lagoons are biological treatment units in which the oxygen demand is met by mechanical aeration equipment. The continuous oxygen supply permits the aerated lagoon to treat more waste water per unit volume per day. The design of an aerated lagoon can be approximated by the kinetics of a completely mixed biological reactor. The lagoons normally operate without solids recycle and can achieve from 50–90% BOD removal depending upon loading, required effluent quality, and whether the solids in the effluent are removed prior to discharge.

Oxidation Ponds

General

The facultative oxidation pond was developed more or less by accident. The first ponds were intended merely as waste-holding devices until suitable treatment or irrigation facilities could be developed. It was observed that as a result of the biological activity in the impounded waters, some of the waste was oxidized. These ponds are relatively shallow, diked structures with a large surface area to maintain aerobic conditions. Their purpose is to produce a suitable effluent for discharge to surface waters. In areas where the land is relatively flat, inexpensive, and available, oxidation ponds will be more economical than other types of aerobic biological treatment. Their predominate use has been in rural areas with adequate sunlight, wind action, and available land.

The term "aerobic" does not completely describe the biochemical reactions taking place in the pond. While ample dissolved oxygen may exist and aerobic metabolism may take place in the upper portion of the pond, there may be little or no dissolved oxygen in the lower depths. The solids layer on the bottom of the pond is devoid of oxygen and anaerobic conditions prevail. Oxidation ponds are so designed and loaded that these anaerobic conditions have little noticeable effect on the quality of the effluent from the ponds. To control the biological processes involved and to adequately design oxidation ponds, attention must be given to the factors that affect the process: temperature, light, organic loading, size and shape, and hydraulic considerations.

Oxidation ponds approach natural purification more closely than any other treatment process, however, the objectives of natural purification and waste treatment are different. The goal of natural purification is to recycle nutrients and organic matter. Treatment has the goal of removing pollutants. Oxidation ponds tend to recycle rather than remove nutrients.

Biochemical Reactions

Bacteria and algae are the key microorganisms in an oxidation pond. Heterotrophic bacteria are responsible for the stabilization of the organic matter in the pond. A portion of the entering BOD settles and undergoes anaerobic fermentation in the bottom sludges. This fermentation will reduce the sludge volume if temperatures are adequate but will release the products of fermentation to the liquid layer. Under uniform conditions, a state of equilibrium will be reached in which the BOD deposited in the sludge is equal to the BOD released during fermentation. Once equilibrium conditions are attained, there is an annual variation of sludge accumulation in the pond because of changes in temperature between summer and winter. The gaseous products of anaerobic fermentation represent a major loss of carbon from oxidation ponds. Ponds designed to allow methane fermentation may result in the release of 56 lb BOD/acre/day at 16°C ranging to 1120 lb BOD/acre/day at 35°C as methane gas (1).

As the soluble organic wastes entering the pond and released from the bottom sludges are metabolized by the bacteria, end products such as carbon dioxide, ammonium and nitrate ions, and phosphate ions become available for the growth of algae. Solar energy furnishes the energy for the growth of the algae. As the autotrophic algae produce new protoplasm, oxygen is an end product and can be used by the heterotrophic bacteria. A simplistic sketch of the microbial relationships in an oxidation pond is noted in Fig. 6.1. The generation of oxygen will be in proportion to the carbon converted into algal protoplasm. If BOD removal is a primary objective of an oxidation pond, the design of the pond must provide for carbon removal by methane fermentation or conversion of carbonaceous material to algae with provision for removal of the algal cells from the effluent.

Bacteria are responsible for the oxidation and reduction processes that take place in the ponds. Algae play an important role in using excess carbon dioxide and creating available oxygen. At low algal densities, oxygen production may be insufficient to satisfy more than a small fraction of the BOD in the pond. In most oxidation ponds, there are enough algal nutrients from bacterial metabolism and in the influent wastes to support an adequate algal

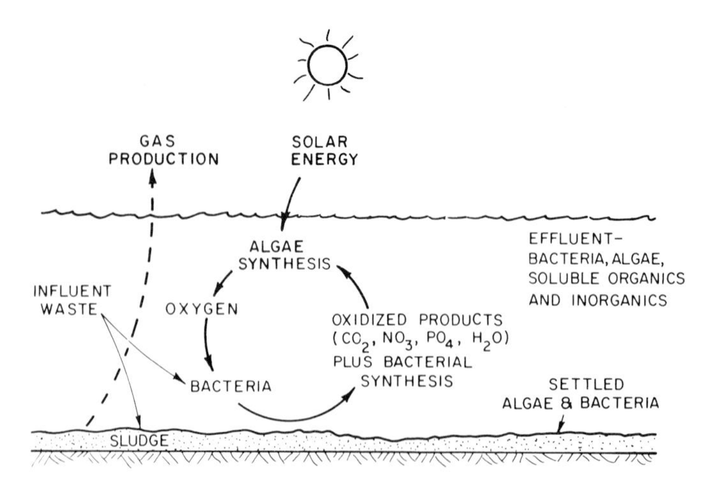

Fig. 6.1. Schematic sketch of biological interactions in an oxidation pond.

concentration. These algae can produce oxygen in excess of the bacterial oxygen demand. Such excess oxygen production accounts for the supersaturated oxygen concentration observed in many ponds.

Surface reaeration cannot meet the oxygen theoretically needed for the oxidation of organic matter in an oxidation pond. Photosynthetically produced dissolved oxygen in the pond can be a barrier against oxygen input by reaeration. As the concentration of dissolved oxygen increases, the rate of reaeration decreases until at saturation no reaeration can occur. In a pond containing growing algae, the dissolved oxygen usually approaches or exceeds saturation during daylight hours. Photosynthesis is the major source of oxygen in all aerobic and facultative ponds.

Satisfactory performance of an oxidation pond depends upon the balance between the bacteria and the algae. An excess of bacterial activity over algal activity, such as caused by high waste loads or inhibition of algal metabolism, will lead to oxygen depletion, to odor nuisance, and poor quality effluent from the pond. Excess algal activity, such as caused by excess algal nutrients and environmental conditions favorable to algal growth, will lead to excessive algal cells in the pond effluent.

The concept that organic wastes are stabilized or oxidized in an oxidation pond is valid only in the sense that the original waste organics are converted into a more stable organic form, algal cells. Oxidation ponds are organic matter generators since algal cells are produced. Mixing, temperature, and radiation are the important factors affecting the growth and concentrations of algae in oxidation ponds. In properly functioning oxidation ponds, more algal cells are produced than are generated solely from the carbon dioxide from bacterial metabolism since the carbon dioxide in the pond waters also can be used. The net result is an increase in organic matter in the system. The removal of carbon dioxide from the pond waters results in an increase in the pH of the waters. In active oxidation ponds, the pH has been known to rise above 9.0.

The production of algal cells can be illustrated by Eq. (6.1).

$$106\,CO_2 + 90\,H_2O + 16\,NO_3 + PO_4 + \text{light energy} \rightarrow C_{106}\,H_{180}\,O_{45}\,N_{16}\,P_1 + 154.5\,O_2 \qquad (6.1)$$

The equation indicates that the weight of oxygen produced per unit weight of algae synthesized is approximately 2.0. The oxygen generating capacity of algae varies from specie to specie and the growth of 1 gm of algae in waste water is associated with 1.3–1.8 gm of oxygen production. A value of 1.6 is commonly used in estimating oxygen production. The oxygen demand in an oxidation pond should be equal to or less than the photosynthetic oxygen production if anaerobic conditions are not desired. However, anaerobic reactions play a major role in the stabilization of BOD in an oxidation pond. Anaerobic conditions in an oxidation pond, such as those that might occur in the lower portions and bottom of the pond, are not undesirable unless odors, nuisance conditions, or lowered efficiency result.

It can be observed that 1 gm of carbon will produce about 1.9 grams of organic matter, 1gm of nitrogen will produce about 12 gm of organic matter, and 1 gm of phosphorus will produce about 75 gm of organic matter as algal cells. These relationships indicate the importance of nitrogen and especially phosphorus in the eutrophication of surface waters since the same relationship will hold in surface waters other than oxidation ponds.

Carbon, nitrogen, and phosphorus are the key nutrients in oxidation ponds, Rapidly growing algae are about 55% carbon as C, 9% nitrogen as N, and 1% phosphorus as P. Based on these percentages or using Eq. (6.1), it is possible to estimate the quantity of algae that a given waste will support and to estimate which of the nutrients will limit algal growth. In low rate ponds and ponds with long retention times, algal growth and oxygen production is related to the carbon dioxide available. Because of the small amounts of nitrogen and phosphorus required, it is rare that these elements become limiting factors, especially with municipal and most agricultural wastes. Certain food processing wastes may be incapable of supporting satisfactory algal growth due to a deficiency of a critical nutrient. Examples might be wastes from sugar or starch production or from certain canning operations where a low nitrogen to carbon ratio may exist or sugar beet wastes which may have a shortage of phosphorus. Where no experience is available with a particular waste water suspected of nutrient deficiency, laboratory and pilot plant studies may be warranted before large scale design.

Kinetic relationships for oxidation pond design have been developed, compared to operating ponds, and found to be a rational approach to design (2).

Well-designed oxidation ponds have demonstrated the ability to handle fluctuating loads. Because of large liquid volume and long detention times, the active bacterial mass in the pond will be low, under 50 mg/l with wastes approximating the strength of domestic sewage. Long detention times compensate for the low bacterial mass and high degrees of BOD removal can be achieved.

Oxygen Relationships

The net oxygen transferred into an oxidation pond is the sum of the algal oxygen production and the quantity added from or lost to the atmosphere. Because sunlight is the energy source for algal growth, algal growth is a function of the penetration of sunlight into the pond which in turn is a function of the turbidity of the water. The efficiency of light conversion by algae grown in waste water has been found to seldom exceed 10–12% and is usually much less. Oxidation ponds are designed with a shallow water depth, generally 3 to 5 ft, to permit maximum light penetration and algal growth. The light available at locations having no ice cover is more than adequate to support the required photosynthetic activities. Ice cover and cold weather will reduce both light and temperature to the point that oxygen production by the algae is negligible or zero.

The depth of an oxidation pond is determined by the depth of light penetration since the efficiency with which algae convert light energy to cellular materials depends upon the intensity of light. In laboratory studies with common algal species, a light intensity of 30 ft-c was the lowest at which a measurable overall algal efficiency was obtained (3). Daylight penetrated only 12 inches on a culture having a concentration of 140 mg/l and growing in an outdoor pond before it was diminished to 20 ft-c.

A pond 4 ft deep allowing light equivalent to 15,000 cal/l/day will produce about 50 lb of oxygen per acre per day with an increase of pH to 9.5. Successful pond designs currently are limited to loading rates of 40–50 lb BOD/acre/day. Extensive mixing would be required to utilize the maximum photosynthetic oxygenation rate of 100–200 lb/acre/day noted in completely aerobic ponds (4).

The oxygen produced by algae is contained in the upper layer of the pond and must be mixed throughout the pond to be of benefit. Wind is the general source of mixing although liquid recirculation has been tried. With active algal growth, supersaturated oxygen concentrations may exist in the top layer of the pond and some oxygen loss to the atmosphere will occur. If unsaturated conditions exist in the top layers, the wind action will assist in transferring the oxygen from the atmosphere to the pond. Good mixing increases the capacity of the pond to handle increased pollution loads.

Vertical mixing of nutrients and oxygen in an oxidation pond by diffusion is negligible compared to that from wave and wind action or by recirculation. In the conventional oxidation pond, mixing is one of the important factors which is able to be controlled yet it rarely is incorporated in practice. Wave action is a problem due to bank erosion and few ponds are equipped for recirculation. Without special mixing by artificial or natural means, atmospheric reaeration will introduce less than 40 lb of oxygen per acre per day even under anerobic conditions. An aerobic pond will absorb less oxygen. Atmospheric reaeration does not contribute rapidly to the oxygen resources of unmixed ponds. The value of reaeration is excluded by supersaturation conditions which can accompany vigorous photosynthesis.

The beneficial effect of algae occurs only when the energy from the sun is available. During nighttime, the respiration of the algal cells will represent an oxygen demand on the system. Normal oxidation ponds may produce supersaturated oxygen conditions during daylight hours which generally are sufficient to keep the ponds aerobic during the night. *In situ* measurements in an aerobic oxidation pond loaded at about 60 lb BOD/acre/day and with a solar input of 200 cal/cm^2/day indicated the algal respiration rate was 0.47 mg/l of oxygen/hr and the mean net photosynthetic rate at the surface was 1.16 mg/l of oxygen/hr (5). The gross surface photosynthetic rate including respiration was 1.62 mg/l of oxygen/hr or for the pond in question 300 lb of oxygen/acre/day. In this pond the photosynthetic efficiency of the algae was about 2.3% of the solar energy imput but was more than sufficient to handle the BOD loading since supersaturated oxygen conditions resulted.

Oxygen production in waste water oxidation ponds can vary from 2.1 gm/m^2/hr at midmorning to 0.03 gm/m^2/hr in the early evening (6). A schematic relationship of oxygen production and oxygen demand is illustrated in Fig. 6.2. Wind action and oxygen transfer from the atmosphere may be sufficient to maintain aerobic conditions during nighttime. The rate of oxygen demand of algal cells without light energy is essentially that of endogenous respiration. Diurnal pH as well as oxygen relationships occur in oxidation ponds.

Temperature Relationships

Temperature affects the rate of metabolism of microorganisms in an oxidation pond, the loading rate of a pond, and its size. The maximum rate of metabolism will occur in late summer and early fall when the temperature

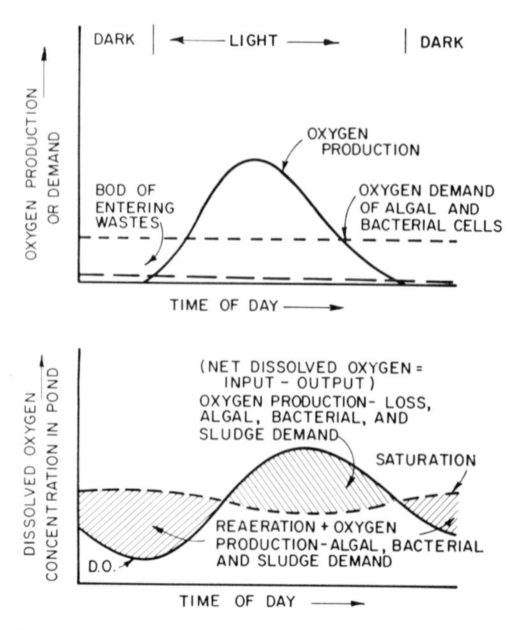

Fig. 6.2. Schematic sketch of typical oxygen production, oxygen demand, and oxygen concentration in an oxidation pond.

of the liquid is at a maximum. No attempt is made to control the temperature in the oxidation pond and it will fluctuate in response to ambient temperatures. The design of an oxidation pond will reflect the expected temperature conditions. Feasible pond loading rates, in terms of pounds of BOD/acre/day, are less in colder climates than in warmer climates.

The temperature effect on the waste degradation rate in oxidation ponds has been shown to be similar to that of other microbial systems, i.e.,

$$K/K_{20} = \theta^{(T-T_{20})} \tag{6.2}$$

where K represents the rate reactions and T the liquid temperature in terms of °C. A θ value of 1.072 was found to fit available data. The equation was valid between 3° and 35°C. These temperatures represent the practical limit caused by decreased activity as the temperature approached freezing and approached the thermal inactivation level of many algae species, respectively (7).

In cooler climates where ice cover may occur during portions of the winter, temperatures under the ice cover are only slightly above freezing. Biological activity is minimal and the accumulation of influent solids occurs. With warmer temperatures in the spring, increased biological activity resumes. Odors are often noted in oxidation ponds during the transition period between winter and spring before spring temperatures and sunlight have restored the algal population and aerobic conditions are again established. During cold weather, oxidation ponds serve more as solids sedimentation units than as biological treatment units.

The fate of the settled organic matter on the pond bottom is a function of the conditions existing at the bottom. If the temperature is near 4°C or if the pH is below 5.5, decomposition of the organic matter is slow and the organic matter accumulates. If the bottom temperature is higher, acid decomposition occurs and if the environmental conditions are correct, methane fermentation may occur.

The pattern of solids accumulation on the bottom of oxidation ponds, anaerobic ponds, and incompletely mixed aerated lagoons is cyclic, increasing in the winter when there is little solids degradation and decreasing in the summer when the accumulated volatile solids are degraded (Fig. 6.3). There will be a net solids accumulation in these ponds and lagoons due to a buildup of inert solids in the influent wastes and undecomposed organic matter.

Gas production in terms of cubic feet produced per pound of BOD applied was found to be maximum when the pond water temperature was greater than 19°C (3). At less than this temperature gas production was less than the theoretical amount expected from solids decomposition. The gases from these solids contained 60–75% methane. At temperatures less than 18°C methane fermentation slowed and organic matter accumulated. These solids will decompose when the temperature becomes greater than 18–19°C (Fig. 6.4). The data from this study also indicated that gas production from solids decomposition did not occur until the water temperature was greater

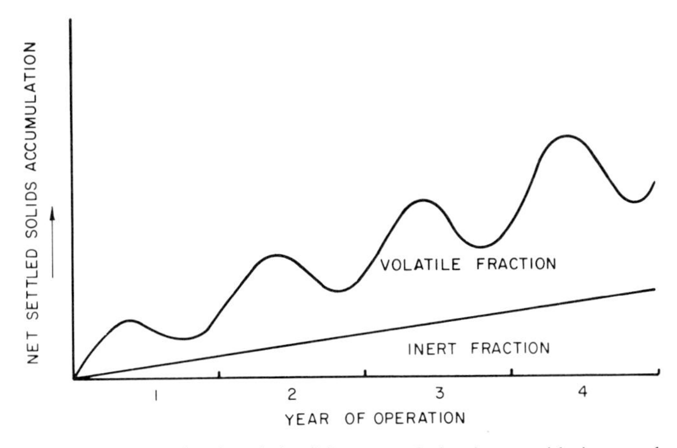

Fig. 6.3. Schematic of settled solids accumulation in an oxidation pond.

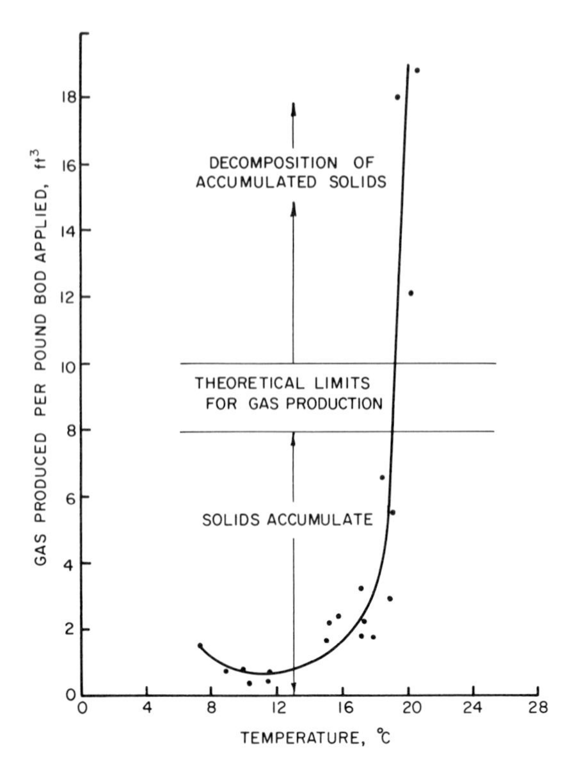

Fig. 6.4. Gas production as a function of temperature and BOD loading in a pond or lagoon (3).

than 15°C (Fig. 6.5). The relationship for methane fermentation was noted as in Eq. (6.3).

$$G = 450\,(T - 15) \tag{6.3}$$

where G is gas production in cubic feet/acre/day and T is the temperature in °C (1). For adequate methane fermentation, the sludge zone should be protected from intrusion of oxygen and the temperature maintained above 15°C. In facultative ponds with quiescent conditions, a liquid level exists below which there will be no free dissolved oxygen. This level is a function of the relative rates of photosynthesis and respiration.

Faculative ponds should be designed to include solids settling and decomposition. Design for optimum algal growth can be secondary since in these ponds, the algae in the surface layers serve only to maintain an aerobic zone and to control odors.

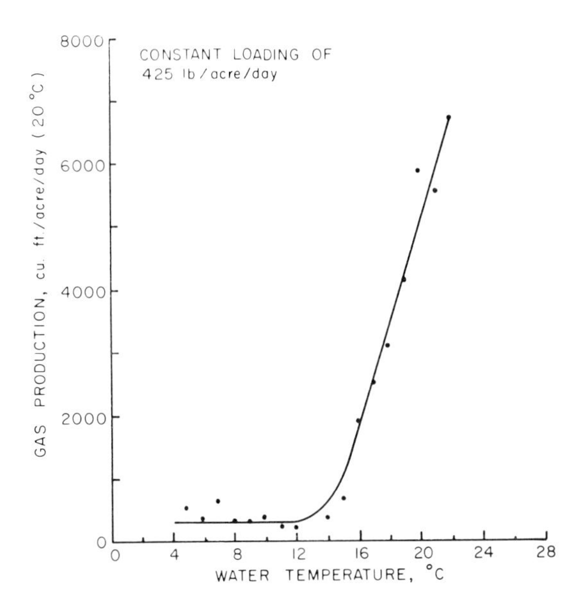

Fig. 6.5. Mean gas production as a function of water temperature at a constant loading in a pond or lagoon (3).

Loading Rate

The environment within an oxidation pond is unable to be controlled by the design engineer. Only physical factors such as size and loading rate can be controlled. A wide fluctuation in environmental conditions is natural and the design must be such that the process will operate satisfactorily within these conditions. Two types of loading rates are important, the hydraulic and the oxygen demand (BOD) loading rates. The effective hydraulic loading is determined by the detention period and the volume of the pond used for the algal production or other biochemical reactions. The actual hydraulic loading rate is determined by the detention period and the total pond volume. The BOD loading rate is determined by the rate of bacterial activity in decomposing the organic matter and the rate at which algae can produce the oxygen.

Rational design of an oxidation pond includes determining the ultimate influent BOD, the desired effluent quality, the quantity of oxygen that must be produced for BOD satisfaction, the quantity of algae that must be synthesized to produce the required oxygen, and the solar energy required to sustain algal growth under actual conditions. The required surface area of the oxidation pond would be equal to the total energy required per day to sustain algal growth divided by the energy in the algal cells at a given conversion of solar energy per surface area. An oxidation pond loading rate should be related to the ratio of BOD load to the oxygen production by algae.

The above approach is feasible although it has rarely been applied to the design of oxidation ponds. Empirical loading rates, approximating the above approach, have been used for design. The loading rate for oxidation ponds is usually expressed in terms of pounds of BOD per surface area per day or population equivalent per surface area per day in recognition of the fact that the primary energy input is sunlight entering through the pond surface. Population equivalents are of lesser importance with agricultural wastes. Design of oxidation ponds for agricultural wastes should be done on the basis of the pounds of BOD in the raw wastes.

Pond performance is affected by environmental conditions such as temperature, solar radiation, and wind speed. Since these vary at a given geographical location, the design criteria reflect the influence of prevailing climatological conditions. The allowable organic loading rate on a pond is a function of the rate at which biological processes can satisfactorily decompose the organic matter without creating nuisance conditions. Design loading rates range from 20 to greater than 50 lb BOD_5/acre/day. The lower figure is common in northern areas where cooler temperatures and ice coverage is common. Higher loadings are used in southern and warmer areas. (Table 6.1).

During the summer considerably higher loading rates than the noted average rates can be treated in oxidation ponds. The practice is to design oxidation ponds for conditions when oxygen production will be the lowest. In most of the United States the policy has been to reduce allowable loadings to the range of about 20–50 lb BOD/acre/day so that problems of the

Table 6.1

Oxidation Pond Design Criteria in the United States[a]

Design criteria	Region		
	North	Central	South
Organic loading (lb BOD/acre/day)			
Mean	26	33	44
Range	16.7–40	17.4–80	30–50
Detention time (days)			
Mean	117	86	31
Range	30–180	25–180	20–45

[a] From reference 8.

transition period may be minimized. Loading rates up to 100 lb BOD/acre/day can be used in tropical and semitropical climates to obtain BOD and COD removal greater than 90 and 80%, respectively (8).

Studies on oxidation ponds in Texas noted that loadings could be 150–200 lb BOD/acre/day providing that the effluents were used for irrigation or diluted with adequate stream flow (9). BOD removal ranged from 85–98%. In other tests, 70% removal was obtained with doubled detention times and lower loadings. A considerable increase in algal population was the reason for the lower BOD removal.

The relatively low organic loadings used for design require large surface areas with resultant long liquid detention times. For example, a waste with a BOD_5 of 300 mg/l and a flow of 1 mgd would require a pond having a surface area of 50 acres, assuming a permissible loading rate of 50 lb BOD/acre/day. If the pond averaged 5 ft in depth, the liquid detention time would be about 34 days. If land is available, oxidation ponds can be satisfactory for high volume, low BOD concentration wastes.

Pond loadings of 20–40 lb BOD/acre/day have demonstrated satisfactory performance throughout the north and north-central regions. The lower loading can result in a lower volume of liquid per unit area and can enhance difficulties of maintaining satisfactory liquid depth, particularly where there is significant seepage, high evaporation, or low waste water volumes. Where these limits are exceeded, nuisances can be expected to occur in direct proportion to the overload even though treatment efficiency, measured in percent removal, may be good.

A minimum liquid depth of about 3 ft has been found desirable in an oxidation pond to reduce the growth of aquatic plants and mosquito propagation. Because of the necessity to maintain this depth, oxidation ponds may be unsatisfactory for relatively low volume wastes having high BOD concentrations. Some agricultural waste waters, especially those from animal production and food processing operations fall in the latter category. Consider a waste having a BOD_5 of 1000 mg/l and a flow of 0.2 mgd. The surface area required would be 33 acres assuming a loading rate of 50 lb BOD_5/acre/day. If the average evaporation plus seepage loss from the pond exceeded 0.23 inches/day, the loss from these sources would be greater than the liquid input per day and maintenance of a satisfactory liquid depth would be difficult. If the organic loading were 20 lb BOD_5/acre/day, the surface area required would be 83 acres, and an average evaporation plus seepage loss of only 0.1 inch/day would be greater than the liquid input per day. Greater difficulties would exist for higher strength, low volume wastes. The required surface area, average liquid detention time, and the ability to maintain an adequate liquid depth should be evaluated critically for high strength, low volume wastes.

Lagoons should not receive significant amounts of surface runoff. Provision should be made for directing surface water around ponds. For new installations and for installations where maintenance of satisfactory water depth may be a problem, diversion structures which permit variable input and exclusion of surface water as desired may be useful.

Effluent Quality

The objective of using oxidation ponds is to treat the incoming wastes without causing nuisance conditions and to produce an effluent of acceptable quality. In adequately designed oxidation ponds, the effluent will contain a negligible amount of the influent waste and on this basis high removal rates are possible. The effluent will contain bacterial and algal cells which will exert an eventual oxygen demand in the waters receiving the effluent. In active oxidation ponds, the algal concentration is considerably larger than that of the bacteria. The oxygen demand of an oxidation pond effluent essentially will be in proportion to the concentration of algae in the effluent.

In low rate oxidation ponds in which the applied organic loading is low, the algal disposal problem may be low since the algae are in low concentrations. Many of the algae are consumed by crustacea such as *Daphnia* or settle slowly and become part of the complex material at the lagoon bottom. Where anaerobic conditions exist, digestion occurs. A large portion of the suspended algae is entrained in pond currents and discharged into the receiving streams. Because of their low settling velocities, unicellular algae may not settle in the ponds. When a pond is in balance, natural processes solve the problem of algae disposal with excess carbon, nitrogen, and energy being dispersed into the atmosphere and water.

In high rate ponds, photosynthesis is practically the sole source of oxygen for the aerobic biological decomposition of the entering waste and settled solids. The algal production is greater than in low rate ponds, and may not be able to be disposed of by natural processes. Suitable provisions must be made for ultimate disposal of the algal crop. Removal techniques that have been used where excess algal cells are in the pond effluent include chemical precipitation and microstraining. Unless the excess algae are removed from the pond effluent prior to discharge, the ultimate waste load may not have been decreased significantly in an oxidation pond. Discharging large concentrations of algae into a receiving stream or lake can be objectionable under some conditions. Although algae may not settle at normal stream velocities or readily die and undergo decomposition, in large concentrations they can lower the oxygen resources of a stream at night and contribute to its total oxygen demand.

The fecal coliform densities in the effluent from many single cell oxidation ponds can be of such magnitude as to violate contact recreation uses unless appreciable dilution capacity is available. Series oxidation ponds produce better bacterial quality than single cell ponds. The organic loading rate has no appreciable effect on the total or fecal coliform quality. True bacterial detention time is more important. Effluent bacterial densities in a single cell pond were observed to vary by an order of magnitude from one moment to the next (10).

Removal of phosphates, nitrates, and other plant nutrients from the pond effluent will be high only if the algal cells are removed from the effluent. Controlled algal production and separation from oxidation pond effluents can be a feasible process for the removal of nitrogen and phosphorus in certain locations where tertiary treatment is required. Methods of algal removal for this purpose, such as centrifugation and flocculation, have been studied but are rarely applied.

Design and Operation

The oxidation pond site should be convenient to the source of the waste but where prevailing winds will not cause any odors to drift to residential areas. Where ice coverage occurs, resultant anaerobic conditions will create some odor potential during the transition period from ice cover to warmer open waters. The pond should be located in areas where wind action is likely to be maximum. Ravines and sheltered valleys are not appropriate locations.

The topography of the area and the soil type will influence the site selection. The site should be on impervious soils or on soils suitable for sealing. The pond bottom should be compacted with clay or other material to avoid excess seepage which can cause difficulties in maintaining adequate water depth and possible contamination of ground water. Contaminants in groundwater near oxidation ponds with relatively high infiltration rates have been investigated (11). The nitrogen in the pond percolation waters was found to be mainly ammonia nitrogen. Little nitrate contamination was observed. Nitrate nitrogen was in the 0.1–0.5 mg/l concentration range while ammonia nitrogen concentrations ranged from 1.0 to 20.0 mg/l. Phosphate concentrations were not significant in the groundwater surrounding the ponds, ranging from 0.2 to 1.0 mg/l 10 ft from the edge of the ponds. Phosphate is adsorbed by soils over a wide pH range.

Pond side slopes are influenced by the nature of the soil and size of the installation. For outer slopes, a ratio of 3:1 horizontal to vertical or greater has been found satisfactory. Flat inner slopes have the disadvantage of providing shallow areas conducive to emergent vegetation. Wave action is more severe for larger installations, warranting flatter sloper to minimize erosion.

Liquid depth should be more than 2–3 ft since shallower depths result in growth of aquatic weeds which in turn increase mosquito breeding. Vegetation-free banks are important in the control of mosquito breeding. Vegetation on the banks should be kept cut, should be removed after cutting, and should not be allowed to float on the water to provide harborage for the mosquito larvae. Covering the banks of a pond at the waters edge with solid material such as plastic, asphalt, cement, etc., to prevent growth of vegetation is useful to control mosquito breeding and to minimize erosion. Mosquito larvae are seldom found in the open water of a pond. The shape of the ponds should be uniform and essentially circular or rectangular with few areas where floating material may accumulate. The ponds should have adequate freeboard above the maximum water line, generally 3 ft for ponds having large surface areas. Weeds, straw, and grass clippings should be excluded from the pond since they will increase the organic load in the pond.

Studies have delineated the oxygen demand and nutrient regeneration caused by the decay of aquatic weeds (12). The maximum potential quantity of oxygen that would be utilized during the decomposition of a variety of aquatic weeds was 1.30 lb BOD/lb volatile suspended solids. On the average 24% of the organic material remained after decomposition. The range was 11 to 46%. Variable nitrogen (0–100%) and phosphorus regeneration (0–100%) occurred during decomposition. In general, phosphorus regeneration was greater than nitrogen regeneration. The effect of the aquatic weeds during decay is twofold; consumption of oxygen and regeneration of nutrients to stimulate further aquatic growth.

Temporary odor problems, such as caused during the spring transition from anaerobic to aerobic conditions, can be controlled by the addition of nitrates. The nitrates will act as hydrogen acceptors in the absence of oxygen to prevent reduced sulfur-containing compounds from being generated. Sodium or ammonium nitrate are common sources of hydrogen acceptors. Mechanical aeration also can be used during such periods.

Continuous odor problems indicate that the oxygen demand in the oxidation pond exceeds the supply. Solutions to the problem include reducing the BOD loading or using mechanical aeration to convert the oxidation pond to an aerated lagoon.

Settling tanks can be used to remove sludge and scum from entering an oxidation pond and adding to the organic load. These tanks should be designed for easy cleanout. Disposal of the scum and solids from the settling tanks should be done often enough to avoid solids carry-over to the oxidation pond and should be accomplished without causing other pollution problems.

Multiple oxidation ponds cells should be interconnected so that they can be operated in parallel or series. Recirculation of the effluent in a single pond back to the influent end is of questionable value except for the mixing effect. Interpond recirculation can be used for altering dissolved oxygen levels or algal population levels to meet requirements imposed by waste loading and climate.

The effluent from oxidation ponds should discharge to surface waters only if the discharge will not result in violation of state water quality standards. The effluent may receive further treatment or be disposed of by liquid spreading systems, irrigation systems, or other land application measures. The disposal area should have proper soil characteristics and surface cover to prevent further pollution problems. Application rates should prevent surface runoff and if possible the nutrients should be utilized by the surface vegetation.

If the pond is located where a safety hazard may exist, the oxidation pond should be fenced to prevent access of animals and children.

Application to Agricultural Wastes

This section is not meant to be an exhaustive literature study. Rather the references were chosen to illustrate oxidation pond fundamentals as applied to agricultural wastes and to indicate the practical application of oxidation ponds as a method of treating agricultural wastes. Both single and multiple cell oxidation ponds have been used to treat agricultural wastes.

Table 6.2

Type and Performance of Industrial Waste Ponds in the United States (1963)[a]

Type of waste and industry	Average loading (lb BOD/acre/day)	Average BOD removal (%)	Odor problems
Aerobic			
Meat and poultry	72	80	Some
Canning	139	98	Many
Wine	221	—	Some
Dairy	22	95	Some
Hog feeding	356	—	Some
Potato	111	—	Some
Anaerobic			
Meat and poultry	1260	51	Many
Canning	392	80	Many
Aerobic–anaerobic			
Meat and poultry	267	94	Few
Canning	617	91	Few

[a] From reference 13.

A summary of ponds and lagoons treating industrial wastes has been prepared and the data for ponds treating agricultural wastes is abstracted in Table 6.2 (13).

The aerobic ponds noted in Table 6.2 are loaded at rates higher than those recommended for conventional ponds with some of the average loadings in the range considered for facultative and anaerobic ponds. Odors were a problem for most of these systems.

Studies on mixed fruit and vegetable canning wastes have indicated that oxidation ponds can be a successful method of treating these wastes. A uniform loading of 100 lb/acre/day was recommended. The average BOD removal in the study was 80%. The use of series ponds was felt to be justified when the BOD loading was higher than 100 lb/acre/day or when large shock loads were common (14).

Meat packing wastes have been treated in a combination of anaerobic-aerobic ponds. Oxidation ponds treating the effluent from the anaerobic ponds achieved a BOD reduction of 59% with a detention time of 18 days. The loading rate averaged 130 lb/BOD/acre/day (15).

Wastes from a milk processing plant were treated in pilot plant oxidation ponds with a BOD loading of 220 lb BOD/acre/day and a 10-day detention period. BOD reductions of 80–90% were obtained. Average pond temperatures ranged from 17° to 30°C (16).

High rate oxidation ponds for poultry processing wastes were loaded at the rate of 214 lb BOD/acre/day. BOD removal efficiencies of 70–96% were obtained. The treatment facilities consisted of primary sedimentation, an equalization tank and two oxidation ponds operated in parallel (17). A two-stage oxidation pond system has been used to obtain 80–95% overall BOD removal for poultry processing wastes (18). The first stage was a high rate pond 8 ft deep loaded at a rate of 935 lb BOD/acre/day with a detention time of 7.9 days. The first pond was anaerobic but achieved 72.5% BOD removal. The second stage was a shallow photosynthesis pond, 2 ft deep loaded at 74 lb BOD/acre/day. BOD removal efficiencies in the second stage were 9% when the algae were not removed from the effluent and 70% when the algae were filtered out.

A three-stage oxidation pond successfully treated bean cannery waste during short (10-week) packing seasons (19). The first and second stages were heavily loaded and were anaerobic. The loading on the total pond area was 260 lb BOD/acre/day. BOD reduction was about 78% and suspended solids removal about 77%. Temporary odorous conditions were corrected by adding sodium nitrate to the pond influent and aerating the influent. The effluent was spray-irrigated on a rye grass field at a rate of 67 lb BOD/acre/day without odors, ponding, or contamination of an adjacent stream.

Data on oxidation ponds treating combined potato and municipal wastes in North Dakota were collected over a 40-month period (20). During summer conditions, the combined wastes were added at rates well above 20 lb BOD/acre/day, in the range of 50–60 lb BOD/acre/day. The net organic load of the wastes accumulated in the winter and added during the summer had to be stabilized during the summer months. The ponds remained aerobic until the BOD in the ponds exceeded 200 mg/l. Once anaerobic conditions were established, the ponds did not return to aerobic conditions until the BOD fell below 100 mg/l. Surface reaeration was reduced by the effect of potato organic matter in the pond water.

Using average design criteria for oxidation ponds, typical pond sizes for animal wastes are indicated in Table 6.3. These sizes should be regarded only as guidelines since different agricultural operations will have different waste characteristics. The most important factors affecting the design and performance of oxidation ponds are the BOD loading and the liquid flow rates.

Aerobic Ponds

Oxidation ponds have more treatment capacity in terms of loading rate than that recommended by most state regulatory agencies, especially if it is realized that the entire volume of the pond is available for treatment. The engineering challenge is to utilize this additional capacity.

Table 6.3

Possible Oxidation Pond Sizes for Animal Wastes[a]

			Pond area (acres)	
Item	Size of operation	Pounds of BOD/day	50 lb BOD/ acre/day	20 lb BOD/ acre/day
Chickens[b]	10,000 birds	180	3.6	9
Swine[b]	100 head	30	.6	1.5
Dairy cattle[b]	100 head	140	2.8	7
Ducks	1,000 birds	30	.6	1.5
Milking parlor[c]	100 head	80	1.6	4
Poultry processing	1,000 birds	27	.53	1.35
Meat packing	per 1,000 lb live weight processed	10	.2	.5

[a] Estimated from average waste and plant data. Individual operations may have waste characteristics widely different from those assumed.

[b] Assumes adequate water to permit the use of an oxidation pond.

[c] Assumes a minimum amount of fecal matter and waste feed is included in this waste.

The work by Oswald and co-workers over the years has established the basic factors involved in oxidation ponds (21, 22). These studies identified the role of photosynthesis and algal age in conventional ponds treating sewage. The loading rate to which a conventional pond can be subjected is a function of the active algal cells in the pond capable of participating in the necessary photosynthetic reactions and of the settling in the pond of some of the influent organic matter. As noted earlier, a conventional pond is characterized by both aerobic and anaerobic decomposition and by a low concentration of active algal cells.

By reducing pond detention times and using shallow depths to foster actively growing algal cells, 95% BOD removals were obtained with BOD loads of up to 225 lb/acre/day in summer and 110 lb/acre/day in the winter (21). These removals could be obtained only when the large quantities of algae were removed from the pond effluent. If the algae were not removed, the BOD removal was 7%.

The high rate aerobic oxidation pond is able to transform the unstable influent organic matter into algal protoplasm. The treatment process is incomplete unless the algal cells are removed from the effluent again emphasizing the need for heterotroph–autotroph balance unless solids separation is contemplated with an oxidation pond.

This high rate process is feasible in climates with continuous mild temperatures and sunlight, where adequate management of the pond is available, and where the separated algal cells can be disposed of adequately and perhaps profitably. The use of the concentrated algal cells as animal feed has been contemplated.

Anaerobic Lagoons

General

Anaerobic lagoons are only similar to aerobic oxidation ponds in that both are impoundments used for holding and treating liquid wastes. Anaerobic lagoons are units that are loaded such that surface reaeration and photosynthetic activity cannot maintain aerobic conditions. Many lagoons currently labeled anaerobic lagoons are overloaded aerobic lagoons. An anaerobic lagoon bears only a superficial resemblance to aerobic lagoons, has a different purpose, and should be designed on a different basis.

The purpose of anaerobic lagoons is the destruction and stabilization of organic matter and not water purification. They can be used as primary sedimentation units to reduce the load on subsequent treatment units. They differ from primary sedimentation units in that the settled solids are not routinely removed but are left in the unit to degrade. A gradual buildup of solids occurs, the rate of buildup being a function of the solids loading rate, the characteristics of the raw waste, and the rate of the solids stabilization. Periodic solids removal is necessary. The biodegradable fraction of the solids undergoes anaerobic decomposition. Considerable gas may be evolved with a resultant decrease in BOD and COD of the lagoon contents.

An anaerobic lagoon is a simple treatment process that can achieve solids separation of a waste as well as biological stabilization of a portion of the waste. Anaerobic lagoons can function as liquid or solids holding units where surge capacity is needed. They can be useful for holding animal wastes prior to field spreading. Anaerobic lagoons can accept variable BOD and solids loadings from accidental spills, process variations, or intermittent operation such as no waste input over weekends or longer, without adverse effects. Like all biological treatment units, an anerobic lagoon performs best if the wastes are added continuously rather than at long intervals.

Size

There is no need for a large surface area to promote surface aeration and or obtain adequate light energy for photosynthesis. Anerobic lagoons require less land area than for an equivalent aerobic lagoon since they are more heavily loaded. The depth of the lagoon is not restricted by light penetration. Anaerobic lagoons should be built with a small surface area and as deep as possible consistent with construction difficulties and groundwater conditions. In practice, anaerobic lagoons have been from 5 to 15 ft in depth. The small surface area promotes anaerobic conditions and decreases the needed land area. Long liquid detention times are not required. Detention periods as short as three to five days have been successful. Longer times also have been successful.

In anaerobic lagoons there is a relatively solids-free liquid layer above a layer of settled solids. A floating scum layer may occur depending on the type of waste. With a small surface area, the scum can form an effective floating cover to minimize surface reaeration and to provide some insulation of the lagoon contents during cold weather.

The depth of a lagoon can be restricted by soil conditions and the temperature that exists within the lagoon. Lagoon temperatures may decrease with depth and may reach a point where biological reactions occurring in the settled solids are inhibited. The volume of the lagoon is dictated by the organic loading rate which is influenced by the desired frequency of solids removal. Capacity should be provided to hold the settled solids between times of solids removal.

Loading

Anaerobic lagoons are comparable to single-stage, unmixed, unheated digesters. Loading values should be based on pounds per volume per time as is done for other anaerobic systems. Loadings of 300–2000 lb BOD_5/acre/day have been reported for anaerobic lagoons illustrating the difference in loading rates between anaerobic lagoons and oxidation ponds. Loadings of 0.36 to 10.4 lb VS/1000 ft^3/day have been reported for lagoons treating a variety of animal wastes (23–25). Studies also have shown that high loading rates, 130–320 lb VS/1000 ft^3/day can be used successfully (26–28). The advantage of anaerobic lagoons lies with wastes that are highly concentrated and in high loading rates. A minimum loading rate of 15–20 lb BOD/1000 ft^3/day should be satisfactory for most wastes. Using the waste characteristics in Chapter 4, appropriate pond sizes can be estimated for a variety of agricultural wastes.

The design of anaerobic lagoons should be on the basis of the BOD or solids loading rate since these factors more clearly represent the factors affecting the microbial reactions in the lagoon. Loadings and designs based on number of animals or pounds of food processed are subject to differences in animal feed, types of animal housing, process efficiency, and management efficiency such as waste feed in the manure, liquid and solid wastes separation at the source in food processing operations, and other in-plant waste control measures.

In terms of percent BOD removal, high loading rates can provide satisfactory treatment. These high rates have the disadvantage of requiring sludge removal more often. The high efficiencies of BOD removal in anaerobic lagoons is dependent upon the development of satisfactory methane fermentation conditions in the settled solids layer. There is little advantage of operating anaerobic lagoons in series since only minor increases in BOD removal efficiency have been observed.

Inadequacies and failures of anaerobic lagoons at a variety of loading rates have been reported. The establishment and maintenance of conditions suitable for optimum biological metabolism plays a larger role in the success or failure of anaerobic lagoons than does the loading rate. Attention should be paid to control of pH, alkalinity, temperature, and mixing.

Anaerobic digestion of complex waste material occurs in a series of steps, the two most important ones being the conversion of complex organics to simple organic materials such as organic fatty acids (volatile acids) and conversion of these compounds to gaseous end products, chiefly methane and carbon dioxide. As long as the system is in balance and the acids are converted to gases or neutralized by the buffer capacity in the system, the anaerobic lagoon functions satisfactorily. Unbalances can occur if too few methane bacteria are present or if unfavorable environmental conditions cause a decrease in the activity of the methane bacteria.

Unbalanced conditions frequently occur during the startup of an anaerobic lagoon and when environmental factors abruptly change, such as when the excess solids are removed from the lagoon or when the lagoon contents warm in the spring. Under these conditions, it is important to control the environment in the lagoon until a proper balance between the two groups of bacteria is established.

Since anaerobic conditions are not inhibited if adequate buffer capacity, i.e., alkalinity, is present, additional alkalinity can be added until an optimum environment is created. Lime has been a common chemical for this purpose, although other chemicals such as sodium bicarbonate, ammonium carbonate, and anhydrous ammonia can be used. The additional alkalinity should be mixed throughout the lagoon contents to avoid localized pH variations that may continue to inhibit optimum biological reactions. Figure 6.6 (29) illustrates initial start-up conditions in an anaerobic lagoon treating beef cattle feedlot wastes. The lagoon was placed in operation in early August. The data indicates that, even with warm temperatures, attention to proper environmental conditions was vital to proper lagoon performance. Alkalinity, as lime, was added to the lagoon as noted until equilibrium was achieved.

With a fixed detention time, the alkalinity in an anaerobic unit is in proportion to the organic loading to the digester. Higher loading rates increase

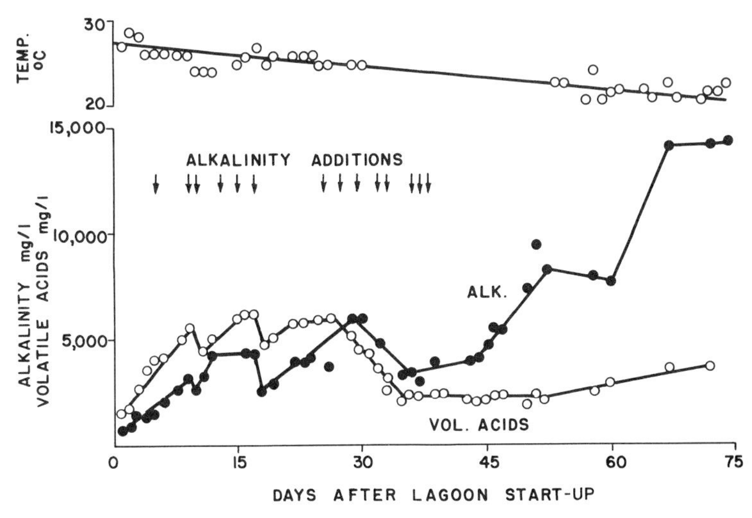

Fig. 6.6. Start-up conditions in an anaerobic lagoon treating beef cattle feedlot wastes (29).

the buffering capacity of anaerobic lagoons and make them less sensitive to operational and waste variations. Lagoons that are lightly loaded not only have inadequate alkalinity to buffer significant volatile acid variations but also have a low population of methane bacteria to readily metabolize any variations of volatile acids.

When an anaerobic lagoon is in biochemical balance, odors are at a minimum and rarely are more than those of the surrounding area. With a high sulfur content of the wastes or waste water, an odor problem results as hydrogen sulfide is produced. Other odors occur when the lagoon is out of balance. Nuisance conditions can arise as a unit is placed into operation, after it is cleaned, after intermittent mixing, and as warm temperatures are established in the spring. Every attempt should be made to minimize the time a lagoon is out of balance.

Despite odor problems, anaerobic ponds can be satisfactory for many wastes if the ponds are located in sufficiently remote areas. Odor problems depend upon variables such as topography, weather, wind direction, and proximity of objectors.

Oxygen and oxidized material such as nitrates should not be added to anaerobic lagoons since they will be used as hydrogen acceptors in preference to oxidized organic matter and carbon dioxide, with the result that anaerobic metabolism can be inhibited if enough quantity of such oxidized material is added. Whenever chemical additives are required, they should be added in the most reduced forms.

Mixing

When the solids in the anaerobic lagoon are actively digesting, considerable quantities of gas are evolved. The evolution of the gas serves to mix the contents of the lagoon making the organic material more readily available to the active organisms. Gasification rarely mixes the lagoon contents completely and is limited to the time of the year when the liquid temperature is warm, generally above 15°C. When thick scum layers accumulate on the surface of the lagoon, it may be necessary to mix the lagoon contents on an intermittent basis.

For maximum biological degradation of the wastes, the content of the anaerobic lagoon can be mixed by mechanical units. If considerable oxygen enters the lagoon during mixing, anaerobic action can be inhibited and the lagoon will tend to approximate an aerated lagoon. Highly loaded lagoons are less subject to this problem.

Better results can be expected if the raw influent is mixed with the digested solids. Higher BOD_5 removals (85%) were obtained when digested sludge was recirculated with the influent (15). Improved results have been noted when a mixer was located near an inlet and other areas of an anaerobic lagoon were mixed manually (30). Odors at an anaerobic lagoon have been reduced by mixing the solids with a pump.

Gas Production

Conversion of the wastes into gas, discharge of the lagoon contents, and seepage into the ground are the only ways by which waste material is removed from an anaerobic lagoon. Seepage through the lagoon bottom and sides should be minimized by adequate sealing to avoid pollution of subsurface waters. Of the remaining methods, gasification removes the greatest quantity of waste material from the lagoon. The theoretical quantity of gas produced is 8 to 10 ft^3 per pound of biodegradable solids added to the unit. High rates of gas production in anaerobic lagoons occur only when the temperature of the lagoon contents are warm.

The gases generated will consist of 60 to 70% methane with the remainder being carbon dioxide and inert gases. This gas represents a potential source of power if collected.

Temperature

Temperature is one of the most important factors affecting the performance of anaerobic lagoons. Because the lagoon is built with the ground for insulation and is generally uncovered, it is subject to temperature fluctuations that can affect the biological system. Anaerobic lagoons function better in warmer climates and are less effective in the colder climates. Low temperatures have been responsible for lagoon failure where the temperature on the bottom of the lagoons never exceeded 15°C anytime throughout the year (31). At liquid temperatures less than 13° to 14°C, gas production is minimal. Maximum decomposition and gas production takes place when the temperature is higher than 17°–19°C (Fig. 6.4). This relationship was verified in full scale pond studies (3) including one involving the use of anaerobic lagoons for the treatment of beet sugar flume wastes (5).

When the temperature in the lagoon is low, an anaerobic lagoon becomes little more than a sedimentation tank. Solids increase on the bottom of the tank during cold weather and undergo decomposition when the warmer temperatures occur. Solids holding capacity should be available in the lagoon during cold weather.

New lagoon installations should avoid penetration into cold ground areas, and they should be placed in operation and be cleaned during the spring and summer when temperatures are more favorable for the development of efficient digestion processes.

Effluent Quality

Anaerobic lagoons offer a possible approach for the treatment of concentrated organic wastes. The lagoons provide excellent settling units to intercept and separate heavy solids from the liquid flow. BOD reductions in anaerobic lagoons can be respectable, 60 to 90%; however, such results are obtained at warm temperatures.

For anaerobic lagoons, an equation of the type of

$$\frac{S_e}{S_o} = \frac{1}{1 + kt} \tag{6.4}$$

has been used to relate effluent quality to waste input. S_e is the effluent BOD concentration, S_o is the influent BOD concentration and t is the detention time in days. The rate constant k will be a function of the type of waste and environmental conditions. For the chemical industry, k has been assumed to be 0.042 (32).

Due to the high loading rates, the effluent from anaerobic lagoons is unlikely to be suitable for discharge to surface waters even with high BOD removals. The effluent will contain significant concentrations of oxygen-demanding material, solids, and nutrients. The quality of the effluent is decreased during start-up operations and when low temperatures exist in the lagoon. The effluent quality from lagoons treating animal wastes contains considerable pollutional material (33).

Anaerobic lagoons are followed by an aerobic unit, generally an oxidation pond or an aerated lagoon if an effluent suitable for discharge to a stream is desired. The degree of treatment that occurs in an anaerobic lagoon may be satisfactory if the contents are to be discharged to the land for disposal, although odors may be a problem during distribution of the contents on the land.

Mixing of an anaerobic pond in an anaerobic-aerobic system is unlikely to occur in practice. An unmixed anaerobic lagoon will remove solids which will accumulate during cold weather unless the lagoon contents are heated or remain above a minimum of about 15°C. As temperatures increase in the spring, the rate of biological decomposition of the settled solids will increase and may result in the effluent from the anaerobic lagoon containing more organic matter than the influent. As a result the BOD load applied to a following aerated unit will vary with climate and ambient temperature. This variation in organic load from the anaerobic lagoon should be considered in the design of the aeration system for the subsequent aerobic unit.

Solids Removal

The rate of decomposition of solids entering an anaerobic lagoon depends on environmental factors such as the temperature of the lagoon and the degree of mixing that takes place. At low temperatures, the quality of the settled solids is similar to those that entered the lagoon. Little decomposition takes place.

The amount of decomposition will depend on the biodegradability of the entering solids. Studies with beef cattle manures have demonstrated a volatile solids reduction of between 40 and 55% under equilibrium conditions. In an anaerobic lagoon treating similar wastes, at least about 50% of the entering wastes can be expected to accumulate in the period between lagoon cleanings even under optimum conditions.

The period between solids removals can be estimated if the volatile solids content of the raw waste and the volatile solids reduction in the lagoon can be estimated, and the resultant solids concentration on the bottom of the lagoon is approximated. Since no decomposition of inert or nonvolatile solids takes place in the lagoon, the introduction of excessive amounts of inert matter should be minimized since it results only in a more rapid filling of the lagoon and a greater frequency of solids removal from the lagoon.

All solids should not be removed from an anaerobic lagoon when the lagoon is cleaned. At least one-quarter to one-third of the active solids should remain so that an optimum anaerobic environment can be maintained to minimize adverse conditions when the lagoon is returned to use. Solids removal from an anaerobic lagoon in cold weather affected the resultant BOD and volatile acid levels in the liquid layer of the lagoon throughout the winter and spring (Fig. 6.7). It was not until the middle of the

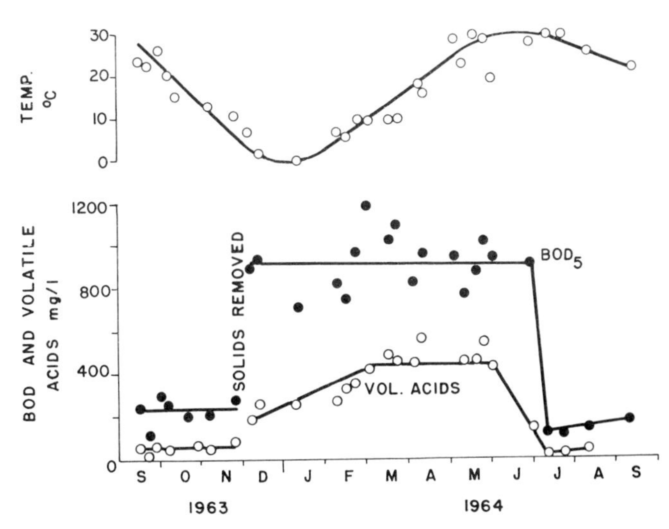

Fig. 6.7. Changes in the characteristics of the liquid effluent from an anaerobic lagoon due to solids removal (34).

following summer that conditions returned to what they were before cleaning.

When necessary, the excess solids in an anaerobic lagoon should be removed in late spring and early summer. The warmer temperatures during this period and the remainder of the summer will permit equilibrium conditions and optimum microbial reactions to be more rapidly established. Removal of excess solids in the fall and winter will delay reestablishment of equilibrium conditions and lengthen the period of poor effluent quality from the anaerobic lagoon.

In an anaerobic lagoon treating the milking parlor wastes (34), the quality of the effluent generally paralleled the quality of the settled solids. A better quality effluent was produced during warm weather and a poorer quality existed during cooler weather in the winter and spring. The settled solids resembled the raw solids of the influent during cold weather but resembled digested solids during the warmer weather.

Data from a number of studies have emphasized the need of adequate environmental conditions and active organisms in the lagoon. It took 6 to 12 months for an anaerobic lagoon in Louisiana to reach maximum efficiency (35). The quality of the effluent from an anaerobic lagoon improved with the buildup of solids and therefore an increase in microbial solids (15).

The settled solids that are removed periodically from active anaerobic lagoons have undergone degradation, stabilization, and concentration. They are less pollutional than the entering untreated solids. However, their quality is such that they should not enter receiving waters. Land disposal offers an acceptable method of disposal for these solids.

Under optimum conditions, i.e., warm temperatures and other optimum environmental conditions, anaerobic lagoons can be loaded at high rates. Adequate solids decomposition will occur. However, the higher the solids loading, the greater the accumulation of nonbiodegradable solids, and the greater the frequency of solids removal.

Application to Agricultural Wastes

Not all of the literature dealing with anaerobic lagoon treatment of agricultural wastes is noted in this section. References have been chosen to illustrate the broad application of this process and to indicate actual results that have been obtained.

Lagoons for livestock waste are operated anaerobically because it requires considerable water to dilute the wastes to the required level for aerobic treatment and because the land required for aerobic lagoons would be large because of the high BOD load of the raw wastes. In many areas, anaerobic lagoons can provide the necessary storage to provide the desired flexibility of handling prior to land spreading of the accumulated wastes. In warm climates, the lagoons can provide both storage and waste decomposition. Anaerobic units have been used as manure holding units prior to land disposal by farmers and animal producers. Active digestion may or may not take place in these units. When the contents are agitated prior to pumping, odors occur. The units are cleaned whenever the operators can get onto the fields.

Anaerobic lagoons have been used with oxidation ponds to treat milking parlor wastes. The milking parlor waste from an 80-cow dairy was treated in an anaerobic lagoon loaded at 9 lb BOD_5/day/1000 ft^3. BOD reductions averaged 85% for the summer with liquid temperatures of 85°F and 20% in the winter when the liquid temperature was 35°F. The average BOD of the effluent from the anaerobic unit during summer was about 100 mg/l (34). The anaerobic unit acted as a sedimentation unit during the winter. The settled solids were degraded during the spring with the result that the BOD of the effluent from the anaerobic unit during the spring was greater than that entering the unit during this period. The relationship of the number of cows to the lagoon size was approximately 0.7 cows/1000 ft^3 of anaerobic lagoon capacity.

An anaerobic lagoon for slaughterhouse waste was loaded at 15 lb BOD/1000 ft^3 of lagoon volume/day (36). The lagoon exhibited the ability to handle intermittent loadings without significant loss of treatment efficiency. Lagoon BOD efficiencies ranged from 52 to 90%. Minor odor problems were noted.

An anaerobic-aerated lagoon system provided overall BOD removals of 99%, suspended solids removals of 98%, and grease removals of 98% from packinghouse wastes. The anaerobic lagoon had a water depth of 15 ft, a BOD design loading of 15 lb BOD/1000 ft^3/day and averaged 65% BOD removal. The actual loading was 12.3 lb BOD/1000 ft^3/day (37). A full-scale study based on the previous report resulted in an anaerobic lagoon loaded at about 29.3 lb BOD/1000 ft^3/day or 195% of design capacity. Liquid detention time was about 5 days. Removals that occurred in this anaerobic lagoon averaged 82% for BOD, 68% for COD, 78% for grease, 69% for volatile solids, and 59% for total suspended solids (38). The organic ammonia was converted to ammonia nitrogen in the lagoons resulting in an average concentration of 120 mg/l in the effluent, an increase of about 200%. The pH remained about 7.0 throughout the year. The temperature ranged from 60° to 75°F with an annual average of 69°F.

Hog processing wastes were successfully treated in an anaerobic pond having an average design detention time of 5 days. A cover of grease kept the temperature above 80°F at all times with a maximum of 94°F during the summer (39). The pond was loaded at 14.7 lb BOD_5/1000 ft^3/day and achieved a 78% BOD_5 removal and a 90% suspended solids removal. An anaerobic pond loaded at 11.2 lb BOD/1000 ft^3 treating meat packing wastes produced an effluent BOD of 820 mg/l and achieved a BOD removal efficiency of 65%. The detention time of the pond was 4.6 days (15).

Anaerobic ponds have been used with success in the meat packing and processing industry since these wastes contain high solids and BOD concentrations. In many cases such ponds are used to pretreat the wastes before discharge to a municipal plant or to an aerobic unit for further treatment. State regulatory agencies in certain states have accepted design loadings of 15 lb BOD/1000 ft^3/day for anaerobic ponds and have allowed for 60% BOD removal when the anaerobic ponds are followed by some sort of aerobic treatment (40).

Beet sugar wastes have been treated in an anaerobic unit followed by either a facultative pond or an algae pond. In this instance, the anaerobic lagoon was loaded with screened and settled flume water up to 2060 lb ultimate BOD/acre/day. A portion of the oxygen demand in the anaerobic lagoon was met by a floating aerator capable of transferring about 300 lb of oxygen/day into the lagoon water when the dissolved oxygen was zero (5). In the anaerobic lagoon, BOD removal was a linear function of the BOD loading up to a loading of 2000 lb of ultimate BOD/acre/day. Removal was about 80% of the applied load. At loadings above about 1000 lb of ultimate BOD/acre/day, odors caused by sulfides became an increasingly severe problem. Although the anaerobic lagoon accepted high loadings and accomplished high removals, the effluent had a high BOD and was unsuitable for discharge without further treatment.

For secondary treatment of this anaerobic lagoon effluent, facultative ponds or high rate algal ponds were found successful. Low soluble BOD resulted in the effluent as illustrated by the low residual BOD, about 10–20 mg/l, when the solids were removed by filtration (5). The system was effective in converting soluble BOD to insoluble BOD and had solids separation been applied to the final treated effluent, BOD removal would have been 98%.

Covered anaerobic ponds have been used to treat potato wastes (20). A pond loaded at 23 lb BOD/100 ft^3/day with a detention time of 4 days produced BOD removals of 17%. A second pond having a detention time of 20 days and loaded at 4.6 lb BOD/1000 ft^3/day produced BOD removals of 22%. The ponds were not mixed and the temperature of the ponds ranged from 60–70°F. The low removal rates were caused by the type of degradation that took place with the large starch and protein compounds breaking down into smaller organic compounds such as acids, aldehydes, and alcohols. Small BOD reductions take place under these conditions. Nutrient limitations did not appear to be a problem.

Anaerobic lagoons treating a combination of domestic sewage and vegetable cannery waste, citrus waste, and fruit cannery waste averaged 75–90% BOD removal when the organic loading ranged between 105 and 360 lb BOD/acre/day (41). A BOD removal of 55% occurred when the loading was 600 lb BOD/acre/day. Sludge solids accumulated during the peak fruit canning season and usually were not completely digested until the following spring. Gas production ranged between 0.2 to 2.0 ml of gas produced/day/gram of volatile solids after equilibrium conditions were established. Higher gas production rates were measured in earlier stages.

A four-day detention time anaerobic lagoon treating swine waste loaded at 280 lb BOD/1000 ft^3/day accomplished only 10% BOD removal and 5% COD removal. Gas formation in the unit mixed the contents and soluble organic compounds were produced from the waste fed to the unit. An anaerobic lagoon having a 20-day detention time and loaded at 55 lb BOD/1000 ft^3/day achieved 60% BOD removal, 65% total volatile solids reduction, and 55% COD removal. The removals primarily were caused by solids sedimentation. The study was conducted in laboratory units at 20–23°C.

Anaerobic lagoons used for beef and hog packinghouse wastes have been loaded at 9–60 lb BOD/1000 ft^3/day. The average loadings were in the 12–15 lb BOD/1000 ft^3/day range. BOD removal efficiencies were consistently above 70%. The lagoons in the 12–15 lb loading rate range averaged 85% BOD removal. The anaerobic lagoon loaded at the 60 lb BOD/1000 ft^3/day rate had a 72% BOD removal with a 3.4-day detention time (42). In these ponds, the sludge accumulation was 7 inches/year and averaged 55–65% volatile solids. Fermentation of the solids proceeded satisfactorily when the volatile acids were low, below 200 mg/l.

In Louisiana over 50 packinghouses installed anaerobic ponds to treat their wastes during 1960–70. The ponds were favored because low installation costs, ability to handle shock loads and dependable nuisance free performance (43). A design of about 300 lb BOD/acre-ft/day (6.4 lb BOD/1000 ft^3/day) resulted in BOD removals of 85% or more. The mean annual temperature in this location was 68°F.

Dairy manure applied to anaerobic lagoons at a rate of 20 lb BOD/1000 ft^3/day (70 lb VS/1000 ft^3/day) resulted in 88% BOD removal (44).

Full-scale systems of anaerobic lagoons followed by either aerated or aerobic lagoons have been used to obtain 80–99% removal of BOD, suspended solids, and grease (Table 6.4) (45). Nutrient removals were not as great. Organic nitrogen removals ranged from 46 to 94%, ammonia nitrogen increased rather than decreased, and soluble phosphate changes ranged from an increase to removals of over 50%. Soluble nitrogen and phosphorus will result from microbial metabolism in the anaerobic unit and are not removed effectively in an aerobic lagoon. Control of these nutrients

Table 6.4

Combined Anaerobic Lagoon–Aerobic Lagoon Treatment of Meat-Packing Wastes[a]

Plant number	BOD	Suspended solids	Grease	Nitrogen: Organic	Nitrogen: Ammonia (NH_3–N)[b]	Nitrogen: Total Kjeldahl	Nitrogen: Nitrate	Phosphorus: Total (P)	Phosphorus: Soluble (P)
Percent reduction (%)									
1	98	95–99	95–99	93–95	+	26–46	—	16–49	+–34
2	82–99	77–95	99	84–90	+	0–32	—	+–56	+–72
3	89–97	83–91	94–98	71–85	+	+–17	—	+–28	+–48
5	99	99	99	46	+	—	—	58	73
6	99	90	97	91	+	46	—	48	25
7	98	94	87	90	+	57	—	26	+
Final effluent concentration (mg/l)									
1	30–250	12–144	13–39	8–20	62–120	—	0–1.2	9–20	6–18
2	200–400	106–124	15–24	14	84	—	0	7.7	4–9
3	40–90	110–127	15–20	18–20	59–25	—	0–3.8	7–13	3–9
5	17–20	16–68	4–17	4–5	50–60	—	0.5–0.8	6–10	5–8
6	30	308	61	34	86	—	0	3–15	7–10
7	20	52	—	7	37	—	—	—	—

[a] From reference 45.
[b] + = increase.

will require additional, specific treatment processes. Even though BOD, suspended solids, and grease removals were high, the effluents from these combined lagoon systems contained considerable concentrations of these contaminants in addition to high concentrations of nutrients.

While a combination anaerobic-aerobic system can handle the scraped solids as well as the runoff from a feedlot, there is little advantage to handling the solids on the feedlot in this manner. The feedlot solids should be handled as a semidry or dry material and transported to the land for final disposal.

A summary of the BOD removal and loading rates for the above and other anaerobic lagoons treating agricultural wastes is shown in Fig. 6.8.

Aerated Lagoons

General

An aerated lagoon differs from an oxidation pond in that aerobic conditions are maintained by mechanical or diffused aeration equipment. Aerated lagoons consist of an aeration device in a square or rectangular excavated basin. Usually, the lagoon is an earthen basin with some protection on the banks for wave action caused by the aeration unit (Fig. 6.9).

A major advantage of the aerated lagoon is the continuous oxygen transfer caused by the aeration equipment. With oxidation ponds, the algae produce oxygen only during daylight hours. The microorganisms require

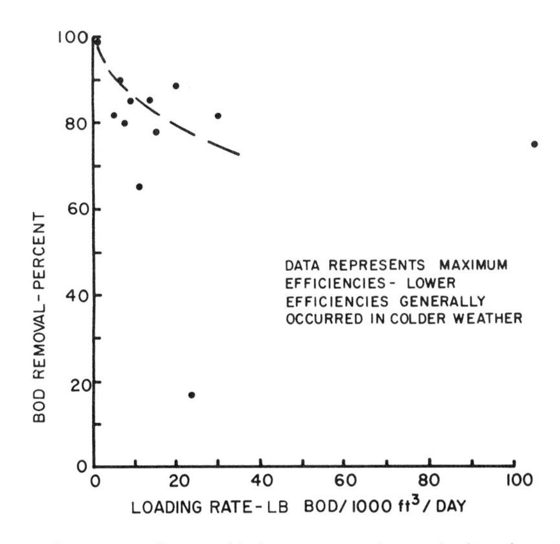

Fig. 6.8. Performance of anaerobic lagoons treating agricultural wastes.

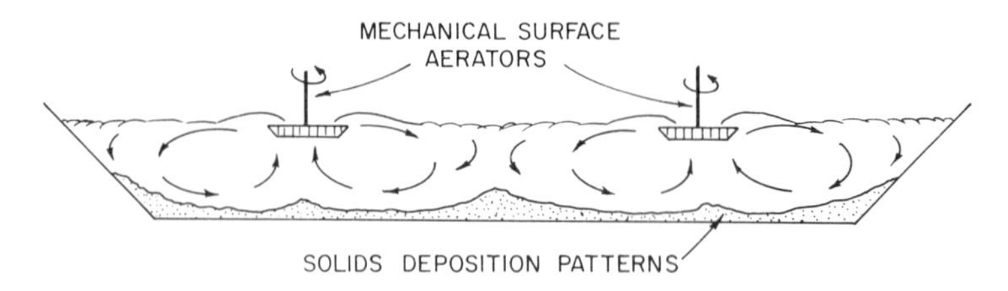

Fig. 6.9. Schematic of an aerated lagoon with surface aerators.

oxygen continuously and sufficient oxygen must be stored during the daylight hours to meet the demand during the nighttime. The continuous oxygen supplied by the aeration equipment in an aerated lagoon permits this unit to treat more waste water per unit volume per day. Aerated lagoons have been used to improve the effluent quality of overloaded oxidation ponds, to alleviate nuisance conditions in such units, to permit a more economical use of land, and to treat wastes without prior treatment.

The principal variables which must be evaluated for design are rate of biological reaction in the aerated lagoon, effect of temperature on the reaction rate, oxygen requirements, synthesis and oxidation of solids in the lagoon, mixing requirements, and pH and nutrient balance.

Certain of these may be evaluated by bench scale investigations while others result from prototype studies.

Operation and Design

The microbial characteristics of an aerated lagoon are similar to those of an activated sludge unit rather than an oxidation pond. Algae will not be important and generally are not found in the lagoon mixed liquor because of the turbulence and turbidity. The aeration equipment will be required to satisfy the entire oxygen demand.

The aerated lagoon is a dilute, well-mixed, biological treatment unit operating without solids recycle and having detention times in the order of 1–10 days, depending upon loading and desired effluent quality. Mixing is sufficient to distribute oxygen throughout the unit but may not be sufficient to keep all of the solids in suspension. The aeration equipment should be designed to deliver the required quantity of oxygen and to provide a minimum of solids deposition in the unit. The aeration equipment normally runs 24 hr a day.

For aerated lagoons, the loading rate should be expressed in terms of the pounds of oxygen demand per unit volume per day (lb BOD/1000 ft^3/day) rather than pounds per surface area per day. The total volume will serve as a biological reactor and the amount of surface area is of lesser importance. Because the biological reaction rates are affected by temperature, design detention time to achieve a given effluent quality should be expressed as a function of temperature.

A common approach to the design of aerated lagoons has been to consider them as completely mixed with respect to both dissolved oxygen and to uniform solids concentrations. Mathematical models developed with this assumption have been applied to the design of aerated lagoons (46). These assumptions represent an ideal situation that may be only approximated in practice. Figure 6.6 represents a typical aerated lagoon and indicates that solids will settle out where velocities are low. The velocity profile is determined by the size and pumpage of the aerator, liquid depth, placement of the aerator, and size of the lagoon. The settled organic solids will become anaerobic and release soluble organic end products and gases to the aerobic upper layers. Because both aerobic and anaerobic areas exist in the aerated lagoon, it may more properly be referred to as a facultative lagoon. If the aerated lagoon is not completely mixed, empirical relationships such as Eq. 6.5 have been used to approximate the BOD removal in these units (47).

$$\text{BOD removal } (\%) = 100Kt/[1 + Kt] \tag{6.5}$$

where K is the BOD removal rate and t is the detention period in days. Values for K are indicated as 0.5/day for cannery wastes, 0.8/day for packinghouse wastes (48), and 0.22/day at 25°C for wastes from fruit and baby food production (49). A value of 0.087/day was obtained with an aerated lagoon treating winery wastes (50). This lagoon had a detention time of 35 days and the coefficient was calculated at 50°F. The constants in the equation are specific for each waste and installation and do not recognize the effect of microbial solids in the process reactions. With wastes with constant characteristics, the microbial solids concentration variation may be small and empirical relationships such as Eq. (6.5) may be adequate for design. Laboratory and/or pilot plant studies should be used to determine the K value to be used for design rather than using K values reported in the literature.

Equations which consider BOD removal as a linear function of effluent BOD_5 and mixed liquor volatile suspended solids also have been applied to aerated lagoons (51) (Eq. 6.6).

$$\frac{L_e}{L_o} = \frac{1}{(1 + S_a \cdot K^1 \cdot t)} \tag{6.6}$$

L_e and L_o are the effluent and influent BOD concentrations, respectively, t is the detention time in days, S_a is the average suspended (or volatile suspended) solids concentration in the body of the aerated lagoon, and

K^1 is the BOD removal rate under these conditions. Equation (6.6) can be used to estimate either soluble BOD or total BOD removals depending on which parameter is used. The total BOD will include the soluble and the particulate oxygen demand in the respective samples. K^1 should be evaluated from studies with the actual waste.

If the aerated lagoon is well mixed such that there is reasonably uniform distribution of dissolved oxygen and suspended solids in the lagoon, the design can be estimated by the kinetics of a completely mixed biological reactor. Results from a comparative study of design methods indicated that field results may not agree with theoretical relationships (52). The constants in these relationships were developed on the assumption that retention times would be short and turbidity sufficient to minimize algal growth. The presence of algae in the aerated lagoons can alter the results as predicted by the theoretical or empirical equations and generally indicate an incompletely mixed aerated lagoon.

The degree of mixing has a large influence on the performance of aerated lagoons. At high mixing levels all of the suspended solids are in suspension and the effluent suspended solids and BOD concentrations are larger than at low mixing levels. At the lower mixing levels, solids are not kept in suspension and some settle to the bottom.

The effluent solids concentration from an aerated lagoon will be higher than for an oxidation pond or comparable aerobic unit since there is little natural opportunity for solids separation. The effluent suspended solids concentration can be reduced by baffling a portion of the aerated lagoon to permit a quiescent zone where suspended solids may settle. A tube settling unit can be installed in the effluent end of the lagoon to reduce the solids content of the effluent. An aerated lagoon can be followed by an unmixed oxidation pond, by a quiescent settling pond, or by a final clarifier to produce a low solids effluent.

Under all conditions, the soluble BOD in the effluent will be low and the total BOD will be related to the effluent suspended solids concentration. Data from full scale aerated lagoons indicated that during the winter a considerable portion of the effluent BOD_5 was in the soluble form, whereas during the summer, the larger portion of the effluent BOD_5 was associated with solids in the effluent. The difference in effluent characteristics was caused by the degree of microbial metabolism that took place at the different temperatures. BOD removals of 50 to 70% can be obtained if suspended solids are not separated from the effluent, whereas 90–95% removals can be obtained if a solids separation unit is used. Figure 6.10 illustrates the relationship that was obtained for the effluent BOD and suspended solids resulting from the treatment of potato processing wastes in a completely mixed aerated lagoon (53). Similar results can be expected from

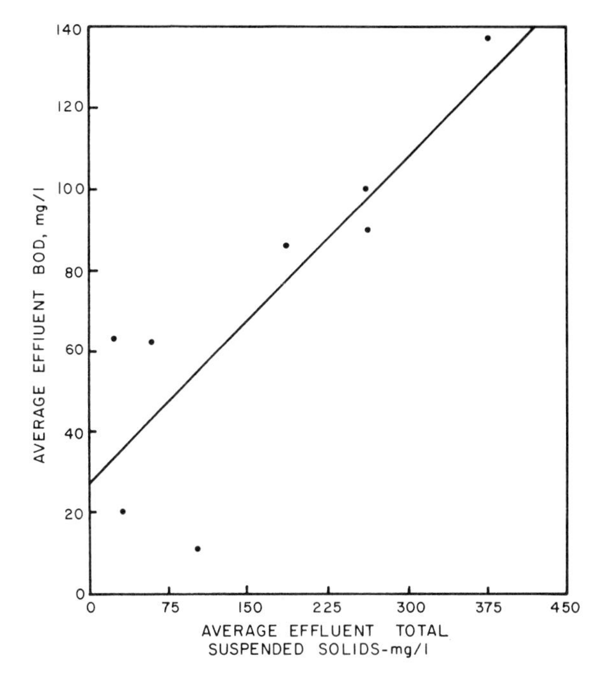

Fig. 6.10. Effluent BOD and suspended solids relationships for an aerated lagoon treating potato processing wastes (53).

mixed aerated units where natural solids separation does not occur. Where very high BOD removals are required, effluent solids removal is required.

The aerated lagoon does not need operational control from a biological standpoint. Biological equilibrium will be established, based on the actual loading conditions, and will adjust automatically to various changes in loads. The microbial reactions are related to the temperature in the unit and the performance will be sensitive to large temperature changes.

Complex maintenance is not required, the usual maintenance being lubrication of the motor and periodic inspections. The capital and operating costs are higher than oxidation ponds because of the need for the aeration equipment and the power used.

Temperature

The liquid temperature in an aerated lagoon depends upon the mixing that takes place and the rate at which heat is lost. The temperature of an aerated lagoon can be estimated from a heat balance incorporating the influent liquid temperature, T_i, the effluent liquid temperature, T_e, the mean lagoon temperature, T_w, and the ambient air temperature, T_a. If heat loss takes place primarily through the lagoon surface, the heat loss will be proportional to the temperature differential between the air and liquid and the surface area of the lagoon, A (54, 55).

$$(T_i - T_e)Q = fA \cdot (T_w - T_a) \tag{6.7}$$

In this equation, all temperatures should be in °C, area in square feet, and Q in MGD. The proportionality factor f or overall heat transfer coefficient has been found to have a value of 8×10^{-6} and 12×10^{-6} (54, 55).

The following equation was developed to predict aerated lagoon temperatures (56)

$$T_L = \frac{8.34 \cdot Q \cdot T_i + 145 \cdot A \cdot (T_A - 2)}{145 \cdot A + 8.34 \cdot Q} \tag{6.8}$$

T_L is the average weekly aerated lagoon temperature (°F), T_i the average weekly influent temperature (°F), T_A the average weekly air temperature, Q the waste water flow (gallons/day), A the surface area of the aerated lagoon (feet^2), and 145 the average heat exchange coefficient (BTU/feet^2-day-°F). The term $(T_A - 2)$ approximates the equilibrium temperature of the aerated lagoon when exposed to air and assuming no liquid influent or effluent. The results obtained from Eq. (6.8) were relatively insensitive to the magnitude of the heat exchange coefficient.

In an aerated lagoon operating in cold weather, the temperature near the surface of aerated lagoons was 1.0° to 1.5°C less than the temperature at a depth of 10 ft. In this location, heat loss to the ground accounted for 5–10% of the total loss from the lagoon (54). BOD removal followed lagoon temperature decreasing from a maximum in July to a minimum in January. The removal decreased 28% as the temperature decreased from 33° to 21°C during this period. For another lagoon, BOD removal increased about 20% as the temperature increased from 14° to 24°C (54).

Ice accumulation can occur on floating aerators during subzero temperatures. The accumulation can result in increased power usage as the aerators settle deeper in the lagoon, and in overturning or sinking of the aerators. Electronic overload switches, electric heaters, and extra buoyancy can minimize such problems.

Aerated lagoons can function satisfactorily and are capable of over 80% BOD removal even in cold climates (57) if attention is given to the fundamentals of the process when it is designed. Diffused aeration and aeration guns were more satisfactory in cold climates than were surface aerators.

Aeration Units

Aeration equipment for aerated lagoons is designed to supply adequate oxygen to meet the oxygen demand of the wastes entering the lagoon and to maintain a residual dissolved oxygen concentration. To avoid oxygen limiting conditions, a dissolved oxygen concentration in the lagoon of 1–2 mg/l should be maintained. The furnishing of oxygen to the unit beyond the amount required to maintain this minimum concentration will not improve BOD removal efficiency.

Because an aerated lagoon is not completely mixed, the dissolved oxygen concentration will vary throughout the unit. The degree of variation is determined by the oxygen uptake rate and the overall velocity profile. If velocity variation is small and the oxygen uptake low, the dissolved oxygen concentration variations may be small. If the lagoon is large and if the recirculation time through the aeration zone of the aerator is large, then a concentration gradient will develop with the highest concentrations existing near the aerator and decreasing with increasing distance from the aerator.

The average liquid detention time in an aerated lagoon is much longer than in an activated sludge unit and the solids concentration is much lower. Oxygen uptake rates are typically in the less than 10 mg/l/hr range as opposed to 25–50 mg/l/hr range for conventional activated sludge units. Oxygen uptake rates have been from 0.1 to 2.6 mg/l/hr in an aerated lagoon treating municipal sewage (58) and 0.5–0.7 mg/l/hr in a tertiary aerated lagoon (59). The oxygen demand in the liquid of a facultative aerated lagoon will be lower than that in a comparable completely mixed aerated lagoon due to the removal of solids by sedimentation.

Two aeration systems can be used for aerated lagoons: (a) diffused air units in which compressed air is forced through perforated plastic pipe or an aeration gun near the bottom of the lagoon and (b) surface aerators which mechanically create turbulence at the liquid surface such as with a series of blades that are partially submerged. Examples of an aeration gun are illustrated in Figs. 6.11 and 6.12 while examples of surface aerators are presented in Figs. 6.13 and 6.14. In cold climates such as Alaska and northern Canada, compressed air introduction of oxygen may be preferable since ice buildup on exposed surface aerators can be excessive (60). The choice of an aeration system is based on factors such as capital and operating costs, flow and characteristics of the waste water, maintenance and replacement needs, and the designers experience.

Mechanical surface aerators pump large quantities of water causing oxygen to be dissolved in the water. The rate of pumping and the type of air–water interface that is created influence the performance of the units.

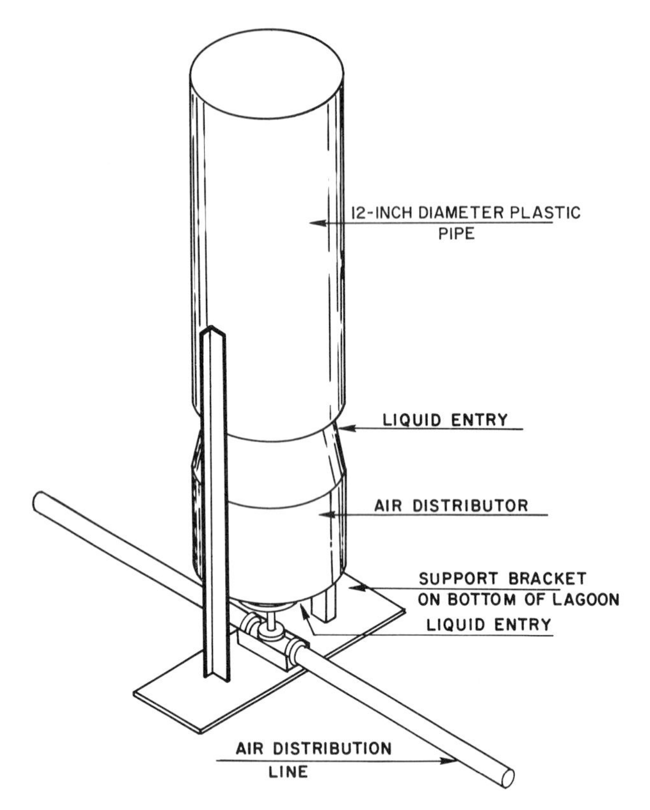

Fig. 6.11. Schematic of air gun used to aerate lagoons and lakes using compressed air (courtesy Aero-Hydraulic Corp.).

Low speed units provide maximum pumping but low turbulence and high speed units provide less pumping and greater turbulence. Performance must be determined in large scale units or under actual conditions.

Early types of mechanical aerators were attached to fixed supports in an aeration basin. These units had the disadvantage of being sensitive to variations in water levels which could alter the oxygen transfer characteristics of the aerators. The development of floating surface aerators has removed this problem and has allowed the aerator to perform uniformly as the liquid level in the lagoon fluctuated. The simplicity and ease of maintenance of surface aerators, especially floating units, has caused them to be successful with agricultural waste waters.

The aeration equipment is expected to mix the lagoon as well as transfer the necessary oxygen. In large systems with a high oxygen demand rate, if the aeration units can transfer enough air to meet the oxygen demand of

Fig. 6.12. Mixing patterns produced by air guns in an aerated lagoon treating municipal and food processing wastes (courtesy Aero-Hydraulics Corp.).

the system, adequate mixing normally will occur. In low rate systems or small systems, the mixing requirements may control and the aeration equipment will be designed to keep the solids in suspension. Under these conditions, adequate oxygen will be available to meet the oxygen demand of the microorganisms. If it is desired to have complete mixing, the liquid velocity at the farthest point in the aeration unit will determine the suspension of the particles and the ultimate power requirements. A velocity of about 0.5 ft per sec will maintain microbial solids in suspension.

If solids settling does occur, the organic matter will undergo anaerobic decomposition on the bottom of the lagoon with a slow release of organics to the upper layers. Gas production will cause solids to rise and increase the oxygen demand of the upper layers. Under these conditions, the aeration equipment should be designed to handle the incoming waste load plus any load caused by the solids that may have settled in the aeration lagoons.

The power requirements for mixing usually are specified in terms of the horsepower required to either maintain specified minimum fluid velocities in the lagoon or to maintain a homogeneous, suspended solids concentration in the lagoon mixed liquor. Values of HP/1000 gal have meaning only for a specific aeration unit since the pumpage of liquid from various designs of surface aerators can vary by factors of 2–10. A more meaningful factor to keep solids in suspension would be the pumpage that an aerator produces per 1000 gal or 1000 ft^3 of basin volume since it is the pumpage that distributes the oxygen and keeps the solids in suspension. The transfer of oxygen is related to the quantity of fluid passing through a rotor or surface aerator.

In an aerated lagoon with detention times greater than about one-day detention time, the horsepower for aeration is a function of the oxygen demand and tends to be represented by a first-order reaction of horsepower and detention time. The horsepower for mixing is a function of the volume of the lagoon and will vary directly with the detention time since for a given flow rate, the volume increases directly with detention time (61). Figure 6.15 illustrates these relationships. Facultative aerated lagoons will require less power for mixing since they are not completely mixed.

Where mechanical surface aeration is used to supply sufficient mixing to attempt uniform dissolved oxygen concentrations throughout the basin

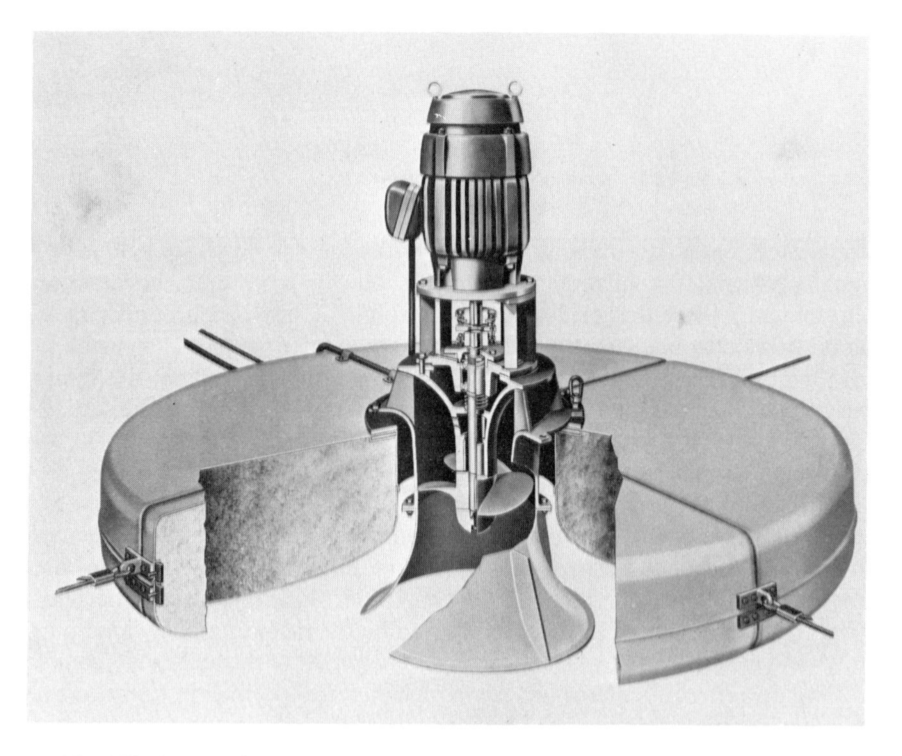

Fig. 6.13. Cross-section of a floating surface aerator (courtesy Rexnord Envirex, Inc.).

Fig. 6.14. Surface aerators in an aerated lagoon treating brewery wastes (courtesy Rexnord Envirex, Inc.).

but not necessarily suspension of all the solids, power levels of 0.015–0.02 HP/1000 gal of basin volume have been suggested (62). Evaluation of two full-scale aerated lagoon systems treating municipal wastes indicated that from 0.2 to 0.33 horsepower per 1000 ft^3 (.027–.044 HP/1000 gal) of lagoon capacity was required if adequate mixing was maintained (63). The mechanical aeration units were capable of transferring from 1.75 to 1.80 lb of oxygen per horsepower hour at 20–25°C and zero dissolved oxygen in the lagoon contents. The oxygen transfer relationships of mechanical aerators under process conditions range from 1.5 to 2.0 lb of oxygen per horsepower hour. Data from municipal waste aerated lagoons demonstrated that power levels of 0.03–0.05 HP/1000 gal would completely mix the lagoons (56) and keep the suspended solids in suspension. Below 0.03 HP/1000 gal of lagoon capacity, complete mixing was not accomplished.

Power requirements to meet the oxygen demand should be related to the oxygen demand (BOD) added to the system per day or to the oxygen uptake rate that will exist in the aerated lagoon. These relationships should be utilized for design purposes when these factors control the aerator design. When mixing requirements control, horsepower relationships per lagoon volume are adequate.

If the aeration system in an aerated lagoon is inadequate to meet the oxygen demand, the dissolved oxygen may be zero and the microbial reactions may be oxygen limited. Under these conditions, increasing the horsepower or aeration capacity may increase BOD removal efficiency. When oxygen limitation is avoided, increases in aeration capacity or horsepower have little effect on the BOD removal efficiencies of an aerated lagoon.

The oxygen transfer efficiency is largely determined by whether the air is used primarily for metabolism and synthesis or for mixing. In low rate systems, such as extended aeration systems, the transfer efficiency may be of the order of 3% while in conventional activated sludge systems it may be of the order of 5–7% and in high rate systems on the order of 15–20%. For comparison of aeration equipment, oxygen transfer efficiency should be expressed at standard conditions which are for clean water at 20°C with zero dissolved oxygen at a barometric pressure of 29.92 inches of mercury.

Surface aerators rarely can be selected without a good knowledge of the process operating conditions, basin geometry, and aerator spacing as well as the need for the aerator to both mix and aerate. Where it is expected that these factors as well as capital and operation costs are critical, pilot and full-scale studies are valuable to select proper aeration equipment.

Application to Agricultural Wastes

A number of studies describing the aerated lagoon treatment of agricultural wastes are noted in this section to illustrate the ubiquity of application, the application of fundamentals to full-scale operations, and individual

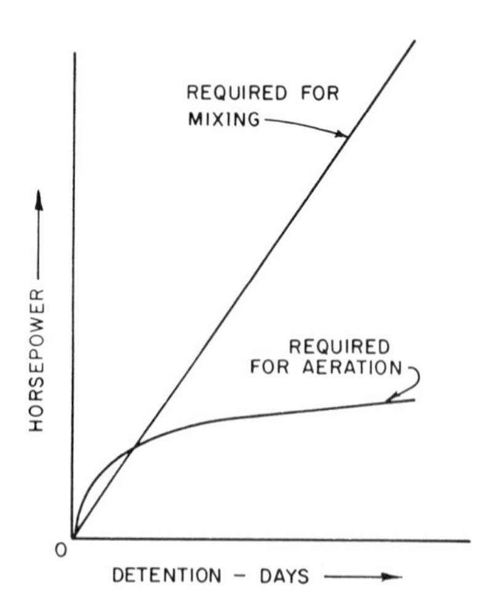

Fig. 6.15. Power relationships required for aeration and mixing in an aerated lagoon.

design information. Aerated lagoons have proved useful for a number of agricultural wastes and have been studied by a number of investigators. Mechanical aeration of potato processing waste and domestic sewage had difficulty in treating a waste mixture with a pH of 11.6. The high pH was the result of the lye used in potato peeling. BOD removal efficiency of 25% was achieved (20). Some form of pH control may be necessary to permit biological treatment of high pH wastes.

Pea processing wastes have been treated in an aerated lagoon having a detention time of 5.5 days. Four 50-HP surface aerators were used and the BOD loading rate ranged from 7.0 to 10.3 lb/1000 ft^3/day. Reductions in total BOD and COD by the aerated lagoon averaged 76 and 59%, respectively (64). Dissolved BOD and COD in the waste was reduced by 95 and 82%, respectively, in the aerated unit. Inorganic nutrients were not reduced appreciably by the aerated lagoon. The major portion of the BOD in the effluent from this system was associated with the suspended solids. The BOD equivalent of the suspended solids in the effluent was 0.24. The endogenous respiration rate of the mixed liquor solids was 0.027 per day.

Aerated lagoon treatment of pear, peach, and apple processing wastes provided greater than 70% BOD removal. Suspended solids in the effluent were the principal source of effluent BOD and COD (65). Nutrient addition was necessary to achieve successful treatment. Nutrient deficient sludge did not settle. It was noted that nutrient savings could be made by increasing the quantity of biological solids in the systems. Endogenous respiration of the solids returned nutrients to the system. Maintaining a higher solids concentration for this purpose could not be accomplished by an aerated lagoon unless effluent solids separation and recycle were practiced. Surface aerators accomplished adequate mixing at 0.04 HP/1000 gal and transferred 2 lb oxygen/HP/hr under operating conditions. The cost of the aerated lagoon treatment was estimated to be $0.04/lb BOD removed and $0.035/lb COD removed.

A series aerated lagoon system achieved 96% BOD reduction with wine wastes (50). Aerators having a capacity of 60 HP were installed in the first lagoon which was loaded at 1200/lb/BOD/acre/day. Three 5-HP aerators were used in the second lagoon. The overall BOD removal decreased 5–10% during the winter. BOD removal of 75% was accomplished in the first lagoon and an additional 84% in the second lagoon.

A food processing waste consisting primarily of carbonaceous wastes required additional nitrogen and phosphorus to obtain satisfactory treatment in an aerated lagoon and good settling characteristics of the resultant microbial solids. At a temperature of 15°–22°C and with adequate nutrients, a loading factor of 0.3 to 0.4 lb BOD/lb MLSS consistently provided 80–90% BOD removal (66).

Aerated lagoons also have been successful with animal wastes. A diffused aeration system obtained 95% BOD removal and 93% grease removal from the effluent of an anaerobic lagoon treating slaughterhouse wastes (37). The aerated lagoon had a 40-day detention time and air was supplied at an average of 0.5 cfm/lb BOD applied.

Aerated lagoon treatment of duck waste water have produced an average of 85% BOD removal. The removals on the 26 farms using this treatment ranged from 60 to 95% (67). The aerated lagoons were designed to have an average detention time of 5 days. Because of variable flows, the actual detention times varied considerably from the average. The aerators used in the lagoons had power relationships that varied from 0.008 to 0.04 name-plate HP/1000 gal lagoon capacity.

Dairy wastes have been treated by a combination of aerated lagoons and irrigation. The aerator was sized at 0.10 HP/cow. Odor was minimum although the dissolved oxygen concentrations were not measurable at all times (68). The lagoon mixed liquor solids were kept between 2 and 3% to accomplish sprinkler irrigation of the lagoon contents.

Laboratory aerated lagoons treating poultry wastes indicated that the minimum value to achieve high BOD reductions, 85+%, was 60 ft^3/lb of applied BOD/day (69). The system minimized odors and reduced the solids load on the ultimate disposal system by at least 50%. At the above volume relationship, over 80% suspended solids destruction was observed.

An unpublished study by the author indicated that aerated lagoons followed by solids removal units treating swine wastes could obtain 98% BOD removal with a 10-day detention time and 90% BOD removal in a 5-day unit. The study was conducted at 20°–22°C. The BOD of the clarified effluent averaged 200–300 mg/l under equilibrium conditions. The units were intermittently fed once a day. Oxygen uptake data in these units illustrate the value of continuous waste loading. Very high oxygen demands, greater than 100 mg/l/hr were exerted in the first few hours after feeding. The oxygen demand decreased to 20–30 mg/l/hr prior to the next feeding.

A comparison of aerated lagoons treating combined domestic and dairy product wastes, poultry wastes, and domestic wastes plus seasonal winery wastes noted that mechanical surface aeration, perforated tubing diffused air, and aeration gun diffused air systems would satisfactorily treat these wastes (52). Data from these lagoons indicated that the temperature relationship, θ, was not constant but increased with decreasing temperatures and decreasing suspended solids. At 20°C, θ was 1.03 at 20 mg/l suspended solids and 1.02 at 300 mg/l suspended solids. At 10°C, the θ values were 1.06 and 1.035 at the two solids levels, respectively.

Kinetic studies with the aerated lagoon treatment of combined citrus processing wastes and domestic wastes indicated an average BOD removal rate coefficient of 1.46 and an average temperature coefficient (θ) of 1.05

(70). About 0.4 to 0.6 lb of waste sludge resulted per pound of influent BOD. This was significantly higher than for comparative studies using domestic sewage alone. Sludge wastage accounted for the greatest portion of overall nutrient removal from the system. The aerated lagoon process for these wastes obtained 91% BOD removal when the detention time was controlled at 7.9 days.

The organic matter present in potato processing wastes was reduced by 90% or more using either surface aerated aerobic lagoons or anaerobic plus aerated lagoons following primary clarification (71). Both systems were economically feasible with the choice depending upon the costs and conditions. Chemicals were not needed for pH control or for inorganic nutrient adjustment. The BOD/P and BOD/N ratios for the clarifier effluent were 130 to 1 and 19 to 1, respectively. Monod relationships were used to express the organic removal patterns and appeared to adequately represent the removal patterns. The BOD equivalent of the effluent volatile suspended solids (C) ranged from 0.25 to 0.6 and increased as the detention time (t) of the aerated lagoon decreased. The following equation was found to express this relationship at a temperature of about 45°F

$$C = \frac{1.68 - \log(t)}{2.51} \tag{6.9}$$

The effluent soluble and total BOD decreased as the aerobic pond detention time increased. Similar results were observed with the effluent from the anaerobic pond.

References

Oxidation Ponds

1. Oswald, W. J., Meron, A., and Zabat, M. D., "Designing Waste Ponds to Meet Water Quality Criteria," Proc. 2nd Int. Symp. Waste Treat. Lagoons, pp. 186–194. Federal Water Quality Administration, Kansas City, Missouri, 1971.
2. Marais, G. v. R., and Shaw, V. A., A rational theory for the design of sewage stabilization ponds in Central and South Africa. *Trans. S. Afr. Inst. Civil Eng.* **3,** 205–227 (1964).
3. Oswald, W. J., Golueke, C. G., Cooper, R. C., Gee, H. K., and Bronson, S. C., Water reclamation, algal production and methane fermentation in waste ponds. *Int. J. Air Water Pollut.* **7,** 627–648 (1963).
4. Oswald, W. J., "Stabilization Pond Research and Installation Experiences in California," Proc. Symp. Waste Stabilization Lagoons, pp. 41–50, Missouri Basin Engineering Health Council, Kansas City, Missouri, 1960.
5. Oswald, W. J., Golueke, C. G., Cooper, R. C., and Tsugita, R. A., "Anaerobic-Aerobic Ponds for Treatment of Belt Sugar Wastes," Proc. 2nd Nat. Symp. Food Process. Wastes, pp. 547–598. Pacific Northwest Water Lab., Environmental Protection Agency, Denver, Colorado, 1971.

6. Bartsch, A. F., and Allum, M. O., Biological factors in treatment of raw sewage in artificial ponds. *Limnol. Oceanogr.* **2,** 77–84 (1957).
7. Hermann, E. R., and Gloyna, E. F., Waste stabilization ponds. III. Formulation of design equations. *Sewage Ind. Wastes* **30,** 963–975 (1958).
8. Canter, G. W., Englande, A. J., and Mauldin, A. F., Loading rates on waste stabilization ponds. *J. Sanit. Eng. Div., Amer. Soc. Civil Eng.* **95,** SA6, 1117–1129 (1968).
9. Gloyna, E. F., and Hermann, E. R., Some design considerations for oxidation ponds. *J. Sanit. Eng. Div., Amer. Soc. Civil Eng.* **82,** SA4 (1956).
10. Little, J. A., Carroll, B. J., and Gentry, R. E., "Bacteria Removal in Oxidation Ponds," Proc. 2nd Int. Symp. Waste Treat. Lagoons, pp. 141–150. Federal Water Quality Administration, Kansas City, Missouri, 1971.
11. Preul, H. C., Contaminants in groundwater near waste stabilization ponds. *J. Water Pollut. Contr. Fed.* **40,** 659–669 (1968).
12. Jewell, W. J., Aquatic weed decay: dissolved oxygen utilization and nitrogen and phosphorus regeneration. *J. Water Pollut. Contr. Fed.* **43,** 1457–1467 (1971).
13. Porges, R., Industrial waste stabilization ponds in the United States. *J. Water Pollut. Contr. Fed.* **35,** 456–468 (1963).
14. Dunstan, G. H., and Smith, L. L., Experimental operation of industrial waste stabilization ponds. *Pub. Works* , 93–95 (1960).
15. Sollo, F. W., Pond treatment of meat packing plant wastes. *Proc. Purdue Ind. Waste Conf.* **15,** 386–398 (1961).
16. El-Sharkawi, F. M., and Moawad, S. K., Stabilization of dairy wastes by algal-bacterial symbiosis in oxidation ponds. *J. Water Pollut. Contr. Fed.* **42,** 115–125 (1970).
17. Anderson, J. S., and Kaplovsky, A. J., Oxidation pond studies in evisceration wastes from poultry establishments. *Proc. Purdue Ind. Waste Conf.* **16,** 8–21 (1961).
18. Nemerow, N. L., Baffled biological basins for treating poultry plant wastes. *J. Water Pollut. Contr. Fed.* **41,** 1602–1612 (1969).
19. Nemerow, N. L., and Scott, S. D., Lagoons plus spray irrigation for effective bean cannery waste treatment. *Proc. Purdue Ind. Waste Conf.* **27,** (1972).
20. Guttormsen, K., and Carlson, D. A., "Potato Processing Waste Treatment—Current Practices," Water Pollut. Contr. Res. Ser. DAST-14. Federal Water Pollution Control Administration, Dept. of Interior, 1969.

Aerobic Ponds

21. Oswald, W. J., Fundamental factors in stabilization pond design: Advances in biological treatment. *Int. J. Air Water Pollut.* **5,** 357–393 (1963).
22. Oswald, W. J., Gotaas, H. B., Golueke, C. G., and Kellen, W. R., Algae in waste treatment. *Sewage Ind. Wastes* **29,** 437–455 (1957).

Anaerobic Ponds

23. Dornbush, J. N., and Andersen, J. R., Lagooning of livestock wastes in South Dakota. *Proc. Purdue Ind. Waste Conf.* **19,** 317–325 (1965).
24. Hart, S. A., and Turner, M. E., Lagoons for livestock manure. *J. Water Pollut. Conf. Fed.* **37,** 1578–1596 (1965).
25. Clark, C. E., Hog waste disposal by lagooning. *J. Sanit. Eng. Div., Amer. Soc. Civil Eng.* **91,** SA6, 27–42 (1965).
26. Loehr, R. C., and Agnew, R. W., Cattle wastes—pollution and potential treatment. *J. Sanit. Eng. Div., Amer. Soc. Civil Eng.* **93,** SA4, 72–91 (1967).

27. Jeffrey, E. A., Blackman, W. L., and Ricketts, R. L., Aerobic and anaerobic digestion characteristics of livestock wastes. *Mo. Univ. Bull.* 57 (1964).
28. Hart, S. A., Digestion tests of livestock manure. *J. Water Pollut. Contr. Fed.* **35,** 748–757 (1963).
29. Loehr, R. C., "Treatment of Wastes from Beef Cattle Feedlots—Field Results." Proc. Agr. Waste Manage. Conf., pp. 225–241. Cornell University, Ithaca, New York, 1969.
30. Canham, R. A., Anaerobic treatment of food canning wastes. *Proc. Purdue Ind. Waste Conf.* **5,** 145–155 (1950).
31. Berry, E. C., "Requirements for Microbial Reduction of Farm Animal Wastes," Proc. Nat. Symp. Anim. Waste Manage., Publ. SP-0366, pp. 56–58. Amer. Soc. Agr. Eng., 1966.
32. "The Cost of Clean Water and Its Economic Impact," Vol. IV. Cyrus W. Rice and Co. Federal Water Pollution Control Administration, Dept. of Interior, 1969.
33. Loehr, R. C., Effluent quality from anaerobic lagoons treating livestock wastes. *J. Water Pollut. Contr. Fed.* **39,** 384–391 (1967).
34. Loehr, R. C., and Ruf, J. A., Anaerobic lagoon treatment of milking parlor wastes. *J. Water Pollut. Contr. Fed.* **40,** 83–94 (1968).
35. Coerver, J. F., Anaerobic and aerobic ponds for packinghouse waste treatment in Louisiana. *Proc. Purdue Ind. Waste Conf.* **19,** 200–215 (1965).
36. Enders, K. E., Hammer, M. J., and Weber, C. L., Field studies of an anaerobic lagoon treating slaughterhouse wastes. *Proc. Purdue Ind. Waste Conf.* **22,** 126–137 (1968).
37. Wymore, A. H., and White, J. E., Treatment of a slaughterhouse waste using anaerobic and aerated lagoons. *Proc. Purdue Ind. Waste Conf.* **23,** 601–618 (1969).
38. Baker, D. A. and White, J. E., "Treatment of Meat Packing Waste Using PVC Trickling Filters," Proc. 2nd Nat. Symp. Food Process. Wastes, pp. 287–312. Pacific Northwest Water Lab., Environmental Protection Agency, Denver, Colorado, 1971.
39. Niles, C. F., and Gordon, H., Operation of an anaerobic pond on hog abattoir wastewater. *Proc. Purdue Ind. Waste Conf.*, **25** (1971).

Aerated Lagoons

40. Steffen, A. J., Waste disposal in the meat packing industry. *Water Wastes Eng.* , C1–C4 (1970).
41. Parker, C. D., and Skerry, G. P., "Cannery Waste Treatment by Lagoons and Oxidation Ditch at Shepparton, Victoria, Australia," Proc. 2nd Nat. Symp. Food Process. Wastes, pp. 251–270. Pacific Northwest Water Lab. Environmental Protection Agency, Denver, Colorado, 1971.
42. White, J. E., "Current Design Criteria for Anaerobic Lagoons," Proc. 2nd Int. Symp. Waste Treat. Lagoons, pp. 360–363. Federal Water Quality Administration, Kansas City, Missouri, 1971.
43. Coerver, J. F., "Anaerobic Lagoon Treatment of Packing Wastes in Louisiana," Proc. 2nd Int. Symp. Waste Treat. Lagoons, pp. 354–359. Federal Water Quality Administration, Kansas City, Missouri, 1971.
44. Bhagat, S. K., and Proctor, D. E., Treatment of dairy waste by lagooning. *J. Water Pollut. Contr. Fed.* **41,** 785–795 (1969).
45. Crandall, C. J., Kerrigan, J. E., and Rohlich, G. A., Nutrient problems in meat industry wastewaters. *Proc. Purdue Ind. Waste Conf.* **25,** (1971).
46. Marais, G. v. R., and Capri, M. J., "A Simplified Kinetic Theory for Aerated Lagoons," Proc. 2nd Int. Symp. Waste Treat. Lagoons, pp. 299–309. Federal Water Quality Administration, Kansas City, Missouri, 1971.

47. O'Connor, D. J., and Eckenfelder, W. W., Treatment of organic wastes in aerated lagoons. *J. Water Pollut. Contr. Fed.* **32,** 365–374 (1960).
48. Water Pollution Control Federation, "Aeration in Wastewater Treatment," MOP-5 Water Pollut. Contr. Fed., Washington, D.C., 1971.
49. Eckenfelder, W. W., and O'Connor, D. J., "Treatment of a Cannery Waste in an Aerated Lagoon," Report prepared for Duffy-Mott Corp., 1958.
50. Tofflemire, T. J., Smith, S. E., Taylor, C. W., Rice, A. C., and Hartsig, A. L., Unique dual lagoon system solves difficult wine waste treatment problems. *Water Wastes Eng.* F1–F5 (1970).
51. Eckenfelder, W. W., Theory of biological treatment of trade wastes. *J. Water Pollut. Contr. Fed.* **39,** 240–247 (1967).
52. Townshend, A. R., Unsal, S., and Boyko, B. I., Aerated lagoon design methods—an evaluation based on Ontario field data. *Proc. Purdue Ind. Waste Conf.* **24,** 327–348 (1970).
53. Richter, G. A., "Aerobic Secondary Treatment of Potato Processing Wastes," Proc. 1st Nat. Symp. Food Process. Wastes, pp. 39–71. Pacific Northwest Water Lab., Environmental Protection Agency, Portland, Oregon. 1970.
54. McKinney, R. E., and Edde, H., Aerated lagoon disposal for suburban sewage disposal. *J. Water Pollut. Contr. Fed.* **33,** 1277–1285 (1961).
55. Timpany, P. L., Harris, L. E., and Murphy, K. L., Cold weather operation in aerated lagoons treating pulp paper mill wastes. *Proc. Purdue Ind. Waste Conf.* **26** (1972).
56. Malina, J. F., "Design Guides for Biological Wastewater Treatment Processes," Final Rep., Proj. 11010 ESQ. Environmental Protection Agency. 1971.
57. Pick, A. R., Burns, G. E., Van Es, D. W., and Girling, R. M., "Evaluation of Aerated Lagoons as a Sewage Treatment Facility in the Canadian Prairie Provinces." *In* Proc. Inter. Sympos. on Water Pollution Control in Cold Climates. Water Pollut. Cont. Res. Ser. 16100EXH. Environmental Protection Agency, 1971.
58. Eckenfelder, W. W., "Industrial Water Pollution Control." McGraw-Hill, New York, 1966.
59. Wahbeh, V. N., and Weller, L. W., "Tertiary Treatment by Aerated Lagoons," Proc. Int. Symp. Waste Treat. Lagoons, pp. 293–299. Federal Water Quality Administration, Kansas City, Missouri, 1971.
60. Dawson, R. N., and Grainge, J. W., Proposed design criteria for wastewater lagoons in artic and sub-artic regions. *J. Water Pollut. Contr. Fed.,* **41,** 237–240 (1969).
61. Benjes, H. H., "Theory of Aerated Lagoons," Proc. 2nd Int. Symp. Waste Treat. Lagoons, pp. 210–217. Federal Water Quality Administration, Kansas City, Missouri, 1971.
62. Eckenfelder, W. W., and Wanielista, M. P., Theoretical Aspects of Aerated Lagoon Design. Paper presented at the Sanit. Eng. Div., Spec. Conf. Amer. Soc. Civil Eng., 1965.
63. McKinney, R. E., and Benjes, H. H., Evaluation of two aerated lagoons. *J. Sanit. Eng. Div., Amer. Soc. Civil Eng.* **91,** SA6, 43–56 (1965).
64. Dostal, K. A., "Aerated Lagoon Treatment of Food Processing Wastes," Final Rep. Proj. 12060. Water Quality Office, Environmental Protection Agency, 1968.
65. Esvelt, L. A., and Hart, H. H., Treatment of fruit processing wastes by aeration. *J. Water Pollut. Contr. Fed.* **42,** 1305–1326 (1970).
66. Dunbar, R. F., Velzy, C. R., and Wicklund, G. W., Pretreatment of a food processing waste. *Proc. Purdue Ind. Waste Conf.* **26** (1972).
67. Loehr, R. D., and Schulte, D. D., "Aerated Lagoon Treatment of Long Island Duck Wastes," Proc. 2nd Int. Symp. Waste Treat. Lagoons, pp. 249–257. Federal Water Quality Administration, Kansas City, Missouri, 1971.
68. Dale, A. C., Ogilvie, J. R., Chang, A. C., Douglass, M. P., and Lindley, J. A., "Disposal of Dairy Cattle Wastes by Aerated Lagoons and Irrigation," Proc. Agr. Waste Manag. Conf., pp. 150–159, Cornell University, Ithaca, New York, 1968.

69. Vickers, A. F., and Genetelli, E. J., "Design Parameters for the Stabilization of Highly Organic Manure Slurries by Aeration," Proc. Agr. Waste Manage. Conf., pp. 37–44. Cornell University, Ithaca, New York, 1969.
70. Coca-Cola Company Foods Division, "Treatment of Citrus Processing Wastes," Final Rep., Proj. WPRD 38-01-67. Water Quality Office, Environmental Protection Agency, 1970.
71. Dostal, K. A., "Secondary Treatment of Potato Processing Wastes," Final Rep., Proj. 12060. Water Quality Office, Environmental Protection Agency, 1969.

7

Aerobic Treatment

Introduction

Aerobic biological treatment is utilized to avoid odor problems during waste storage and/or to achieve an effluent quality suitable to meet stream standards. The processes described in this chapter are of two types, suspended growth and fixed film or adhered growth processes. Activated sludge is an example of a suspended growth process. The microorganisms are mobile, move with the liquid flow, and are able to flocculate and separate from the liquid in a solids separation unit. Processes with and without solids recycle are possible. Mixing of the microbial solids and wastes can be done by mechanical mixers, diffused aeration, or both.

The trickling filter is an example of a fixed film or adhered growth process. In this process, the microorganisms adhere to solid media. The organic wastes flow over the media and the adhered microbial mass. Aeration is achieved by natural air movement through the media and no controlled mixing of the microorganisms and the wastes occur. Solids recycle is not a normal part of a trickling filter process although recirculation of a portion of the filter effluent through the filter is possible.

Both processes are classified as secondary treatment processes because they generally follow a primary solids separation step and because they achieve greater organic removal than a primary unit. Certain aerobic treatment processes treat raw wastes directly. BOD and suspended solids removals of over 80% can be obtained by aerobic processes.

The liquid wastes from a number of agricultural industries, especially food processing, are seasonal, complicating the problem of achieving adequate secondary treatment of these wastes by conventional processes. Some of the aerobic systems described in this chapter are suitable primarily for dilute liquid wastes. Others are suitable for more concentrated wastes but few are designed to function at high solids concentration such as the 8–10% range.

The number of possible objectives of handling agricultural wastes in an aerobic liquid system include: control of objectionable odors, improve the characteristics of the wastes for easier handling, reduce the pollutional content of the wastes prior to discharge to the environment, and reduce the nutrient content prior to land or surface water disposal. Both laboratory and field studies have demonstrated that animal waste slurries, effluents from anaerobic systems, and runoff from confinement operations can be treated aerobically. Comparable results have been demonstrated with a variety of food processing wastes.

Aeration Systems

Introduction

Oxygen is only slightly soluble (7–10 mg/l) at normal ambient temperatures. In a waste treatment unit, the dissolved oxygen concentration will be depleted rapidly by the microbial metabolism unless sufficient oxygen is added to the system to meet or exceed the oxygen demand of the microorganisms in the unit. The aeration system is designed to furnish the necessary oxygen and in most cases to mix the aerobic unit. Diffused aeration, mechanical aerators, and combinations of both are the aeration processes in common use. The key to a successful aerobic treatment process is the design and application of a satisfactory aeration system.

Basic Relationships

The problem of supplying sufficient oxygen is one of oxygen transfer from air bubbles or the atmosphere through a liquid media to the microbial cells. Dissolved oxygen is utilized at a rate proportional to the rate of organic removal by the microorganisms. The basic relationship used to estimate the rate at which oxygen must be supplied by an aeration system is

$$\frac{dC}{dt} = K_L a(C_S - C_L) \tag{7.1}$$

where

$\frac{dC}{dt}$ = rate of change of dissolved oxygen concentration with time, mass/volume-time
K_La = overall gas transfer coefficient, time^{-1}
C_S = oxygen saturation concentration for given atmospheric composition and pressure, mass/volume
C_L = actual oxygen concentration at time t, mass/volume

The numerical value of the overall transfer coefficient, K_La, is a function of the gas–liquid interfacial area, the liquid turbulence, and the diffusivity of the oxygen in the liquid, and is dependent upon the temperature of the liquid, the relative humidity, and the concentration of organic material in the waste water. The oxygen is added to the liquid from the atmosphere by surface aeration or by bubbles during diffused aeration. The microbial floc in the aerobic treatment system will utilize the dissolved oxygen in the liquid as a hydrogen acceptor (Fig. 7.1). Where the microbial floc is attached to air bubble, direct transfer of the oxygen is possible.

The saturation value of dissolved oxygen C_S, is affected by temperature, dissolved inorganic material, and the air pressure at the depth the air is introduced into the waste water.

Equation (7.1) is adequate where the respiration rate of the microorganisms is zero and must be modified to include the oxygen utilization rate of the microorganisms, R_r.

$$\frac{dC}{dt} = K_La(C_S - C_L) - R_r \tag{7.2}$$

The microbial oxygen utilization rate in a biological treatment system is dependent upon the available food, number of active organisms, and temperature. In low rate, conventional aerobic treatment systems, R_r can be in the range of 5–50 mg O_2/l/hr. The uptake rate, R_r, can be in the range of

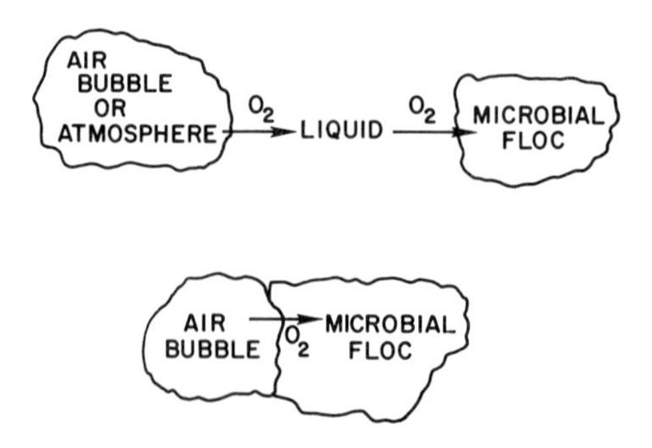

Fig. 7.1. Pathways of oxygen transfer in aerobic biological systems.

50–100+ mg O_2/l/hr in a high rate system. On a mass basis, R_r can range about 5–40 mg O_2/hr/gVSS for conventional units such as an oxidation ditch and activated sludge process.

The oxygenation capacity of an aeration unit or system is a measure of the rate of input of oxygen that can be achieved under given operating conditions. It is defined as the mass of oxygen absorbed in unit time at a specific temperature and pressure. Usual units are pounds per hour or milligrams/liter per hour. Aeration efficiency is the ratio of the quantity of oxygen absorbed to that blown in for diffused aeration systems or the ratio of the oxygen absorbed into the mixed liquor per unit of energy of the mechanical equipment for mechanical and diffused aerators.

The oxygenation capacity of an aerator will govern the number of aerators that will be required for the treatment of a given flow of waste at a given detention period. The aeration efficiency will influence both capital and operational costs since it governs the size of compressor or motor and the rate of consumption of energy required to achieve a given rate of oxygen absorption.

The oxygen capacity of aeration equipment usually is determined in water at zero dissolved oxygen and adjusted to 20°C. Under these conditions the oxygenation capacity can be expressed as:

$$N_o = (K_L a_{20}) \cdot C_S \cdot W \cdot 10^{-6} \tag{7.3}$$

N_o = oxygenation capacity, pounds/hour.
$K_L a_{20}$ = oxygen transfer coefficient in water at 20°C., hours^{-1}
W = weight of liquid under aeration, pounds.

A true measure of W is important. W should be the volume of the liquid that is under the direct mixing and pumping influence of the aeration systems. If the aeration unit is not completely mixed, the magnitude of W will be less than the liquid basin volume. Estimation of W is difficult when incomplete mixing occurs.

Under operating conditions, the oxygen demand of the aerobic system is a function of the oxygen utilization rate of the microorganisms:

$$N = R_r \frac{W}{10^6} \tag{7.4}$$

For aerobic biological treatment, the oxygenation capacity of the aeration system should equal or exceed the oxygen demand of the aerobic system. Under steady state conditions in an aerobic biological unit this means that

$$N = R_r = K_L a(C_S - C_L) = N_o \tag{7.5}$$

Under operating conditions, if N_o is greater than N, the aeration system will transfer more than enough oxygen to meet the demand and a residual dissolved oxygen concentration will be maintained. The dissolved oxygen concentration, C_L, will adjust until the system is in balance. If N_o is considerably greater than N, C_L may be considerably above zero and can approach the saturation conditions for that mixed liquor.

Aeration equipment is rated on the basis of the oxygen transferred in tap water under standard temperature and barometric conditions with zero dissolved oxygen. This rating is rarely the quantity of oxygen transferred under process conditions since the dissolved oxygen level normally is kept at a level of at least 1–2 mg/l to assure aerobic conditions. The effect that the actual dissolved oxygen concentration (C_L) will have on the pounds transferred is illustrated in Table 7.1. As noted, the difference from standard conditions, i.e., zero dissolved oxygen, can be significant. It is important to recognize that aerator ratings reflect conditions that are unlikely to occur under process conditions.

Because standard conditions rarely exist in practice, modifications to Eq. (7.5) must be made to reflect the aerator oxygenation capacity under actual process conditions. These modifications include factors to adjust for different K_La, C_S, temperature, and pressure conditions:

$$R_r = N_T = \alpha K_L a_{20} \cdot (\beta C_S - C_L) \cdot \theta^{(T-20)} \cdot \frac{P}{14.7} \tag{7.6}$$

where

N_T = oxygenation capacity of the aeration system under process conditions at temperature T.

α = the ratio of K_La in the mixed liquor at temperature T to K_La in water at temperature T.

K_La_{20} = K_La in water at 20°C.

β = the ratio of oxygen saturation in mixed liquor at temperature T to the oxygen saturation in water at 20°C.

C_S = the oxygen saturation in water at 20°C.

θ = the temperature correction factor for the system.

C_L = dissolved oxygen concentration in the aeration tank under steady state conditions.

P = atmospheric pressure at the treatment plant operation, psi.

The use of this equation includes many simplifying assumptions and can best be considered as an idealized, approximate approach to the design of aeration systems.

Aeration systems are designed before the aerobic biological system is in operation and before N or R_r are known. By establishing α, β, and θ or the actual K_La and applying them in Eq. (7.6), the oxygenation capacity of the system can be estimated and the aeration system designed. Since it is not practical to determine these parameters at full-scale conditions prior to construction of the treatment facilities, laboratory or pilot plant scale

Table 7.1

Oxygen Transfer as Affected by Residual Dissolved Oxygen Concentration[a]

DO (mg/l)	DO deficit $(C_S - C_L)$ (mg/l)	Oxygen transfer rate (mg/l/hr)	Pounds of oxygen transferred per hour[b]
0	8.8	17.6	147
1	7.8	15.6	130
2	6.8	13.6	113
3	5.8	11.6	97
4	4.8	9.6	80

[a] $K_L a$ assumed as 2 mg/l/hr/mg/l. The uptake rate was calculated using Eq. (7.2) and assuming that the system was operating at equilibrium. α and β assumed to be 1.0.

[b] Basin of 1,000,000-gal capacity assumed.

conditions resembling estimated process design conditions are used to estimate the above parameters that may occur in the full-scale aeration units.

A temperature correction value (θ) of 1.024 is commonly used for most aeration systems. Values of from 1.016 to 1.047 have been reported with 1.024 fitting most experimental data except in very cold water. Based upon testing of surface aerators, a value of 1.012 has been suggested as being more correct where the water temperature is between 5° and 25°C (1).

Temperature affects both the oxygen saturation concentration and the oxygen transfer coefficient. As the temperature increases, C_S decreases and $K_L a$ increases. The overall oxygenation capacity is not as widely affected by temperature as are these two factors (Fig. 7.2) and varies about 3–5% over a temperature range of 10°–30°C. The waste characteristics affect α and β directly and influence the oxygenation capacity of an aeration system to a greater degree than does temperature.

$K_L a$ can vary with the conditions of the test and the actual conditions in

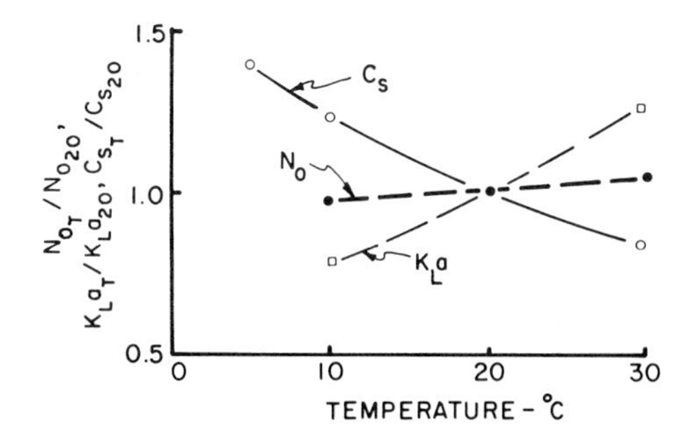

Fig. 7.2. Effect of temperature on $K_L a$, C_S, and overall oxygen transfer.

practice. It is assumed that K_La is a constant regardless of time and dissolved oxygen level. At low dissolved oxygen levels, K_La may be dependent on the dissolved oxygen level. Experiments have shown that in diffused aeration plants treating municipal wastes, K_La was little affected by suspended solids in the range of 1000–6000 mg/l while K_La was reduced considerably by the presence of detergents in the mixed liquor. Reductions of up to 50% in the K_La value were noticed at detergent concentrations greater than 3–4 mg/l (2). High mixed liquor solids concentrations can reduce K_La by altering the viscosity of the aerating medium. The presence of 10,000 mg/l sludge solids was noted to reduce absorption in the aerating unit to only 0.2 of that in the pure waste (3). Data on K_La values in a contact stabilization unit treating a combination of municipal and meat packing wastes illustrate the changes in the values under varying conditions (Table 7.2).

Alpha (α) is a gross term for the factors that affect K_La and usually is used to indicate factors other than temperature. Alpha varies with the waste type and strength. Values of α generally are less than 1.0 with most values in the range of 0.5 to 0.9. Occasionally, the characteristic of the waste may be such that α is greater than 1.0 but this is the exception rather than the rule. Alpha will vary with temperature, basin geometry and size, intensity of mixing, and organics in the waste water under aeration. In a specific aeration basin, alpha is primarily a function of the unmetabolized organic material in the system. The greater the organic material in the system, the lower the alpha value. The knowledge of alpha is an important aspect of the evaluation of aerator equipment.

Beta (β) values also are usually less than 1.0 as many of the characteristics of waste waters decrease the oxygen saturation capacity of a mixed liquor.

Table 7.2

Aeration Data at a Contact Stabilization Plant

	Temperature	
	15°C	24°C
Oxygen uptake data (mg/l/hr)		
Contact tank	40–55	40–70
Reaeration tank	50–70	80–150
K_La (per hour)		
Tap water	22–25	40–60
Contact tank	10–15	14–24
Reaeration tank	13–18	22–30
Alpha (α)		
Contact tank	0.51	0.39
Reaeration tank	0.63	0.58

Values of β generally are in the range of 0.7 to 0.9. The saturation of oxygen in a waste water or mixed liquor sample can be obtained by permitting the sample to over aerate until it has reached a saturation dissolved oxygen level. The ratio of C_S in the sample to C_S in tap water at standard conditions is defined as beta. Standard conditions for tap water have been defined as sea level barometric pressure and zero chlorides.

Determination of the α, β, and $K_L a$ values should be based on conditions that exist or are expected in the biological unit and not in the raw waste since it is the conditions that exist in the aeration unit that will influence the performance of the aeration equipment. Where wastes are not available to use in laboratory or pilot plant experiments to establish design parameters, it may be possible to estimate the parameters from on-going systems using similar wastes or from existing published values. The best method is to use the specific waste in question.

Dissolved Oxygen Relationships

While aerators may be rated at zero dissolved oxygen to establish a common base, aeration systems are designed to maintain a specific residual in the biological unit. The residual should be above the critical dissolved oxygen concentration that will limit the respiration of the microorganisms. Estimates of critical oxygen concentration for nonnitrifying bacteria vary from 0.0003 (4) to 2.4 mg/l (5). For nitrifying bacteria the critical concentration for *Nitrosomonas europea* at 30°C was reported to be 1 mg/l and for *Nitrobacter winogradskyi* at 30°C was reported (6) to be about 2 mg/l.

The rate of respiration of nitrifying sludges normally is independent of the concentration of dissolved oxygen above about 1 mg/l. Below 1 mg/l the rate of nitrification falls and with it the rate of consumption of oxygen. Nitrification ceases below about 0.5 mg/l dissolved oxygen (DO) and at still lower concentrations the rate of carbonaceous matter oxidation also is reduced. The rate of consumption of dissolved oxygen in mixed liquor appears to be substantially independent of the dissolved oxygen concentration above about 0.5 mg/l.

The dissolved oxygen concentration in the aeration unit need not be maintained higher than the minimum critical dissolved oxygen concentration since the oxygenation capacity and the size of the aeration system is a function of the dissolved oxygen deficit that is maintained [Eqs. (7.5) and (7.6)]. The larger the deficit, the more efficient will be the aeration equipment in transferring oxygen. Since the power required to achieve a given rate of input of oxygen is inversely proportional to the oxygen deficit, there appears to be no advantage in maintaining a minimum dissolved oxygen concentration much above the critical dissolved oxygen concentration.

Von der Emde (7) observed that when the oxygen concentration fell below 1.5 mg/l purification of the waste waters was affected adversely and concluded that under normal conditions, oxygen should be regarded as the most important limiting factor of aerobic biological units. He recommended that the oxygenation capacity of aeration equipment be not less than 1.5 lb of oxygen per pound of applied BOD to fully treat waste waters and to avoid oxygen limitations. This ratio does not include an oxygen demand caused by nitrification.

In practice, waste loads, characteristics, and temperature vary. The aeration equipment should be able to produce an acceptable effluent under maximum conditions. To accomodate these variations, aeration units usually are designed to maintain a minimum dissolved oxygen concentration in the mixed liquor (C_L) of about 2 mg/l. Furnishing oxygen to the mixed liquor beyond that amount required to satisfy the oxygen demand for a dissolved oxygen in the system of 1–2 mg/l will not give improved treatment. The utilization of more pounds of oxygen requires a process with more oxygen demand.

Determination of Aeration Parameters

The engineer must understand the wastes and the treatment system with which he will work. He must specify the operating conditions for which the system is to be designed, i.e., specify the dissolved oxygen concentration to be maintained under aeration for a given oxygen uptake rate, at a fixed temperature, in a tank of definite dimensions. The manufacturer will need an idea of α to select the proper equipment. The equipment manufacturer should be responsible for providing accurate information on the K_La value of the aeration equipment in tap water since he is most familiar with his equipment under standard conditions and in most basins. Changing waste water characteristics may preclude an accurate prediction of α and β. It should be the responsibility of the design engineer to determine the above critical factors that are valid for the expected aeration conditions since he is most familiar with the facility, its waste water characteristics, its operation, and expected future changes.

The capital cost of a treatment facility is largely a function of the volume of the treatment units such as the clarifiers and aeration tank. A reduction in the flow rate without decreasing the quantity of oxygen demanding material will reduce the cost of the clarification system only. A reduction in oxygen demanding material only without a reduction in flow rate will only reduce the size of the aeration unit. Operating cost is mainly a function of the quantity of waste to be oxidized and stabilized. The major operating cost is likely to be the power necessary to drive the aeration equipment. Because agricultural wastes are in general low volume, high strength wastes, operating costs tend to be a larger proportion of the total treatment cost than would be the case for municipal waste treatment. The efficiency of the aeration system, i.e., quantity of oxygen transferred to the waste under aeration per unit of power per time unit, is of considerable importance. Therefore it is important to understand how aeration equipment are evaluated and rated.

One of the most complex aspects of designing an aerobic biological waste treatment unit is the selection of suitable aeration equipment. The only sure way to determine that aeration equipment will perform satisfactorily is to test it after installation. While estimates of oxygen transfer coefficients and oxygen uptake rates can be obtained using small scale laboratory studies, more exact and useful data is obtained in full-scale aeration basins especially with the wastes to be treated. When full-scale units are not available, small scale studies offer a way to develop useful estimates of parameters for design of actual aeration systems and equipment. Turbulence is the main difference between full-scale and small scale aeration studies of a specific waste. The basic oxygen transfer processes are small scale and are influenced by the smallest turbulence which produces diffusion of oxygen to the micro-organisms. Even with similar power inputs per unit volume, different turbulence can result in units with different geometry, aerator location, and type of aerator. Small scale laboratory tests are not indicative of the turbulence in a full-scale aeration basin unless the degree of agitation is described quantitatively.

The common method of measuring overall energy input such as power/volume/time has little significance in comparing aeration basins or extrapolating small scale or pilot studies since it does not characterize the turbulence that is generated which in turn controls the transfer of oxygen. The differences in turbulence account for some of the differences reported between equipment and scale of aeration studies. An appropriate method of relating input oxygenation energy to turbulence energy remains to be developed.

In lieu of relationships of energy input to turbulence regimes, empirical relationships based upon power input per unit volume have been developed from full-scale aeration systems to provide reasonable design parameters for mixing and solids suspension in aerobic biological treatment systems. Power levels of 0.015–0.02 HP/1000 gal of basin volume have been suggested to obtain uniform dissolved oxygen concentrations but not necessarily all solids in suspension in aerated lagoons (8). Data from two full-scale aerated lagoons indicated that from 0.027 to 0.044 HP/1000 gal was required if adequate mixing was maintained (9). European data has indicated that in activated sludge basins the power input was 0.04–0.07 HP/1000 gal to keep solids in suspension whereas in oxidation ditches, 0.015–0.023 HP/1000 gal was sufficient (10).

The use of performance specifications has been suggested to assure that the aeration equipment will perform as expected and designed (11). These specifications should define and describe clearly the analytical tests to be conducted, the methods to be employed, the number and location of samples, the time of sampling, and expected results at varying temperatures. With the aeration parameters estimated or known, Eq. (7.6) can be used to prepare a table of the expected results for the aeration equipment.

Performance specifications can be useful in assuring that the aeration equipment meets the design requirements. They also can have valuable use as a predictive tool. The conditions that exist when the aeration equipment meets the specifications can be used as the base conditions against which the effects of changing waste water characteristics, operating conditions, α, β, and K_La factors, and alteration in basin configuration and mixing patterns can be determined. Routine aeration unit evaluation by the plant operator using the techniques in the specifications provides the operator with another tool to understand his plant better and to predict and prepare for an increase in aeration capacity that may be needed in the future.

Techniques commonly used to measure the oxygen transfer coefficient of aeration systems are (a) the unsteady state method incorporating sulfite oxidation, (b) the steady state method with microbial solids, and (c) the unsteady state method with microbial solids. An assumption inherent in the use of these techniques is that the test liquid is completely mixed and of uniform composition at any instant of time. A significant deviation from this assumption will result in erratic or erroneous data.

The unsteady state method incorporating sulfite oxidation is useful in determining the oxygen transfer relationships in tap water. In the presence of metallic catalysts such as copper and cobalt, sulfite ion and oxygen react rapidly and irreversibly. The procedure is to add an excess of sulfite ion and a small amount of catalyst to the water. Approximately 0.8 lb of dry sodium sulfite per 1000 gal will be adequate per test and will provide an excess to compensate for oxidation during initial mixing. The aeration equipment is started and the dissolved oxygen concentration is monitored either chemically or with a dissolved oxygen analyzer. The latter is preferred because of the minimum response time and avoidance of analytical errors.

Once the excess sulfite ion is oxidized, the dissolved oxygen concentration will increase (Fig. 7.3). A plot of values of $C_S - C_t$ against t on semilog paper should produce a straight line, the slope of which is K_La (Fig. 7.4). The results of the test are a function of the catalyst concentration. Studies have indicated that the catalyst concentration of cobalt should be less than 0.05 mg/l as Co. When repeated tests are run in the same tank, there is no need to continually add catalyst. If the catalyst is continually added, the

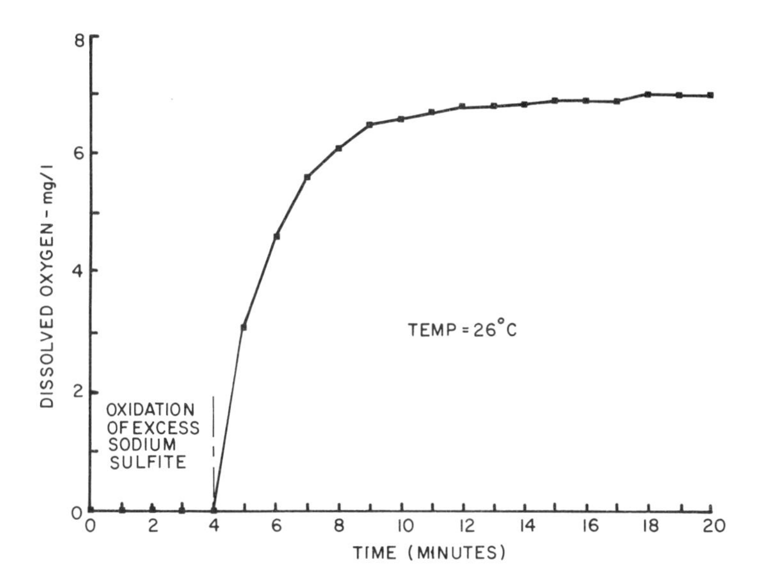

Fig. 7.3. Aeration of the tap water in an oxidation ditch to be used for the treatment of poultry wastes—evaluation of K_La in tap water.

catalyst concentration will increase to the point where the concentration may affect the test results.

The performance of the aeration equipment under process conditions is of considerable interest and the sulfite oxidation method is not suitable for these conditions. The value of K_La obtained under process conditions will depend upon the composition of the liquid that is aerated as well as the geometry of the unit and other process variables. In practice both the steady state and unsteady state methods are used with microbial solids since a convenient means of eliminating the oxygen demand of the mixed unit without altering its physical properties or making difficult the estimation of dissolved oxygen has not been found.

Equation (7.2) is used with the steady state method. At equilibrium, dC/dt is zero and the equation becomes

$$R_r = K_La(C_S - C_t) \tag{7.7}$$

C_t is the dissolved oxygen concentration that exists in the aeration unit under the test conditions and should be above the critical dissolved oxygen concentration. In large units that may not be completely mixed, the tests should be run at a number of points in the unit. C_S is determined by saturating a sample of the mixed liquor with oxygen and measuring the dissolved oxygen at a given temperature. In Eq. (7.7), C_S is equal to $\beta \cdot C_{S_{20}}$.

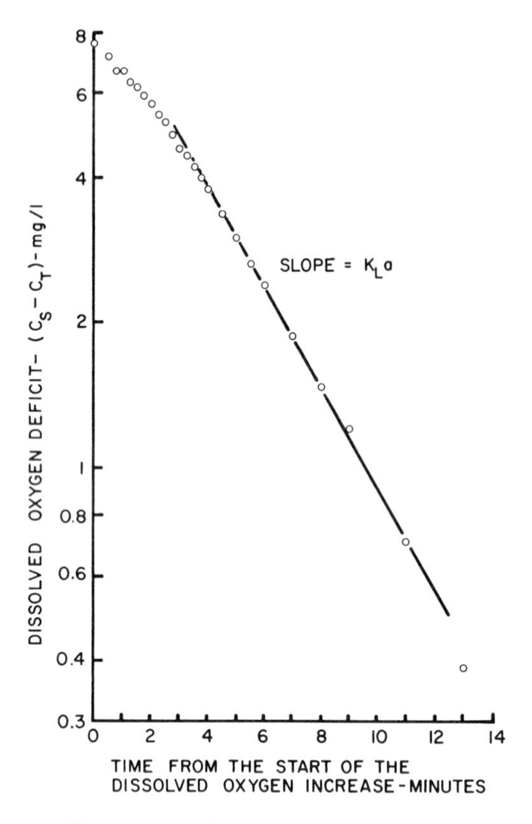

Fig. 7.4. Determination of K_La in tap water.

The oxygen uptake rate, R_r, is determined by aerating a sample of the mixed liquor, sealing it from the atmosphere, providing continuous mixing, and observing the decrease in dissolved oxygen with time. A plot of dissolved oxygen with time should result in a straight line throughout most of the range. The slope of the straight line is R_r. Variation from a straight line may occur in the beginning of this test when equilibrium conditions are established and at the end of the test when the dissolved oxygen concentration becomes limiting.

When using the unsteady state method with microbial solids, dC/dt is not zero. Equation (7.2) is used by expressing it as:

$$\frac{dC}{dt} = (K_La \cdot C_S - R_r) - (K_La \cdot C_L) \tag{7.8}$$

Equation (7.8) has the form of a straight line. By plotting dC/dt against C_L a straight line with a slope equal to K_La will result. The K_La factor thus can be determined without analytically measuring the oxygen uptake rate, R_r, or C_S.

The oxygen uptake rate is affected by the degree of turbulence in the aeration unit. Since the uptake measurement done under laboratory conditions with a sample of the mixed liquor may not exactly describe the uptake rate in the aeration unit, Eq. (7.8) permits the determination of K_La without introducing this type of possible error and permits the determination of the uptake rate directly in the aeration unit.

This method involves turning off the aeration equipment and noting the decrease in dissolved oxygen that occurs with time. Since little oxygen is entering the aeration unit, the rate of dissolved oxygen decrease will be caused by microbial utilization of the oxygen and will be the microbial oxygen uptake rate, R_r (Fig. 7.5a). In most cases this rate will deplete the oxygen within a short time after the aerator is turned off and little solids settling will occur in this period. The dissolved oxygen probe should be agitated in the mixed liquor so that a suitable velocity, generally one foot per second, occurs in the area of the probe.

The aeration equipment remains off until a zero or low dissolved oxygen level is obtained. The aeration rate is then returned to normal and the dissolved oxygen concentration determined at frequent intervals until steady state conditions are approached. The dissolved oxygen concentration (C_L) are plotted with time. Tangents to this curve are plotted at various values of C_L to establish values of dC/dt (Fig. 7.5a). These values are plotted against C_L as determined above (Fig. 7.5b) to determine K_La of the system under actual process conditions. The intercept on the y axis is equal to the quantity $(K_La \cdot C_S - R_r)$. Because R_r has been determined when no aeration occurred, the intercept on the y axis permits C_S to be determined since both K_La and R_r are now known.

An advantage of this method is that K_La is determined independently of R_r or C_S and therefore is not influenced by any errors in their measurement. Any errors in measuring R_r will cause C_S to be in error. This method also permits information on K_La, C_S, and R_r under actual process conditions rather than taking aeration liquor samples and analyzing them for R_r or C_S under laboratory or other conditions that may not represent process conditions. Figure 7.6 illustrates the use of this approach with oxidation ditch mixed liquor treating poultry wastes.

If process conditions are unstable, the analysis cannot be applied. If a straight line is obtained from a plot of dC/dt versus C_L, the system is stable and the parameters R_r and K_La are constant. An unstable system will give a curve when dC/dt is plotted against C_L and indicates that R_r and K_La are variable. Several factors contribute to the instability of an aeration system, the most important of which is lack of dissolved oxygen. The lack of dissolved oxygen at points in the system can create instability throughout the entire system, even though sufficient dissolved oxygen is present in specific parts of the system. If the equilibrium dissolved oxygen concentration under conditions of equilibrium falls below 1.0 mg/l the evaluation of the unit by any analysis is questionable.

Aeration Equipment

Three equipment systems are presently available for the aeration of waste waters in biological reactors: (a) diffused air systems in which compressed air is forced into the liquid through a porous ceramic material or perforated pipe at a liquid depth of 10–15 ft or less, (b) the sparged air-turbine system in which compressed air is forced through a circular diffuser placed directly beneath a turbine type mixer which is placed at a significant depth in the liquid, and (c) a surface entrainment aerator which mech-

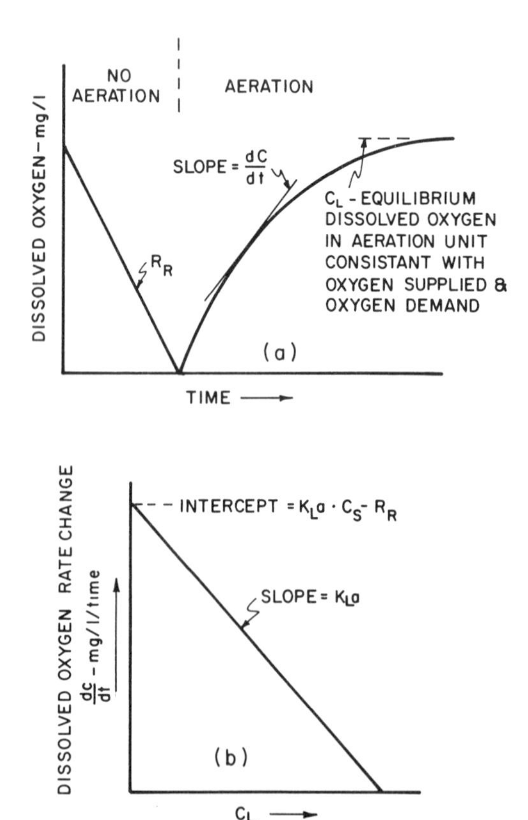

Fig. 7.5. Determination of K_La and C_S using nonsteady state methods.

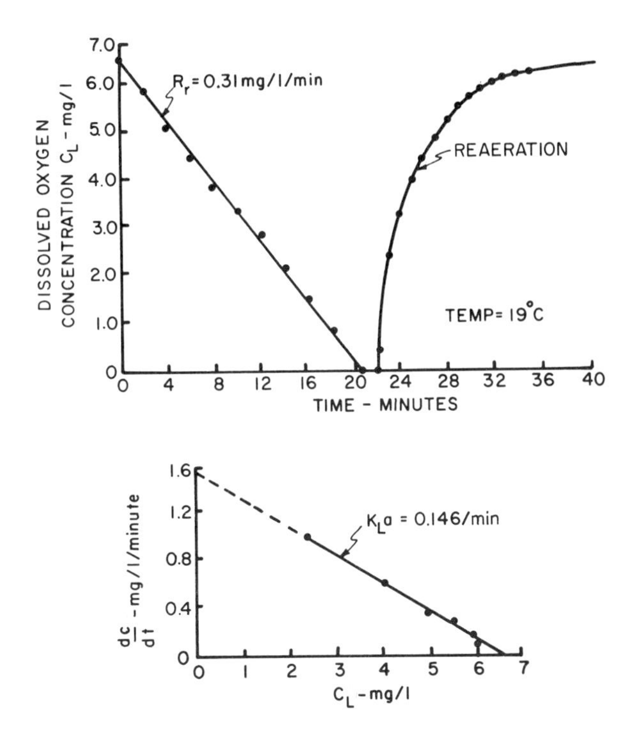

Fig. 7.6. Oxygen transfer and uptake characteristics—mixed liquor of oxidation ditch treating poultry wastes.

anically creates turbulence at the liquid surface with a series of circular blades mounted so as to be partially submerged. The choice of an aeration system is based on such factors as capital and operating costs, flow and strength characteristics of the waste water, maintenance and replacement experience, and designer's preference. The trend in agricultural waste water practice is toward the use of surface aerators in aerated lagoons because of the availability of land adjacent to agricultural operations.

Design of an aeration system for a specific treatment situation is based on the computational or experimental techniques described earlier. The aeration equipment is expected to mix the aeration unit as well as transfer the necessary oxygen. In large systems and high rate systems, if the aeration units can transfer enough air to meet the oxygen demands of the system, adequate mixing normally will occur. In low rate systems or small systems, the mixing requirements may control and the aeration equipment will be designed to keep the solids in suspension. Under these conditions, adequate oxygen will be available to meet the oxygen demand of the microorganisms.

The operating characteristics of an aerobic biological system reflect the dynamic nature of influent waste characteristics, flow, temperature, solids recycle, and similar system components. In most aeration units the dissolved oxygen concentration at any point in the unit is not constant. Variations of as much as 0.5 mg/l of dissolved oxygen can occur with time at a single point in the aeration unit.

Diffused aeration, in which compressed air is put into a liquid through porous diffusers, has been a widely used aeration method. These units are installed such that the rising bubbles create turbulence and mixing. Figure 7.7 illustrates the aeration and mixing patterns in a large activated sludge unit using diffused aeration to treat combined municipal and meat-packing wastes. The rate of oxygen transfer by these systems is dependent upon the air–water interfacial area produced and the degree of mixing and turbulence generated. Because small bubble size provides a greater air–water interfacial area, early diffusers had small openings. Problems of clogging of these openings by chemical precipitates and biological growths

Fig. 7.7. An activated sludge unit using a diffused aeration system.

prompted the development of diffusers with large openings and less maintenance. For standard conditions, the oxygen absorption efficiency of various types of diffuser systems will range from 3 to 10%.

In a submerged turbine system, a diffuser (sparger) ring delivers compressed air beneath the turbine through relatively large holes (Fig. 7.8). The turbine pumps a large quantity of water with the entrained air bubbles and as this mixture passes through the turbine, the bubbles are sheared creating a large air–water interfacial area. The turbine generates considerable turbulence and mixing to aid in the transfer of the oxygen from the air bubbles. The diffuser ring supplies the oxygen and the turbine supplies the mixing. In this manner, the two functions of an aeration system, to meet the oxygen needs and to keep the solids in suspension, can be met by separate pieces of equipment. When the oxygen demand is low, the turbine can keep the solids in suspension while the air supply is decreased. Although oxygen absorption efficiencies of up to 50% may be obtained, high turbine speeds are required (12). From the practical standpoint an average efficiency of about 15–20% is accomplished with most waste waters. These efficiency values are for standard conditions and must be corrected for the waste characteristics, temperature of the liquid, pressure, and desired dissolved oxygen in the liquid.

Submerged turbine aerators can satisfy high oxygen uptake rates that are beyond the capacity of slow speed surface aerators. Submerged turbine aerators are independent of normal liquid level variations and icing problems in cold climates.

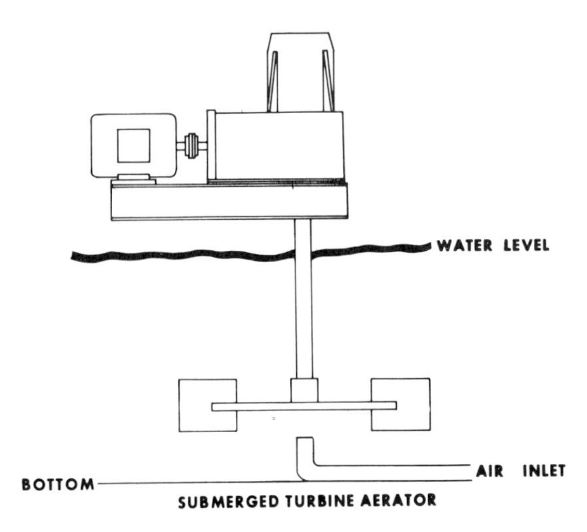

Fig. 7.8. Schematic of a submerged turbine aerator (courtesy of Mixing Equipment Co.).

Fig. 7.9. Fixed level surface mechanical aerator in an aerated lagoon.

Surface aerators are generally slow speed with a high pumping capacity to move large quantities of liquid efficiently. Many early surface aerators were attached to fixed supports in an aeration basin (Fig. 7.9). The unit suffered the disadvantage of being sensitive to slight variations in water levels which could alter the oxygen transfer characteristics of many mechanical aerators. In one investigation (13), the required power increased by 20% (4 kW) for a two-inch increase in water level. The advent of floating surface aerators has removed this problem and has allowed the unit to perform uniformly as the liquid level in the aeration system fluctuated. The simplicity and ease of maintenance of surface aerators, especially floating units, (Fig. 7.10) have caused them to be popular in treating many agricultural waste waters.

At least four types of surface aerators are manufactured, the plate type, an updraft type, a downdraft type, and a brush type. The plate type is a rotating plate with vertical blades that create surface agitation to accomplish a high degree of oxygen transfer. The updraft type depends upon pumping large quantities of water at the surface with relatively low pumping energy. The downdraft type is an aerator where oxygen is supplied by air brought in by a negative head produced by a rotating blade or propeller. The brush type consists of a horizontal revolving shaft with blades attached to the shaft and extending below the water surface. The brush type is used extensively in oxidation ditches (Fig. 7.11).

Fig. 7.10. Float-mounted surface aerator for a large aerated lagoon (courtesy of Mixing Equipment Co.).

Surface aerators achieve gas–liquid contact by means of an impeller or turbine located in the liquid surface of the water to be aerated. All types of surface aerators described above promote gas–liquid contact in this manner. The rate of oxygen transfer is influenced by the total air to water interfacial area produced and the mixing and turbulence generated. Oxygen transfer occurs by: (a) spraying the liquid in the air, (b) incorporation or dispersion of air into the liquid, and (c) generation of a high level of surface turbulence. The turbulence aerated by the impeller blades disperse the air in the liquid discharged from the aerator. The discharge of the liquid from the aerator generates relatively high liquid surface velocities in the immediate vicinity of the aerator which assist in oxygen transfer.

These actions are noted in Fig. 7.12. In the aeration zone, the dissolved oxygen concentration is increased from DO_1 to DO_2 where DO_1 and DO_2 are the average dissolved oxygen concentrations of the liquid entering and leaving the aeration zone, respectively. The oxygen transferred is related to the quantity of water moving through the aeration zone and the dissolved oxygen difference between discharged and inlet concentrations, i.e.,

$$R = \frac{Q_A \cdot (8.34)(DO_2 - DO_1)}{10^6} = \frac{L(DO_2 - DO_1)}{10^6} \tag{7.9}$$

where R is the oxygen transferred, pounds/hour, and Q_A and L are the flow

Fig. 7.11. Brush type surface aerators in an oxidation ditch (courtesy of Lakeside Engineering Corp.).

rates of liquid through the aeration zone, in terms of gallons per hour and pounds per hour, respectively.

In a surface aerator, $K_L a$ is a function of the characteristics of the aerator, i.e., shape and height of blades, revolutions per minute, and geometry of the basin. The dissolved oxygen concentrations in the aeration basin will be determined by the dissolved oxygen concentrations leaving the aerator,

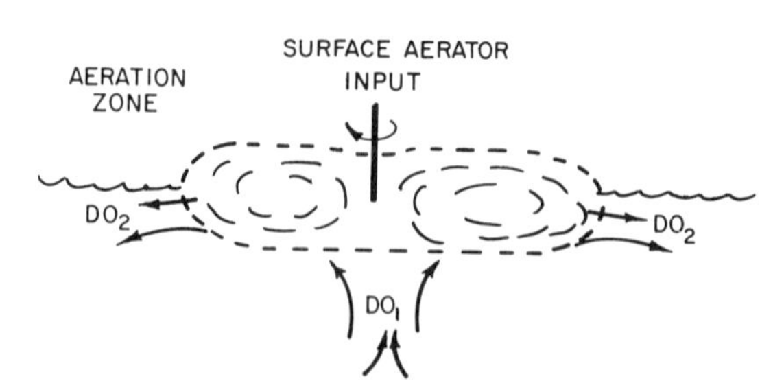

Fig. 7.12. Change in dissolved oxygen concentration through the aeration zone of a surface aerator.

the pumping rate of the aerator, the oxygen uptake rate, the configuration of the basin, and the location of the aerators in the basins. Surface aerators are essentially pumping units installed at the liquid surface and must pump a relatively large quantity of liquid to keep the driving force ($DO_2 - DO_1$) as high as possible, distribute the oxygen throughout the liquid, and keep solids in suspension.

Surface aerators mix effectively in their own zone of influence but not effectively throughout an aeration basin if there is more than one in the same basin. Hydraulic translation of liquid from the zone of influence of one aerator to that of another aerator is slow. The transition between two zones of influence acts as a partition wall slowing intermixing in the basin. Differences in oxygen demand per zone of influence can be observed, especially in large field installations.

Control of oxygen transfer capacity of surface aerators is related to aerator design. Submerged aerators that disperse compressed air at the suction of the aerator can control oxygen transfer by varying air flow. Entrainment type surface aerators (Fig. 7.8) control oxygen transfer by changing the speed of the aerator or the submergence of the aerator. These latter measures adjust the horsepower of the aerator and thus the oxygen transfer capacity.

The efficiency of a surface aerator is expressed in terms of the pounds of oxygen transferred at standard conditions per hour per brake horsepower. In general, available surface aerators have an efficiency range of 2–4 lb O_2 transferred under standard conditions per hour per horsepower (14). Oxygen transfer relationships in actual waste treatment systems will be lower than data quoted in equipment manufacturers literature since the standard conditions do not occur under process conditions.

With a mechanical aerator, maximum oxygen transfer is at the aerator rather than through the surface of the aeration unit. When the surface of an aerated lagoon was covered to evaluate oxygen transfer at the aerator versus through the surface, the cover reduced the oxygen transfer rate by only about 10% (15). Other investigators have estimated the oxygen transfer through the surface as low as 2–5% during diffused aeration (2).

The determination of the oxygen input capacity of an aerator is not simple. With localized aerators, it is difficult to have a uniform value of dissolved oxygen throughout a unit. Since uptake is influenced by turbulence, and since such turbulence varies throughout the basin, the uptake rate for oxygen determined in a laboratory on a mixed liquid sample may not be indicative of the average uptake in the aeration unit.

Surface aerators rarely can be selected without a good knowledge of the process operating conditions, basin geometry, and aerator spacing as well as the need for the aerator to both mix and aerate. Where it is expected that these factors as well as capital and operating costs are critical, pilot and full scale studies are valuable to select proper aeration equipment.

Activated Sludge Processes

General

Activated sludge processes are aerobic biological processes that can be used to treat many wastes. The processes are versatile, flexible, and an effluent of any desired quality can be produced by varying the process parameters. High BOD removal efficiencies can be achieved routinely. These processes require less land but more competent operation than do simpler processes such as oxidation ponds and aerated lagoons. A number of activated sludge processes exist, however only those that have potential for agricultural wastes are discussed.

The term, activated sludge, is applied to both the process and to the biological solids in the treatment units. Activated sludge is the complex biological mass that results when organic wastes are aerobically treated. The sludge will contain a variety of heterotrophic microorganisms including bacteria, protozoa, and higher forms of life. The predominance of a particular microbial species will depend upon the waste that is treated and the way in which the process is operated.

In practice, most activated sludge plants operate in the stationary phase of microbial growth and are continuous flow systems. The biological growths will aggregate into flocculant masses which can be kept in suspension while the activated sludge unit is mixed but which will settle when the mixing is stopped. These properties are important since they permit the microbial solids to be distributed throughout the biological unit yet separate readily in a separation unit. A well-treated effluent containing few biological solids will result. Depending upon the activated sludge modification, the settled solids are recycled back to the biological unit or wasted to a solids handling and disposal system, or both.

For optimum treatment, the raw waste must be balanced nutritionally. If the microbial solids undergo extensive endogenous respiration in the same unit in which the influent BOD is metabolized, a portion of the microbial nutrients will be released to the mixed liquor. A tradeoff exists between the addition of nutrients and the length of aeration period for nutrient deficient wastes. Activated sludge produced from nutrient deficient wastes tend to have poor settling and filtering characteristics, especially if nitrogen is in a short supply.

The total oxygen demand of an activated sludge unit is related to the quantity of untreated BOD, the quantity of activated sludge present, and the time of contact between the sludge and BOD. The rate of oxygen demand is a function of the phase of growth of the bacteria.

After endogenous respiration occurs, further treatment results in no greater removal of raw waste BOD but does reduce the quantity of sludge solids for ultimate disposal. Since nearly all the BOD entering a system is converted into cell mass, the effluent BOD will be caused almost entirely by the endogenous respiration of the organisms contained in the effluent suspended solids. Separation of microbial solids will clarify the final effluent, provide a concentrated quantity of active organisms to return to the aeration basin or to waste as desired, and will reduce the volatile fraction of the effluent solids. The ratio of VSS/TSS in the clarified effluent from an activated sludge system treating potato wastes was 0.52 while the ratio in the return sludge was 0.74 and that in the aeration basins was 0.70.

In activated sludge systems or similar aerobic treatment units, the total oxygen demand and the oxygen uptake rate is a function of the solids retention time in the system. The total oxygen demand will be lower at short SRT values since synthesis is the predominant oxygen demand and will increase slowly as the SRT increases because of increased endogenous demand. The oxygen uptake rates are rapid at low SRT values since there is a high food to mass ratio and rapid microbial growth. The uptake rates rapidly decrease as SRT increases because of greater dilution of the influent wastes and a lower food to mass ratio. The oxygen demand at long SRT values is caused by the endogenous oxygen demand.

The settling characteristics of activated sludge floc also have been correlated with the SRT of a system (16). Waste activated sludge solids from a unit with a SRT greater than 4 days were found to have better solids separation in a final clarifier. The settling characteristics of the floc was not correlated with either the soluble COD in the unit or the F/M ratio.

The Basic Process

The basic process was developed in England with E. Arden and G. J. Lockett generally given the credit for the experimentation that led to the process as it is now known. Their work was done in 1914. However, experiments in the late 1800's demonstrated the possibility of the process (17).

A flow diagram for the basic process includes two units, an aeration unit in which the aerobic treatment occurs and a sedimentation tank in which the solids are separated from the liquid. These two important functions of the system are separated so that each process can be controlled individually to yield optimum results.

The biological solids in the aeration tank are known as the aeration tank solids or mixed liquor suspended solids (MLSS). The volatile fraction of the mixed liquor (MLVSS) frequently is used as a measure of the active organisms in the tank. However, it is only a gross estimate of the active mass. At activated sludge loadings normally encountered in treating domestic wastes, the percentage of total carbon, total and protein nitrogen, lipid, and total carbohydrates in MLVSS was observed not to vary significantly (18). The consistency of gross MLVSS makes it possible to use MLVSS as the basis on which to express measurements of activated sludge viability and activity.

The return of a portion of the settled solids permits a higher MLSS and hence active mass to be kept in the aeration tank. The larger concentration of active mass purifies the wastes in a shorter time, permits a smaller aeration tank, and permits the solids retention time to be different from the hydraulic rentention time. Without sludge return, the activated sludge process would be approximated by an aerated lagoon with its larger size and smaller active mass.

Diffused aeration at about 6–8 psi is used to supply the oxygen to keep most activated sludge systems aerobic and to keep the aeration unit well mixed (Fig. 7.7). The air generally is supplied along one side of the tank with the result that the mixed liquor will have a spiral roll pattern as it moves through the tank. The basic process treating municipal wastes has a liquid detention time of 6–8 hr and a MLSS concentration of 1500–3000 mg/l in the aeration tank. Other average operational parameters are a BOD loading rate of 20–50 lb BOD/1000 ft^3/day, a lb BOD/lb MLSS ratio of 0.25–0.5, a solids retention time of three or more days, a return sludge rate of 25–50% of the influent flow, and approximately 2% sludge wasting per day. Average air flow rates are 1000–1500 ft^3/lb BOD removed/day. These values will differ for different activated sludge modifications and waste concentrations.

The basic system functions as a plug flow system in that the untreated waste and the return sludge enter the aeration tank at one end of the tank and move as a unit to the other end of the tank. The incoming wastes and solids are not mixed uniformly with all of the contents of the aeration tank. To reduce the oxygen demand in the aeration tank, settleable solids in the raw wastes are removed in a primary sedimentation unit. With municipal wastes, the BOD of the raw wastes is reduced about 30–35% in the primary sedimentation unit while the suspended solids are reduced about 50–60%. Other wastes will have different removal relationships since the percent removals in the primary units are related to the type of solids in the raw wastes. The dissolved oxygen concentration in the aeration unit should be kept above 0.5 to 1.0 mg/l. Activated sludge systems can be used for soluble and colloidal wastes as well as mixtures.

Aeration tanks with a large length to width ratio generally are designed as plug flow units in which the mixed liquor does not totally intermix. Baffles may be placed along the tank length to prevent short-circuiting. The biological reactions proceed more rapidly near the inlet of the tank where the concentration of organic matter is the largest. The highest oxygen demand exists at the inlet end of a plug flow aeration unit. The aeration units should be placed to meet the oxygen demand as closely as possible. Aeration in the plug flow tank can be tapered to approximately parallel the oxygen demand.

The food to microorganism ratio when the wastes enter the aeration tank is high, causing higher oxygen demand rates and microbial synthesis at the influent end of the tank. By the time the waste has reached the effluent end of the tank, the organic matter has been metabolized and the organisms may be in the endogenous phase. Ratios of inlet to outlet demand rates of 5–10 to 1 are not uncommon. In a conventional activated sludge plant treating domestic wastes, the oxygen demand when nitrification does not occur may be as high as 100 mg/l/hr in mixed liquor near the inlet of the aeration unit and fall rapidly to about 40–60 mg/l/hr after aeration for about an hour and then more gradually to about 10–20 mg/l/hr after 5–6 hr aeration. If full nitrification of domestic sewage is achieved the average demand as a whole may be about 20 mg/l/hr higher and the aeration capacity must be able to supply this demand as well as the carbonaceous demand.

Because of the plug flow effect, the activated sludge MLSS in the basic process are on an alternating feed–nonfeed cycle which can affect the ability of the microorganisms to handle variable hydraulic and organic loads. The microbial concentration is considerably less than the MLVSS concentration. The viable organism content of a conventional activated sludge unit amounted to only about 15% of the MLVSS (18). Ranges of viable organisms were $0.1–2 \times 10^8$ cells/mg VSS. The viable aerobic heterotrophic microorganism content of conventional and low rate activated sludge units is between 10 and 20%. The concentration of nitrifying organisms was estimated at about 5% of the bacterial mass even where the activated sludge was nitrifying completely (19). The rates of oxygen demand in conventional systems have been shown to range from 7 to 37 mg/l DO/gm VSS/hr (2) and 15–35 mg/l DO/gm SS/hr with higher rates associated with nitrifying sludges (20).

Solids production is an important aspect of aerobic biological treatment since they are a large fraction of the solids ultimately to be handled and disposed of from these facilities. Equations in Chapter 5 permit the estimation of sludge production in a biological treatment unit.

When understood and properly operated, conventional activated sludge plants can produce high (90 +%) BOD removals. The key to the quality of the effluent lies in a positive solids removal system. Compared to other aerobic biological treatment processes, the advantages of the activated sludge process include small land area, positive control over the degree of treatment, and flexibility of treatment to meet varying waste loads. Disadvantages include higher capital costs and the need for competent and constant control. The basic process can be sensitive to variations in BOD and hydraulic loads, varying air requirements throughout the aeration tank, a feed–nonfeed cycle for the microorganisms since the wastes are introduced only at one point in the tank, and inability to control nitrification.

The conventional activated sludge process can be used with liquid agricultural wastes, especially those from canning operations either separately or with municipal wastes (21, 22). The characteristics of animal wastes or waste waters have led design engineers to favor other aerobic biological processes for these wastes.

Modifications

Numerous modifications of the basic activated sludge process have been developed to overcome shortcomings of the basic process. Detailed description of the basic activated sludge process and of the modifications are available in many summaries (3, 23, 24). Three modifications are particularly useful for treating agricultural wastes which generally have high and variable strengths and which may be seasonal. These modifications are contact stabilization, complete mixing, and extended aeration. The selection of a given process depends on the characteristics of the waste water, the effluent quality required, and factors such as land availability.

Contact Stabilization

Contact stabilization, or biosorption (Fig. 7.13) is a process which was developed at a municipal treatment plant (25) and rapidly used with industrial wastes (26) and in package plants (27). In this process the raw wastes are mixed and aerated with activated sludge for a short time and then settled in a sedimentation tank. The contact unit is used to adsorb the particulate and colloidal organic matter to the activated sludge floc. The detention time in the contact unit will depend upon the time necessary for adsorption and can be as low as 20–30 min. Many designers have increased the detention time to 1–2 hr to be conservative. This added time permits some of the adsorbed material to be solubilized and released into solution where it will be part of the effluent from the sedimentation tank.

Excess sludge may be wasted from the process either from the sedimentation tank or from the effluent of the reaeration tank. Solids wasted from the

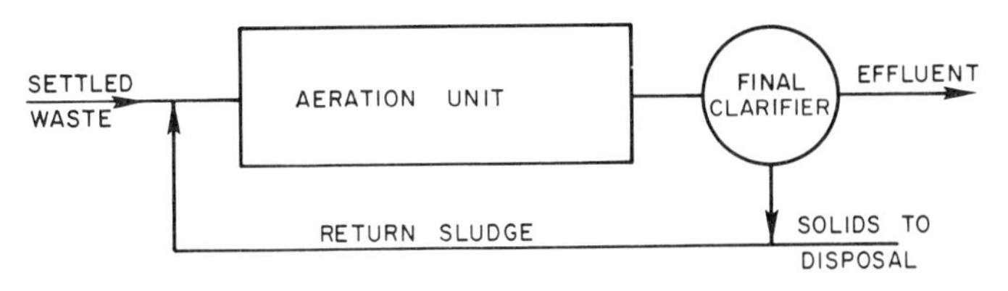

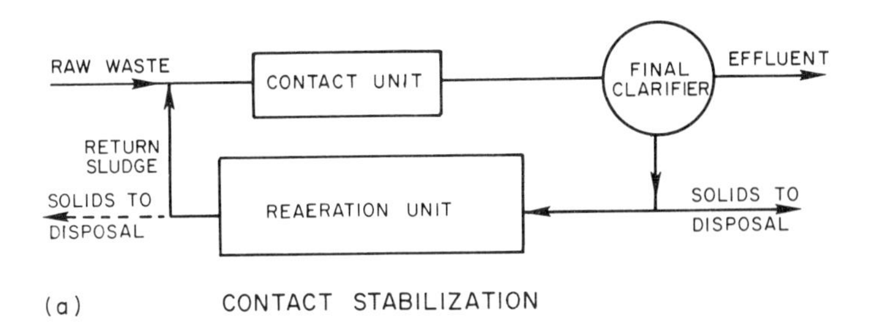

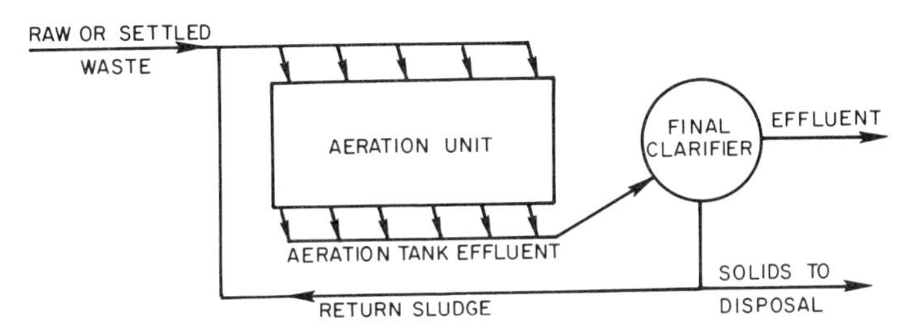

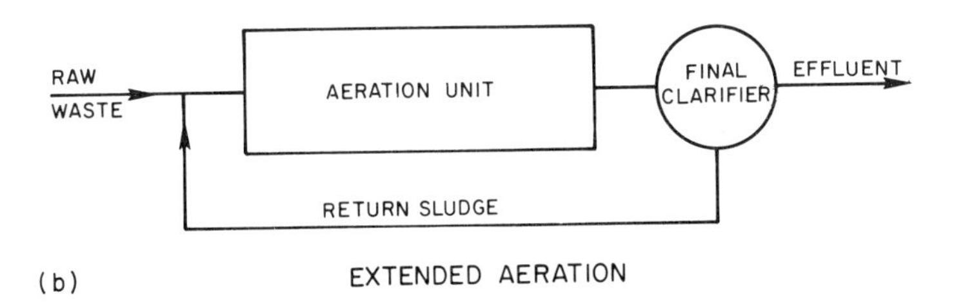

Fig. 7.13. Flow diagrams of activated sludge modifications.

sedimentation tank will have a higher oxygen demand than those wasted from the reaeration unit since the latter have been oxidized in the reaeration unit where endogenous respiration is predominant. During periods of light loadings, aerobic digestion will take place in the reaeration unit.

The microbial solids from the reaeration unit will be devoid of food for a period prior to their return to the contact unit. The new food which they receive encourages rapid growth. The resultant sludge has good settling characteristics.

The essential difference between the contact stabilization and the conventional activated sludge treatment process is that the adsorption phase is separated from the biological oxidation phase. The sludge in the reaeration tank normally will range between 6000 and 8000 mg/l. The solids in the contact tank can range from 3000 to 5000 mg/l.

The detention time in the reaeration unit is from 4 to 6 hr. The reaeration unit volume is less than the volume in a conventional activated sludge plant since the tank is designed on the basis of the flow containing the settled solids and not the entire raw waste flow. Aerating concentrated solids in these units rather than more dilute solids, as in the conventional activated sludge process, reduces the aeration volume while providing high BOD reduction. Oxygen requirements for the system are about 40% for the contact tank and 60% for the reaeration tank. A higher oxygen demand rate will occur in the contact tank as the waste and microorganisms are mixed.

Generally accepted design criteria for contact stabilization systems indicate that as long as the average daily BOD_5 load applied does not exceed approximately 50 lb per 100 lb of mixed liquor suspended solids in the system, a reduction of approximately 90% or greater of the applied BOD_5 can be expected. The SRT of the system will be comparable to that in a basic activated sludge system.

Evaluation of sludge production indicated (28) that gross sludge production, including solids lost in the effluent, was 0.35 lb dry solids/lb BOD applied/day at temperatures ranging from 6 to 10°C. The sludge production was directly proportional to the plant loading. The temperature fluctuations (6°–10°C) and the MLSS concentrations (5000–5600 mg/l) in the contact and stabilization zones did not appear to affect the sludge production. The study concluded that plant loadings less than 0.1 lb BOD/lb sludge/day should be avoided if a good settling sludge was to be maintained.

The process is less sensitive to fluctuating or toxic loads since it has a high concentration of microbial solids contacting the raw wastes. Additional concentrations of microbial solids are separated from the raw wastes and unaffected by transitory toxic conditions. The smaller contact tank permits better mixing to occur and a smaller plug flow effect.

Primary sedimentation rarely is a component of this process since contact stabilization can handle the particulate material in the raw wastes. One of the main disadvantages for the process is that soluble wastes may not be adequately treated due to the short detention time of the wastes and the microbial solids in the contact tank.

Conventional activated sludge systems can be converted readily to contact stabilization when increased treatment capacity is required. Prefabricated package plants for small waste flows are available from a number of manufacturers (Fig. 7.14).

Contact stabilization has been applied for the treatment of agricultural wastes. It has been used with wastes from the killing and packing of broilers (29). At an aeration tank loading of 30–33 lb BOD/1000 ft^3/day the BOD was reduced from 1220–1720 mg/l in the influent to 14–22 mg/l in the effluent. Suspended solids removal was 94%.

Sixty-five percent removal of the BOD in the wastes from the production of powdered and sweetened condensed milk was obtained with contact stabilization (30). The plant had detention times of 0.8 hr in the contact unit and 3.3 hr in the reaeration unit. Loadings varied from 0.65 lb BOD/lb MLSS/day at 20°C to 1.16 lb BOD/lb MLSS/day at 25°C.

In a comparative study of systems for the treatment of fruit processing wastes, both conventional activated sludge systems and contact stabiliza-

Fig. 7.14. A contact stabilization plant treating combined municipal and cheese plant wastes. The sedimentation section is in the middle with the contact and reaeration basins on the periphery of the circular unit.

tion provided greater than 90% removal of the organic load when the systems were loaded at low rates, less than 0.4 mg COD removed/day/mg MLVSS (31). The resulting solids were removed satisfactorily in clarifiers. Completely mixed aeration basins provided effective buffering capacity to avoid pH fluctuations in the aeration systems caused by abrupt changes in the pH of the waste flow.

Complete Mixing

The ideal activated sludge treatment system would include the following: ability to handle shock organic and hydraulic loads, uniform waste feeding, uniform air utilization, complete waste mixing, uniform organic level in the aeration system, and simplicity of operation. Complete mixing activated sludge contains these attributes. Complete mixing is a process in which the incoming wastes are completely mixed with the entire contents of the aeration tank within a few minutes after the wastes enter the tank (Fig. 7.13). In small tanks the wastes and return sludge can be introduced at a single point but in large tanks multiple inlets are required. There are many variations of complete mixing including combination aeration-sedimentation units, and separate aeration and sedimentation tanks with or without separate sludge wasting. The system can be operated at any point in the growth cycle that is desirable to provide the required BOD reduction and solids oxidation.

The aeration tank acts as a surge tank to level out any fluctuations in organic or hydraulic load. The microbial solids will respond to changes in the waste characteristics and adjust themselves accordingly to reestablish equilibrium conditions. The use of the entire mass of activated sludge results in a more uniform oxygen demand and uniform loading. A detailed analysis of the process has been provided by McKinney (32) who was a leader in developing and utilizing the process. In recent years the process has been utilized at a number of municipal and industrial treatment facilities. The oxidation ditch below animals in confined production facilities can be considered as an example of a long detention time complete mixing activated sludge unit.

In the complete mixing process, adequate mixing will produce direct contact between the microorganisms and the wastes and will reduce short circulating and untreated waste loss to a minimum. A modification of the complete mixing system permits the return sludge and the untreated waste to enter the unit at separate points. This approach is useful when there is the possibility of toxic material in the unit.

In a study of the process used to treat a combination of municipal and meatpacking wastes (33), a high degree of treatment was possible with a relatively low quantity of oxygen, approximately 0.5 lb O_2/lb BOD_5 applied. It appeared that the organic load was such that synthesis reaction was normal but that endogenous respiration was reduced, being about 0.1/day rather than the 0.48/day previously reported. Air flow rates were as low as 400 ft^3/lb BOD applied with an average rate of 560 ft^3/BOD applied. The aeration tank loading was 116 lb BOD/1000 ft^3/day. The oxygen uptake relationship ranged from 35 to 70 mg/l/hr.

A complete mix activated sludge process satisfactorily treated potato processing waste water, accomplishing over 90% BOD removal. BOD removals of 70–80% were obtained when the solids in the effluent from the aeration basin were not removed in a final clarifier (34). The high pH values, up to 11.5 at times, of the influent waste water were buffered in the aeration basins and were not detrimental to the treatment efficiencies. Oxygen requirements were related to the BOD removal rate, S_R, and the average MLVSS, X_D, according to the following equation:

$$O_2(\text{lb/day}) = 0.48\ S_R(\text{lb BOD removed/day}) + 0.03\ X_D(\text{lb MLVSS}) \quad (7.11)$$

Total annual activated sludge treatment costs for secondary treatment only and not disposal of waste activated sludge solids were estimated as $0.038/lb BOD applied and $0.021/lb COD applied.

Citrus wastes required nitrogen and phosphorus additions to accomplish satisfactory complete mix activated sludge treatment. The recommended design loading rate was 0.3 lb BOD/lb MLSS and the capital and operating expense for the treatment system was about 1.5 cents/box of fruit processed or about 3.2 cents/lb BOD removed (35).

Extended Aeration

Where wastes can be held for long detention periods, extended aeration offers the opportunity to provide a high degree of treatment and to oxidize the entering and synthesized solids. This process includes long-term aeration, at least 24 hr for municipal wastes and longer for stronger wastes (Fig. 7.13). The solids are settled in the sedimentation tank and returned directly to the mixing tank generally with separate wasting, producing a long SRT. Any excess sludge in the effluent has been aerobically digested in the aeration tank and should have little residual BOD. The fundamental research on extended aeration was initiated on studies of the treatment of dairy wastes (36).

The correct basis for design of extended aeration plants is the BOD loading per unit of activated sludge solids in the system, i.e., lb BOD/lb MLSS. Operational characteristics for the extended aeration process are 4000–6000 mg/l MLSS, a loading rate of 2–20 lb BOD/1000 ft^3/day, a lb BOD/lb MLSS ratio of 0.05–0.10, a solids retention time of 10–20 + days, and an air flow rate of 1500–2000 + ft^3/lb BOD treated/day. The active mass may range from 7 to 10% of the MLSS.

In order to have this process operating in the declining growth and endogenous respiration range, a low food to microorganism ratio must be maintained. With this process, the mass of organisms produced is equal to the mass of organisms lost in endogenous respiration, in the effluent, and by wasting. The wastes solids will be the inorganic and nonbiodegradable sludge residue plus a small fraction of active microorganisms.

Extended aeration units commonly are small units which are completely mixed as evidenced by little suspended solids or dissolved oxygen gradients in the units. Since they are completely mixed, they are more resistant to load or flow fluctuations.

As measured by effluent quality, overall performance of extended aeration plants depend on the ability of the system to retain suspended solids. The discharge of excess suspended solids frequently results from hydraulic overload of the sedimentation section of the plants. Although extended aeration units are unsophisticated, they do require operational and maintenance attention to achieve the high BOD removals (90 +%) that they can produce.

Because of the long detention times, a high degree of nitrification may occur. This process will have a higher total oxygen requirement since the oxygen is used for both synthesis and oxidation of the microbial solids and for nitrification.

Plant performance is affected by the degree of nitrification. Long-term BOD tests on the influent and effluent from an extended aeration plant have demonstrated that significant differences can occur when BOD reduction efficiencies are based on 5-day 20°C BOD values without consideration of the nitrogenous oxygen demand of the effluent (37).

An extended aeration system for poultry processing wastes achieved 90% removal. The loading of the process was about 15 BOD/day/100 lb MLSS and the MLSS were 89% volatile. The average hydraulic detention time was 24 hr. BOD loadings approaching 40 lb BOD/day/100 lb MLSS over a 12-hr period were successfully handled on an individual day basis. The air supplied was 1500 ft^3/lb BOD removed (38).

Citrus wastes were subjected to an extended aeration process. BOD removals ranged from 90 to 98% with an average of 94% (39). Mixed liquor suspended concentrations of 3000–5000 mg/l were held in the system when the above removals were achieved. The system was designed for a loading of about 7 lb BOD/day/100 lb MLSS.

Dairy cattle manure have been treated in an aerobic treatment system having a long detention time. At a temperature of 65°F, a volatile solids reduction of 70% was obtained with a detention period of 74 days. At a temperature of 48°F or lower, only 45% reduction in volatile solids was obtained (40). In general, the percent volatile solids reduction decreased with decreasing temperature at a fixed detention time thus following the general temperature relationships of biological systems.

Completely mixed activated sludge systems and aerated lagoons can operate as extended aeration units. The key is the long detention time and the opportunity to oxidize the microbial solids. Total oxidation of the microbial solids will not occur since portions of the microbial solids resist degradation and will be in the effluent. Extended aeration solids can be dewatered on sand beds and can be disposed of on the land.

Prefabricated complete mix and extended aeration plants are marketed by numerous equipment manufacturers. Extended aeration plants are useful for wastes from an expanding operation since they can be converted into complete mixing units with only slight additions to handle the waste sludge.

Aerobic Digestion

Introduction

Waste treatment processes essentially are separation and concentration processes. Biological treatment processes use microorganisms to metabolize the organic matter and concentrate it in the form of synthesized microbial solids. Settling units can be used to separate the inert and microbial solids from the waste flow before discharging the effluent to the environment. Each separation and concentration process produces waste solids which require further handling and disposal.

A treatment process that can dispose of the organic solids and not produce any residue is desirable but has not yet been developed. Aerobic digestion, however, is a process which, if operated correctly, can produce a minimum amount of biodegradable solids. The residue from this process is a stable sludge with low oxygen demand, good settling characteristics, and no offensive odor. Aerobic digestion is an aerobic biological process designed to have a long solids retention time. The gross oxidation in the system includes the direct oxidation of any biodegradable matter by the active mass of organisms, and endogenous respiration, the oxidation of microbial cellular material.

A residual organic, insoluble, and relatively nonbiodegradable fraction of the bacteria will remain as described in Chapter 5. This inert polysaccharide material will accumulate in the system at the rate of about 11–15%

of the ultimate BOD removal. In aerobic digestion, endogenous respiration is the predominant metabolic reaction. Most applications of aerobic digestion have been used to treat waste-activated sludge but the process can be used for other concentrated organic wastes.

Factors Affecting Design

The factors affecting the design of aerobic digestion systems include the rate of sludge oxidation, oxygen requirements, temperature, and characteristics of the residue and supernatant. The sludge reduction possible in aerobic digestion systems is related to the sludge age or solids retention time in any prior treatment system, such as in an activated sludge unit, and in the aerobic digestion system.

Sludge oxidation rates have been found to vary considerably. These rates will vary depending upon the microbial content of the sludge, the characteristics of the wastes used to grow the sludge, the sludge age, and the temperature. In addition, the different rates of sludge oxidation and oxygen utilization in aerobic digestion studies are due to different starting points in the particular research. As indicated above, in the aerobic digester, the percent volatile solids reduction of a waste sludge having undergone previous biological treatment, i.e., waste-activated sludge, will be less than that of waste solids that have not been treated as yet, i.e., untreated animal wastes, cannery wastes, or primary sewage solids. The primary criteria for solids reduction is the total solids retention time, i.e., the SRT of the solids in previous biological units plus the SRT in the aerobic digester. Percent volatile solids reductions and oxygen demands should be related to the total solids retention time rather than to only that in the aerobic digester.

In any biological system net protoplasm accumulation can be expressed in terms of the solids increase by synthesis and solids loss by endogenous respiration. This relationship can be used to understand the microbial reactions in an aerobic digestion unit. The factors a and b in the following equation represent the fraction of BOD resulting in synthesized microbial solids and the endogenous degradation of microbial active mass, respectively.

$$\text{Sludge accumulation (lb active mass/day)} = a(\text{lb BOD removed/day}) - b(\text{lb degradable biological solids}) \quad (7.12)$$

or

$$\Delta M_a = aF - b \cdot M_a \quad (7.13)$$

When aerobic digestion is used with waste solids that have had previous treatment, the microorganisms in these solids have removed the organic matter in the influent waste flow, have metabolized any stored and adsorbed food, and microbial synthesis is minimal. Therefore,

$$\Delta M_a = -bM_a \tag{7.14}$$

Solving Eq. (7.13) yields

$$M_{a_t} = M_{a_o}(e^{-bt}) \tag{7.15}$$

The net decrease in sludge accumulation that occurs is caused by endogenous respiration.

Equation (7.13) illustrates an important point relative to aerobic digestion. The decrease in solids that occurs in an aerobic digester will be caused only by a decrease in the active mass in the digester and that decrease is an exponential onc. The inorganic and nonbiodegradable organic solids will not decrease. Although percent volatile solids reduction is convenient to measure aerobic digestion process efficiency, it is not the fundamental relationship. The reduction in active mass would be a better parameter to describe the performance of an aerobic digester.

Reported values for b have usually been based on mixed liquor volatile solids rather than on the active mass which is the biodegradable biological solids, since the active mass is difficult to determine directly. Thus, reported values for b have varied considerably. The key to an accurate determination of the endogenous respiration constant lies in an accurate determination of the active mass.

During the initial phase of endogenous respiration, the rate equation may be approximated by monomolecular kinetics and b can be expressed as a constant. However, as aeration continues, b will decrease as the more easily metabolized cellular components are oxidized. Under the above conditions, the net change of sludge mass in an aerobic digester will be a function of the decrease of biodegradable solids, the accumulation of biologically undegradable solids remaining from endogenous respiration, and the accumulation of inert solids present in the original sludge.

When the waste solids contain significant organics that have not yet been biologically treated, microbial synthesis cannot be neglected and the synthesis term, aF, must be included when using Eq. (7.13).

The oxygen requirements in aerobic digestion systems also are related to synthesis and endogenous respiration.

$$\text{Oxygen utilization} = cF + b_1 M_a \tag{7.16}$$

When synthesis in aerobic digestion systems is minimal, oxygen requirements are primarily related to endogenous respiration. Therefore,

$$d\text{O}/dt = b_1 M_a \tag{7.17}$$

Investigations to determine endogenous respiration oxygen uptake rates have produced widely varying results. Few of these results have been reported in terms of the active mass in the system. Most have been reported in terms of mixed liquor volatile solids which will contain both active and inactive fractions.

Ideally with prolonged aeration, the active sludge mass would continue to oxidize itself to carbon dioxide and water so that no net sludge accumulation would occur in the system. However, approximately 20–25% of the microbial solids synthesized in a biological treatment system are relatively inert to biological oxidation and accumulate in the system.

The rate of volatile solids destruction and the oxygen uptake rate will decrease as the SRT in the aerobic digester increases. The uptake rates of waste sludges will reflect the prior treatment of the sludge (Table 7.3).

Table 7.3

Oxygen Uptake Rates of Waste Solids from Aerobic Biological Treatment Units[a]

Waste sludge	Uptake rate ($mg\ O_2/gm\ VSS/hr$)
Conventional activated sludge plant	10–15
Contact stabilization plant	10 or less
Extended aeration plant	5–8
Primary clarifiers	40 +

[a] From reference 41.

Uptake rates in the range of 1.0 or less mg O_2/gm VSS/hr are associated with well-stabilized sludges. This rate is reached at a total sludge retention time of about 120 days (41). It may be impractical to provide this degree of aerobic digestion prior to final disposal. Because the oxygen uptake rate is low in an aerobic digester, the air requirement is more often a function of the amount of air required for tank agitation than of the oxidation requirements. In general an air supply of 15–20 cfm/1000 ft of digester capacity is enough to keep the solids in suspension (42).

When the waste solids contain significant amounts of unmetabolized organic matter, such as in primary sewage sludge or untreated agricultural wastes, microbial synthesis, cF, cannot be neglected in determining the oxygen requirements. Under these conditions, the air requirement will be a function of both synthesis and endogenous respiration and may exceed that needed for tank agitation. The largest air requirement for the system, i.e., for either oxygen demand or for agitation, will control and be used for design purposes. In one example the oxygen requirements for the aerobic digestion of mixed primary and activated sludge were about six times that for the aerobic digestion of activated sludge alone (43).

Temperature affects the rate of metabolic activity of the microorganisms in the system. In many aerobic digestion systems this change in the sludge oxidation and endogenous respiration rates can be masked by the variability of sludge age and microbial content of the sludge. Temperature has an appreciable effect at short aerobic digestion detention times, less than 15 days, but this effect is decreased considerably as the time is lengthened or the loading rate decreased. For long detention times, beyond a total sludge retention time of about 120 days, temperature has little effect since digestion is essentially complete at all temperatures.

With increasing sludge age in an aerobic digestion unit, the volatile solids reduction approaches a limit. The limit that can be expected will be a function of the waste fed to the system. Activated sludge from domestic and industrial wastes undergoing aerobic digestion has approached limits of about 40–60% volatile solids reduction.

Little change in the cellulosic material occurs during aerobic digestion. The nitrogenous matter is the most susceptible to degradation. As aerobic oxidation proceeds, cellular nitrogen is broken down and released to solution in the form of ammonia and the ammonia is further oxidized. Nitrates in excess of 900 mg/l were obtained in one study (44). Parameters related to nitrification can be used to note the progress of aerobic digestion since the oxygen supply is in excess of the demand and primarily is used for mixing. These parameters are a pH and alkalinity decrease and a nitrate increase. The pH during aerobic digestion can be in the 5–6 range depending upon degree of nitrification and the buffer capacity of the units.

Drying of aerobically digested sludge solids on sand beds can be an efficient method of disposal if the treatment facility is operated properly and the sludge drying beds designed adequately. Well-conditioned aerobically digested sludges have been shown to drain more rapidly than anaerobically digested sludges (45). The drainability of sludges digested for short periods (5 days) was poorer than nondigested samples, but for digestion periods greater than 10 days, the drainability was improved. The digested sludges dried satisfactorily with no development of obnoxious odors. The BOD of the supernatant liquid generally was below 80 mg/l. COD values ranged from 360 to 670 mg/l for the various times and temperatures studied with the longer detention times generally giving the better results. The BOD of the supernatant may have been due to nitrification.

The effluent from full-scale field aerobic digesters has been evaluated. The average BOD of the decant liquor was 10 mg/l (42). Aerobically digested sludge spread at a thickness of 6 inches dried to a spadable condition in 5 days with a terminal thickness of 1 inch.

The residual solids requiring disposal from aerobic digestion systems will equal the sum of inert solids in the raw waste, residual nonbiodegradable organic solids, and any remaining unmetabolized influent organic solids. The latter fraction should be very small. The residual solids have been shown to be very stable with no odor and readily filterable. The effluent liquid is well oxidized and should not be in need of further treatment before discharge to the environment. The BOD of this liquid is due more to nitrification than to any unoxidized organic matter.

Aerobic digestion offers advantages, when compared to anaerobic digestion, which include the lack of any need for insulation, added heat, or special covers, the ability to handle low waste sludge concentrations eliminating the need for sludge thickening devices, and the production of solid residue and liquid effluent with a low oxygen demand. The disadvantages include the cost of supplying air, the lack of any usable end product such as methane which is produced in anaerobic digestion, and the somewhat larger sludge volumes that will be produced in aerobic digestion. In an anaerobic digester less synthesis takes place since considerable energy is unavailable to the microorganisms.

Application to Agricultural Wastes

The greatest application of aerobic digestion has been with the waste solids from municipal waste water treatment plants. There have, however, been a few aerobic digestion studies that involved agricultural wastes. Aerobic digestion was effective for the oxidation of sludges developed from whey wastes (46). Volatile solids reductions of 18–22% were obtained with 72-hr aeration and COD reductions of 10–15% were obtained with 24-hr aeration. The aerobic digestion of waste activated sludge produced from a domestic waste comprised of one-third pretreated meatpacking waste and two-thirds domestic sewage was studied. Reductions in volatile solids of 14–20%, 34–36%, and 39–53% at 5, 15, and 30 days detention time, respectively, at a temperature of 20°C were reported (47).

Figure 7.15 indicates the use of surface aerators in an aerobic digestion lagoon used for the excess solids from the aerobic biological treatment of food processing waste.

The fundamentals of aerobic biological treatment, as described in this section can be applied to other situations where aerobic digestion can be used with agricultural wastes. The economics of the situation such as size, mixing requirement, and aeration requirements would determine the value of aerobic digestion systems for specific agricultural wastes.

Fig. 7.15. Aerobic digestion of excess biological solids from an aerobic system treating food processing wastes.

The Oxidation Ditch

Introduction

Oxidation ditches are feasible aerobic biological treatment systems for agricultural wastes whether they are liquid such as food processing wastes or slurries such as livestock wastes. The largest number of ditches used for agricultural wastes have been installed in confined animal production facilities. The majority of the ditches have been utilized for swine operations although they have been incorporated in dairy and beef cattle as well as in poultry facilities. The reasons for the use of oxidation ditches in livestock confinement operations include: (a) odor prevention and control, (b) labor saving, (c) waste treatment, and (d) ease of incorporation in confinement housing.

Both in-house and external oxidation ditches are possible. Oxidation ditches are shallow and can be incorporated in most confined livestock facilities. External oxidation ditches are similar to aerated lagoons. The choice between an aerated lagoon and an external oxidation ditch will be related to the shape and quantity of available land and to the oxygenation and power requirements of the surface aerators. For specific waste and performance requirements, external oxidation ditches will require more land than aerated lagoons since the ditches are shallower.

With in-house oxidation ditches, labor will be saved by avoiding moving the animal wastes to a treatment system and in the ability to use pumps to transport the liquid wastes from the ditch. Because of the evaporation by the rotor, the total waste volume will be less than if the wastes were not so aerated. The effect of evaporation loss will be greater for low volume, low moisture wastes. In these systems water may have to be added to keep the water level in the ditch at proper levels.

Because the oxidation ditch is an aerobic biological treatment system, the odors associated with anaerobic conditions will not occur. The system is not odor-free but the odors that exist are rarely considered objectionable. Ammonia emission will occur if the oxygen supply is inadequate or if nitrification does not exist. An earthy smell common to aerobic biological systems will exist when the aeration capacity is sufficient to meet both nitrogenous and carbonaceous oxygen demand. Both of these odors will not be detected after being exhausted from a confinement building or if an external oxidation ditch is used.

Operational problems occur but with proper maintenance and an understanding of the microbial processes occurring in the ditch, they can be held to a minimum. The resulting waste management scheme is low in odors, low in manual labor, and is convenient for the operator. Effluent from a continuous overflow oxidation ditch must be kept in aerated holding units or disposed of in a short time. Residual organic content and microbial activity in this effluent can cause anaerobic conditions and odors if the effluent is not aerated while being held prior to disposal.

The oxidation ditch was developed by the Institute of Public Health Engineering in the Netherlands. The basic process is known by a variety of names such as the aeration ditch or the Pasveer ditch after the name of an early advocate of the process. The system is an aerobic biological waste treatment system with a long liquid detention time and adequate mixing. The basic system is economical for small communities or industries and requires a minimum of supervision. The characteristics of the oxidation ditch include low capital cost, ease of operation, and minimum maintenance.

The oxidation ditch is similar to the aerated lagoon in that a surface aerator is used to supply the necessary oxygen. The oxidation ditch has a long, narrow shape with a center partition and can be modified to fit into available space. The key components of the system are a continuous open channel and a surface aeration rotor to mix the ditch contents as well as to supply the oxygen (Fig. 7.16). Depending on specific wastes and modes of operation, other components may be added and many alternatives exist for further treatment or ultimate disposal (Fig. 7.17). The components will vary de-

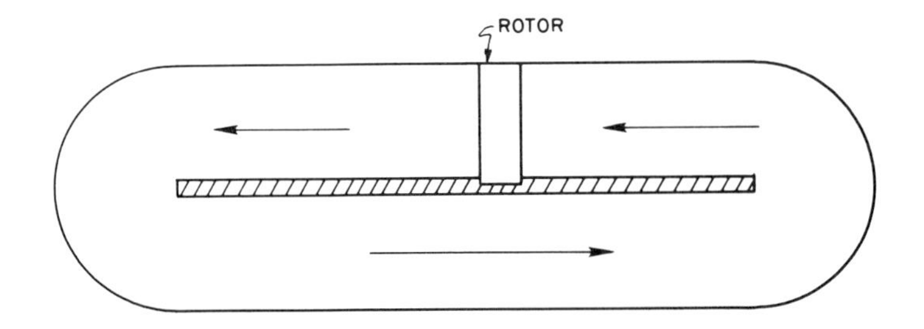

Fig. 7.16. Diagram of the basic oxidation ditch.

pending upon the objectives of a particular system. By a suitable choice of conditions it is possible to obtain both purification of the waste water and stabilization of the influent and synthesized solids. Ditches for animal wastes can be used to provide partial treatment under slatted floors with additional treatment in an external oxidation ditch or other aerobic unit. They can also be used to provide containment and partial treatment until

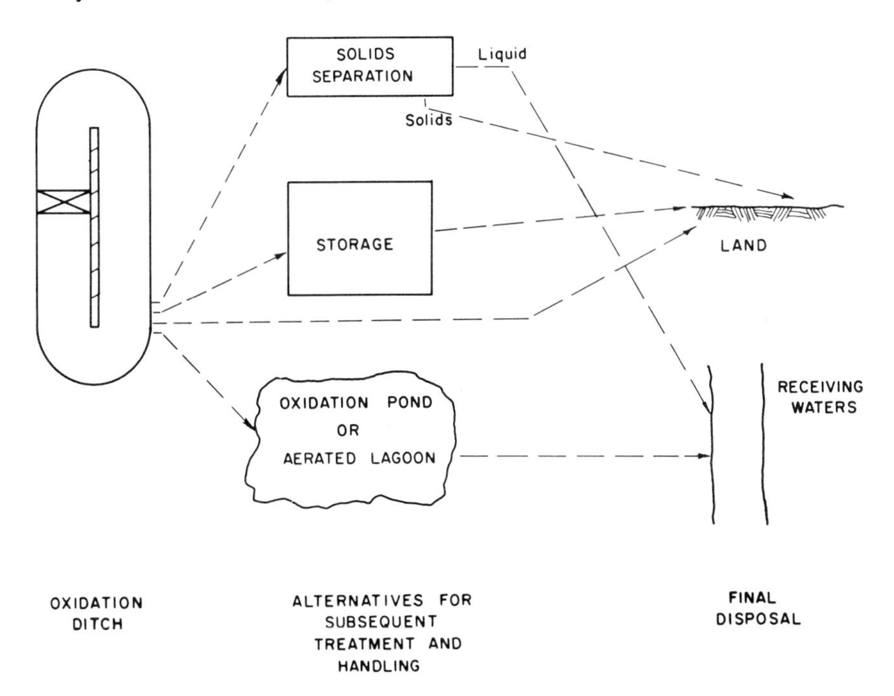

Fig. 7.17. Integration of an oxidation ditch with additional treatment and disposal alternatives.

the waste can be removed and integrated with crop and land management methods. This aspect of storage reduces the possibility of surface water and groundwater pollution since the pollutional strength of the wastes, i.e., COD, BOD, solids, and nitrogen, can be reduced prior to land disposal.

Attempts to have construction costs as low as possible have caused some oxidation ditches to be designed both as an aeration unit and as a sludge separation unit. This is accomplished by cyclic operation of the rotor. While the rotor is off, the mixed liquor solids can settle, and a portion of the clarified liquid is removed. This procedure can be effective with municipal wastes or other high volume, low strength wastes. It also is effective with animal wastes except when the mixed liquor in these ditches reaches high levels, from 1 to 5 +% total solids, and solids separation by sedimentation does not occur. The construction of ditches may vary from unlined earth to concrete walls depending upon local soil conditions and whether the ditch is of an external or in-house type. A butyl rubber lining has been used for an oxidation ditch treating hog wastes (48).

Untreated wastes can be added directly to the ditch. Experience has indicated that prior treatment such as solids separation is not needed. Depending upon the quality of the untreated wastes, the effluent may be of suitable quality for discharge to surface waters. Such may be the case for municipal wastes or for some food processing wastes, especially if the solids in the effluent are removed before the effluent is discharged. The effluent from oxidation ditches treating livestock wastes is not suitable for discharge to surface waters because of the color, mineral salts, and residual BOD in the effluent. The effluent is suitable for land disposal. Odors will be a minimum with the process if sufficient oxygen is supplied by the rotor.

Experience with municipal wastes as well as with animal wastes has shown that BOD reductions of 80–90% can be obtained. The advantages of the process include no odors, a treated effluent, and oxidation of waste solids. The large volume of the ditch, as compared to the influent flow, and the long detention time make the oxidation ditch less sensitive to changes in the quality and quantity of the untreated waste.

The low capital cost of the oxidation ditch is attractive for many municipal and industrial waste operations. Over 400 of these systems are in operation treating animal waste waters with equal numbers treating municipal wastes in Europe and the United States. The process has value for seasonal and intermittent wastes and where there will be minimum operation and maintenance.

Process Design

Two types of oxidation ditches are in common use: (a) the in-house ditch commonly used in livestock confinement facilities and (b) the external ditch that is separate from the waste production site. The in-house ditch is located beneath self-cleaning slotted floors of swine and cattle confinement buildings or below cages in confined poultry buildings. The majority of the in-house ditches have been used for swine but the process has been used with beef cattle, young dairy cattle, and poultry. The in-house ditch takes advantage of the continuous and uniform waste-loading to the unit, the controlled temperature in the confinement building, and the continuous mixing and aeration to produce a near ideal biological waste treatment process. The in-house ditch also eliminates the equipment that otherwise would be required to transfer the livestock wastes from the source to the ditch since the animals continuously push the waste through the slots to the ditch.

The external ditch can function as well as the in-house ditch and is the preferred method for municipal and other wastes where the treatment system cannot be installed in the waste production area. The external ditch is uncovered and exposed to ambient temperatures. In addition the wastes may be added on an intermittent basis as they are pumped from the point of production. As a result of variable temperatures and loading, operational problems such as shock loads, foaming, and variable efficiencies may result. If the input is small compared to the ditch volume, these variations may not be significant.

Since animal wastes are added directly to the oxidation ditch under confinement operations, considerably smaller water volumes are used compared to systems in which water is used to flush the waste to the treatment or disposal site. Because treatment facilities are designed on the basis of flow rate as well as BOD loading, the smaller water usage can mean smaller treatment systems. Because oxygen demand is related to BOD loading, smaller water usage will not decrease the size of the required rotos. Animal wastes have higher BOD loading than many other wastes and rotor capacity per unit of flow will normally be greater for these wastes than for food processing or municipal wastes.

In an oxidation ditch, the volume of the ditch per animal or per case of food packed is of less importance than the oxygenation capacity of the rotor. The rotor must supply the required oxygen to meet the oxygen demand of the wastes entering the ditch. If adequate oxygen is not supplied, the ditch becomes oxygen-limited, poorer process efficiencies result, and odors can occur.

Greater ditch volumes per animal or case of food packed will dilute the wastes but the same quantity of oxygen will be required for a given quantity of animal or food waste regardless of the dilution. With animal wastes, the volume of the ditch plays a role by providing space to accumulate the nonbiodegradable fraction remaining after oxidation. A larger ditch volume will permit the ditch to have greater storage capacity and perhaps lower solids concentration which in turn will permit more efficient oxygen transfer relationships since the oxygen transfer efficiency decreases as the total solids concentration increases (Fig. 7.18) (49). The figure also illustrates the need for routine maintenance of the rotor. The rotor is less efficient when it is not operating satisfactorily such as when belts are slipping.

For in-house oxidation ditches, the size of the ditch is controlled to a large extent by being able to fit the ditch into a confinement building. Ditch volume is important since it must be large enough to permit adequate detention time for treatment, stabilization, and the storage that is desired. The degree of solids stabilization and waste treatment purification that occurs is a function of the loading rate of the ditch. With low volume wastes such as animal wastes, long detention times are the rule.

Loading

The ditch design should be in terms of pounds of BOD per quantity of active organisms to be maintained in the ditch. Suspended solids commonly are used as a best estimate of active organisms. Although data on the characteristics for various animal and food processing operations are available in the literature, it is desirable to determine the actual BOD load for a given agricultural operation when designing an oxidation ditch for

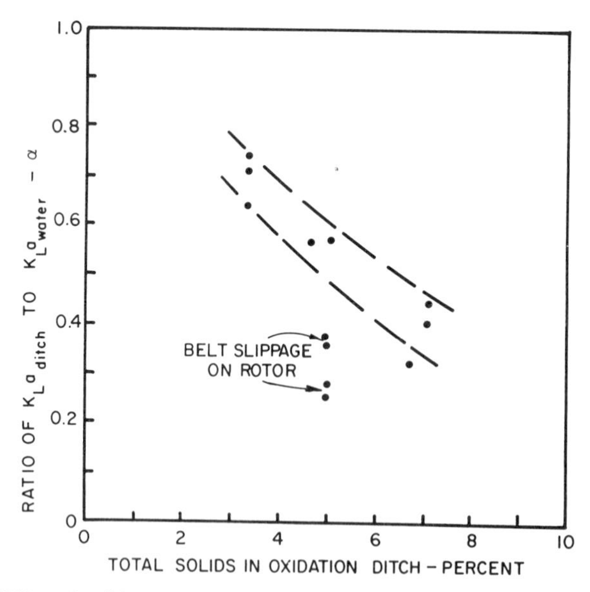

Fig. 7.18. Effect of solids content on the value of α in an oxidation ditch treating poultry wastes (49).

specific wastes. Because operational characteristics of oxidation ditches for agricultural wastes have only recently been studied, little data is available to indicate the proper suspended solids level to be maintained for different wastes. Consequently, many ditches for these wastes are designed on the basis of pounds of BOD or number of animals per liquid volume. This relationship is empirical and unrelated to the microbiological processes that are fundamental to the success of the process. It is hoped that current and future research will permit the ditch loading to be put on a more fundamental basis.

The following loading rates have been suggested for animal wastes to provide adequate treatment: 55 ft^3 per 800 lb fattening calf, 80 ft^3 per calf and cow (50), and 6–10 ft^3 per finishing hog (51). Design recommendations for in-house oxidation ditches for a variety of livestock wastes are noted in Table 7.4 (52). Additional investigations are necessary to verify the recommendation for dairy and beef cattle, sheep, and poultry since the majority of work has been done with swine wastes. The data in Table 7.4 provide a reasonable estimate of equipment and power requirements for these units. The rotor and ditch volume requirements per animal are based on the type of rotors currently in use. Future improvements in rotors will alter these relationships should better oxygen transfer efficiencies result. The data in Table 7.4 have been corrected to 80% of the oxygenation capacity of the rotor in tap water, i.e., an alpha value of 0.8. Such values may or may not represent the actual conditions in an oxidation ditch. The oxygenation capacity of a rotor in an actual system is most important and an estimate of the transfer relationships, i.e., α and β, in a specific waste slurry must be known to correctly size a rotor.

For dilute wastes, such as food processing or municipal wastes with a 24-hr retention time, a loading rate of about 10–15 lb BOD_5/1000 ft^3/day is recommended. With a mixed liquor suspended solids concentration of about 3000–4000 mg/l, the food to organism relationship is in the range of 0.03–0.06 lb BOD/lb MLVSS/day. Because of the low food to organism ratio, endogenous respiration is predominant. The net growth of microbial solids is low and active organisms are in low concentration. Data from experiments at loadings of 11–156 lb BOD/1000 ft^3/day have shown that BOD reductions of 90–87% were obtained (53). The ditch loading is not restricted to low levels. However, the effluent quality will be less at high loadings and may not be acceptable if discharge to a surface stream is contemplated. Low loadings will occur with animal wastes because of the high ditch volumes provided per animal.

In oxidation ditches treating animal wastes, the mixed liquor total solids concentration can be high, above 1%. The concentration can be higher depending on the type of wastes, ditch volume per unit of waste, and whether

Table 7.4

Design Recommendations for In-The-Building Oxidation Ditches[a]

Animal	Weight (lb per unit)	Daily BOD_5 (lb per unit)[b]	Daily required oxygenation capacity (lb per unit)[c]	Number of animals per foot of rotor (units per foot)[d]	Ditch volume (ft^3 per unit)[e]	Daily power requirement (kWh per unit)[f]	Daily cost (cents per unit)[g]
Swine							
Sow with litter	375	0.79	1.58	16	23.7	0.83	1.66
Growing pig	65	0.14	0.28	91	4.2	0.15	0.30
Finishing hog	150	0.32	0.62	41	9.6	0.33	0.66
Dairy cattle							
Dairy cow	1300	2.21	4.42	6	66	2.33	4.66
Beef cattle							
Beef feeder	900	1.35	2.70	10	40	1.42	2.84
Sheep							
Sheep feeder	75	0.053	0.11	230	1.6	0.06	0.12
Poultry							
Laying hen	4.5	0.0198	0.0396	650	0.6	0.021	0.042

[a] From reference 52 (Jones, D. D., Day, D. L., and Dale, A. C. "Aerobic Treatment of Livestock Wastes," Bull. No. 737. Univ. of Illinois, Urbana, Agricultural Experiment Station in cooperation with Purdue Univ., Revised Bulletin, April 1971).

[b] Use specific production data when known.

[c] Twice the daily BOD_5.

[d] Based on 25.5 lb of O_2 per ft of rotor per day.

[e] Based on 30 ft^3 per lb of daily BOD_5.

[f] Based on 1.9 lb O_2 per kWh.

[g] Based on electricity at 2 cents per kWh.

the ditch is the overflow or storage type. Ditches that have high evaporation have had total solids concentrations in the 5–8% range before being pumped out for disposal. Because of the high nonbiodegradable volatile solids in such systems the food to mass ratio, measured by soluble BOD/MLVSS would be quite low, below 0.01.

The detention times in oxidation ditches range from one or more days for continuous systems treating dilute wastes to months for essentially storage systems treating animal wastes. Because of its large capacity relative to in influent flow, the oxidation ditch acts as a surge tank to level out peaks of loading.

Oxygenation Relationships

Aeration is accomplished by a surface rotor rotating about a horizontal shaft (Fig. 7.11). Except for a portion of the rotor blade immersed in the liquid, the entire mechanism is above the liquid. Usual immersion depths are from 2 to 8 inches. Common rotors consist of iron plates or angles attached perpendicular to a rotating shaft or a circular cage with rectangular pieces of metal attached to the periphery. The amount of oxygen required depends upon the microbial oxygen demand. Design relationships for rotors require an estimate of the effectiveness of transferring oxygen to a slurry over a range of rotor speeds, diameters and configurations, liquid

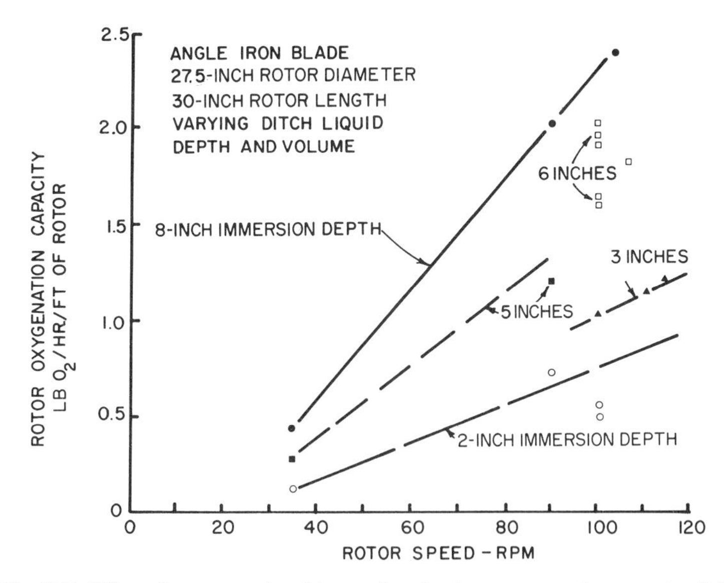

Fig. 7.19. Effect of rotor speed and immersion depth on oxygenation capacity (54).

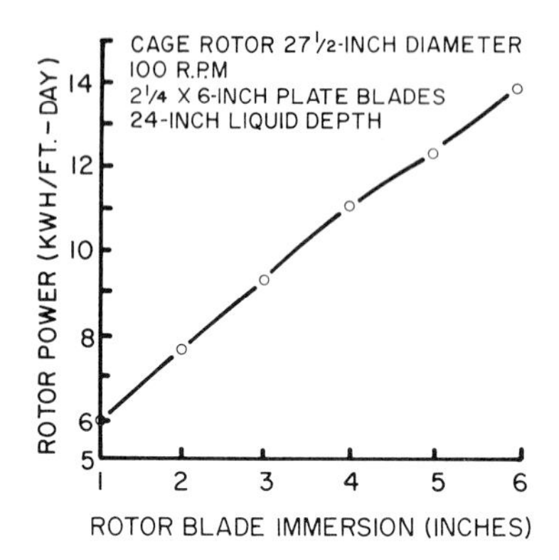

Fig. 7.20. Variation of power requirements of a cage rotor as a function of rotor immersion depth (52).

waste properties, and desired dissolved oxygen levels in the wastes. The oxygen transfer coefficient, K_La, and the pounds of oxygen transferred to the ditch contents increase as the rotor immersion increases (Fig. 7.19) (54) while the oxygen transferred per kilowatt hour decreases as the immersion increases. The power consumption for a given rotor increases as immersion of the rotor blades increases (Fig. 7.20). Similarly the pounds of oxygen transferred increases as the rotor speed increases although the oxygen transferred per kilowatt hour decreases with an increase in speed. Multiple rotors can be installed in ditches where oxygen demand is high. Oxygen uptake rates of about 10 mg/l/hr (50) and 15 mg/l/hr at 40°F (55) have been reported in oxidation ditches treating beef cattle wastes. Uptake rates of 25–50 mg/l/hr have been reported as common in ditches treating swine wastes (50). A study of the oxidation ditch treatment of poultry waste obtained uptake rates in the 5–10 mg/l/hr range at 10°–12°C and up to 50–70 mg/l/hr at 17°–20°C (49). The uptake rates are a function of the BOD loading rate and the temperature of the mixed liquor. Oxygen uptake rates will be greater if nitrification occurs.

A comparison of commercial and privately manufactured rotors indicated that the oxygenation capacity of these rotors ranged from 1.3 to 2.1 lb oxygen transferred/hour/foot of rotor at 6-inch immersion depths and 100 rpm (50). The variations were caused by changes in liquid volume. The oxygenation capacities were not linear as the immersion depth and rotor speed increased. For example, the oxygen transferred at 2-, 5-, and 8-inch immersion depth for an angle iron rotor running at 100 rpm was 0.8, 1.5, and 2.3 lb oxygen/hr/ft of rotor. The above oxygenation capacities of the rotors were determined under standard conditions using tap water with zero dissolved oxygen. The conditions existing under operating conditions must be known or reasonably estimated and the oxygenation capacities corrected accordingly to account for decreased transfer relationships under process conditions. Available information indicates that K_La values are a function of the total solids in the systems. For livestock wastes, α values of 0.5 and less, and β values less than 1.0 have been recorded at high solids concentrations. The oxygenation capacity of a rotor operating in a livestock building was observed to be less than half that expected from a rotor operating with municipal wastes (56). The above data indicate that the oxygenation capacity of a rotor functioning with livestock wastes can be considerably less than that of a rotor operating in less concentrated wastes and less than that expressed by equipment manufacturers under standard conditions.

Because of the low loading and long liquid detention time in the ditch, nitrification can take place if dissolved oxygen concentrations above 2.0 mg/l are maintained. The oxygen demand of nitrification increases the total oxygen demand of the system. The rotor must be of sufficient size to supply both the carbonaceous and nitrogenous demand. In this way the oxidation ditch differs from an aerated lagoon since in the latter process nitrification rarely occurs. A usable rule of thumb is that the oxygenation capacity should be twice the BOD_5 loading to the ditch. Although it may not be realistic to design for the maximum possible oxygen demand including nitrification, this demand can occur in units with a long detention time.

Experience with a full-scale oxidation ditch noted evidence of both nitrification and denitrification in the unit. With adequate oxygen, up to 50 mg/l of nitrate nitrogen was observed in the effluent from a unit treating municipal wastes (57). If nitrification is not required, the oxygenation capacity of a rotor can be decreased by reducing the rotor immersion and/or hours of aeration so that the oxygen transferred to the unit matches the carbonaceous oxygen demand while not supplying adequate oxygen for the entire nitrogenous demand. Both of these approaches will promote denitrification of the nitrites or nitrates that do result.

If the oxygenation to load ratio is greater than five, complete nitrification of a municipal waste water will occur (58). When the oxygenation capacity to load ratio is greater than two, nitrification can be expected. If the ratio falls to less than two, denitrification can occur in the mixed liquor (59). Although nitrification did occur in an oxidation ditch treating hog wastes, over 90% of the nitrogen in the system disappeared (60). The phenomenon has been attributed to denitrification in the mixed liquor under aerated conditions since ammonia stripping at the pH levels in the ditch, 6.5–7, was not significant. The ditch had effluent nitrite concentrations from 0 to 700 mg/l, nitrate concentrations from 0 to 700 mg/l, and ammonia concentrations of 5–400 mg/l during continuous operation.

In some livestock systems where land disposal of the wastes is to be practiced, the purpose of the oxidation ditch is to reduce the pollution potential of the wastes prior to disposal and to minimize odors. High effluent quality is not necessary. The quantity of oxygen and therefore rotor selection solely to minimize odors in these systems is not known but is under study.

The rotor is designed to provide adequate mixing as well as to aerate the ditch contents (Fig. 7.21). As a general rule, adequate velocity and oxygenation will occur when the immersion of the rotor is approximately one-fourth to one-third the liquid depth in the ditch. Oxidation ditches are designed as shallow units having depths of from 15 to 30 inches. When rotor immersion is less than that suggested above, lower velocities will result and solids sedimentation can occur. Velocities of 1.0 to 1.5 ft/sec should be maintained in the ditch to minimize settling of the solids. The aeration process takes place primarily within the immediate vicinity of the rotor. Additional aeration takes place during the movement around the ditch but this aeration is minor compared to that that occurs at the rotor. The amount of aeration at the rotor depends upon the degree of turbulence which is a function of the power and immersion depth of the rotor. The dissolved oxygen of the mixed liquor will decrease from a maximum value immediately after the rotor. Depending upon the oxygen demand of the mixed liquor

Fig. 7.21. Aeration and turbulence caused by rotors in an oxidation ditch (courtesy of Lakeside Engineering Corp.).

and the oxygen supplied by the rotor, the dissolved oxygen concentration can be reduced to zero prior to the liquid again passing through the rotor.

In lightly loaded systems, such as the oxidation ditch, the power requirements needed to keep the solids in suspension may exceed the power needed for oxygen transfer. This leads to a high power use per BOD removed.

Power usage and oxygen transfer of the rotor are functions of the rotor immersion depth. A uniform water depth is necessary for best results. Routine attention will be required to maintain a constant overflow if there is excess liquid or to add water routinely if evaporation exceeds input. The use of a floating rotor would be of benefit to handle liquid level variations and to permit certain ditches to be used as aerated storage units without overflow. No floating rotors for oxidation ditches are operating as yet.

Operational Characteristics

Reports from Europe and Canada have indicated that the oxidation ditch will operate successfully in cold weather. The rotor and surrounding area should be enclosed and heated where necessary. The ditch may have an ice cover in parts and a decrease in efficiency will occur. However, the long detention time in the ditch can partially compensate for the lower efficiency per organism in the colder weather.

Evaporation losses will reduce the quantity of effluent to be handled from an oxidation ditch. With high volume wastes such as municipal or food processing wastes, the effect of the evaporation may be negligible. Livestock wastes, especially sheep and poultry wastes, have a low moisture content and do not contribute large volumes of water to the ditch. In these cases evaporation can be significant. Experiments using an oxidation ditch with poultry and beef wastes have indicated that tap water may have to be added on a periodic basis to the ditches to keep the water level constant. The quantity of evaporation that takes place in oxidation ditches is not well established. Evaporation from a ditch will add to the moisture in animal confinement operations. Environmental control and ventilation in these buildings must take the increased moisture load into account when the buildings are designed.

There are certain problems associated with oxidation ditches. These include foaming, humidity, odors, and rotor maintenance. Foaming can be a nuisance and in some cases objectionable. Even in a properly operating ditch, some foam is present because of the agitation by the rotor and surface active agents resulting from degradation of the organic wastes. Antifoam agents such as kerosene or proprietary chemicals can relieve excess foam problems. Excess foam usually occurs during start-up conditions and when the ditch is overloaded. When the ditch is designed and operated properly, and when there is an adequate microbial population in the ditch, foaming problems are a minimum. The total depth of the ditch should include free-board of at least a foot to allow for some foam during start-up conditions.

Problems with bearings, improperly designed rotors, motors, and belts have caused many problems with earlier rotors and drive units. The experience gained from these problems has been incorporated into current equipment. Equipment breakdown is less frequent. It is important that the maintenance instructions provided by the equipment manufacturer be followed for maximum equipment life.

Typical odors from the ditch are related to its operation and to microbial fundamentals. Septic or anaerobic odors will result from an inadequate oxygen supply. An earthy smell is normal with a well operating ditch. During start-up operations, ammonia will be generated as the organic nitrogen is oxidized to ammonia. The agitation by the rotor will release a portion of the ammonia to the atmosphere. The ammonia odor will disappear when nitrification occurs. If nitrification does not occur either because of inadequate oxygen or a lack of nitrifying organisms, ammonia production and release will persist. In confined livestock buildings under these conditions, the ammonia concentration in the atmosphere can be a nuisance to humans who work in the facility.

Confined livestock facilities should be well ventilated when an oxidation ditch rotor is started up in a ditch that has been collecting waste for a period of time. The gases released during the initial aeration period can be a mixture of carbon dioxide, methane, ammonia, as well as other volatile compounds. The start-up of an oxidation ditch should take place with a minimum of waste in the ditch but with a sufficient quantity of active organisms as seed to ensure rapid degradation of the untreated wastes. After a ditch has been in operation for some time, it will become necessary to remove sludge to avoid excessive amounts in the ditch, maintain better oxygen transfer, and have adequate mixing. A constant overflow or intermittent solids removal can be used.

The BOD in the effluent of an oxidation ditch is related to the suspended solids in the effluent. In the effluent from an oxidation ditch treating beef cattle, the ratio of effluent BOD to effluent suspended solids was 0.17 on the first day of operation but increased to about 0.90 after 4 months (55). Toward the end of the experiment the BOD_5 was in the form of microbial cells. When effluent oxidation ditch suspended solids have been removed by sedimentation, BOD removals in the range of 95–97% have been reported.

Application to Agricultural Wastes

The oxidation ditch has been applied successfully to many liquid agricultural wastes. The application has been greatest with livestock wastes but also has occurred with food processing wastes. Tomato wastes having an influent BOD of 800 mg/l were treated in an oxidation ditch. BOD removals of 89% were obtained with a liquid detention time of 2 days (61). An oxidation ditch used to treat seasonal cannery wastes, municipal sewage, animal slaughtering wastes, and wastes from a butter factory exhibited a BOD removal of about 30 lb BOD/day/ft of rotor length. The power requirement was 0.4–0.5 kWh/lb BOD removed (62).

Hog waste treatment has been accomplished in a full-scale oxidation ditch with an oxygenation capacity to load ratio of 1.8 (60). Foaming was excessive at the start. Effluent BOD and COD values after settling were about 15 mg/l and 600 mg/l, respectively, although the raw pig wastes had BOD and COD concentrations of about 30,000 and 80,000 mg/l, respectively. Sludge accumulation was about 18% of the total BOD loading per day. The ditch was of reinforced concrete and included a final sedimentation tank. The annual operating cost was about 10–15% of the dollars invested.

The solids in another study with hog wastes were noted to be of two types, the usual sludge particles and a fibrous cellulose material which was about 10% of the total solids (48). Velocities of 1–1.5 ft/sec kept the fibrous material in suspension although they settled rapidly if the flow stopped.

When an oxidation ditch was operated as a nonoverflow unit with beef cattle wastes, the COD increased at a linear rate reaching about 50,000 mg/l after 160 days. The BOD remained low and relatively constant ranging from 900 to 2500 mg/l during this period. A high cellulose ration was fed to the animals and the COD increase was caused by the nonbiodegradable fraction in the waste. In one experiment, the total solids reached 11.8% and in another, 7.3% before the solids had to be removed. The solids entering the ditch averaged 5.1 lb/day/head and 34% solids reduction was obtained in 143 days of treatment. The rotor operating cost was estimated at about 2.5 cents/1000 lb animal/day or about 1 cent/lb gain/animal/day (55). A similar full-scale study with beef cattle showed that the BOD of the mixed liquor was less than 1000 mg/l with supernatant BOD values usually less than 200 mg/l (50).

If a batch or storage type oxidation ditch is used, the COD and solids will increase almost linearly until solids are removed. BOD values and volatile solids levels will remain low and constant or slightly increase since the organic demand and the solids are being oxidized. In an oxidation ditch with an overflow, the COD and solids levels will build up to equilibrium levels consistent with the volume and dilution of the ditch.

A comparison was made of an oxidation ditch and an under-the-cage storage pit for the wastes from laying hens (63). The lack of dissolved oxygen in the ditch indicated that the rotor was not able to transfer enough oxygen to meet the demand. In spite of this, there was a definite decrease in the level of objectionable odors. Egg production and feed conversion with both systems were comparable and mortality was slightly less in the room with the oxidation ditches. Foaming was a problem and was thought to be partially caused by shock loadings that occurred with the intermittent cleaning of dropping boards in this operation.

An oxidation ditch has been used to maintain aerobic storage conditions for poultry wastes from 936 birds (64). Storage capacity of 1.8 ft^3/bird resulted in emptying the oxidation ditch twice annually. The slurry was odorless and easily removed by vacuum tanker. The ditch ran continuously until the rotor was not able to move the ditch contents sufficiently to keep the solids in suspension. At the time of emptying, the ditch had a total solids content of 5.2%. Evaporation was sufficient to permit an overflow of only 1400 gal of liquid in 9 months. The temperature in the ditch ranged from 7° to 11°C. Foaming, floating feathers, and failure of rotor bearings were the main problems encountered.

An oxidation ditch treating poultry wastes can be used either as a continuous input, constant overflow unit, or as a batch, no overflow aerated holding tank. When an oxidation ditch was used directly beneath laying hens, evaporation resulting from the normal ventilation system and the rotor action was more than adequate to balance the water input each day (49). No overflow resulted and the system functioned as a nonoverflow, aerated holding system for over a 9-month period. Water had to be added to the system each day to provide the proper immersion depth for the rotor. In this system treating 250 birds, the average evaporation rate was about 7.5 gal/day. Because of evaporation, the final weight of liquid removed when the ditch was cleaned out was only slightly more (29%) than the weight of manure added to the ditch.

Approximately 53% total solids, 62% volatile solids, 33% BOD, and 31% total nitrogen loss occurred in this system over the 9 months of operation. Because the rotor provided adequate oxygen to meet the oxygen demand, no odors were generated except ammonia which persisted until nitrification occurred. Under normal operation, the odor level was comparable to that found in offices and homes. The solids increased linearly, since there was no overflow, until at cleanout the ditch contained 8.4% total solids (Fig. 7.22). The α value for the rotor decreased in what appeared to be a linear fashion as the solids increased (Fig. 7.19). At 7% total solids, α was about 0.4.

Rotating Biological Contactor

Introduction

The rotating biological contactor (RBC) is analogous to a rotating trickling filter. The disks, connected to a common shaft and spaced a short

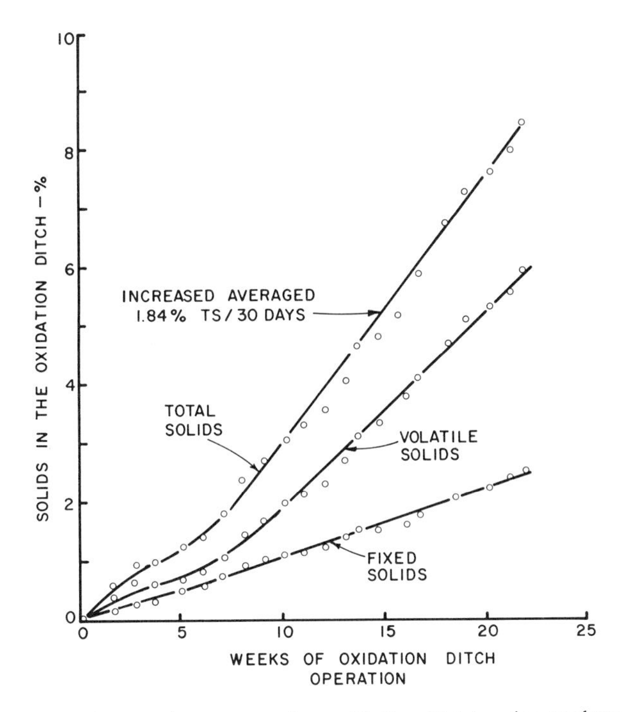

Fig. 7.22. Solids increase in a nonoverflow oxidation ditch treating poultry wastes (49).

distance from each other, rotate in a semicircular tank through which the liquid waste flows. The large diameter, light weight plastic disks are submerged about 35–45% in the tank. A biological film develops on the disk surface in a manner similar to that developed on the surface of a trickling filter. When submerged in the waste water, the microbial film absorbs organic matter. In rotation the discs carry a layer of waste water into the atmosphere where it trickles down the disk surfaces and the microbial film absorbs oxygen. Organisms on the disk surface use the oxygen and the organic matter for growth, reducing the oxygen demand of the waste water. A schematic cross-section of a typical unit is illustrated in Fig. 7.23 and a large scale unit is shown in Fig. 7.24. Disk speed can be varied and generally is in the range of 2–5 rpm.

The rotating disks serve to provide support of a fixed biological growth, contact of the growths with the waste water, and aeration of the waste water. The growths continuously shear off the disks. The mixing action of the system keeps these solids in suspension until the treated waste water carries them out of the unit. The aerated mixed liquor in the unit also serves to treat the waste water. The discharged waste water contains metabolic products of degradation and the excess solids synthesized in the unit. The

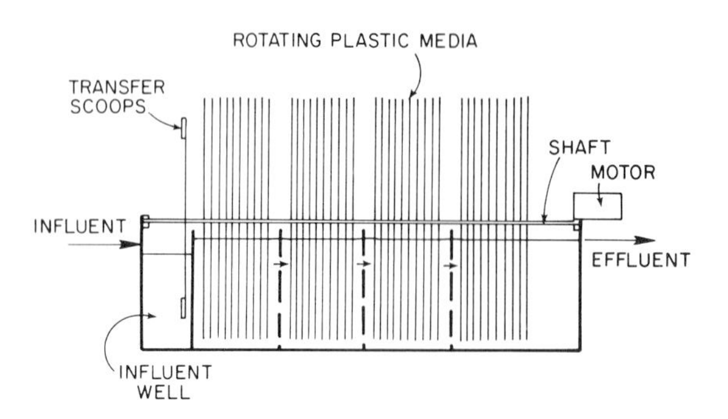

Fig. 7.23. Schematic cross-section of a rotating biological contactor.

solids settle well in a final clarifier. The process can be used independently or in conjunction with other treatment processes.

The system has been used in Europe since 1960 (65, 66) with a large number of commercial units installed in western Europe. Research and development on the process in the United States has been conducted since about 1965. Several full-scale units have been installed for municipal and industrial wastes. The process is currently under investigation with poultry, swine, and meatpacking wastes. When a number of sequential stages are used, successively increasing BOD removals result with the removal following a first-order pattern (Fig. 7.25) (67). Depending upon the operating method and loading, a range of processes such as high rate and low rate systems, roughing systems, and including nitrification and denitrification can be achieved. Ferric chloride and ferrous sulfate have been added to raw wastes treated in these units to remove phosphorus (68). The disks can be mounted in equal size stages so that the system functions as a plug flow or series type reactor.

The fundamentals of aerobic biological waste treatment can be applied to this process to estimate sludge production and oxygen inputs and demands. The organic loading rate should be based on the pound of BOD applied per pound of microbial solids in the system. Estimates of microbial solids on the rotating disks are difficult and empirical relationships based upon pounds of BOD applied per surface area of disk or per unit liquid volume in the unit are useful as design relationships. Design parameters based upon hydraulic detention time or liquid flow per surface area of the disks are likely to be satisfactory only with wastes of consistent quality and quantity. These latter parameters are not fundamental to the biological treatment processes and are difficult to extrapolate to wastes of different characteristics. Although it has been rarely tried, it may be possible to return a portion of the solids separated from the effluent to the unit to increase the concentration of the active organisms in the mixed liquor. Such return sludge could increase the flexibility of the process and its ability to function under variable loads.

Application

In cold climates, freezing of the disk surfaces is a possibility. Contact with outside air can be limited by a simple covering with closed side walls. Insulation is not required to avoid freezing because the warm waste water should keep the ambient air around the unit above freezing if it is covered. As expected, the temperature of the liquid in the unit will affect the efficiency of the microorganisms and be reflected in the treatment efficiency. If the unit is enclosed, adequate ventilation through the enclosure should be assured to provide adequate oxygen for the microorganisms.

A laboratory rotating biological contactor was used to study the process with wastes comparable to that of domestic sewage (69). The unit was loaded at a low rate, liquid detention time was 5 hr, and 98.6% BOD removal was achieved. The carbonaceous demand was satisfied in one-fifth of the unit

Fig. 7.24. Full-scale rotating biological contactor module (courtesy Autotrol Corp.).

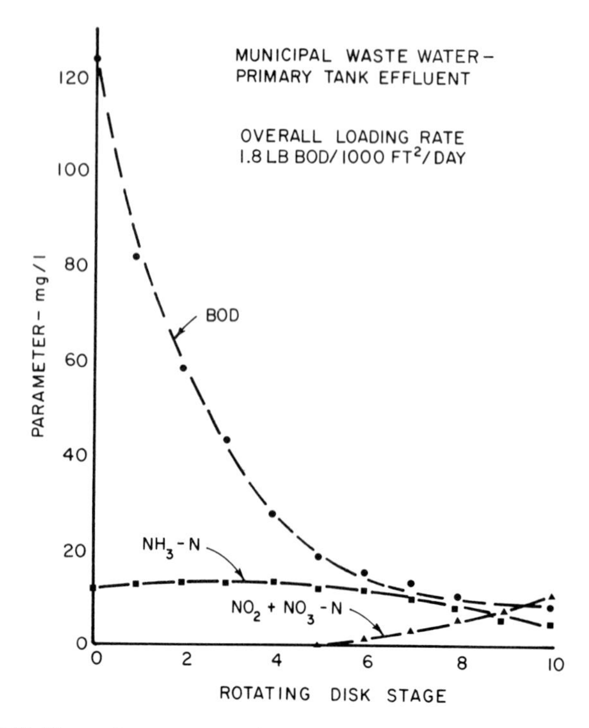

Fig. 7.25. Change in waste water characteristics with a rotating disk unit (67).

with algae growth and nitrification predominating in the rest of the unit. Assuming that the oxygen demand was satisfied in the first of the five disk stages, the loading on this stage was 0.67 lb BOD/day/1000 ft^2 of disk area. The total solids in the unit were removed at the end of the experiment and measured. Net solids production in this unit was 0.16 lb solids/lb applied BOD. The low solids production is indicative of a lightly loaded system with considerable endogenous respiration of the microbial solids on the disks. About 50% nitrogen loss occurred which was attributed to denitrification in the mixed liquor and on the disks.

Wastes containing high concentrations of milk and cheese wastes have been treated by these units. Using a two-stage unit, the BOD removals through the unit ranged from 92 to 98% with influent wastes having concentrations from 200 to 1350 mg/l (70). The disk loading rates ranged from 0.41 to 2.3 lb BOD/day/1000 ft^2 of disk area. Another unit treating milk and dairy processing wastes achieved an average of 96% BOD removal through the unit and its final clarifier. The influent BOD averaged 1060 mg/l (71).

The operating power requirements in these units is for rotation of the disks. Because of the low density of the plastic media, the buoyancy of the disks helps offset the weight of the attached biomass and power requirements are low. In a full-scale study using dilute wastes, the power requirements were a function of disk speed. At 3 rpm the requirements were approximately 0.05 HP. At 6 rpm, the requirements were 0.3 HP (72). At a disk speed of 3.2 rpm, 6 lb BOD were removed/HP-hr for 87% BOD removal.

A three-stage, pilot plant unit was used to treat the wastes from an anaerobic lagoon receiving beef cattle processing wastes (73). The BOD reduction through the three stages was 50% when the unit was loaded at 5.5 lb BOD/day/1000 ft^2 of disk area and 83% when the loading was 1.8 lb BOD/day/1000 ft^2 of disk area. Further details on the performance of the unit are noted in Table 7.5.

A rotating biological contactor and an extended aeration unit were compared on wastes from a vegetable processing operation (74). The contactor removed about 96% of the influent over a loading rate of 3 to 10.5 lb COD/day/1000 ft^2. These removals occurred in spite of a temperature variation of 7°C, an influent COD variation of 1:3, and a liquid detention time variation of 1:2.5. The contactor removed the same amount of COD with less than 50% of the power requirements of the extended aeration unit. The contactor appeared to recover from heavy shock loadings more quickly than the aeration tank.

Winery wastes were shown to be biodegradable by an RBC six-stage unit. At a loading rate of about 1 lb BOD/1000 ft^2/day, 96% BOD removal was obtained and the resultant sludge settled rapidly (75). Efficiency decreased to about 83% BOD removal at a loading rate of about 1.8 lb BOD/1000 ft^2/day.

BOD removal in a rotating biological contactor is related to the pounds of BOD applied, hydraulic retention time, disk speed, and biodegradability

Table 7.5

Performance of a Rotating Biological Contactor with Partially Treated Meat-packing Wastes[a]

Trial	rpm	Hydraulic loading (gpd/ft^2/stage)	BOD loading (lb/day/1000 ft^2)	BOD removal (%) (3 stages)	D.O. in effluent (mg/l)
1	3	8	5.5	50.2	0
2	6	8	5.0	64.5	trace
3	6 (stage 1) 3 (stages 2,3)	4	1.8	83.2	1–1.5

[a] From reference 73.

of the waste. Different removal patterns may occur in each study. Data from many studies illustrated a general removal pattern (Fig. 7.26) when related to the loading rate. Better and more fundamental parameters for design will result from future studies.

Trickling Filters

General

Trickling filters are designed to aerobically treat dilute liquid wastes. Wastes with a high concentration of particulate matter, such as liquid animal manures, are inappropriate wastes to be treated by trickling filters. High concentrations of organic or inorganic solids will result in clogging, reduced efficiencies, and increased maintenance problems if wastes containing such solids are applied to trickling filters. Pretreatment to remove these solids is required if trickling filters are to be used to treat such wastes.

Trickling filters are not filters but are aerobic oxidation units which absorb and oxidize the organic matter in the wastes passing over the filter media. Trickling filters remain one of the more common biological treatment methods for municipal waste waters. The media used in trickling filters generally is crushed stone or rock of large size, generally 2 to 4 inches in size, or plastic media of various configurations. Horizontal redwood

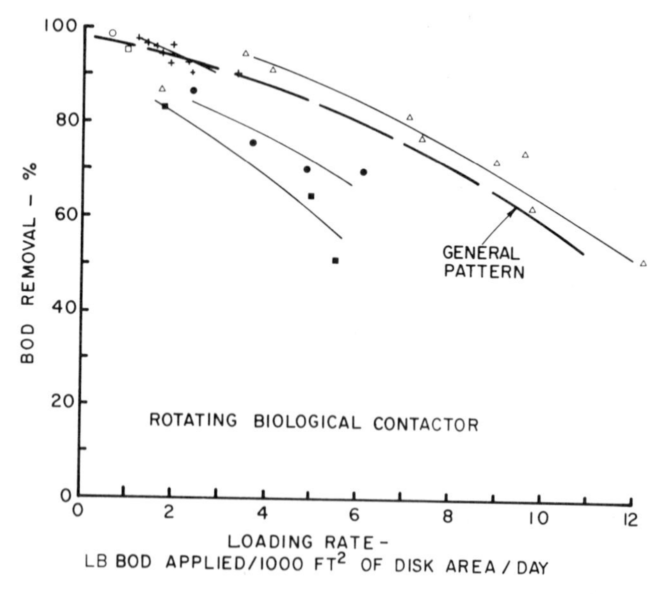

Fig. 7.26. BOD removal in an RBC unit as a function of loading rate.

Table 7.6

Physical Properties of Common Filter Media[a]

Media	Nominal size (inches)	Weight (lb/ft^3)	Surface area (ft^2/ft^3)	Void space (%)
Granite	1–3	90	19	46
Granite	4	90	13	60
Slag	3	68	20	49
Plastic	21 × 38	6	27	94

[a]From reference 76 (with permission of Microform International Marketing Corporation exclusive copyright licensee of Pergamon Press journal back files).

slats and other inert material have been used as a trickling filter support media. The properties of common filter media are noted in Table 7.6 (76).

The waste to be treated is distributed either intermittently or continuously over the top of the media and flows through the media. The rotary distributor is the most common type of liquid distribution system used with trickling filters today (Fig. 7.27). Fixed nozzle distribution systems, in which the distribution nozzles were permanently positioned in the filter media, were used in the early part of this century but are no longer used because of clogging and poor liquid distribution problems. The waste water is discharged above the trickling filter by the rotary distributor permitting some aeration of the liquid to take place before contact with the media. Further aeration takes place as the liquid flows over the media.

The media is the body of the filter and ranges in height of 4 to 7 ft for stone or rock to 10 to 40 ft for plastic media. The media surface serves as support for the microorganisms that metabolize the organic matter in the waste. The filter should have as small a media as possible to increase the surface area in the filter and the active organisms that will be in the filter volume. However, the media must be large enough to have enough void space to permit liquid and air flow and remain unclogged by the microbial growth. Large size media such as 2–4-inch rock and plastic media function satisfactorily.

The microbial slime layer and the water flowing over the media will considerably increase the weight of the material in the filter. For plastic media, the combined weight of the slime, water, and media may be 4–5 times more than the weight of the media alone. The support structure for the filter should be designed to hold such loads.

The filter media sits on an underdrain system which collects the liquid from the filter and transports it to a final sedimentation basin. The underdrain systems (Fig. 7.28) assist in keeping the filter aerobic by facilitating air movement through the filter. Solids are removed from the system con-

Fig. 7.27. Circular trickling filter indicating size of typical rock media, rotary distributor, and liquid distribution pattern.

tinuously from the final sedimentation unit. With common trickling filter systems, settled solids are not returned to the trickling filter.

Heterotrophic facultative bacteria are the largest population of microorganisms in a trickling filter. Protozoa and higher forms of animal life will be present in a filter. In an overloaded filter which may be clogged or oxygen-limited, protozoa will be absent because of the lack of dissolved oxygen. Algae will grow on the surface of filters that are not overloaded but will not grow below the surface since sunlight cannot penetrate. A gray to brown microbial growth will exist within the filter even though the surface may be green due to algal growth. The algae and the higher forms of animal life contribute little to the efficiency of the filter.

The organic matter in the waste water stimulates biological growth on the surface of the media. The growth is established first in areas where the flow does not wash it from the media and will spread throughout the media. Even in warm weather when microbial growth is most rapid, it takes a considerable period of time for the microbial growth to become established in a filter and for equilibrium performance conditions to occur. This period can be from 4 to 6 weeks. It is important that the microbial growth not be killed by toxic conditions in a waste since the filter will not function at design efficiency until the growth is reestablished which can be another lengthy period. Where seasonal wastes are to be treated, such as at some canning plants, trickling filters may not be a satisfactory solution to the waste treatment problem because of the length of time required to build up the necessary microbial population and produce the desired effluent.

After the microbial layer on the media is established, the liquid flows over the layer rather than through it. The liquid waste flows down the media as a wave-creating turbulence between the waste and the liquid film in the microbial surface (Fig. 7.29). The organic matter in the waste is transferred to the liquid film and waste products of metabolism are transferred from the liquid film to the waste. This transfer is continuous through the filter depth.

The outer layers of microorganisms are exposed to the fixed liquid layer

Fig. 7.28. Clay tile underdrain system for a rock media trickling filter. The influent pipe and riser to the rotary distributor is shown (courtesy W. S. Dickey Clay Mfg. Co.).

and metabolize most of the waste. Aerobic metabolism is maintained by continuous transfer of oxygen from the void spaces in the filter to the fixed liquid layer. The rate of oxygen transfer is related to the oxygen differential between the air and the fixed liquid layer. For a given size of filter and type of filter media, the available surface for oxygen transfer is established and there is a fixed quantity of oxygen that can be transferred into the filter per unit of time. This fixed quantity will limit the maximum organic loading that can be applied to the filter and still maintain aerobic conditions.

Most filter loading rates have been established such that the rate of metabolism is not limited by the oxygen transfer. In trickling filters the organic matter limits the rate of microbial metabolism. Trickling filters are food-limited systems operating in the declining growth phase.

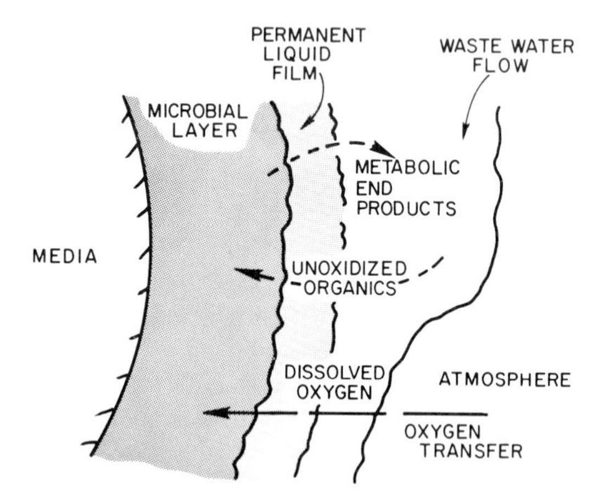

Fig. 7.29. Schematic of processes in a trickling filter.

The thickness of the aerobic microbial layer has been estimated at about 0.005 cm while the actual thickness of the entire microbial layer is much greater. Because only the surface of the microbial layer obtains most of the food and oxygen, the microorganisms attaching the microbial growth to the media die and are carried from the media by the waste water flow. Microbial growth becomes reestablished in areas where older growths have been removed. This cycle is continuous in a trickling filter.

The effluent from the filter contains unmetabolized organic matter in the applied waste water, biological solids separated from the media, and metabolic end products. The biological solids must be separated from the trickling filter effluent before the liquid is discharged from the treatment plant if a high quality effluent is desired. This is done in a final sedimentation tank which is an integral part of a trickling filter system. If any liquid

Table 7.7

Types of Trickling Filters Based on Loading Rates

Loading rate	Standard rate filter	High rate filter	Roughing filter
Hydraulic (MGAD)	Less than 4	10–40	> 100
Organic (lb BOD/1000 ft^3/day)	Less than 15	30–100	> 100

is used for recirculation in a trickling filter, it is generally a portion of the effluent from the sedimentation tank.

Filter Types and Loadings

Loadings to trickling filters are expressed in terms of the amount of oxidizable matter applied to the filter media, i.e., pounds of BOD_5 per 1000 cubic feet of filter media per day, or in terms of hydraulic flow applied to the surface of the media, i.e., millions of gallons applied per acre of filter surface (MGAD). Using these loading rates, three general types of filters can be described: standard rate, high rate, and super rate or roughing filter. A comparison of these types is shown in Table 7.7. One of the two parameters will control the design of the filter. A roughing filter is used for industrial wastes either as a pretreatment device or where high treatment efficiencies are not needed. Process flow patterns for various types of trickling filters are illustrated in Fig. 7.30.

In standard rate filters, the waste water is applied to the media, and passes through the media to the final sedimentation tank (Fig. 7.34). Effluent recirculation may be done during low flow periods to keep the media moist and to provide some organic matter for the microorganisms but it is rarely done continuously.

A high rate filter generally uses continuous recirculation (Fig. 7.31) to produce an acceptable effluent since the applied loading rates are higher. Recirculation equalizes variable waste loads, reduces waste loads, and keeps a thinner microbial layer on the media.

The light weight of the plastic media (Table 7.6) reduces the need for heavy supporting structures and permits tall units reducing land requirements. The large void space permits higher application rates without clogging and free passage of air to supply the necessary oxygen. A honeycomb type structure commonly is used with plastic media to increase surface area and porosity. When placed in the trickling filter, the layers of the media are placed at angles to each other so that the possibility of waste water falling through the media without contacting the media is very small (Fig. 7.31). The larger surface area per unit volume increases BOD removals and can permit larger hydraulic loads.

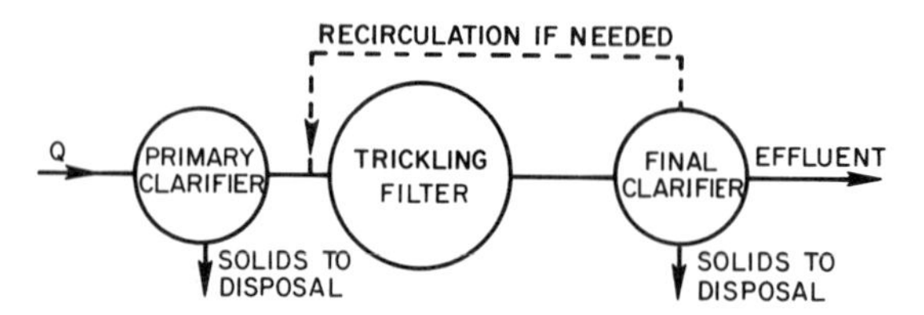

SINGLE STAGE, STANDARD RATE

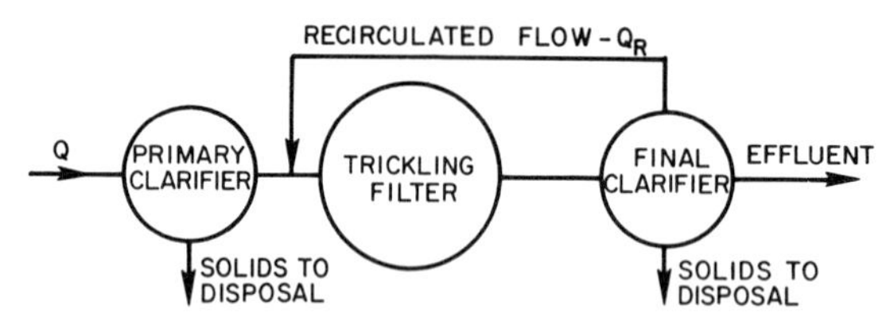

SINGLE STAGE, HIGH RATE

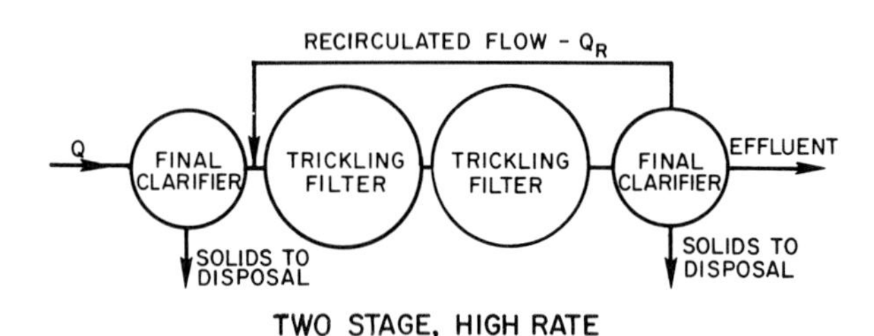

TWO STAGE, HIGH RATE

Fig. 7.30. Flow diagrams of trickling filter treatment processes.

Redwood horizontal slats have proven functional as a trickling filter media showing no deterioration after 9 years of use (77). The configuration of these slats in a trickling filter provides a larger void space to surface area ratio than conventional rock filter media. Aeration is achieved by the splashing action of water droplets which fall from slat to slat. This media has been used as both a high rate trickling filter and as an active biofilter. This type of a biofilter is a process in which the influent waste, filter underflow, and secondary clarifier sludge are mixed together and passed through a tower filled with the horizontal slats. The mixed liquor suspended solids function with the filter slime to provide a larger quantity of active microorganisms.

Performance

The removal of organic matter in trickling filters results from the adsorption and metabolism as the waste water passes through the filter. The BOD removal can be expected to be proportional to both the liquid residence time in the filter and the active mass of organisms in the filter. For a given filter design, the active mass of organisms in the filter will be fixed and is considered not to be a variable parameter. Under these assumptions, and because a trickling filter is a food-limiting process, first-order reaction kinetics are used to relate the BOD removal,

$$dL/dt = kL \tag{7.18}$$

where L is the BOD in the filter at liquid residence time t, and k is a coefficient incorporating the surface area of active microbial film per unit volume. Upon integration, this becomes

$$L_e = L_o e^{-kt} \tag{7.19}$$

where L_e is the BOD in the filter effluent, L_o is the BOD applied to the filter, and t is the liquid residence time of the filter. L_e does not represent the total BOD in the filter effluent, only that fraction of the applied BOD that remains. The total BOD in the filter effluent is L_e plus the BOD of the biological solids separated from the media. The BOD of the effluent from the final sedimentation tank is L_e plus that portion of the biological solids not removed in the sedimentation tank.

The liquid residence time, t, is related to the depth and hydraulic loading

Fig. 7.31. Plastic media trickling filter and rotary distributor (courtesy Dow Chemical Co.).

of the filter and to the physical characteristics of the filter media. The following empirical equation has been found satisfactory to estimate the liquid residence time in a trickling filter.

$$t = \frac{CD}{Q^n} \tag{7.20}$$

C and n are constants that will vary with the type of filter media and hydraulic characteristics, D is the depth of filter, and Q is the rate of flow without recirculation in terms of MGAD. With t defined in the above manner, Eq. (7.19) becomes

$$L_e = L_o e^{-KDQ^{-n}} \tag{7.21}$$

where K reflects both constants, k and C.

Values of n have ranged from 0.4 to 0.6 for rock filters, 0.67 for large spherical balls and screen filters, 0.5 for plastic media, and 0.76 for 0.25-inch granite media (76, 78–80). The numerical value of n decreases as the slime thickness increases. The value of n should be evaluated under process conditions in which a slime layer has been developed.

Values of K have ranged from 0.045 to 0.088 for plastic media (79). The value of K is related to the type and size of the trickling filter media, temperature, and the treatability of the waste.

The values of n and K are related to process conditions and can be determined experimentally. With actual data, a logarithmic plot of BOD remaining (L_e/L_o) versus filter depth can be prepared at different flow rates (Fig. 7.32a). The slope of the resulting lines, K_F, is equal to $K \cdot Q^{-n}$ and will vary with the flow rates. Mathematically manipulating this relationship by taking the logs of all factors, i.e.,

$$\log K_F = \log K - n \log Q \tag{7.22}$$

these values can be plotted as a function of flow rate (Fig. 7.32b). The value of n is the slope of the resultant line. The value of K can be determined once n is known. In addition, when Q is 1.0, $\log Q$ is zero and K equals K_F. The numerical value of K is dependent upon Q and will have the units of liquid flow per unit of depth. n is dimensionless. With these constants known, the depth of the filter to remove a specific amount of applied BOD can be calculated.

The above equations provide a basis of determining the efficiency of single pass, standard rate trickling filters. With modifications, these equations also can be useful when recirculation is practiced. The hydraulic load to the filter is increased by the recycled flow. With recirculation, the BOD applied to the filter, L_m, is a combination of that in the pretreated waste,

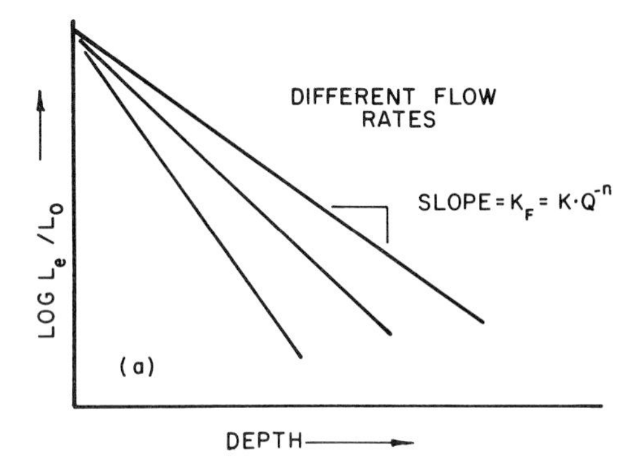

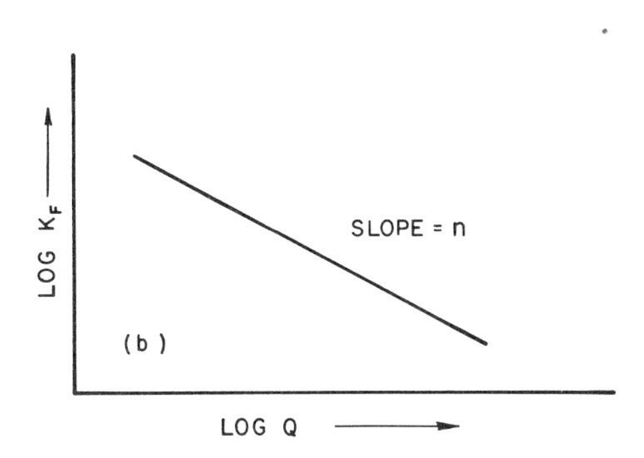

Fig. 7.32. Experimental evaluation of the constants K and n.

L_i, and that in the recirculated waste water, L_{e_R}. In Eq. (7.23), R is the recirculation ratio, i.e., the ratio of the recycled flow,

$$L_m = \frac{L_i + RL_{e_R}}{R + 1} \tag{7.23}$$

Q_R, to the flow of the pretreated waste water, Q. By defining Q_2 as the hydraulic load in MGAD including recirculation flow, Eq. (7.21) can be written as

$$L_e = L_m e^{-KD/Q_2^n} \tag{7.24}$$

In this case, the final effluent is used for recirculation (Fig. 7.30) and is mixed with the incoming waste directly ahead of the filter. The filter is considered to operate in a manner comparable to a standard filter except that with recycle the hydraulic load is increased and the efficiency is related to the BOD of the mixed flow. The effect of recirculation is to decrease the concentration of BOD added to the filter. The increased flow rate through the filter will decrease the residence time which in turn decreases the removal in a single pass through the filter. Recirculation permits additional treatment of the BOD which is not removed in a single pass.

A number of flow patterns for trickling filters have been proposed for specific wastes and by different manufacturers. Two-stage filtration (Fig. 7.31), using rock media, frequently is used for strong wastes to permit adequate effective filter depth for the necessary degree of treatment. Tall plastic media filters permit single stage units with strong wastes.

Typical treatment efficiencies for trickling filters treating municipal wastes range from 85 to 95% BOD removal for standard rate filters and 75–85% BOD removal efficiencies for high rate filters. Superrate or roughing filters will have lower BOD removal efficiencies.

Lightly loaded trickling filters can produce a high degree of nitrification. They have been suggested as polishing treatment units where the nitrogenous oxygen demand is of concern in the receiving waters. Studies on nitrification in trickling filters have indicated that up to 100% of the available ammonia can be nitrified (81). As the hydraulic load increased, the depth of the filter needed to be increased to achieve the degree of nitrification. In this study, the constants K and n in Eq. (7.21) were 0.65 and 0.49, respectively, at 30°C. A good correlation between the values of K for different media and the specific surface of the media (feet^2/feet^3) was observed. A temperature correction factor (θ) of 1.07 was found satisfactory. High rate and roughing filters are unlikely to produce nitrification.

As with all biological treatment units, the performance of trickling filters is influenced by the temperature of the liquid. Metabolic reaction rates are greater as the temperature increases. The liquid temperature determines the rate of microbial metabolism while the air temperature differential between the air in the media and that outside the media determines the air flow through the filter. Since currently designed trickling filters are food rather than oxygen limited systems, induced aeration through the filters by fans is not effective.

The movement of air through trickling filters is essential to the maintenance of aerobic conditions and will be upward or downward depending upon the relative temperature of the liquid and the ambient air. When the liquid is colder than the air, air movement will be downward and outward through the underdrain system or ventilation ports. The draft causing a movement of air in a filter is a function of the depth of the filter, the average air densities inside and outside the filter, and should be directly related to the difference between the temperature of the air and the liquid.

Liquid temperature in the filter will decrease during the winter, decrease the rate of microbial metabolism, and reduce the efficiency of the filter.

There is no opportunity in a trickling filter to compensate for decreased metabolism rates by increasing the active mass in the system. The efficiency of a trickling filter will be less in the winter.

Pretreatment

The biological growth on a filter removes the organic material in a waste water by adsorption and metabolism and the filter is intended to be aerobic. The waste water applied to the filter should consist primarily of dissolved and colloidal organic matter. Removal of large particles that may clog a filter is necessary to avoid closing of the voids in the media which would prevent air movement and oxygen transfer to the microorganisms. Clogging can also result from excessive microbial growth filling or bridging the void space.

When clogging causes a filter to become anaerobic and/or if the removal efficiency decreases, the problem may be handled by better removal of the particulate matter and by reducing the organic concentration applied to the filter by incorporating or increasing effluent recycle. The solids causing the clogging frequently can be removed by increased hydraulic flow rates or by high pressure water applied at the clogging site.

Sedimentation is the process most commonly used to remove the particulate material that can cause clogging problems. With certain wastes centrifuges, dissolved air flotation, and fine rotary screens may be effective in removing the particulate material.

Application to Agricultural Wastes

Trickling filters have been used more with food processing wastes than with any other type of agricultural waste. Depending upon the type of waste and the effluent quality required, a trickling filter treatment system can include primary sedimentation to remove settleable solids in the raw waste, recirculation pumps, and lines, and final sedimentation units. When the fundamental relationships for design and use of trickling filters are not understood, difficulties and poor performance can result.

Trickling filters have been explored for the treatment of potato starch processing wastes. Standard rate rock filters, loaded up to 30 lb BOD/1000 ft^3/day obtained BOD reductions of about 90%. The effluent pH was generally between 7 and 8 although the wastes had an acid pH when they were applied to the filters. Loadings up to 70 lb BOD/1000 ft^3/day in a high rate filter, with a recirculation ratio of 10, also obtained above 90% BOD removals. At higher BOD loadings, the filter, which contained media of about 0.75 inches in diameter, became clogged with the biological solids sloughed from the media (82). Media with larger openings may have permitted higher loadings without clogging problems.

Plastic media did not exhibit clogging problems when loaded up to rates of 130 lb BOD/1000 ft^3/day of potato processing wastes. BOD reductions were about 70% prior to the final sedimentation unit and about 85% after the sedimentation unit (83). The recirculation ratio was 6.

A pilot plant using plastic media was found to work satisfactorily at a loading rate of 400 lb BOD/1000 ft^3/day. Approximately 75% BOD removal occurred (84) when potato lye peel process water was treated. No pH adjustment of the wastes was necessary although addition of nutrients was advantageous. Excess solids production was about 0.6–0.8 lb/lb BOD removed at the above loading. These solids were concentrated to 3% solids by gravity and to greater concentrations by other methods.

Settled peach and pear canning wastes were applied to plastic media trickling filters at loading rates from about 300 to 1800 lb BOD/1000 ft^3/day (83). The raw waste and effluent pH values were 10.5 and 6.2, respectively. Recycle ratios were 1 and 2. At 316 lb BOD/1000 ft^3/day, 86% BOD removal occurred. When the loading was 1030 and 1760 lb BOD/1000 ft^3/day, BOD removals were 33 and 36%, respectively. Addition of nitrogen and phosphorus to these wastes resulted in a BOD removal of 45% when the loading rate was 1510 lb BOD/1000 ft^3/day.

A two-stage system, using clay blocks with openings of 1 inch, obtained BOD removals of above 94% when the loading rate was 150 lb BOD/1000 ft^3/day and the recycle ratio was 10 (84). The acid pH of the milk waste was neutralized before treatment and sufficient nitrogen and phosphorus were added prior to treatment to obtain a BOD:N ratio of 20:1 and a BOD:P ratio of 75:1.

The wastes from the canning of peaches and fruit cocktail have been treated by a pilot plant trickling filter using plastic media (85). A heavy slime growth, composed primarily of fungi, became established when the raw wastes were treated. This growth grew so thick that anaerobic conditions occurred. An evaluation of the wastes indicated that they were nitrogen deficient. The addition of nitrogen in the raw wastes to obtain a BOD:N ratio of 20:1 changed the characteristics of the filter growth. A thin growth comprised mainly of bacteria became established. This growth sloughed continuously and settled rapidly. The filter acted as a roughing filter. The BOD removal was 53% at a loading rate of 640 lb BOD/1000 ft^3/day and 20% at a loading rate of 950 lb BOD/1000 ft^3/day. BOD removals increased after the nitrogen was added. A BOD removal of 39% was accomplished at a loading rate of 1160 lb BOD/1000 ft^3/day after the addition of nitrogen. These removals represent data on the influent and effluent of the trickling filter. Greater BOD removals resulted if the solids in the filter effluent were removed.

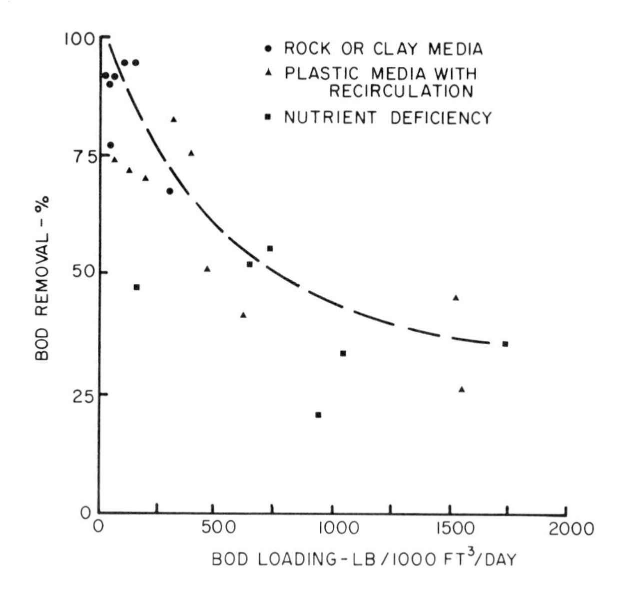

Fig. 7.33. BOD removal patterns obtained by treating agricultural wastes by trickling filters.

A combined system of anaerobic lagoons followed by trickling filters have been used to treat meat-packing wastes (86). The anaerobic lagoon served to remove the gross solids in wastes. The plastic media trickling filters were loaded at about 73 lb BOD/1000 ft^3/day at a hydraulic load of 0.8 gpm/ft^2 of surface area. The BOD and COD removal averaged 74 and 73%, respectively. The wastes received adequate aeration in the filter with the dissolved oxygen in the filter effluent averaging about 4 mg/l. Grease removal was about 69%. No phosphate removal occurred although about 20% nitrogen removal resulted, chiefly as a decrease in ammonia concentration. Some nitrification occurred in the filter although denitrification took place in the final clarifier. The above removals refer to changes that took place in the combined trickling filter–final clarifier system.

Yeast plant effluent was treated in a pilot plant plastic media trickling filter. BOD removals averaged from 25 to 85% due to organic loading fluctuations that ranged as much as 300% at a specific hydraulic loading (87). Influent BOD variations of 2500–8000 mg/l were the result of normal production operations in the yeast plant. Although there was a scatter in the resultant data, a removal rate constant of about 15 mg/l BOD removed/ft^3 of media/gpm of waste flow resulted. A zero order removal model fit this data.

Trickling filters have been tried with dilute animal wastes. When coagulated cattleshed wastes were applied to laboratory filters, effluents containing 20 mg/l BOD could be obtained (88). Dilution and recirculation of filter effluent was necessary to treat the wastes successfully without coagulation. The influent BOD was about 570 mg/l. Field trickling filter experiments at lower temperatures, variable application rates, and lesser initial dilution were less successful. In these cases the effluent BOD, after settling, ranged from 40 to 140 mg/l.

Settled farm waste waters were applied to a trickling filter at rates from 30 to 52 lb BOD/1000 ft^3/day (89). BOD removal efficiencies were in the range of 75–90%. The effluent was well nitrified and infrequent ponding occurred.

Attempts have been made to treat dairy manure with trickling filters preceded by sedimentation tanks. Loadings ranged from 6 to 22 lb BOD/1000 ft^3/day (90). Effluent BOD concentrations ranged from 80 to 750 mg/l and varied with temperature and loading. Scum and sludge removal was not practiced in the sedimentation unit which became overloaded and released solids to the trickling filter. It is important that the settled solids be removed continuously from primary sedimentation units if subsequent trickling filters are to be successful.

A summary of the trickling filter studies treating agricultural wastes is presented in Fig. 7.33. The BOD removal is related to BOD loading. A number of factors become apparent from a comparison of this type: (a) the percent removal decreases logarithmically as the BOD loading increases, (b) wastes deficient in nutrients have lower BOD removals, and (c) even at very high loading rates, sizable BOD removals will occur, between 25 and 50% as observed from this data. Roughing filters can be used at high loading rates to accomplish conversion of the soluble wastes components into microbial mass and reasonable BOD removals, especially if the filter solids are removed from the effluent before discharge. Although a broad trend results from this data, the removal efficiencies of each waste will be related to the characteristics of that waste, the operational procedure of the trickling filter system, and the ambient environmental conditions.

References

Aeration Systems

1. Landberg, G. G., Graulich, B. P., and Kipple, W. H., Experimental problems associated with the testing of surface aeration equipment. *Water Res.* **3,** 445–456 (1969).
2. Downing, A. L., and Boon, A. G., Oxygen transfer in the activated sludge process. *Int. J. Air Water Pollut.* **5,** 131–148 (1963).
3. Eckenfelder, W. W., and O'Connor, D. J., "Biological Waste Treatment." Pergamon, Oxford, 1961.

4. Warburg, O., and Kuborvitz, E., Atmong bei sehr Klenen Sauerstoffdrucker. *Biochem. Z.* **214,** 5–18 (1929).
5. Kempner, W., The effect of oxygen tension on cellular metabolism. *J. Cell. Comp. Physiol.* **10,** 339–363 (1937).
6. Schoberl, P., and Engel, H., Das Verhalten der Nitrifizierenden Backterium Gegenuber Golostern Sauerstoff. *Arch. Mikrobiol.* **48,** 393–400 (1964).
7. von der Emde, W., Aspects of the high rate activated sludge process. *Int. J. Air Water Pollut.* **5,** 299–318 (1963).
8. Eckenfelder, W. W., and Wanielista, M. P., "Theoretical Aspects of Aerated Lagoon Design." Presented at the Sanit. Eng. Div., Spec. Conf. Amer. Soc. Civil Eng., 1965.
9. McKinney, R. E., and Benjes, H. H., Evaluation of two aerated lagoons. *J. Sanit. Eng. Div., Amer. Soc. Civil Eng.* **91,** SA6, 43–56 (1965).
10. Zeper, J., and de Man, A., New developments in the design of activated sludge tanks with low BOD loadings. Presented at the *5th Int. Water Pollut. Res. Conf.*, San Francisco, 1970.
11. Benjes, H. H., and McKinney, R. E., Specifying and evaluating the performance of aeration equipment. *J. Sanit. Eng. Div., Amer. Soc. Civil Eng.* **93,** SA6, 55–64 (1967).
12. Kalinske, A. A., Economic evaluation of aerator systems. *Environ. Sci. Technol.* **3,** 229–234 (1969).
13. Novak, R. G., Techniques and factors involved in aerator selection and evaluation. *J. Water Pollut. Contr. Fed.* **40,** 452–463 (1968).
14. Water Pollution Control Federation, "Aeration in Wastewater Treatment," MOP-5. Washington, D.C., 1971.
15. McKinney, R. E., and Edde, H., Aerated lagoon for suburban sewage disposal. *J. Water Pollut. Contr. Fed.* **33,** 1277–1285 (1961).

Activated Sludge

16. Bisogni, J. J., and Lawrence, A. W., Relationships between biological solids retention time and settling characteristics of activated sludge. *Water Res.* **5,** 753–763 (1971).
17. Martin, A. J., "The Activated Sludge Process." MacDonald & Evans, London, 1927.
18. Weddle, C. L., and Jenkins, D., The viability and activity of activated sludge. *Water Res.* **5,** 621–640 (1971).
19. Downing, A. L., Painter, H. A., and Knowles, G., Nitrification in the activated sludge process. *Inst. Sewage Purif., J. Proc.* Part 2, pp. 130–154 (1964).
20. Downing, A. L., Factors to be considered in the design of activated sludge plants. *In* "Advances in Water Quality Improvement" (E. F. Gloyna, and W. W. Eckenfelder, eds.), pp. 190–202. Univ. of Texas Press, Austin, 1968.
21. National Canners Association, "Liquid Wastes from Canning and Freezing Fruits and Vegetables," Final Rep., Proj. 12060 EDK. Environmental Protection Agency, 1971.
22. Dallas, Oregon, "Combined Treatment of Domestic and Industrial Wastes by Activated Sludge," Final Rep., Proj. 12130 EZR. Environmental Protection Agency, 1971.
23. Sawyer, C. N., Milestones in the development of the activated sludge process. *J. Water Pollut. Contr. Fed.* **37,** 151–162 (1965).
24. McKinney, R. E., and O'Brien, W. J., Activated sludge—basic design concepts. *J. Water Pollut. Contr. Fed.* **40,** 1831–1843 (1968).
25. Ulrich, A. A., and Smith, M. W., The biosorption process of sewage and waste treatment. *Sewage Ind. Wastes* **23,** 1248–1257 (1951).
26. Eckenfelder, W. W., and Grich, E. R., High rate activated sludge treatment of cannery wastes. *Proc. Purdue Ind. Waste Conf.* **10,** 549–553 (1955).
27. Baker, R. H., Package aeration plants in Florida. *J. Sanit. Eng. Div., Amer. Soc. Civil Eng.* **88,** SA6, 75–86 (1962).

28. Moore, M. E., and Todd, J. J., Sludge production in the contact stabilization process. *Eff. Water Treat. J.* **8,** 551–556 (1968).
29. Dart, M. C., The treatment of meat trade effiuents. *Eff. Water Treat. J.* **7,** 20–33 (1967).
30. Boon, A. G., The role of contact stabilization in the treatment of industrial waste water and sewage: A progress report. *J. Inst. Water Pollut. Contr.* **69,** 67–84 (1969).
31. Esvelt, L. A., and Hart, H. H., Treatment of food processing wastes by aeration. *J. Water Pollut. Contr. Fed.* **42,** 1305–1326 (1970).
32. McKinney, R. E., Mathematics of complete mixing activated sludge. *J. Sanit. Eng. Div., Amer. Soc. Civil Eng.* **88,** SA3, 88–113 (1962).
33. McKinney, R. E., Benjes, H. H., and Wright, J. R., Evaluation of a complete mixing activated sludge plant. *J. Water Pollut. Contr. Fed.* **42,** 737–757 (1970).
34. R. T. French Co., "Aerobic Secondary Treatment of Potato Processing Waste," Final Rep., Proj. WPRD 15-01-68. Water Quality Offlce, Environmental Protection Agency, 1970.
35. Winter Garden Citrus Products Cooperative, "Complete Mix Activated Sludge Treatment of Citrus Process Wastes," Final Rep., Proj. 12060 EZY. Environmental Protection Agency, 1971.
36. Hoover, S. R., Jasewicz, L., Pepinsky, J. B., and Porges, N., Dairy waste assimilation by activated sludge. *Sewage Ind. Wastes* **23,** 167–178 (1951).
37. Morris, G. L., VanDenBerg, L., Culp, G. L., Geckler, J. R., and Porges, R., Extended aeration plants and intermittent watercourses. *U.S., Pub. Health Serv., Publ.* 999-WP-8 (1963).
38. Teletzke, G. H., Chickens for the barbecue—wastes for aerobic digestion. *Wastes Eng.* 134–138 (March, 1961).
39. Coca-Cola Co., "Treatment of Citrus Processing Wastes," Proj. WPRD 38-01-67. Final Rep., Water Quality Offlce, Environmental Protection Agency, 1970.
40. Nye, J. C., Dale, A. C., and Bloodgood, D. E., Effect of temperature in the aerobic decomposition of dairy cattle manure. *Annu. Meet., Amer. Soc. Agr. Eng.* Paper 69-926 (1969).

Aerobic Digestion

41. Ahlberg, N. R., and Boyko, B. I., Evaluation and design of aerobic digesters. *J. Water Pollut. Contr. Fed.* **44,** 634–643 (1972).
42. Drier, D. C., Aerobic digestion of solids. *Proc. Purdue Ind. Waste Conf.* **18,** 123–140 (1963).
43. Loehr, R. C., Aerobic digestion: Factors affecting design. *Water and Sewage Works.* Reference Edition, R 169–R 180, 1965.
44. Jaworski, N., Lawton, G. W., and Rohlich, G. A., Aerobic sludge digestion *Int. J. Air Water Pollut.* **5,** 93–102 (1963).
45. Randall, C. W., and Koch, C. T., Dewatering characteristics of aerobically digested sludge. *J. Water Pollut. Contr. Fed.* **41,** R215–R238 (1969).
46. Jasewicz, L., and Porges, N., Aeration of whey wastes. I. Nitrogen supplementation and sludge oxidation. *Sewage Ind. Wastes* **30,** 555–561 (1958).
47. Lawton, G. W., and Norman, J. D., Sludge digestion studies. *J. Water Pollut. Contr. Fed.* **36,** 495–504 (1964).

Oxidation Ditch

48. Baxter, S. H., Pontin, R. A., and Watson, J. S., "Development of a Prefabricated Feeding Piggery with Waste Treatment in Pasveer-Type Oxidation Ditches," Farm Building Report, Scottish Farm Buildings Investigation Unit, Aberdeen, December, 1966.

49. Loehr, R. C., Anderson, D. F., and Anthonisen, A. C., An oxidation ditch for the handling and treatment of poultry wastes. *In* "Livestock Waste Management and Pollution Abatement," Publ. PROC-271, pp. 209–212. Amer. Soc. Agr. Eng., 1971.
50. Jones, D. D., Day, D. L., and Garrigus, U. S., Oxidation ditch in a confinement beef building. *Winter Meet., Amer. Soc. Agr. Eng.* Paper 69-925 (1969).
51. Day, D. L., Jones, D. D., Converse, J. C., Jensen, A. H., and Hansen, E. L., Oxidation ditch treatment of swine wastes—summary report. *Winter Meet., Amer. Soc. Agr. Eng.* Paper 69–924 (1969).
52. Jones, D. D., Day, D. L., and Dale, A. C., "Aerobic Treatment of Livestock Wastes," Bull. No. 737. University of Illinois, Urbana, Agricultural Experiment Station in Cooperation with Purdue University, Revised Bulletin, April 1971.
53. Schmidtke, N. W., "Low Cost Pollution Control with a Ditch Oriented Modified Brush Aeration System," 1st Annu. Symp. Water Pollut. Res. University of Waterloo, Waterloo, Ontario, 1966.
54. Day, D. L., Jones, D. D., and Converse, J. C., "Livestock Waste Management Studies—Termination Report," Agr. Eng. Res. Rep., Proj. 31-15-10-375. University of Illinois, Urbana, 1970.
55. Moore, J. A., Larson, R. E., Hegg, R. O., and Allred, E. R., Beef confinement systems—oxidation ditch. *Annu. Meet., Amer. Soc. Agr. Eng.* Paper 70-418 (1970).
56. Jones, D. D., Day, D. L., and Converse, J. C., Oxygenation capacity of oxidation ditch rotors for confinement livestock buildings. *Proc. Purdue Ind. Waste Conf.* **24,** 191–208 (1969).
57. Guillame, F., "Evaluation of the Oxidation Ditch as a Means of Treatment in Ontario," Res. Publ. No. 6. Ontario Water Resources Commission, Ontario, 1964.
58. Pasveer, A., Research on activated sludge. V. Rates of biochemical oxidation. *Sewage Ind. Wastes* **27,** 783–790 (1955).
59. Scheltinga, H. M. J., Aerobic purification of farm waste. *Inst. Sewage Purif., J. Proc.* Part 6, pp. 585–588 (1966).
60. Scheltinga, H. M. J., Farm wastes. *J. Inst. Water Pollut. Contr.* (*Brit.*) **68,** 403–413 (1969).
61. Webster, L. F., Oxidation ditch treat tomato wastes. *Water Pollut. Contr.* (*Canada*) **105,** 75–78, 1967.
62. Parker, C. D., and Skerry, G. P., "Cannery Waste Treatment by Lagoons and Oxidation Ditch at Shepparton, Victoria, Australia," Proc. 2nd Nat. Symp. Food Process. Wastes, pp. 251–269. Pacific Northwest Water Lab., Environmental Protection Agency, 1971.
63. Walker, J. P., and Pos, J., "Caged Layer Performance in Pens with Oxidation Ditches and Liquid Manure Tanks," Proc. Agr. Waste Manage. Conf., pp. 249–253. Cornell University, Ithaca, New York, 1969.
64. Stewart, T. A., and McIlwain, R., The aerobic storage of poultry manure. *In* "Livestock Waste Management and Pollution Control," Publ. PROC-271, pp. 261–263. Amer. Soc. Agr. Eng., 1971.

Rotating Biological Contactor

65. Hartmann, H., Entwicklung und Betreib von Tauchtropfkorpern. *Gas- Wasserfach* **101,** 281–285 (1960).
66. Hartmann, H., The bio-disc filter. *Oesterr. Wasserwirt.* **17,** 264–269, 1965.
67. Torpey, W. N., Heukelekian, H., Kaplovsky, A. J., and Epstein, R., Rotating disks with biological growths prepare wastewater for disposal or reuse. *J. Water Pollut. Contr. Fed.* **43,** 2181–2188 (1971).
68. Bretscher, U., Die Phosphat—Elimination mit Tauchtropfkorpen. *Gas- Wasserfach* **110,** 538–542 (1969).

69. Kolbe, F. F., A promising new unit for sewage treatment. *Die Siville Ingenieur in Suid-Afrika,* (South Africa) pp. 327–333. (1965).
70. Bretscher, U., Small sewage treatment plant with RBC's for the purification of dairy wastewater. *Soc. Swiss Sewage Spec.,* Report No. 9614, November, 1967.
71. Birks, C. W., and Hynek, R. J., Treatment of cheese processing wastes by the bio disc process. *Proc. Purdue Ind. Waste Conf.* **26,** (1971).
72. Antonie, R., Application of the BIO-DISC process to treatment of domestic wastewater. Paper presented at the *43rd Water Pollut. Contr. Fed. Conf.,* Boston, 1970.
73. Chittenden, J. A., and Wells, W. J., Rotating biological contactors following anaerobic lagoons. *J. Water Pollut. Contr. Fed.* **43,** 746–754 (1971).
74. Burm, R. J., Cochrane, M. W., and Dostal, K. A., "Cannery Waste Treatment with RBC and Extended Aeration Pilot Plants," Proc. 2nd Nat. Symp. Food Process. Wastes, pp. 227–249. Pacific Northwest Water Lab., Environmental Protection Agency, 1971.
75. LaBella, S. A., Thaker, I. H., and Tehan, J. E., Treatment of winery wastes by aerated lagoon, activated sludge process, and rotating biological contactors or RBC. *Proc. Purdue Ind. Waste Conf.* **27,** (1972).

Trickling Filters

76. Bryan, E. H., and Moeller, D. H., Aerobic biological oxidation using Dowpac. *Int. J. Air Water Pollut.* **5,** 341–347 (1963).
77. Schaumberg, F., "Secondary Treatment with Del Pak Horizontal Media—An Engineering Report on the 1970 Pilot Plant Study at Corvallis, Oregon." Cornell, Howland, Hayes, & Merryfield, Corvallis, Oregon, 1971.
78. Eckenfelder, W. W., Trickling filter design and performance. *J. Sanit. Eng. Div., Amer. Soc. Civil Eng.* **87,** SA4, 33–46 (1961).
79. Germain, J. E., Economical treatment of domestic wastes by plastic medium trickling filters. *J. Water Pollut. Contr. Fed.* **38,** 192–205 (1966).
80. Meltzer, D., Investigation of the time factor in biological filtration. *Inst. Sewage Purif. Proc. J.* Part 2, pp. 196–175 (1962).
81. Balakrishman, S., and Eckenfelder, W. W., Nitrogen relationships in biological treatment processes. II. Nitrification in trickling filters. *Water Res.* **3,** 167–174 (1969).
82. Buzzell, J. C., Caron, A. J., Ryckman, S. J., and Sproul, O. J., Biological treatment of protein water from the manufacture of starch. *Water Sewage Works* **111,** 327–330, 360–365, 1964.
83. Guttormsen, K., and Carlson, D. A., "Potato Processing Waste Treatment-Current Practice," Water Pollut. Contr. Res. Ser. DAST-14. Federal Water Pollution Control Administration, Dept. of Interior, 1969.
84. Hatfield, R., Strong, E. R., Heinsohn, F., Powell, H., and Stone, T. G., Treatment of wastes from corn industry by pilot plant trickling filters. *Sewage Ind. Wastes* **28,** 1240–1246 (1956).
85. National Canners Association, "Waste Reduction in Food Canning Operations," Rep. Grant WPRD 151-01-68. Federal Water Quality Administration, Dept. of Interior, 1970.
86. Baker, D. A., and White, J., "Treatment of Meat Packing Wastes Using PVC Trickling Filters," Proc. 2nd Nat. Symp. Food Process. Wastes, pp. 237–312. Pacific Northwest Water Lab., Environmental Protection Agency, Denver, Colorado, 1971.
87. Quirk, T. P., "Biological Treatment of High BOD Yeast Wastes," Proc. 2nd Nat. Symp. Food Process. Wastes, pp. 367–408. Pacific Northwest Water Lab., Environmental Protection Agency, Denver, Colorado, 1971.
88. Painter, H. A., Treatment of wastewaters from farm premises. *Water Sanit. Eng.* **7,** 353–358, 1957.

89. Wheatland, A. B., and Borne, B. J., Treatment of farm effiuents. *Chem. Ind.* (*London*) pp. 357–362 (1964).
90. Bridgham, D. O., and Clayton, J. T., "Trickling Filters as a Dairy Manure Stabilization Component," Proc. Nat. Symp. Anim. Waste Manage., Pub. SP-0366, pp. 66–68. Amer. Soc. Agr. Eng., 1966.

8

Anaerobic Treatment

General

Anaerobic treatment is one of the least understood processes in waste treatment. As a result, the application of anaerobic waste treatment has been both narrow and slow. The inability to utilize the advantages of anaerobic treatment rests in a lack of understanding of the fundamentals of the process. In addition, emphasis on attaining required treatment plant effluent quality has been greater than on solids destruction. Only recently have anaerobic treatment processes undergone the research and developmental changes that allow the designer to utilize the processes in a more optimum manner.

Almost all naturally occurring organic matter and many synthetic organic compounds can be fermented, or digested anaerobically. If the process is carried to completion, the gaseous products are methane and carbon dioxide. If partial digestion occurs, intermediate compounds may be produced, including many that are odorous. The purpose of anaerobic treatment is to stabilize the organic compounds in the untreated wastes.

A comparison of aerobic and anaerobic methods indicated that anaerobic digestion became economical at influent concentrations of greater than 4000 mg/l COD. The advantages of anaerobic digestion were most obvious at higher COD concentrations, above 20,000 mg/l. At this concentration anaerobic methods cost about a fourth of equivalent aerobic treatment. As a rough guide, it was suggested that below COD concentrations of 4000 mg/l aerobic methods should receive priority and above this anaerobic methods would be preferable (1). The actual anaerobic process to be adopted will b determined by the desired effluent quality and operating conditions.

Many reasons exist for the utilization of anaerobic processes for waste treatment, such as: (a) higher loading rates than are possible for aerobic treatment, (b) useful end products such as digested sludge and/or combustible gases, (c) stabilization of organic matter, (d) alteration of water-binding characteristics to permit rapid sludge dewatering, (e) solids reduction for easier handling, and (f) low microbial growth which will decrease the possible need for supplemental nutrients with nutritionally unbalanced wastes. The digested solids do not have an offensive odor and odor complaints are unlikely when well-digested waste solids are disposed of on sand-drying beds or on the land for final disposal. Inadequately digested solids are capable of generating odors. The application of anaerobic processes to a specific waste requires an understanding of the basic factors that are involved. Factors that should be evaluated when considering anaerobic waste treatment include: mixing, loading, temperature, solids retention time, nutrient availablilty, and buffer capacity.

Fundamentals

Microbial Reactions

In anaerobic processes, microorganisms are used under anaerobic conditions to convert organic solids to methane, carbon dioxide, and nonbiodegradable solids. The microbial reactions and some of the factors affecting these reactions have been discussed earlier in Chapters 5 and 6.

Buffer Capacity

If volatile acids are produced at a faster rate than they are utilized, adverse conditions will not occur as long as the buffer capacity of the system can neutralize the excess acids. The buffer capacity can be expressed as the alkalinity of the system. With a fixed detention time, the alkalinity in an anaerobic unit will be in proportion to the organic loading to the unit. Figure 8.1 was derived from a number of anaerobic units digesting sewage sludge. Early digestion tanks and some today are designed at loadings of less than 0.1 lb of volatile solids per cubic foot per day (lb VS/cfd). By noting the alkalinity in digesters at these loadings, it is not difficult to understand why variations in volatile acid concentrations caused by intermittent loading conditions and temperature variations can cause these digesters to become inhibited. Anaerobic units that are lightly loaded not only may have inadequate alkalinity to buffer volatile acid variations, but they also will have

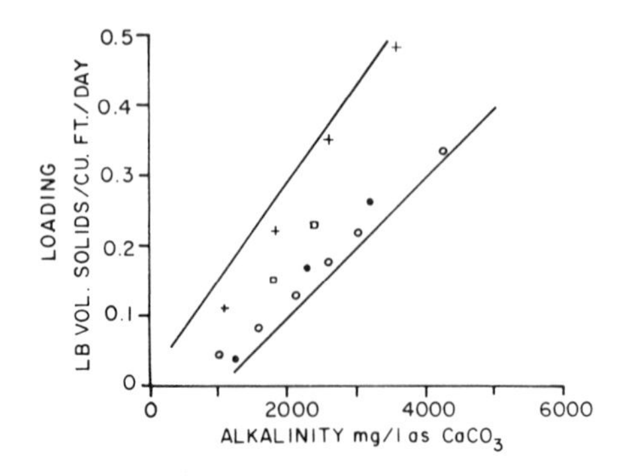

Fig. 8.1. Alkalinity in digesters is a function of the loading.

a low population of methane bacteria to readily metabolize any large increase in volatile acids.

The parameters of pH, alkalinity, and volatile acid content are related. A knowledge of only pH is of little value and must be considered in relationship to the alkalinity (buffering capacity). The volatile acid content of the mixed liquor of a digester is a useful parameter in controlling digestion. Any sudden change from a constant value is an indication of disruption of equilibrium conditions. The inhibition noted in anaerobic units with high volatile acid concentrations has been shown to be the result of the cations associated with the neutralized volatile acids, rather than of the acids or the acid anions (2). As long as cation toxicity does not exist and as long as there is sufficient alkalinity to neutralize the volatile acids that are produced, the anaerobic metabolic reactions proceed and the anaerobic unit performs satisfactorily. High and increasing volatile acid concentrations indicate that the methane bacteria are being limited or inhibited. The cause of the limitation or inhibition should be determined and corrected as rapidly as possible. Low and/or decreasing volatile acid concentrations are indicative of satisfactory anaerobic metabolism. Signs of anaerobic process failure include decreasing alkalinity, decreasing pH, increasing volatile acids, and increasing carbon dioxide content of the digester gas.

The data used in developing Fig. 8.1 were obtained from anaerobic units with long liquid retention times. Systems with short liquid retention times will have decreased buffer capacity, at comparable solids loading and tank size, because of the greater hydraulic flow through the unit.

Process Inhibition

Excessive volatile acids are not the only items that can be detrimental to anaerobic units. Industrial wastes may contribute inhibitory concentrations of inorganic and organic materials. Alkaline earth metal salts, such as those of sodium, potassium, calcium, or magnesium, which may be in high concentrations in certain wastes, can be the cause of the inefficiency and/or failure of anaerobic systems. The inhibition has been shown to be associated with the cation portion of the salt. The cation effects in anaerobic units are a function of the types and concentrations of all cations present in the environment. Optimum ionic concentrations for maximum efficiency of anaerobic treatment were found to be 0.01 *M* for the monovalent ions sodium, potassium, and ammonium and 0.005 *M* for the divalent ions calcium and magnesium (3). On the molar basis, the order of increasing inhibition of the cations, when in excess of the optimum concentration, was sodium, ammonium, potassium, calcium, and magnesium. Concentrations ranging from 1000 to 5000 mg/l usually have an inhibitory effect on anaerobic processes. Combinations of cations act in a different manner than single ion concentrations. The divalent ions are less toxic if an optimum concentration of a monovalent ion is present.

In anaerobic systems, ammonia will be in equilibrium as the ammonium ion or as dissolved ammonia gas:

$$NH_4^+ \rightleftharpoons NH_3 + H^+ \tag{8.1}$$

As the pH of the system increases, equilibrium shifts to the right and more free ammonia is in solution. The ammonium ions result from either degradation of nitrogenous compounds, such as proteins and urea, or addition of ammonium compounds as an additional source of alkalinity. A natural relationship will exist between the ammonia nitrogen content of the digester contents, the digested sludge volatile solids, and the alkalinity in the digester. Additional information in ammonia inhibition in biological systems is presented in Chapter 11.

The free ammonia concentration determines the ammonia nitrogen toxicity. High pH values will cause a shift to the more toxic free ammonia form. When the free ammonia concentration exceeded about 150 mg/l, laboratory anaerobic units ceased to function. Gas production diminished markedly when the free ammonia concentration reached 140 mg/l (4). Ammonia nitrogen concentrations above 3000 mg/l at any pH have been stated as detrimental (5).

High rate anaerobic treatment operates in a restricted pH range which is controlled by the ammonia nitrogen content which in turn regulates the equilibrium within the treatment unit. An ammonia nitrogen concentration of 1250 mg/l has been indicated as a limiting value (6). Full-scale digestion studies of sewage sludge have indicated that ammonia nitrogen became toxic or inhibitive only when the rate of its formation above a threshold level of 1700–1800 mg/l increased more rapidly than the ability of the methane organisms to acclimate to the ammonia. The rate of ammonia nitrogen formation was controlled by proper feeding of the digester and using the volatile acid–alkalinity ratio as an operational control (7). Digester loadings were 0.11–0.16 lb VS/cfd. Both the ammonium nitrogen concentration and the pH of the digester must be known to determine the free ammonia concentration that might be the cause of the ammonia toxicity in a digester.

A change in environmental conditions may cause adverse volatile acid–alkalinity relationships. Additional buffering capacity can be added to keep the system in balance while the environmental conditions are corrected. The information on inhibition caused by alkaline metal salts indicates that care should be taken when adding more buffering capacity to anaerobic systems for control. Adverse synergistic cation inhibition should be avoided. When anaerobic systems appear inhibited, cation analyses of the system and of the waste are useful. If high concentrations of the inhibitory cations cannot be eliminated from the waste, then dilution of the waste may be needed.

Certain industries may contribute heavy metal salts to wastes. Low concentrations of copper, zinc, chromium, and nickel can be toxic to anaerobic treatment. Studies have indicated that soluble copper and zinc concentrations as low as 10 mg/l have resulted in complete inhibition of gas production. Precipitation of these metals will remove the toxic effect of the metals. Toxic metals should be removed before the wastes enter an anaerobic waste treatment system. Should the metals reach the anaerobic unit in sufficient concentration to cause inhibition, the metals can be precipitated as their sulfides. Sulfates or sulfides can be added to the anaerobic units if adequate quantities are not in the waste or the units. Approximately 0.5 mg/l of sulfide is needed for each 1 mg/l of toxic heavy metal present. The addition of sulfate or sulfide precursors has allowed the presence of more than 10% of heavy metals, on a dried volatile solid basis, without inhibiting the digestion process (3).

Excessive concentrations of sulfides are toxic and should be prevented. Sulfates are the major precursors of sulfides in anaerobic treatment units. Sulfides also can result from the anaerobic degradation of sulfur containing organic and inorganic compounds and from sulfides in the raw wastes. Soluble sulfide concentrations up to 200 mg/l as sulfur exerted no significant effect on anaerobic treatment. Concentrations of soluble sulfides greater than 370 mg/l as sulfur have been shown to produce severe toxic effects and complete inhibition of gas production (3). Insoluble sulfide complexes have had no significant effect on anaerobic treatment up to at least 400 mg/l of sulfide. Excessive concentrations of sulfides can be precipitated to reduce the sulfide toxicity in the anaerobic unit. Iron salts can be used for such precipitation. Coprecipitation of heavy metals and sulfides will decrease the toxicity of both types of ions.

While oxygen is important in aerobic biological units, the addition of oxygen to anaerobic units will be detrimental and should be avoided. The advantages of anaerobic systems such as the production of a usable end product, methane, and minimal biological solids production occur because the degradation of organic matter takes place in the absence of oxygen. As noted in Chapter 5, microorganisms will use metabolic pathways that yield the greatest amount of energy. Satisfactory anaerobic digestion will occur when the microorganisms obtain a minimum amount of energy from the metabolic reactions and when the end product of the reaction contains considerable usable energy, i.e., the methane.

Oxygen and oxidized material such as nitrates should not be added to anaerobic units since they will be used as hydrogen acceptors in preference to oxidized organic matter and carbon dioxide. When nitrates have been added to anaerobic units as a source of nitrogen or in hopes of increasing the buffering capacity of the units, volatile acids have increased and gas production decreased significantly (8, 9). Nitrate inhibition of methane production in sewage digestion studies was found to be small at a continuous dose of 10 mg/l NO_3-N and virtually complete at 50 mg/l NO_3-N. Methane formers could withstand at least 100 mg/l NO_3-N on a continuous basis and 250 mg/l NO_3-N as a slug dose and again produce methane after the nitrate was elimminated. Methane production resumed within a few hours after the nitrate disappeared (10). The organic loading, i.e., hydrogen acceptor demand, and the number of active organisms in a digester are important in evaluating the effect of the addition of any oxidized compound such as nitrate. Different anaerobic processes will have different tolerances.

Whenever chemical additives are required in an anaerobic system they should be added in the most reduced forms. For example, the reduced forms of nitrogen such as ammonium carbonate, chloride, and hydroxide and anhydrous ammonia are suitable sources of nitrogen and buffering capacity while nitrates are not suitable. When chemical additives are introduced to an anaerobic system, the resulting concentrations should be maintained below inhibitory concentrations.

Nutritional Requirements

The nutritional requirements are related to the net growth of microbial cells (Chapter 5), i.e.,

$$\Delta Ma = aF - b(Ma) \tag{8.2}$$

Values of a, based on COD removal or utilization in anaerobic systems, range from 0.05 to 0.18. Values of b, based on mixed liquor volatile solids in anaerobic systems, range from 0.01 to 0.1 per day. These ranges of a and b have resulted from laboratory investigations using synthetic soluble substrates. Confirmatory data are needed on pilot plant and full-scale systems treating complex organic wastes. In anaerobic systems treating nutritionally balanced organic material such as sewage sludge or animal wastes, a deficiency of any of the necessary elements is unlikely. However, when anaerobic systems are designed to stabilize food processing wastes that may not be nutritionally balanced, the possibility of nutrient limitation should be investigated.

Retention Time

The advantage of anaerobic systems is with high solids wastes and high loading rates. An important parameter is the length of time the wastes, especially the solids, remain in the systems. In a continuous flow completely mixed anaerobic unit, the liquid retention time and the solids retention time are equal. In units that are not completely mixed, or that employ recycle, the solids retention time is greater than the liquid retention time. The relative importance of these retention times in biological systems was discussed in Chapter 5.

The biological solids retention time (SRT) represents the average retention time of microorganisms in the system and can be determined by dividing the pounds of volatile solids in the system by the pounds of volatile solids leaving the system per day. If the SRT of the system is less than the minimum microbial reproduction time, they are removed from the system faster than they can reproduce. In anaerobic systems, the microbial growth rate is low and the minimum SRT values are much longer than for aerobic treatment. Minimum SRT values for anaerobic systems have been estimated to be in the range of 2–6 days. Both liquid and solids retention time deserve attention in utilizing anaerobic systems.

Temperature

Temperature will affect the performance of biological systems since it affects the activity of the microorganisms. The optimum temperature of mesophilic anaerobic treatment is 30° to 37°C (11). Although the time to obtain a desired degree of treatment is less with thermophilic treatment than with mesophilic treatment, thermophilic conditions have not been demonstrated as practical or economical to date.

It does not follow that anaerobic treatment *must* occur at optimum mesophilic temperatures. Satisfactory anaerobic treatment can occur if an adequate mass of active microorganisms (*Ma*) and a sufficiently long SRT are provided for the system. The factors *a* and *b* in Eq. (8.2) are temperature-dependent and decrease with decreasing temperature. At reduced temperatures, microbial solids production will be less and there will be less active organisms in the system. At low temperatures, the SRT must increase if comparable degrees of waste treatment are to be accomplished. This can be accomplished by solids recycle. Relationships between temperature and minimum SRT values have not been well established. The minimum SRT values at 25° and 15°C may be about 1.2 to 1.5 and 2.0 to 3.0, respectively, of the values at 35°C (12).

The rate of digestion is not the only factor to be evaluated in selecting proper operating temperatures. Other factors include the amount of heat required to raise the influent to the operating temperature, and the characteristics, especially the settleability, of the anaerobic sludges produced at different temperatures. In practice, the range of 25°–35°C appears to be most convenient and economic. Reasonably constant temperature is important to the process.

Anaerobic treatment of wastes such as sewage sludge and organics that are equally biodegradable can produce gas at the rate of 8–9 ft^3/lb VS added to the anaerobic units. The methane content of the gas will range between 60 and 80%. It may be difficult, if not impossible, to maintain mesophilic temperatures using only the gas produced in the anaerobic treatment of many organic wastes. Supplemental heat can be used to maintain optimum temperatures. An alternative is to compensate for the decrease in rate constants by increasing the active mass and by maintaining the SRT greater than the minimum SRT at the lower temperatures. It is possible to design and operate full-scale systems at ambient temperatures if proper attention is paid to the fundamentals of anaerobic treatment.

Mixing

The maximum rate of biological reaction takes place when the organisms are brought continuously into contact with the organic material. Mixing also serves a number of other functions such as: (a) maintenance of uniform temperatures throughout the unit, (b) dispersion of potential metabolic inhibitors, such as volatile acids, and (c) disintegration of coarser organic particles to obtain with a greater net surface for increased rates of degradation.

The value of mixing has been reported by many investigators. Successful digestion was achieved at solids loadings considerably in excess of conventional design loadings when the digester contents were thoroughly mixed (13). With agitation, approximately 85% BOD reduction occurred in less than 3 days whereas with quiescent digestion, 25 days were required to produce the same results (14). These results have been verified in full-scale installations.

Natural mixing obtained from gasification plus sludge recirculation can keep highly loaded anaerobic units adequately mixed. Modern gas and mechanical mixing devices are able to keep anaerobic units uniformly mixed. The type of solids distribution that occurs in mixed and unmixed anaerobic units is shown in Fig. 8.2. The optimum quantity of mixing per unit volume has yet to be adequately evaluated.

Loading

Recommended loading rates for anaerobic units treating sewage solids and wastes of a similar nature are in the range of 0.04 to 0.07 lb VS/cfd (15). Both laboratory and pilot plant anaerobic systems have demonstrated that higher loading rates are possible if optimum environmental conditions are maintained in the anaerobic unit. Optimum conditions include: adequate mixing, continuous feeding, adequate SRT, and high temperatures. When these conditions have been incorporated into anaerobic units, loading rates between 0.1 and 0.5 lb VS/cfd have been handled satisfactorily (16–20). Because sewage sludges have a low solids concentration, generally between 1 and 5%, higher loading rates are diffIcult to obtain unless special consideration is given to concentration of the sludges before they are sent to the anaerobic digester. Higher loading rates may be feasible for the anaerobic treatment of certain wastes that have a high organic concentration, such as packinghouse waste solids and animal wastes.

Applications

While anaerobic systems have been applied widely to wastes having a high solids content, they also can be applied to more soluble wastes and to

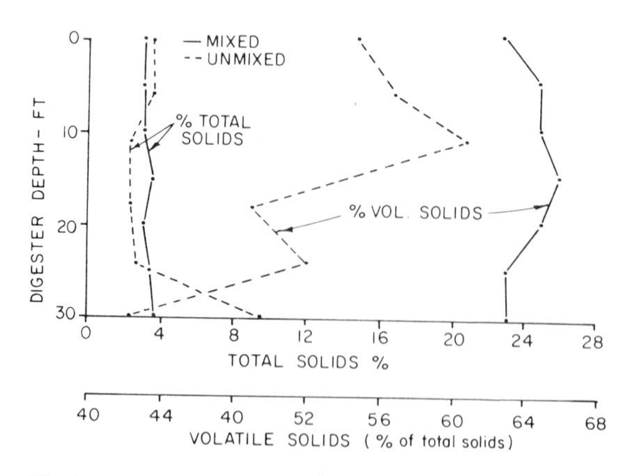

Fig. 8.2. Solids distribution in mixed and unmixed digesters.

wastes with a low organic concentration if proper attention is given to the fundamentals of the process. Anaerobic systems for wastes with a low organic content must maintain a long SRT to accomplish adequate treatment. In a completely mixed anaerobic unit, the liquid and solids retention times are the same. With dilute wastes, the large unit volumes that may be required may be uneconomical. Alternative systems employing solids recycle or solids holding can produce long SRT values while obtaining low liquid retention times and smaller unit volumes. Anaerobic treatment should not be relegated only to wastes with high solids content but also should be evaluated for partially or completely soluble wastes and for dilute organic wastes.

Basic Anaerobic Processes

There are four basic processes that can be employed for anaerobic waste treatment: single stage, two stage, two stage with solids retention, and anaerobic filter.

Single Stage

The single stage anaerobic unit (Fig. 8.3) represents one of the oldest anaerobic processes. Both biological stabilization and solid–liquid separation are to be accomplished in the same unit. Conditions optimum for both factors are difficult to achieve in a single unit. Adequate mixing will inhibit solids separation by sedimentation which is the usual method of solid–liquid separation in anaerobic systems. Quiescent conditions for satisfactory gravitational solids separation will not allow adequate food and microorganism contact for optimum biological action. Liquid detention times are long, on the order of 20+ days, solids retention times are even longer, the active mass is small because of the low loading and poor contact with the wastes, and the system is more sensitive to load and temperature fluctuations than are some of the other anaerobic units. Anaerobic lagoons and animal waste storage units are examples of the practical use of a single stage anaerobic unit for agricultural wastes. The most common use of anaerobic units for agricultural wastes is as a preliminary treatment system prior to further treatment, such as an aerobic unit, or to disposal, such as on the land.

Two Stage

Realization that optimum conditions for both solids separation and biological action are incompatible in a single state anaerobic unit led to the

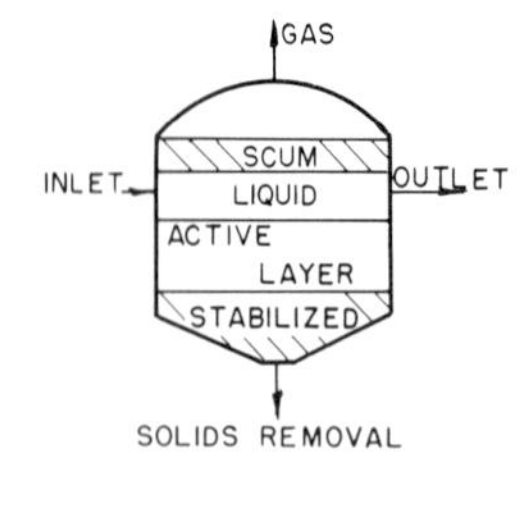

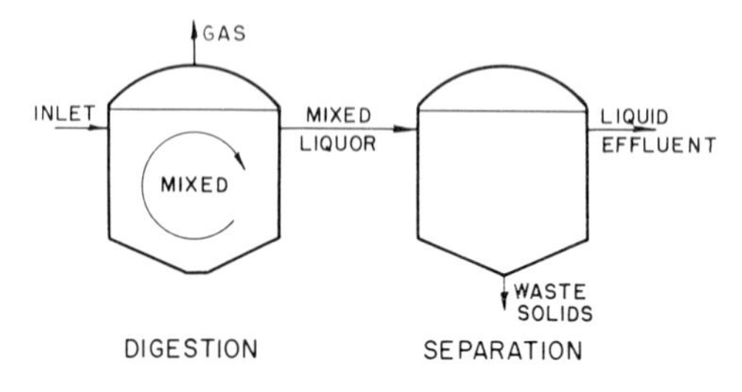

Fig. 8.3. Single and two-stage digestion systems.

development of the two-stage system (Fig. 8.3). The first stage is a continuously fed and completely mixed, temperature-controlled biological unit. The second stage is a solids concentrator and separator. Sedimentation tanks and centrifuges may serve this function. Two advantages accrue from a continuously and completely mixed first stage anaerobic unit: (a) the entire volume of the first stage may be used for biological stabilization of the organic waste and (b) such a unit can be designed for optimum biological action.

The development of two-stage systems has permitted the higher loadings, above 0.1 VS/cfd, to be successful. The solid and liquid retention times in the first stage are the same. The capacity of the first stage is controlled by the liquid retention time which must be more than the minimum SRT. No flexibility to alter the SRT exists after the system is built.

Two Stage with Solids Return

Systems of this type (Fig.8.4) provide greater flexibility in maintaining a minimum SRT, in maintaining a higher active mass of microorganisms in the system, and in permitting the use of smaller first stage tank capacities. With sludge return, the microbial population can be maintained at levels to obtain better efficiency in a unit having greater stability than that found in single stage and two-stage anaerobic units without sludge return. A system with sludge return to a mixed unit would have a small liquid detention time and a large solids retention time. This anaerobic system is analogous to an activated sludge system.

The advantages of sludge recycle in biological systems have been apparent for many years, although it has not been accepted as part of anaerobic systems to the extent that it has in aerobic systems. Laboratory and pilot plant studies on the anaerobic digestion of meat-packing wastes (21) indicated feasibility of high rate, completely mixed anaerobic systems with sludge return for these wastes. A full-scale system has been designed and built based on the pilot plant studies (19). The detention time in the first stage, based on influent flow, was approximately 12 hr and that based on total flow was approximately 3.5 hr. The sludge retention time was greater, a matter of 10–15 days.

Although an anaerobic system with sludge return has advantages for the treatment of any biodegradable organic waste, the greatest advantages are in treating dilute wastes. In systems without solids recycle, the dilute wastes may not support an adequate active microbial mass. Solids recycle permits

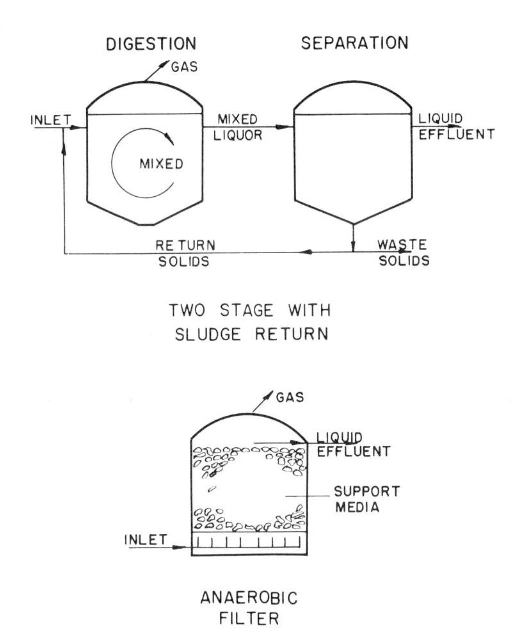

Fig. 8.4. Schematic of a two-stage digestion system with sludge return and of an anaerobic filter.

smaller unit volumes and efficient waste stabilization. The key to proper solids recycle lies in producing a sludge that will separate or settle well.

Anaerobic Filter

The anaerobic filter (Fig. 8.4) represents the newest anaerobic treatment process (22). The unit is useful for the treatment of dilute soluble wastes and in denitrifying oxidized effluents for nitrogen control. The filter consists of a bed of submerged media through which waste flows in an upward direction. The anaerobic organisms grow within the bed and either cling to the media or grow in the voids. The filter has a large capacity for retaining microorganisms. Of the types of media that have been tested, one-inch rounded stones appear to be the best (23).

Wastes with COD concentrations of over 750 mg/l appear well suited for this process. COD removal efficiencies of 90+% have been obtained with hydraulic detention times greater than 18 hr and with wastes having COD concentrations up to 2000 mg/l (23). These results were obtained at 25°C. Effective treatment was obtained at 15°C although the rates were lower than at 25°C.

The microorganisms remain in the filter for a long time, creating a long SRT even though the hydraulic retention time is short. No solids recycle is incorporated in the anaerobic filter. Because of the long SRT, it is possible to provide anaerobic waste treatment at ambient temperatures, at high loading rates, with little power required, and with minimum solids production for further disposal. The filter responds to changing waste loads and has been indicated to operate well on a periodic basis (23). This would be an advantage for wastes such as seasonal food processing wastes.

Difficulties will be encountered with wastes that will clog the filter media and therefore high solids wastes are not appropriate unless solids removal precedes the anaerobic filter. The wastes must have adequate alkalinity to buffer any pH reduction during treatment. Initial start up of an anaerobic filter should be done carefully to establish the proper organisms. Once the methane organisms are established, oxidized conditions should not be permitted to occur in the filter.

Application to Agricultural Wastes

Anaerobic treatment can be feasible with agricultural wastes because of their high organic strength. Many investigations have indicated the potential of the process with different wastes and under a variety of environmental conditions. The studies cited in this section are indicative of the results that were obtained and are presented to illustrate the application of the fundamentals to the anaerobic treatment of agricultural wastes.

Controlled studies with slaughterhouse wastes demonstrated the effect of temperature on the performance of anaerobic digestion systems and on the quality of sludge produced (24). At BOD loading rates of 59–75 lb/1000 ft^3 of digester volume/day and liquid retention times of 1.4–1.8 days, BOD removals from 74 to 95% were obtained. The amount of BOD removed for a given volume of digester increased approximately linearly with the strength of the waste being treated. Temperatures from 25° to 55°C were investigated for their effect on the performance. The more stable conditions occurred at 33°C with a BOD removal of 95% and the sludge in granular form, free of gas bubbles, and able to be removed from the solution by settling. At lower temperatures the sludge became thixotropic, entrapping gas bubbles and forming a layer of scum on the surface.

The above studies were expanded to a larger scale in which the digested solids were separated from the effluent and returned to the digester in a ratio of 1 : 1 by volume. BOD removals of 95% were obtained. The loading rate was 110 lb BOD/1000 ft^3/day with a hydraulic retention time of 30 hr. Operating temperature was 33°C. Interruption of heating and waste supply had little effect on the operation of the digester (20). The economy of the proposed process was linked to the heat energy balance and through it to the strength and temperature of the influent wastes. Because of the warm temperatures and high strength of most slaughterhouse wastes, the heat produced by burning the resultant digestion gases could be enough to maintain the desired temperatures in the system without excessive or perhaps no auxiliary heat or power. Similar heat relationships were noted in a full-scale demonstration of a solids recycle system treating meat-packing wastes (21).

Two-stage anaerobic treatment with solids return was used successfully with pear processing wastes over a loading rate range of 0.10 to 0.46 lb VS/ft^3/day (25). At low liquid retention times of 0.5 and 1 day, treatment efficiency decreased because of difficulties in solids flocculation and separation. COD removal efficiencies of up to 95% were obtained.

The anaerobic filter has been used to treat potato starch wastes. With a hydraulic detention time of 1.6 days, a loading of 5.1 lb COD/1000 ft^3/day, and operating at 37°C, an anaerobic filter obtained 76% COD removal efficiency (26). Potato processing wastes have been treated in pilot scale anaerobic filters. Average organic removals of 70% were obtained at loadings of about 57 lb BOD/1000 ft^3/day. Average suspended solids removal at the loading was about 60%. The filter operated at a temperature of 25°C and the entering waste had an average pH of 10.2. The gas that was produced had a methane content of 70% (26).

Whey has been treated anaerobically. At a loading of 26 lb/day/1000 ft^3

digester capacity and a hydraulic detention time of 80 days, the BOD was reduced 99%. Similar results were obtained at a loading of 130 lb BOD/day/1000 ft^3 with a six-day detention time (27). The gas yield was 5 ft^3/gal of whey fermented and the gas contained 65% methane. The ammonia nitrogen in the digester contents increased from 75 to 450 mg/l as the proteins degraded. No pH control was used and the pH of about 4.3 in the raw wastes was increased to 6.9 in the digester effluent.

Laboratory investigations with beef cattle wastes have demonstrated that loading rates from 0.1 to 0.4 lb of total solids/ft^3/day have been successful (28). Analysis of data from that study indicated that even higher loading rates may have been feasible. Other investigations on the anaerobic digestion of beef cattle manure have demonstrated that between 8 and 9 ft^3 of gas/lb VS added can be expected from digestion systems treating beef cattle wastes (29, 30). Similar studies with dairy cow and sheep manure indicated only 2.5 to 3 ft^3 of gas/lb VS added could be expected (29). The low values that occurred with the cow and sheep manure undoubtedly were caused by the fact that not all of the volatile solids in these manures were biodegradable, demonstrating that the waste characteristics are influenced by the type of food ration and type of animal. The gas produced from digestion of animal wastes generally contains between 50 and 70% methane depending upon the environmental conditions. Carbon dioxide is the other major component of the gas.

Poultry and dairy manure were digested at rates from 0.17 to 0.31 and from 0.13 to 0.22 lb VS/cfd, respectively (30). Both manures can be stabilized with controlled digestion. With dairy manure only about 10–15% of the volatile solids were destroyed. About 50% of the volatile solids in the poultry manure were destroyed.

When poultry manure was digested at a rate of 0.088 lb VS/cfd, the ammonia nitrogen concentration in the digesters reached toxic levels (31). Ammonia toxicity could be a significant problem in the high rate of digestion of concentrated solids with a high nitrogen content, such as poultry wastes.

Controlled conditions were found necessary for the active digestion of duck wastes. Without controlled environmental conditions, the volatile acids concentration increased to the point that bacterial metabolism ceased. Continuous loading rates as high as 0.15 lb VS/cfd were successful (32).

Comparative studies of animal wastes indicated that dairy, beef, swine, and poultry wastes were amenable to anaerobic digestion. Suggested digester loading rates and hydraulic detention times for the three wastes were 0.24 lb VS/cfd and 10 days for dairy bull wastes, 0.24 lb VS/cfd and 10–15 days for swine wastes, and less than 0.18 lb VS/cfd and 10–15 days for poultry wastes (33). The alkalinity in the digesters increased as the loading rate increased. The volatile solids reduction in digestion reflected the prior history and relative biodegradability of the wastes. Volatile solids reductions for dairy bull waste ranged from 18 to 26%, for swine from 49 to 61%, and for poultry 57–68%. The methane concentration of the produced gases was within the normal range: 61–66% for dairy bull waste, 58–61% for swine waste, and 52–58% for poultry wastes.

Digestion of dairy manures produced similar results, i.e., low gas production 1.5–2.0 ft^3/lb VS destroyed, and low reduction in volatile solids, 38–53% (34). The digesters were loaded at rates of 0.10–0.18 lb VS/cfd. The percent methane in the gas ranged from 74 to 79%.

Digestion of swine wastes was found to be inhibited at loading rates of 0.25 and 0.50 lb TS/cfd (35). Volatile acids ranged from 16,000 to 19,000 mg/l and were in excess of the alkalinity. Representative pH values in the digesters were 6.6 to 6.9. Ammonia nitrogen concentrations of 1800 to 2500 mg/l existed and it was suggested that the methane bacteria were inhibited by the high ammonia nitrogen concentrations. Intermittent mixing provided the same results as continuous mixing. Little settling of the digested swine wastes occurred and solids recycle to increase the SRT was not possible. The liquid detention times in the digesters was from 10 to 20 days.

Based upon available data, a loading rate of 0.1 to 0.2 lb VS/cfd for mixed anaerobic digesters treating animal wastes should produce successful digestion and solids stabilization. Such values are only guidelines since temperature, method of feeding, mixing, toxicity problems, and operational control all affect the success of the process. Higher loadings may be possible for dairy cattle wastes because of their lower biodegradable content. Dairy cattle and sheep wastes can be expected to produce smaller quantities of gas per volatile solids destroyed than will poultry, swine wastes, and wastes from beef cattle in confinement since dairy and sheep wastes have undergone considerably more degradation in the animals.

Digestion systems for agricultural wastes will not produce an effluent acceptable for discharge to surface waters. A volume comparable to that of the original waste will remain and the transport problem of the contents or effluent of the digester will be of the same order as that of the untreated waste. If incomplete digestion occurs, an odor problem results.

A summary of the loading rates and anaerobic processes used with agricultural wastes is presented in Table 8.1. Because of the necessary equipment, the high initial cost, the potential operating problems, the need for competent operators, and the fact that further treatment and disposal are necessary, controlled anaerobic digestion systems will not be widely applied to agricultural wastes. The exception could be very large animal production operations where adequate financing, operational control, and utilization of the gases and residue are possible. Single stage digestion systems can be

Table 8.1

Anaerobic Treatment of Agricultural Wastes

Waste	Temperature (°C)	Process loading lb BOD/1000 ft³/day	lb VS/ft³/day	Process	Ref.
Slaughterhouse	25–55	59–75	—	Two stage with solids return	24
Slaughterhouse	33	110	—		21
Pear	35	—	0.10–0.46	Two stage with solids return	25
Potato starch	37	5.1	—	Anaerobic filter	26
Potato processing	25	57	—	Anaerobic filter	26
Whey	37	26	—	Single stage	27
Whey	37	130	—	Single stage	27
Beef cattle	23	—	0.1–0.4	Single stage	28
Poultry manure	—	—	0.17–0.31	Single stage	30
Dairy manure	—	—	0.13–0.22	Single stage	30
Poultry manure	35	—	0.088	Single stage	31
Duck manure	37	—	Up to 0.15	Single stage	32
Dairy bull	—	—	0.24	Single stage	33
Swine	—	—	0.24	Single stage	33
Poultry manure	—	—	0.18	Single stage	33
Dairy manure	—	—	0.1–0.18	Single stage	34
Swine	35	—	0.25–0.50	Single stage	35

important parts of a suitable anaerobic–aerobic treatment system. Anaerobic filters appear to have promise for soluble, high strength food processing wastes.

References

1. Cillie, G. G., Henzin, M. R., Stander, G. J., and Bailtie, R. D., Anaerobic digestion. IV. Application of the process in waste purification. *Water Res.* **3,** 623–644 (1969).
2. McCarty, P. L., and McKinney, R. E., Volatile acid toxicity in anaerobic digestion. *J. Water Pollut. Contr. Fed.* **33,** 223–232 (1961).
3. McCarty, P. L., Kugelman, J. J., and Lawrence, A. W., "Ion Effects in Anaerobic Digestion," Tech. Rep. No. 33. Dept. Civil Eng., Stanford University, Stanford, California, 1964.
4. McCarty, P. L., and McKinney, R. E., Salt toxicity in anaerobic digestion. *J. Water Pollut. Contr. Fed.* **33,** 399–415 (1961).
5. McCarty, P. L., Anaerobic waste treatment fundamentals. 3. Toxic materials and their control. *Pub. Works* November, 41–44 (1964).

6. Albertson, O. E., Ammonia nitrogen and the anaerobic environment. *J. Water Pollut. Contr. Fed.* **33,** 978–995 (1961).
7. Melbinger, N. R., and Donellon, J., Overloading the toxic effects of ammonia nitrogen in high rate digestion. *J. Water Pollut. Contr. Fed.* **43,** 1658–1670 (1971).
8. Cramer, F. J., Curing a sick digester. *J. Water Pollut. Contr. Fed.* **37,** 1317–1318 (1965).
9. Straub, C. P., Effect of addition of nitrogen on digestion of paper pulp. *Sewage Works J.* **16,** 20–41 (1944).
10. Brezonik, P. L., and Lee, G. F., Sources of elemental nitrogen in fermentation gases. *Int. J. Air Water Pollut.* **10,** 145–160 (1966).
11. Fair, G. M., and Moore, E. W., Effect of temperature of incubation upon the course of digestion. *Sewage Works J.* **4,** 589–600 (1932).
12. Lawrence, A. W., and McCarty, P. L., "Kinetics of Methane Fermentation in Anaerobic Waste Treatment," Tech. Rep. No. 75. Dept. Civil Eng., Stanford University, Stanford, California, 1968.
13. Morgan, P. F., Studies of accelerated digestion of sewage sludge. *Sewage Ind. Wastes* **26,** 462–478 (1954).
14. Rudolfs, W., and Trubnick, E. H., Compressed yeast wastes treatment. 2. Biological treatment in laboratory units. *Sewage Works J.* **21,** 100–109 (1945).
15. American Society of Civil Engineers, "Sewage Treatment Plant Design," Manual of Practice No. 36 Amer. Soc. Civil Eng., New York, New York, 1959.
16. Estrada, A. A., Design and cost considerations in high-rate sludge digestion. *J. Sanit. Eng. Div., Amer. Soc. Civil Eng.,* **86,** SA3, 11–27 (1960).
17. Nash, N., and Chasick, A. H., High rate digester performance at Jamaica. *J. Water Pollut. Contr. Fed.* **32,** 526–536 (1960).
18. Torpey, W. N., High-rate digestion and concentrated primary and activated sludge. *Sewage Ind. Wastes* **26,** 479–494 (1954).
19. Steffen, A. J., and Bedker, M., Operations of a full scale anaerobic contact treatment plant for meat packing wastes. *Proc. Purdue Ind. Waste Conf.* **16,** 423–437 (1962).
20. Pettet, A. J., Tomlinson, T. G., and Hemens, J., The treatment of strong organic wastes by anaerobic digestion. *J. Inst. Pub. Health Eng.,* July, 170–191 (1959).
21. Schroepfer, G. J., Fullen, W. J., Johnson, A. S., Ziemke, N. R., and Anderson, J. J., The anaerobic contact process as applied to packinghouse wastes. *Sewage Ind. Wastes* **27,** 460–486 (1955).
22. Young, J. C., and McCarty, P. L., The anaerobic filter for waste treatment. *J. Water Pollut. Contr. Fed.* **41,** R160–R173 (1969).
23. McCarty, P. L., "Biological Treatment of Food Processing Wastes," Proc. 1st Nat. Symp. Food Process. Wastes, pp. 327–346. Pacific Northwest Water Lab., Environmental Protection Agency, Corvallis, Oregon, 1970.
24. Lloyd, R., and Ware, G. C., Anaerobic digestion of wastewaters from slaughter-houses. *Food. Mfr.* **31,** 511–515 (1956).
25. van den Berg, L., and Lentz, C. P., Anaerobic digestion of pear waste: Factors affecting performance. *Proc. Purdue Ind. Waste Conf.* **27** (1972).
26. Guttormsen, K., and Carlson, D. A., "Potato Processing Waste Treatment—Current Practice," Water Pollut. Contr. Res. Ser. DAST-14. Federal Water Pollution Control Administration, Dept. of Interior, 1969.
27. Parker, C. D., "Methane Fermentation of Whey," Proc. 2nd Nat. Symp. Food Process. Wastes, pp. 501–508. Pacific Northwest Water Lab., Environmental Protection Agency, 1971.
28. Loehr, R. C., and Agnew, R. W., Cattle wastes—pollution and potential treatment. *J. Sanit. Eng. Div., Amer. Soc. Civil Eng.* **93,** SA4, 72–91 (1967).

29. Jeffrey, E. A., Blackman, W. L., and Ricketts, R. L., "Aerobic and Anaerobic Digestion Characteristics of Livestock Wastes," Eng. Ser. Bull. No. 57. University of Missouri, Columbia, 1963.
30. Hart, S. A., Digestion tests of livestock wastes. *J. Water Pollut. Contr. Fed.* **35**, 748–757 (1963).
31. Cassell, E. A., and Anthonisen, A., "Studies on Chicken Manure Disposal," Part 1, Laboratory Studies, Res. Rep. No. 12, New York State Dept. of Health, Albany, 1966.
32. Gates, C. D., "Treatment of Long Island Duck Farm Wastes," Res. Rep. No. 4. New York State Water Pollution Control Board, Albany, New York, 1959.
33. Gramms, L. C., Polkowski, L. B., and Witzel, S. A., Anaerobic digestion of farm animal wastes (dairy bull, swine, and poultry). *Annu. Meet. Amer. Soc. Agr. Eng.* Paper 69-462 (1969).
34. Dalrymple, W., and Proctor, D. E., Feasibility of dairy manure stabilization by anaerobic digestion. *Water Sewage Works,* 361–364 (1967).
35. Schmid, L. A., and Lipper, R. T., "Swine Wastes, Characterization and Anaerobic Digestion," Proc. Agr. Waste Manage. Conf., pp. 50–57. Cornell University, Ithaca, New York, 1969.

9

Utilization of Agricultural Wastes

General

Agricultural wastes are excesses of agricultural production that have not been effectively utilized. Most waste management approaches are methods of concentration and/or relocation of wastes, such as source separation, biological waste treatment, incineration, or land disposal. Recycling, reprocessing, and utilization of the wastes in a positive manner offers the possibility of returning the excesses to beneficial use as opposed to the traditional methods of waste disposal and relocation. The keys to successful processes of this nature are a beneficial use, an adequate market, and an economical, although not necessarity profit-making process. Many such processes would be satisfactory if they caused the overall cost of waste management to be less than other alternatives. Any additional steps in utilization should repay extra storage, processing, and distribution costs that are incurred. A return greater than the extra cost of utilization is desirable in that it reduces the total cost of waste management but such reduction may not be sufficient to result in an overall profit for the producer.

The common method of utilizing agricultural wastes has been to return them to the land. Labor costs have risen and convenient land can be at a premium. The cost of handling agricultural wastes may make them not competitive in price with inexpensive chemical fertilizers. Investigations on other utilization methods have increased.

The utilization of waste materials from agricultural production operations can assist in reducing some waste management problems. Many examples can be cited. Fruit and vegetable wastes are being utilized as stock feed. Tomato skins and seeds have been dehydrated and used as part of animal feeds, as have corn husks, cobs, and trimmings. Solid wastes from the canning of peas, corn, grapefruit, oranges, and some solids screened from the liquid wastes of other processes have been converted into a dried cattle feed. Animal manure is being composted, dried, and pelletized for soil conditioners, animal feed supplement, and fertilizer base. Peach pits have been converted into charcoal briquets. The use of horse manure as a medium for growing mushrooms is a specific agricultural by-product utilization.

Other possibilities that have been studied include paper-making from rice or cereal straw, chemicals such as furfural extracted from cereal straw, tartarate from wine grape residue, and monosodium glutamate from bagasse. Several industries including soap, leather, glue, gelatin, and animal feed manufacturing have been based on meat-packing waste products. Biochemicals such as hormones, vitamins, and enzymes also have been produced from packinghouse residues. While these methods offer the possibility of waste utilization, they rarely solve the entire waste problem since the material produced or removed from the waste may be only a small component leaving the problem of managing a large volume of residue.

Seafood processing produces considerable quantities of material that result in usable by-products. The chief by-products are fish meal, fish oils, and condensed fish solubles (1). Fish meal can be the result of waste fish products such as from ground fish, crabs, and shrimp and as a result of the rendering of whole fish such as menhaden alewives, anchovies, mackerel, and herring. Fish oils are nonedible oils and one of the most valuable of the fish rendering products. Condensed fish solubles are the by-products from the stickwater generated in fish meal and oil plants. The dried solubles are used as press cake for meal. Fresh fish wastes also are processed into cooked and canned pet food. In addition, a number of plants process crushed shells for use in poultry feed. Among other by-products are lime and limestone from oyster and clam shells, glue from fish waste and bones, hormones such as insulin, and enzymes.

The amount of residue such as straw, stubble, leaves, and tree limbs from crops and orchards is on the order of hundreds of millions of tons. In some areas drastic measures, such as burning, are used to dispose of troublesome residues and to control plant disease and weeds. Burning as a disposal method is being reduced and more of the material is being utilized by returning it to the soil as a mulch which is later plowed under. A small fraction of the residue, such as straw and peanut hulls, is being used as bedding for farm animals. This material eventually is returned to the soil.

Efforts are needed to develop methods to utilize additional excesses from agricultural production. All activities in this direction must be directed toward effective and economically feasible solutions and toward the development of adequate markets for the usable by-products. Without a sufficient market, the recovered material may be of little economical value.

There are many utilization processes that can be used with agricultural wastes. Processes that have appeared applicable to a variety of wastes and waste waters are discussed in this chapter. The fundamentals of the processes and their real or potential application with agricultural wastes are indicated. The processes are composting, drying and dehydration, by-product development, methane generation, and water reclamation.

Composting

Composting offers an opportunity to recover and reuse a portion of the nutrient and organic fraction in agricultural wastes. Animal wastes can be composted alone but frequently are combined with wastes that may have a high carbon content, such as sawdust, corncobs, paper, and municipal refuse. Enthusiasm for composting in the United States has been caused largely by the possibility of producing a saleable and possibly profitable product. Experience in this country indicates that, except in special cases, composting should be thought of as a treatment process and not as a profit-making operation. Composting is but a component in an overall waste management system.

Because composting is a microbiological process, the fundamentals of biological waste treatment noted in Chapter 5 can be applied. Important factors in the process include intimate mixing of the wastes, small particle size, oxygen for the microbial degradation of the wastes, time to accomplish the composting, and moisture. The composting can be done in open windrows or in enclosed, environmentally controlled units. With the use of the controlled units, composting can be accomplished in 5–7 days while in open windrows, it may take 3–8 weeks or more to produce satisfactory compost.

The process of aerobic composting can be separated into distinct stages of stabilization and maturation. During the stabilization stage, the temperature rises to a thermophilic level where the high temperature is maintained followed by a gradual decrease in temperature to ambient conditions. These conditions can be observed in batch or flow-through type composting operations. As the temperature increases, multiplication of bacteria occurs and the easily oxidized organic compounds are metabolized. Excess released energy results in a rapid rise in temperature. The temperature can be in the 130°–160°F range depending upon the method of operation. At these temperatures, the pathogenic organisms are reduced or destroyed. The ultimate rise in temperature is influenced by oxygen availability. Compost units kept aerobic reach and maintain higher temperatures than do those allowed to be anaerobic.

When the energy source is depleted, the temperature decreases gradually and the fungi, such as actinomycetes, become active. At this stage the organic material has been stabilized but can be further matured. During maturation slow organic matter degradation occurs until equilibrium conditions occur and the volatile matter content reaches about 50%. The final product is a mixture of stable particles useful as a soil conditioner.

Small particle size is important to increase the rate of microbial decomposition. Grinding of the wastes will help reduce the particle size. The more open textured the mixture, the better the aeration that can be achieved. Adequate aeration infers the addition of enough air so that aerobic conditions will exist throughout the compost. Control of the required air may be necessary to avoid rapid drying and cooling of the compost especially in the latter stages of composting.

The moisture content of compost should be in the 50–60% range on a wet weight basis for optimum composting rates. Under uncontrolled conditions, the composting material may dry out increasing the time required to complete the process. The moisture level should be maintained as necessary during the process.

The completion of the composting process can be noted by a marked drop in temperature, no significant increase in temperature when aerated, and a decrease in the rate of volatile solids reduction. Volatile solids reductions of 40–50% can occur in a composting operation depending upon the characteristics of the uncomposted solids. One of the difficulties in composting is obtaining a representative sample of the untreated and treated material.

There are many organisms that can utilize the organic matter in agricultural or municipal wastes and in general these wastes have a multitude of bacteria per gram. There is little need or value in adding enzyme or bacterial cultures to composting systems. The key to a successful composting operation is to have the environmental conditions satisfactory for the organisms that exist.

The major objectives in composting are to stabilize putrescible organic matter, to conserve as much of the crop nutrients and organic matter as much as possible, and to produce a uniform, relatively dry product suitable for use as a soil conditioner and garden supplement or for land disposal.

The operational fundamentals for satisfactory composting have been investigated over the past decades and the following factors elucidated (2): Windrow composting in the open is a simple procedure; recycling of compost to the process for reseeding purposes is of little, if any value, to the process because of the sequence of organisms in the composting operation; aerobic composting is actually a semi-aerobic process requiring less aeration than the term implies; simple turning is an effective way of maintaining needed aeration, and forced aeration is not necessary and can be technologically difficult on a large scale; fly control and destruction of disease vectors occurs in the process; finished compost is a low grade fertilizer, more valuable for its soil conditioning and moisture retaining properties; and it is feasible to reduce the area of land required for disposal of municipal wastes by composting and compaction.

The composting of dairy, beef, swine, and poultry manure has been shown to be technically possible (3–9). Both windrow and mechanical composting of these wastes has produced suitable compost when the fundamentals of the process are followed. A number of studies have illustrated the value of adding dry, large particulate material such as ground corncobs, wood shavings, straw, sawdust, and dried compost to reduce the moisture content, facilitate air movement, and reduce composting time. On-site composting of poultry manure within the poultry house (10) resulted in an odorless, fly-free environment, and was relatively inexpensive. Solid meat-packing wastes also have been successfully composted (11).

The solids from fruit processing operations such as discarded whole fruit, halves, and fragments which may or may not have been lye peeled have been satisfactorily composted. Some pits and a small fraction of leaves and stems were included. Separate peach and apricot wastes were mixed with municipal compost, rice hulls, and other material. Lime added to the mixture helped neutralize the fruit acids that were generated and accelerated the process. Both bin, windrow, and forced air composting were utilized (12).

Under normal agricultural conditions, the costs of applying compost to the land has been greater than the benefits received (13). The value of the increased nutrients in compost are likely to be less than the costs of preparing, transporting, and distributing the compost. With available inexpensive fertilizers, composting is unlikely to be profitable in the money-making sense. The cost of composting and the need to supplement it with chemical fertilizers restrict its marketability to the specialty fertilizer field.

There is no doubt that agricultural wastes can be stabilized by composting. While composting is a process capable of salvaging and recycling inorganic and organic resources in solid wastes, the soil conditioning and resource values of compost are too small to make compost attractive to U.S. agriculture now or in the reasonable future. A suitable market must be available before composting can be attractive as a method for the disposal of agricultural solid wastes. Without a suitable market most of the original dry matter remains for further disposal.

Composting can be feasible for specific agricultural production units and in unique regional situations. As yet it is not a process that can adequately dispose of the volume of agricultural wastes generated throughout the country. Perhaps its most appropriate role in the United States and western Europe is as a process to conserve landfill refuse disposal sites and to serve as a reservoir of organic material to be utilized at a future date.

By-Product Development

General

The meat slaughtering and meat-packing industry is an example of the utilization of techniques that can convert agricultural wastes into useful products. Typical by-products and their uses have been edible fats—tallow, grease; meat scraps and blood—animal feed; bone—bone meal for fertilizers; intestines—sausage casings, surgical thread; glands—pharmaceutical products; and feathers—feather meal for animal feed.

The wastes from vegetable and fiber processing plants have been intensively studied for many years for possible use as animal feeds, as a raw material for manufactured products, and as a source of chemical compounds. More than 150 million tons of agricultural wastes, such as straws, oilseed hulls, and corncobs which are basically cellulosic in nature are produced annually in the United States. This figure does not include animal wastes. Sawdust represents an additional 40 million tons of material. Commercial production of vegetables contributes greater than 300 million tons of additional fresh solid wastes annually (14). When these wastes are in solid form, greater opportunities for processing and utilization occur.

The nature of processing plant operations rarely makes it economical to collect, process, and utilize the solids in liquid wastes. The utilization of these wastes depends largely upon whether the usable fraction can be separated and collected from the water being discharged from a processing plant. The reusable material is present in this water in very dilute suspensions of highly variable composition such as peels, skins, pulps, seeds, and fibers. The liquid fraction may be saline, alkaline, or acidic and contain a large variety of soluble organic compounds. In some processing operations, the separation difficulties are minor and an economical by-product may result. A notable example of such utilization is the production of molasses and pulp from the processing of sugar cane and sugar beets. These by-products can be used as animal feed supplements. The greatest opportunity for by-product recovery exists when wastes are separated into specific fractions at the source during food and meat processing operations.

Milk Product Waste

The processing of fluid milk and the manufacture of milk products results in millions of gallons of waste water containing suspended and soluble milk solids which must be adequately treated before discharge to surface waters. In some areas, much of this waste water is used for irrigation of crops after milk solids are removed. The milk solids can be used as animal feed, feed supplements, a starting material for chemicals such as alcohol, and as a growth medium for microorganisms which can produce pharmaceutical chemicals.

In making of cheese, whey, a liquid waste, results. Liquid whey from cheddar cheese production contains about 54% of the nutrients from milk used and the acid whey from cottage cheese production contains about 73% of the nutrients of the nonfat milk used. Most of the whey is discharged to waste treatment plants and to streams thereby increasing the cost of the cheese production operation and the national costs of water pollution control. The utilization of the whey offers possibilities to minimize pollution and to increase the nutrition of the world. Such possibilities include blending whey powder with basic food materials to produce new and/or less expensive foods such as process cheese food, fruit sherbets, custards, and bakery goods.

Dehydration of whey can be accomplished by roller drying and spray drying. About one-third of the whey now being produced is cottage cheese whey which can not be dried directly by the roller or conventional spray process because of its high acid content. Foam spray drying can be used with all types of whey including that of cottage cheese. After drying, the whey powder contains about 11–13% protein, 70–75% lactose, and 7–8% ash. Cottage cheese whey can contain about 10–11% lactic acid. If the whey can be fractionated, the relatively pure protein, sugar, and other material may be able to command a price considerably greater than that of the whole whey powder. Research on methods of whey separation and utilization is continuing.

The salt content of whey, about 1%, can be a barrier to its use as a by-product. Desalting equipment, such as a combination of reverse osmosis and electrodialysis, may be used to produce a protein supplement grade of whey that may be economically competitive (15). The technical and marketing aspects of this approach remain to be evaluated in depth.

Animal Feeds

In addition to the use of fruit and vegetable processing solids for animal feeds, animal wastes have been considered as a portion of such feeds. Animal wastes contain considerable energy and nutritive value. The gross energy value of chicken feces range from 3.22 to 4.48 kcal/gm of dry matter and the nitrogen content from 0.03 to 0.07 gm of nitrogen/gm of dry matter depending upon the feed ration (16).

The value of dried animal wastes as a part of the feed for other animals is related to the metabolizable energy (ME) rather than the gross energy content of the wastes. The ME represents that portion of the gross energy that does not appear in the feces and urine. Compounds of high ME are readily digestible. The efficiency of feed conversion into product and the amount of feed consumption by the animal both are related closely to the ME content of the diet. The ME of air-dried poultry waste varies depending upon the original diet and has been reported to range from 82 to 560 kcal/lb of material with an average of 360 kcal/lb (17). These are low values compared to standard animal feeds.

The utilization of dried chicken manure as part of the feed for chickens and ruminants as well as the utilization of dried ruminant feces for ruminants and other species has been suggested and experiments carried out to demonstrate the efficacy of this approach. Dried poultry wastes can be successfully used as part of a feed ration for poultry. Optimum values of dried poultry waste as a part of the poultry ration have been indicated as about 10% (18), 12.5% (19), and 15% (20). Laying hens have the ability to adjust feed intake to achieve a constant ME when fed diets of varying energy concentration. Hens fed low energy diets consume more feed. Hens fed dried poultry waste exhibit increased feed consumption and greater quantities of waste per unit of feed intake.

Recycling of dried poultry waste into animal feed does not eliminate the problem of poultry waste handling and disposal. Calculations have indicated that only 30% of the dried poultry waste is utilized the first time it is recycled as animal feed. If the resultant waste is again dried and recycled, less than 10% of the wastes are utilized in the second cycle (17). As a result, considerable amounts of poultry manure accumulate for ultimate disposal.

Dried poultry wastes and broiler litter can be used as part of the ration for ruminants (21, 22). When 32% dried poultry wastes were included in beef cattle rations, relatively poor performance was obtained (23), indicating that this was too high a proportion. The wastes contained only 17% protein. Milk production from cows receiving 20% of their protein from dried poultry wastes was equal to or above that of cows receiving similar portions of protein in silage or soybean meal. It was noted (23), that from management and nutritional considerations, dehydrated animal wastes must contain more than 25% crude protein to economically compete with other supplemental nitrogen sources for nutrients.

Air-dried poultry manure was fed to dairy cattle and steers (24) and was able to be used as the sole source of supplemental nitrogen for these animals when part of low protein-based diets. The nitrogen, calcium, and phosphorous in the dried manure was readily available and utilized by the animals.

The advantages of refeeding poultry and other wastes have been enumerated by a number of studies. Careful handling of the wastes, including drying and pasteurization, and utilization of the wastes in proper balance with other feedstuffs must occur. The literature dealing with the utilization of poultry manure has been reviewed and opportunities described in detail (25).

The use of wastelage, cattle manure which has been removed daily, blended with hay, and stored as silage before use, shows promise as a portion of animal feed. The total protein content of waste from cattle ranges from 9 to 17% of the dry matter with glutamic, aspartic, proline, threonine, leucine, and glycine among the major amino acids in the waste (26). Again, however, the reuse of cattle manure as part of animal feed will not solve the manure disposal problem since a considerable portion of the manure components are resistent to further degradation and will remain for disposal.

Wastes from ruminants are composed of a variety of nitrogen-containing compounds, various vitamins, and undigested feed products composed primarily of cellulose, hemicellulose, and lignin. Where bedding is used, barn wastes also include a large portion of low quality roughage such as straw, wood shavings, or sawdust. The composition of manure, as well as that of the total barn wastes, is dependent on the rations fed and the bedding systems used. If the forage is the sole constituent of the ration, a large portion of the manure, about 60%, may be undigested plant cell walls. These undigested materials are resistant to natural decomposition, but contain a further potential source of energy in ruminant nutrition. If the resistant portion of ruminant wastes were treated and/or processed in a manner to allow subsequent animal digestion by the rumen microbial population, these wastes could become an available source of energy to the animal.

Chemical treatment and use of manures in this manner may offer an expedient method of reuse and volume reduction of otherwise potentially polluting wastes. Studies have shown that the digestibility of the material in animal wastes can be increased by chemical treatment. The chemical treatments that have been explored include use of sodium hydroxide and sodium peroxide which reduced the cellulose, hemicellulose, and lignin to more digestible material, and sodium chlorite which altered the lignin content (27). Feeding experiments have shown that the chemically treated wastes used in a corn silage ration up to 25% of the dry matter were consumed by sheep equally as well as an all corn silage ration.

The use of biological methods to capture the nutrients in agricultural wastes was suggested many years ago when it was proposed that house fly larvae be used to consume wastes and then be harvested as a potential protein source for animal feed. Research has demonstrated that the method can be used with poultry manure to produce a protein source that can replace soybean meal in the diet of the growing chick. Analysis of the dried fly pupae showed that they were high in protein (63%) and fat (15%). In addition, the method deodorized the manure in 213 days, removed more than 50% of the moisture, and reduced the manure volume (28, 29). The large scale feasibility of the method remains to be investigated.

Details of these and other possibilities dealing with the reuse of animal wastes as an animal feed supplement, including problems associated with antibiotics, arsenicals, hormones, larvacides, and other feed additives, have been gathered in a succinct review (30). The review discusses the opportunities of refeeding animal wastes to poultry, swine, and cattle. The practice of incorporating animal and food processing wastes in feed rations for animals is not a new concept, although it has received publicity only recently as a potential method for agricultural waste management.

In general, the nutritive value of incorporating animal wastes in feed rations is greater if the wastes of single stomached animals are added to the feed ration of ruminants and if the ruminant wastes are treated chemically before being added to feed rations. The experimental use of animal wastes in feed rations is not confined to the United States. Research using this approach has been reported in South Africa, Canada, Australia, and England.

When nutritional principles are followed, a portion of animal wastes can be used as a feed supplement for animals. Certain unknowns related to transmittal of drugs, feed additives, and pesticides to the second animal and to the agricultural product, such as eggs and milk remain to be clarified. Some of the specific nonnutritive feed additives that may pose potential problems in recycling animal wastes as feed include: pellet binders, flavoring agents, enzymes, hormones, tranquilizing drugs, antibiotics, arsenicals, antifungals, coccidiostats, and antioxidants. The additives of greatest concern are antibiotics, arsenicals, and hormones.

The U.S. Food and Drug Administration does not sanction the uncontrolled use of animal wastes as a feedstuff for animals. The basic reason is the possible transmittal of feed additives or disease organisms. Available research suggests that refeeding of wastes is a possibility for utilizing a portion of animal wastes. When permitted, the use of these wastes for refeeding will be handled in a manner similar to that used for animal by-products that are used as feed for animals such as blood meal and feather meal. As a minimum, the resultant product would not be permitted to contain harmful levels of pathogenic organisms, toxic metabolites, pesticides, drugs, or other substances shown to be of concern in the animal or product consumer.

It must be recognized that the use of animal and food processing waste in animal feed rations is not an ultimate disposal method since a portion of the waste is not able to be utilized by the animal and will remain for disposal. Recycling of wastes in this manner cannot go on indefinitely. There always will be wastes to be treated and disposed of from animals being refed processed animal wastes as part of their food. A completely closed loop cannot be envisioned when animal wastes are refed.

Other Processes

The world need for protein has led many investigators to look at the vast quantities of cellulosic agricultural wastes as a source of material for the production of single cell protein and yeast. A number of studies have closely examined these opportunities (31, 32) and have concluded that the opportunities are there when it becomes economically feasible to adopt such practices.

Pyrolysis offers the opportunity of the recovery of useful by-products from agricultural wastes. Pyrolysis is defined as the chemical change brought about by heat. It is a process of destructive distillation and is conducted in a closed reactor devoid of oxygen. The process has been used for several hundred years to make charcoal and has been applied commercially to organic compounds to recover by-products such as methanol, acetic acid, and turpentine. The process has been applied to municipal refuse as well as animal wastes. The advantages of pyrolysis for animal wastes are suggested as volume reduction of the total solids so that there is less material for disposal, reclamation of a portion of the organic material as combustible gases for heat energy or as valuable by-products, and use of the resultant char as an adsorbent.

Data from the pyrolysis of municipal refuse indicated that ammonium sulfate, a liquor containing over 94% water as well as low carbon acids, ketones, and aldehydes, light oils containing mainly benzene, and gas consisting of hydrogen, carbon monoxide, methane, and ethylene were produced (33). With a reasonably dry refuse, 40–50% moisture, the energy of the gas was more than sufficient to provide heat for the process. Similar experiments were run with dried sewage sludge and tree leaves. The reaction products were identified as a multitude of chemical compounds (34). This study noted the problems of chemical separation and analysis were formidable. The production of the above by-products by pyrolysis will have to compete economically with similar products produced by more conventional means.

The pyrolysis of animal wastes indicated that the heating value of the gases from dried beef cattle wastes was about 1900 BTU/lb of total solids, from dried poultry wastes was about 1640 BTU/lb total solids, and from dried swine wastes was about 1400 BTU/lb total solids (35). Pyrolysis temperatures of 400°–500°C when used with beef cattle feedlot wastes produced maximum yields of liquid organic compounds (36). A char containing one-third ash was produced but was difficult to burn in the open air although it had some properties as an adsorbent. A combustible gas with a fuel content of 300–400 BTU/ft^3 was obtained and it was estimated that about 2–3 million BTU/ton of dry cattle feedlot manure was recovered in the gas mixture. It was concluded (36) that the pyrolysis process applied to cattle feedlot wastes was uneconomical since the cost of equipment to separate potentially valuable materials from the exhaust stream was not offset by the market value of these materials. In addition, some of the volatile liquid fractions had objectionable odors which would require removal from the gas exhausts.

Drying and Dehydration

Drying and dehydration of agricultural wastes are not utilization processes, but are used to produce a product suitable for utilization as a soil conditioner or as a feed supplement. Dried animal wastes are likely to have more value if they are packaged and sold to the home gardener, florist, or nurseryman. Dried animal wastes have a greater potential for utilization than food processing wastes because the animal wastes generally contain greater nutritive value.

Both of these processes have the same result, removal of moisture and reduction of waste volume. When dried waste is at equilibrium moisture content with the air, it is easily converted into a granular material that improves the handling characteristics of the product. An indication of the handling characteristics of poultry manure as related to moisture content is noted in Fig. 9.1. The equilibrium moisture content is related to the relative humidity of the ambient air. Even for relative humidities greater than 80%, equilibrium moisture contents are in the 15–35% range. A moisture content in the 10–15% range will inhibit mold and microbial growth and minimize the dust problem (37).

The fundamentals of these processes involve the addition of heat and/or air and mixing of the waste. Animal solid wastes usually are treated without the addition of other organic matter. The objectives of the processes are to eliminate odors and flies, to remove moisture, and to facilitate the handling of the manure. The removal of water from poultry manure not only reduces

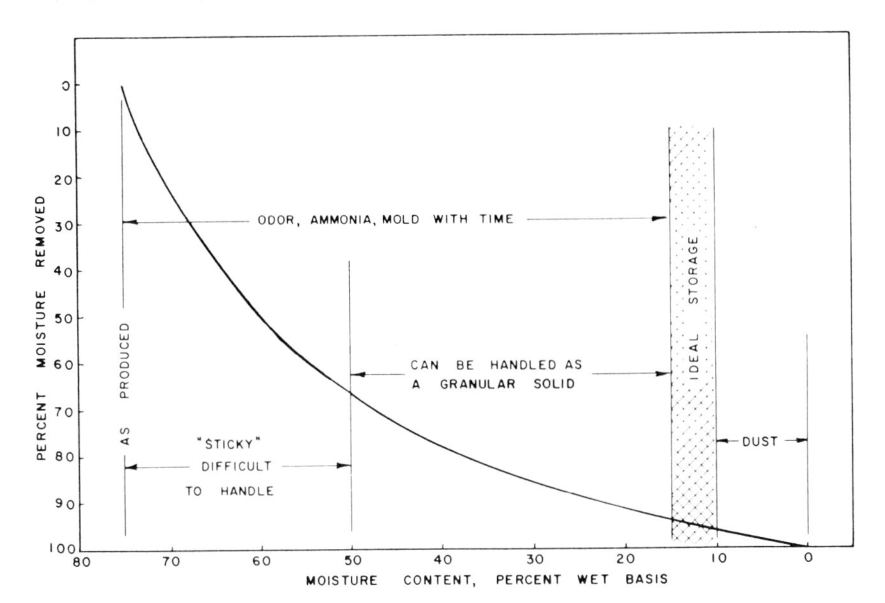

FIG. 9.1. Characteristics of chicken manure as related to moisture content (37).

the weight and volume of the material but also reduces the odor of the material. The offensiveness of the odor of chicken manure decreased with a decrease in moisture content (38). Drying of agricultural wastes should be accomplished with wastes that are not in the process of anaerobic decomposition since odor production will be increased during the drying of anaerobic wastes.

When heat is used to accelerate the drying process, temperatures in the 200–1100°F range have been used. These temperatures will volatilize a number of the organic compounds present in the waste. The gaseous effluent from the high temperature drying can be a source of odors and nuisance complaints. Satisfactory afterburners or other air pollution control equipment may be necessary.

The odor problem at a poultry manure drying plant should be considered in the initial design stage to adequately develop suitable odor control methods. The odors can be controlled by a closed, gas recirculation system within the drying operation and subsequent combustion or scrubbing. The latter produces a liquid waste for disposal which may be undesirable. Suggested methods for odor control at an existing poultry drying plant included a filter to remove solids from the exhaust gas stream, sodium hydroxide scrubbing of the gas stream, and the use of activated charcoal to adsorb the odors compounds (39).

Commercial driers are available and have been used with a number of animal wastes (40, 41). Drying poultry manure in the poultry house has been investigated and shown to reduce the weight of manure to be handled to about $\frac{1}{4}$–$\frac{1}{3}$ of the original as defecated weight. With commercial dryers, a product having a moisture content of 11–20% can be produced. The market potential of dehydrated agricultural wastes is unknown. Without a suitable market, drying of wastes can be uneconomical. Low temperature, natural or forced air drying of certain animal wastes can reduce the volume of wastes to be handled and may be able to be incorporated into feasible land disposal waste management systems.

Methane Production

During the anaerobic digestion of animal wastes, gas containing 60–80% methane can be produced when consistently high rates of digestion are maintained. These gases have been considered as an energy source close to the site where they are generated. Approximately 8–9 ft^3 of gas can be produced per pound of volatile solids added to the digester when poultry, beef cattle, and hog wastes are digested. Lower gas production rates result when sheep and dairy cattle wastes are digested since much of these wastes are no longer biodegradable. A detailed description of anaerobic digestion is presented in Chapter 8.

Controlled process conditions, equipment, and associated costs and problems such as treatment and disposal of the supernatent liquor, disposal of the residual sludge solids, removal of corrosive components such as hydrogen sulfide from the digester gas, and the need for a dependable and constant power supply are a part of a methane production operation. The difficulties of using this method with either animal or food processing wastes can be more than for alternative waste management methods. Consequently, little if any usable gas is produced and utilized from the anaerobic digestion of agricultural wastes in the United States, although the process remains as a potential source of future energy.

Water Reclamation

Agriculture is a major user of water in the United States, primarily for water used in the food processing industry and through the use of irrigation in agriculture. A major reusable by-product of the food and fiber industry is water. These waters can be reused in the processing plants and for irrigation and recharge of groundwater provided some in-plant operational changes are made. These changes can be grouped into two categories: circulation and discharge of highly polluted water separately from the wash and cooling waters, and the elimination of some processes which result in highly polluted water such as the lye-peeling of potatoes.

Irrigation agriculture is changing. Irrigation waters are now better controlled concerning the quantity of water to use and the method to irrigate for maximum efficiency of water use. The reuse of irrigation water can be accomplished by a return pumping system where the normal and irrigated runoff water is caught in shallow ponds and reapplied to other fields. The primary purpose of the method is to increase the efficiency of the use of irrigation waters. A secondary effect is that soluble and solid materials usually lost in runoff are kept on the farm.

Land disposal of agricultural waste waters offers further possibilities for recharge of groundwater and water reclamation. The microbial and plant action in the plant–soil filter removes objectionable organics, most inorganics and bacteria. Upon reaching the groundwater level, the waste water has lost its identity and has been renovated.

Each time water goes through a cycle of domestic or agricultural use, the salt content increases. Upon continual reuse or recycling, the salt increase may be a problem with reclaimed water since high salt concentrations are undesirable for human consumption and crop production. Animal manure also is high in salts. Management techniques are necessary to apply manure and processing waste waters to soils at the proper rate to avoid developing saline soils. Through careful planning and control, spray irrigation of food processing waste can be adapted to various ground conditions and terrain. This method has been used for year-round and for seasonal operation of food processing plants.

Summary

A review of available agricultural waste utilization possibilities indicates that most of these possibilities are approaches whose time has not yet come as far as their ability to solve the agricultural waste management problem or to provide a by-product that can be economically viable. It is true that the state of development concerning suitable methods for utilizing agricultural wastes is in an embryonic stage. Information is becoming available for many processes that permit a reasonable evaluation as applied to agricultural wastes. On the basis of current information, many utilization processes will result in a low level of application until such time as social or economic constraints make their use more appropriate. There are relatively few situations in the agricultural industry where capital investment in production facilities for recovery of by-products from waste materials can be justified. Costs of collecting, transporting, and processing waste materials must all be evaluated in comparison with the size of the market and the selling prices for resultant products.

Detailed information on the costs of the utilization methods must be available before the broad application of the methods can be evaluated. A variety of costs must be delineated: the cost of utilizing the method, the net cost as it affects the profit of the producer, and the ultimate cost to the public. These costs must be developed for different types and sizes of operations. It is important to treat the removal and distribution of agricultural wastes as an expense to the waste-generating operation. By doing so, evaluation of the waste-generating and waste utilization methods is improved and a more intelligent evaluation of alternate methods may be made.

Technical solutions to the utilization of agricultural wastes exist and others will be found. The application of these methods will depend upon developments in the social, legal, economic, political, and in the marketing and public relations areas. The most effective agricultural waste utilization method for the vast quantity of agricultural wastes continues to be land disposal in a crop production cycle with or without prior treatment of the wastes.

References

1. Soderquist, M. R., Williamson, K. J., Blanton, G. I., Phillips, D. C., Law, D. K., and Crawford, D. L., "Current Practice in Seafoods Processing Waste Treatment," Final Rep., Proj. 12060 EVD. Water Quality Office, Environmental Protection Agency, 1970.
2. McGauhey, P. H., "American Composting Concepts," Publ. SW-2r. Solid Waste Management Office, Environmental Protection Agency, 1971.
3. Wiley, J. S., A report on three manure composting plants. *Compost Sci.* **5,** 15–16 (1964).
4. Livshutz, A., Aerobic digestion (composting) of poultry manure. *World's Poultry Sci. J.* **20,** 212–215 (1964).
5. Bell, R. G., and Pos, J., The design and operation of a pilot plant for composting poultry manure. *Annu. Meet., Amer. Soc. Agr. Eng.* Paper 70-419 (1970).
6. Grimm, A., "Dairy Manure Waste Handling Systems," Proc. Agr. Waste Manage. Conf., pp. 125–144. Cornell University, Ithaca, New York, 1972.
7. Martin, J. H., Decker, M., and Das, K. C., "Windrow Composting of Swine Wastes," Proc. Agr. Waste Manage. Conf., pp. 159–172. Cornell University, New York, 1972.
8. Galler, W. S., and Davey, C. B., High rate poultry manure composting with sawdust. *In* "Livestock Waste Management and Pollution Abatement," Publ. PROC-271, pp. 159–162. Amer. Soc. Agr. Eng., 1971.
9. Willson, G. B., Composting dairy cow wastes. *In* "Livestock Waste Management and Pollution Abatement," Publ PROC-271, pp. 163–165. Amer. Soc. Agr. Eng., 1971.
10. Howes, J. R., "On-Site Composting of Poultry Manure," Proc. Nat. Symp. Anim. Waste Manage., Publ. SP-0366, pp. 66–68. Amer. Soc. Agr. Eng., 1966.

11. Nell, J. H., and Krige, P. R., The disposal of solid abattoir waste by composting. *Water Res.* **5,** 1177–1189 (1971).
12. Rose, W. W., Chapman, J. E., Roseid, S., Katsuyama, A., Porter, V., and Mercer, W. A., Composting fruit and vegetable refuse. *Compost Sci.* **6,** 13–25 (1965).
13. Tietjen, C., and Hart, S. A., Compost for agricultural land. *Sanit. Eng. Div., Amer. Soc. Civil Eng.* **95,** SA2, 269–287 (1969).
14. Kohler, G. O., "Animal Feeds from Vegetable Wastes," Proc. 1st Nat. Symp. Food Process. Wastes, pp. 383–386. Pacific Northwest Water Lab., Federal Water Quality Administration, 1970.
15. Goldsmith, R. L., Goldstein, D. J., Horton, B. S., Hossain, S., and Zall, R. R., "Membrane Processing of Cottage Cheese Whey for Pollution Abatement," Proc. 2nd Nat. Symp. Food Process. Wastes, pp. 413–446. Pacific Northwest Water Lab., Environmental Protection Agency, 1971.
16. Pryor, W. J., and Connor, J. K., A note on the utilization of chickens of energy from feces. *Poultry Sci.* **43,** 833–834 (1964).
17. Young, R. J., Evaluation of poultry wastes–feed ingredient and recycling waste as a method of waste disposal. *Tex. Nutr. Conf.,* Texas A & M Univ. **27,** 1–11, 1972.
18. Hodgetts, B., The effects of including dried poultry waste in the feed of laying hens. *In* "Livestock Waste Management and Pollution Abatement," Publ. PROC-271, pp. 311–313. Amer. Soc. Agr. Eng., 1971.
19. Flegal, C. J., Sheppard, C. L., and Dorn, D. A., "The Effects of Continuous Recycling and Storage on Nutrient Quality of Dehydrated Poultry Waste (DPW)," Proc. Agr. Waste Manage. Conf., pp. 295–300. Cornell University, Ithaca, New York, 1972.
20. Nesheim, M. C., "Evaluation of Dehydrated Poultry Manure as a Potential Poultry Feed Ingredient," Proc. Agr. Waste Manage. Conf., pp. 301–319. Cornell University, Ithaca, New York, 1972.
21. Long, T. A., Bratzler, J. W., and Frear, D. E. H., "The Value of Hydrolyzed and Dried Poultry Waste as a Feed for Ruminant Animals," Proc. Agr. Waste Manage. Conf., pp. pp. 98–104. Cornell University, Ithaca, New York, 1969.
22. Fontenot, J. P., Bhattacharya, A. N., Drake, C. L., and McClure, W. H., "Value of Broiler Litter as a Feed for Ruminants," Proc. Nat. Symp. Anim. Waste Manage., Publ. SP-0366, pp. 105–109. Amer. Soc. Agr. Eng., 1966.
23. Bucholtz, H. F., Henderson, H. E., Thomas, J. W., and Zindel, H. C., Dried animal waste as a protein supplement for ruminants. *In* "Livestock Waste Management and Pollution Abatement," Publ. PROC-271, pp. 308–310. Amer. Soc. Agr. Eng., 1971.
24. Bull, L. S. and Reid, J. T., Nutritive value of chicken manure for cattle. *In* "Livestock Waste Management and Pollution Abatement," Publ. PROC-271, pp. 297–300. Amer. Soc. Agr. Eng., 1971.
25. Johnson, T. H., and Mountney, G. J., Poultry manure—production, utilization and disposal. *World's Poultry Sci. J.* **25,** 202–217 (1969).
26. Anthony, W. B., "Utilization of Animal Waste as a Feed for Ruminants," Proc. Natl. Symp. Anim. Waste Manage., Publ. SP-0366, pp. 109–112. Amer. Soc. Agr. Eng., 1966.
27. Smith, L. W., Goering, H. K., and Gordon, C., "Influence of Chemical Treatment Upon Digestibility of Ruminant Feces," Proc. Agr. Waste Manage. Conf., pp. 88–97. Cornell University, Ithaca, New York, 1969.
28. Calvert, C. C., Morgan, N. O., and Martin, R. D., House fly larvae: Biodegradation of hen excreta to useful products. *Poultry Sci.* **49,** 588–589 (1970).
29. Calvert, C. C., Martin, R. D., and Morgan, N. O., House fly pupae as food for poultry. *J. Econ. Entomol.* **62,** 938–939 (1969).
30. Anonymous, Animal waste reuse—nutritive value and potential problems from feed and additives—a review. *U.S., Dep. Agr., ARS* ***ARS*** **44-224** (1971).

31. Meller, F. H., "Conversion of Organic Solid Wastes into Yeast—An Economic Evaluation," Rep. Contract PH-86-67-204. Bureau of Solid Waste Management, Public Health Service, Dept. of Health, Education and Welfare, 1969.
32. Callahan, C. D., and Dunlap, C. E., "Construction of a Chemical-Microbial Pilot Plant for Production of Single Cell Protein from Cellulosic Wastes," Rep. SW-24c. Solid Waste Management Office, Environmental Protection Agency, 1971.
33. Sanner, W. S., Ortuglio, C., Walters, J. G., and Wolfson, D. E., "Conversion of Municipal and Industrial Refuse into Useful Materials by Pyrolysis," Rep. No. M428. Bureau of Mines, Dept. of Interior, 1970.
34. Shuster, W. W., "Partial Oxidation of Solid Organic Wastes," Final Rep., SW-7rg. Bureau of Solid Waste Management, Public Health Service, U.S. Dept. of Health, Education and Welfare, 1970.
35. White, R. K. and Taiganides, E. P., Pyrolysis of livestock wastes. *In* "Livestock Waste Management and Pollution Abatement," Publ. PROC-271, pp. 190–191. Amer. Soc. Agr. Eng., 1971.
36. Garner, W., Bricker, C. D., Ferguson, T. L., Wiegand, C. S. W., and McElroy, A. D., "Pyrolysis as a Method of Disposal of Cattle Feedlot Wastes," Proc. Agr. Waste Manage. Conf., pp. 101–124. Cornell University, Ithaca, New York, 1972.
37. Sobel, A. T., "Moisture Removal," Proc. 1971 Cornell Agr. Waste Manage. Conf. pp. 107–114, Cornell University, Ithaca, New York, 1971.
38. Sobel, A. T., "Removal of Water from Animal Manures," Proc. Agr. Waste Manage. Conf., pp. 347–362. Cornell University, Ithaca, New York, 1969.
39. Hodgson, A. S., The elimination of odor from the effluent gases of chicken manure drying plant. *J. Agr. Eng. Res.* **16,** 387–393 (1971).
40. Shepard, C. L., ed., "Poultry Pollution: Problems and Solutions," Res. Rep. No. 117. Michigan State University, East Lansing, 1970.
41. Niles, C. F., Egg laying house wastes. *Proc. Purdue Ind. Waste Conf.* **22,** 334–341 (1967).

10

Land Disposal of Wastes

Introduction

Of the three locations for the ultimate disposal of wastes, surface waters, atmosphere, and land, the land represents not only an appropriate disposal medium for agricultural wastes but also an opportunity to manage wastes with a minimum of adverse environmental effects. Application of manure, sewage sludge, municipal waste water, and industrial wastes on land for both disposal and fertilizer value has been practiced for centuries. The challenge is to utilize the chemical, physical, and biological properties of the soil as an acceptor for the residues of man with minimum unwanted effects to the crops that are to be grown, to the characteristics of the soil, and to the quality of the groundwater and surface runoff. The land cannot function as a neglected waste sink. The soil and the wastes must be managed carefully as a total system to obtain the best use of this resource.

Waste disposal approaches are essentially those of relocation of wastes from the point of generation to a suitable and nondetrimental site. The soil can serve as a receptor of organic and inorganic residues if land disposal methods are based on an understanding of the applicable scientific and engineering fundamentals and if the actual methods are well designed and managed. It is important to avoid the mistakes made in the use of the air and water as waste receptors when the land is used for waste disposal.

Land disposal of wastes incorporates organic and inorganic recycle and reuse. Less is known about the use of land for water disposal than is known about the use of surface waters for this purpose. Engineers and scientists have been concerned about the assimilative capacity of streams, estuaries, and lakes for numerous decades while comparatively little is known about the waste assimilative capacity of a soil. Each soil will have its own maximum capacity to assimilate and treat wastes and to renovate waste waters. This capacity will be related to the soil characteristics, environmental conditions, and crops to be grown. The maximum assimilative capacity of a soil therefore represents the maximum waste loading to a soil.

The soil assimilative capacity is related to the microbial, chemical, and physical reactions in the soil layers. Dimensionally, the units of the soil assimilative capacity should be the quantity of a critical waste parameter applied per unit time per unit of soil depth or volume of soil infiltrated. Examples would be gallons of waste water, pounds of organic matter, or pounds of nitrogen per cubic foot of soil or per foot of soil depth per week, month, or year. At a minimum, the soil assimilative capacity and hence the soil loading rate should be stated in terms of quantity of waste applied per unit time per surface area. Current application rates generally are in terms of inches of liquid per week or tons of wet manure per acre. While these rates may provide operating disposal rates, they provide little understanding of the fundamental or controlling parameters. Future data on soil loading rates for wastes and waste water should be given in terms of the more fundamental units noted above.

Each disposal site will have its own controlling parameter which will in turn depend upon the characteristics of the waste, characteristics of the soil, and the most important item of environmental concern. Acceptable waste loading rates to the soil and undesirable reactions that may occur in the soil are known only in general terms. Information that will permit engineered use of the land as an acceptor of wastes is only beginning to become available.

The information that does exist has originated with the use of inorganic fertilizers and manures to increase crop yields. Data is available on the quantity of nutrients and trace elements from these sources that becomes part of the crop. The fate of the remaining organics and inorganics has remained obscure. The talents of many disciplines, such as agronomy, soil science, agricultural engineering, and sanitary engineering are needed to develop the criteria that will permit acceptable use of the land as a resource to accommodate the residue of man and agriculture. Opportunities abound for innovative research and waste management approaches using the land as a point of waste disposal.

A schematic diagram of the use of land for crop production and/or waste disposal (Fig. 10.1) illustrates the factors that are involved. The human and animal wastes are residues of crops grown on the land. Disposal of these wastes on the land is often a beneficial recycle. The industrial wastes

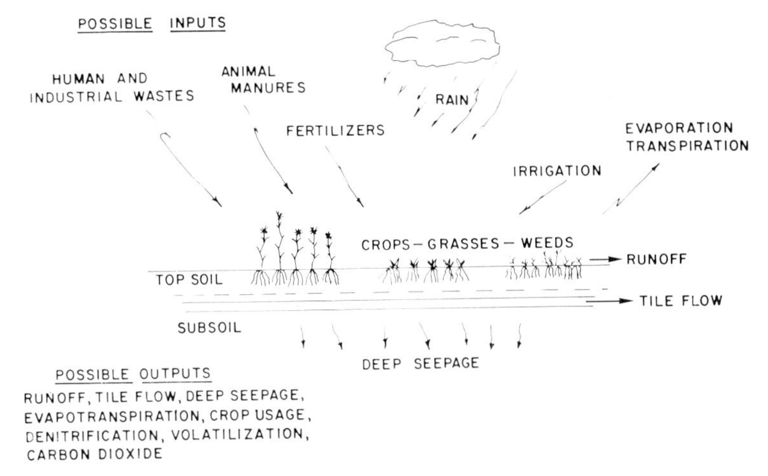

Fig. 10.1. Schematic land use balances.

contained in municipal sludges or disposed of directly to the land include materials mined from deep in the earth or chemical conversions of simple organic compounds. Many of these waste components can be foreign to the upper soil horizons. Control will be needed to avoid biologically toxic concentrations of such material.

In crop production, inorganic fertilizers represent an external input whose source has not been the land. The proper application of this input has produced the foods needed for the world population. Human and animal wastes rarely are returned to the specific land that grew the consumed food. Rather, these wastes are disposed of on land in proximity to their production. They may be applied at loadings greater than that needed for the crops to be grown on the land.

Outputs such as evapotranspiration, denitrification, volatilization, and carbon dioxide production cause no environmental quality problems. Of particular environmental concern are constituents of the runoff, tile drain flow, deep seepage, and accumulations of potentially inhibitory concentrations of minerals in the soil.

Land disposal of wastes can serve a variety of useful purposes. Such a method of waste disposal partially restores the nutrient cycle of food and crops which is interrupted by the disposal of wastes to waters and air. For agricultural wastes, this disposal often is less expensive than other waste treatment and disposal. This method, when properly employed, will replenish groundwater and improve soil. It can be used for untreated wastes and for wastes partially treated by drying, aeration, or removal of nutrients.

Land for disposal of partially treated wastes offers unlimited opportunities to combine treatment and land disposal possibilities. Treatment requirements prior to land disposal are less than those for disposal to surface waters since the land will provide subsequent treatment and recycle. Prior treatment may range from only that required to minimize odors to that necessary to accomplish reduction of BOD and nutrients (Fig. 10.2). The degree of required treatment will depend on type of crop production, types of soil, general land management, desired flexibility of disposal timing, and site selection.

Growth of a crop on the disposal area is important to the success of the use of land for waste disposal. A growing crop is essential to increase the rate of absorption, evaporation, transpiration, and to avoid soil erosion. A crop will decrease the impact of water droplets and provide water storage space with the increased surface area of the crop. Plant transpiration results in a considerable loss (disposal) of water and uptake of nutrients. The crop should be harvested periodically to promote crop growth and obtain continued removal of the nutrients in the applied wastes.

Hydraulic and organic loading rates consistent with the ability of the land to accommodate the wastes are important. Intermittent use of the land for disposal purposes followed by a "resting" or nonapplication period is a common way to avoid overloading the soil. The soil surface must be permitted to drain periodically and the pore spaces permitted to fill with air. This is important not only for adequate root growth of crops but for the use of the soil as a disposal media. Compaction impedes root growth, impairs crop productivity, lowers the percolation capacity of the soil, and results in more rapid soil saturation and water loss by runoff. The use of land as a waste water disposal site is practical on a continuing basis only if the application rate is less than the soil infiltration rate. Tillage can help correct clogging problems.

The timing between waste water application will depend upon the nature of the soil and generally is determined by trial and error. Because of the variability of soil types, climate, and vegetation, general design criteria for land disposal of wastes are difficult to describe. Extrapolation of reported results to other situations also is difficult. Local conditions must be considered carefully. Typical requirements of an engineered land disposal system are noted in Table 10.1.

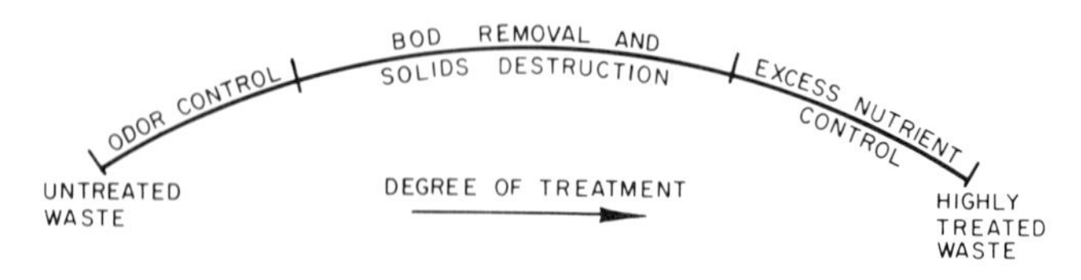

Fig. 10.2. Types of treatment that may be necessary prior to land disposal of wastes.

Table 10.1

Requirements of Feasible Land Disposal Systems

Reliable	Economically feasible
Acceptable from the health standpoint	Minimum environmental problems
Compatible with soil structure and use	Able to be rationally designed
Technically feasible	Compatible with available technology and manpower
Acceptable life expectancy	Able to accept wastes at rates consistent with waste production
Flexible operation	

The characteristics of a soil have been determined by natural geologic processes. Bodies of soil are relatively homogeneous within specific areas. Information on the characteristics of soils is available through the Soil Survey of the United States which describes, classifies, and maps the soils of the nation. These maps are a detailed and accurate soil inventory which can be applied to the selection of soils for waste disposal. The responsibility for defining and naming different kinds of soils, for maintaining standards, and for the completion and publication of a soils inventory for the entire United States is vested by law in the Soil Conservation Service of the U.S. Department of Agriculture. The soil survey description of soils includes texture, structure, porosity, organic matter, chemical composition, vegetation, and other important factors.

Although soil survey information has largely been used for agriculture and crop production, it can be applied for waste disposal and other environmental and sanitation developments. The soil survey enables predictions to be made of soil conditions for an area as small as one acre with an accuracy of 70 to 90%. Greater use of soil survey data should be made in determining sites and rates for the disposal of liquid and solid wastes.

Where agricultural wastes are disposed of on land, the fate of the added organics and inorganics is of concern. Organic compounds will be decomposed to humus which will cause no problems. Nitrogen can be oxidized and reduced or leached away. Other inorganic material such as potassium, sodium, and other elements can increase with continual waste applications to the point of possible concern in subsequent use of the soil. Examples of such concern can be compounds which in excess retard or inhibit the growth of a crop, adversely affect the nutritive quality of a crop, or cause health problems in animals who use the crop for feed created by accumulations of excessive amounts in the feed. The movement of the applied waste components in the soil, especially to groundwaters or surface waters, also is of interest. A number of mechanisms can transport the waste components to locations that result in environmental quality problems. The control of these mechanisms offers a positive approach to minimize such problems that can result from the use of land for waste disposal.

Transformations in the Soil

The ability to develop acceptable land disposal systems for wastes requires an understanding of the reactions and transformations that take place when wastes are applied to the soil. Soil is a composite medium containing inert rock, gravel, and sand, reactive clay minerals, organic matter, living and dead vegetative and animal matter, plus a large variety of soil microorganisms. Potential treatment mechanisms in the soil are many. They include biological oxidation, ion exchange, chemical precipitation, adsorption, and assimilation into growing plants and animals. The biological, physical, and chemical processes in a soil provide a treatment of wastes that incorporates the factors and components of present day biological and advanced waste treatment methods for municipal and industrial wastes. The capacity of a soil to handle complex organics varies with soil properties and climatic conditions. Infiltration rates and types of vegetative cover are major factors in using soil as a means of degrading organic wastes. Such biodegradation requires good soil aeration which in turn is a function of the soil moisture content. Improper drainage and aeration will result in anaerobic conditions and less soil assimilative capacity. Failure to consider the limited infiltration capacity of a soil has led to the failure of many soil disposal systems.

A comprehensive review of the use of the soil as a treatment system (1) emphasizes what is needed to design and utilize an engineered system for septic tank effluents and similar material. The principles are applicable to the disposal of other wastes on the land. Many factors affecting the performance of soil systems for waste disposal are not able to be altered by the engineer and scientist. Design factors that are important include waste loading or dosing patterns, pretreatment of waste, drainage, and vegetation. The use of a soil for waste treatment is determined by the chemical and physical characteristics of the soil which vary from site to site making general conclusions difficult.

The application of significant amounts of waste water annually will alter the equilibrium of chemical movement in any ecosystem. Many waste water constituents will be retained by the soil while other natural constituents may be displaced to groundwater and eventually surface water. With sustained additions of wastes and waste water, ecosystems will reach a new equilibrium state. Water quality as measured by surface water draining a particular watershed will represent the new chemical equilibrium where decomposition of plant material and weathering of soil yields constituents to the soil solution and streams in balance with new rates of biological and chemical processes.

A number of waste constituents are of interest in soil treatment systems. These are oxygen-demanding material, nitrogen, phosphorus and other minerals in the applied wastes. The ecological relationships and microbial transformations that occur in the soil have been reviewed in detail (2). The following paragraphs describe the transformations that are pertinent to the waste assimilative capacity of the soil.

Carbon

The fundamentals of carbon metabolism of organic wastes, as described in Chapter 5 for biological waste treatment systems, also occur in the soil. Organic agricultural wastes are residues of natural plant and animal compounds and as such are able to be degraded by soil microorganisms. Most of the aerobic bacterial activity occurs near the surface of the soil and decreases with depth. When wastes are added to the soil, the more easily degradable materials are metabolized with cellulose, hemicellulose, and lignins persisting for longer periods of time.

The BOD removal efficiency of soil can be affected by the amount of vegetative cover and the infiltration capacity. Anything that adds surface area at the soil–air interface, such as litter or living plants, will increase the biological decomposition capacity of the soil disposal system. At very high rates of waste application, the residence time of the dissolved or particulate BOD may not be great enough for complete biological decomposition to occur. However, net BOD removal efficiency has been observed to be high even in coarse soils and at high infiltration rates. The treatment of waste having a low BOD and a large volume may be limited by the percolation capacity of the soil. A waste with a small volume and a high BOD is more likely to be limited by the oxidative capacity of the microorganisms and sorptive capacity of litter on the soil surface.

The soil has a large capacity to remove organic matter from wastes applied to the soil. When anaerobic livestock lagoon effluent was disposed of on land, the organic load was removed in the soil profile (3). About 50% of the COD was removed in the upper three inches of the soil where biological activity was the greatest. Physical filtration removed particulate matter. Overall COD removal through 30 inches of soil was about 95%. COD removals of 67–91% were obtained when municipal waste treatment plant effluents were applied to sand and silt loam. The lower percent removals occurred at shallow depths (1 ft) and with the sand (4).

From the environmental quality standpoint, our concern is the quantity of material that can have an adverse effect. A materials balance would indicate that the carbon added in a waste ultimately will exist as carbon in the protoplasm of soil microorganisms and in plant growth, as slowly degradable soil humus, and as carbon dioxide released to the atmosphere for continued photosynthetic activity. None of these organic residues can be considered to have an adverse environmental effect. As in more conventional waste treatment systems, carbon is not an element of environmental quality concern. Adverse effects, such as waterlogged soils, inhibition of plant growth, and odor production can occur when the waste assimilative capacity of a soil is exceeded. These conditions can be controlled by proper waste application rates, soil resting periods, and cultivation practices.

Oxygen

Oxygen is an important component of a soil system. If the soil is organically overloaded, oxygen can become limiting and the biological system may become anaerobic, reducing the hydraulic loading capacity of the soil. The magnitude of the changes is related to the type and quantity of the waste and the frequency with which it is added. Anaerobic conditions also will predominate if excessive flooding or waterlogged conditions prevail. Intermittent applications of wastes to soils and cultivation of the soil improve soil aeration and drainage and permit continuing use of a plot of land for waste disposal. Factors such as temperature and pH also affect waste decomposition in the soil and therefore waste-loading rates.

The oxygen content of the soil water or atmosphere can be a measure of an efficient soil disposal process for any waste. When the oxygen content is close to zero, the waste application rate is approaching the maximum for the local environmental conditions.

Nitrogen

Nitrogen is a key nutrient for protein synthesis and crop growth. Until the advent of adequate quantities of inorganic fertilizers, nitrogen management on a farm was one of the major factors limiting crop yields. Nitrogen undergoes transformations involving organic, inorganic, and gaseous compounds. The quantity of excess nitrogen that is not incorporated into plant and microbial growth or held in the soil is of concern in waste management.

The nitrogen processes of interest include mineralization, immobilization, nitrification, and denitrification. Mineralization describes the process in which nitrogen is converted to a form that is both mobile in the soil–water system and available to plants. Organic nitrogen is converted to the ammonium form and then the latter is oxidized (nitrification) to nitrite and nitrate nitrogen. Immobilization describes the process in which the nitrogen is tied up in organic nitrogen forms such as microbial cells. The processes of mineralization and immobilization occur simultaneously and depend upon the relative availability of biodegradable carbonaceous and nitrogenous matter. If nitrogenous matter is in excess, nitrogen is mineralized. If carbon is in excess, nitrogen is converted into cell mass and is immobilized. Under anaerobic conditions and excess carbon, the oxidized nitrogen can be reduced (denitrification) to gases such as nitrous oxide and molecular nitrogen.

With the exception of certain fruit and vegetable processing wastes, most agricultural wastes contain excess nitrogen. When these wastes are added to the soil, the relative rates of mineralization and immobilization will be affected by the ratio of carbon to nitrogen in the soil. If the wastes are added to solids with considerable quantities of crop residues, mineralization may not predominate.

Ammonium ions are positively charged and move slowly with the soil water because of the attractive forces between the ammonium ions and negatively charged clay and organic colloids. As long as the nitrogen stays in the ammonium form, the possibility of nitrogen loss by leaching is low. However, in normal soils ammonium is oxidized to nitrate, a negatively charged ion that does move freely with the soil water. Nitrate leaching can be of concern if large amounts of nitrate are present during periods before a crop is planted, when the crop is not utilizing these nutrients rapidly, and when irrigation or rainfall exceeds soil water storage or crop moisture requirements. Under such conditions, excess water and soluble nitrates will move through the soil. The nitrification rate in soil is affected by rate of nitrogen application, soil aeration, temperature, and soil moisture. Nitrification rates increase as temperature increases. Nitrification is not considered a limit to crop production since many plants can utilize both ammonium and nitrate nitrogen.

Nitrogen can be lost from the soil in several ways: leaching, denitrification, and volatilization. As noted, oxidized nitrogen is soluble and can be leached out of the root zone of the crops. Once the nitrates are below the root zone, little opportunity will exist for utilization by a crop or for denitrification. Hence the nitrates will remain in the soil or move to the groundwater. Denitrification can occur in the soil and can be one of the more important processes accounting for nitrogen losses in soils. Most of the evidence for denitrification losses has originated from nitrogen balances under controlled conditions in which the amount of nitrogen added and that initially in the soil could not be accounted for by the amounts removed by cropping, leaching, and remaining in the soil. Denitrification in the soil occurs where there is poor aeration, high oxygen demand, and residual carbonaceous material. Volatilization of ammonia represents another loss of nitrogen from the soil. Volatilization occurs in alkaline pH ranges with rapid rates occurring when the pH is above 9.5 and when there is considerable aeration or air movement over the soil. With the exception of ammonia volatilization, the above nitrogen transformations are microbial and are controlled by the same environmental factors discussed in Chapters 5 and 11.

Environmental quality concern with nitrogen is related to three factors: (a) adverse health effects that may occur when the nitrate in drinking water exceeds acceptable limits or when nitrates and nitrates in forages and leafy vegetables reach toxic levels, (b) oxygen demand caused by the oxidation of reduced nitrogen compounds in surface waters, and (c) eutrophication of surface waters stimulated by an abundance of nitrogen. Waters containing more than 10 mg/l nitrate nitrogen are not recommended as a source of drinking water for humans and animals because of potential adverse health effects. Ruminants are susceptible to nitrate toxicity because of rumen microorganisms that can reduce nitrate to nitrite. Nitrate poisoning of livestock from consuming forage in excess of 0.3% nitrate nitrogen on a dry weight basis has been reported. Livestock deaths in Canada have occurred at levels between 0.4 and 0.5% nitrate nitrogen in forage following rates of manure application that exceeded recommended rates by five fold in terms of nitrogen (5). High nitrogen application rates also can be detrimental to crop growth and yield. Although environmental quality interest is in the ionic forms of nitrogen, ammonium, nitrite, and nitrate, these forms generally constitute 2% or less of the total soil nitrogen. The remainder is as organic nitrogen.

A schematic diagram of these concerns is shown in Fig. 10.1 in which the input and output from a soil are noted. In waste management, the concern is to avoid the excess amounts of nitrogen that can be lost to surface water and groundwaters and to avoid toxic levels in crops.

The waste loading rate on soil must depend in part on the quantity of nitrogen in the wastes and the relative rates of nitrogen application, degradation, and utilization by the growing plants. Ideally, the nitrogen application rates should be consistent with nitrogen needs and utilization rates so that nitrogen excesses that can adversely affect environmental quality are minimized. A nitrogen balance for Ontario, Canada conditions has shown that crops of continuous corn or grass could be expected to utilize up to 300 lb N from manure per acre. Application rates greater than this rate could lead to a decrease in crop yield and be a potential cause of water pollution (5). Reductions in the amount of applied nitrogen were indicated where the crops were grown on coarse-textured soil. Where heavy waste loads to the soil are contemplated or are necessary because of shortage of land, control of the nitrogen content of the wastes by prior treatment may be necessary.

The ammonium ions in soil water can be adsorbed onto the clay and organic fraction of the soil. The exchange capacity of soils differs widely depending upon the amount of organic matter and clay in the soil. Agricultural soils with a high cation exchange capacity can adsorb significant amounts of ammonia either added by the wastes or resulting from the microbial metabolism of organic nitrogen compounds. Other cations such as calcium and magnesium are adsorbed to the soil and compete with ammonium for exchange sites. The ammonium is not irreversibly adsorbed since it can be oxidized biologically to nitrate in well-aerated soils. If the adsorption takes place under anaerobic conditions, the ammonium will not be oxidized.

The migration of the nitrates is related to the quantity of water that passes through the soil to the groundwaters. Under nonirrigated conditions, water usually moves through the soil to depths below the root zone only during the fall, winter, and early spring months. Leaching occurs during these periods because infiltration is greater than evapotranspiration and runoff. Application of fertilizers and wastes to land during these periods coincides with the increased opportunity for nitrate movement.

Planned denitrification in the soil offers opportunities to avoid excessive leaching of nitrates to groundwaters. Large numbers of denitrifying bacteria that can use nitrates as a hydrogen acceptor in the absence of oxygen exist in the soil. The rate of denitrification is a function of the temperature, pH, and decomposable carbon content in the soil. Both aerobic and anaerobic zones can coexist in the same soil. In soils normally considered to be aerobic, losses of applied nitrogen by denitrification can occur in anaerobic pockets. The anaerobic conditions usually increase after heavy rains and after heavy applications of wastes. Quantitative data on the magnitude of losses by denitrification in the field is fragmentary. Most of the gas evolved during denitrification is N_2 with N_2O and NO being generated at low pH levels.

To establish a planned soil denitrification system the nitrogen must be oxidized to the nitrate form, anaerobic conditions must be created after the nitrates are formed, and a sufficient energy source must be available in the anaerobic zone for the denitrifying bacteria. These conditions can be approximated by alternating waste application and resting cycles. Sequential nitrification and denitrification has been utilized in municipal waste treatment systems for nitrogen control but has yet to be applied to land waste disposal sites in a major way to control potential nitrate leaching to groundwaters. As the land becomes used more extensively for waste disposal, planned denitrification in the soil may provide an opportunity for relatively large nitrogen loadings per area of land while minimizing adverse nitrogen releases to groundwaters. Uncontrolled denitrification has been observed in many soils that have received wastes. In one example up to 80% nitrogen loss reduction was observed when anaerobic livestock lagoon effluent was applied to land (3). The potential for planned denitrification in the soil requires further exploration.

Phosphorus

Phosphorus applied to a soil is converted to water insoluble forms in a short time. When phosphorus in fertilizers, animal wastes, sewage sludge, or any material is added to a soil upon which crops are grown, a certain portion of the phosphorus is utilized by the crop and the remainder accumulates in the soil. The percentage of accumulated phosphorus increases as phosphorus inputs increase because of a nearly constant amount of phosphorus taken up by a crop of equal yield. The accumulation will occur after the input is greater than that necessary to satisfy plant requirements.

Phosphorus immobilization in the soil is related to the mineral constituents of the soil. Phosphorus fixation in acid soils is caused by the formation of insoluble iron and aluminum compounds. In alkaline soils the fixation is due to insoluble calcium compounds. Adsorption of the phosphorus on clay particles is caused primarily by the free iron and aluminum associated with the clay. The result is a low soluble phosphorus concentration and a higher phosphorus concentration in particular form. With such soils, phosphorus movement occurs primarily as the soil moves. Adequate practices to control soil erosion will help control the loss of both organic matter and the nutrients associated with the soil.

Phosphorus immobilization is less in sandy soils than in soils with a high clay content because of the lower adsorption and reactive capacities of sandy soil. The capacity of a soil to adsorb phosphorus is not infinite. Each soil has a phosphorus adsorptive capacity which can be exceeded by prolonged application of large amounts of phosphorus. The phosphorus adsorptive capacity may not be exceeded for decades or centuries in soils having a high clay content. The formation of insoluble phosphorus compounds generally restricts the soluble orthophosphate concentration in soil water solutions to low concentrations, frequently the 0.01 to 0.05 mg/l range. These levels of soluble orthophosphate concentration are comparable to the equilibrium release value of the soil and represent base level concentrations for the respective soils. This phosphate then, is the amount that would normally leach out of a geological substratum whether the soil had or had not received wastes. In the absence of erosion or direct discharge, agriculture should not contribute important controllable amounts of phosphorus to lake or stream eutrophication.

Examination of surface water characteristics before the advent of widespread fertilization or phosphate containing detergents indicates that phosphorus concentrations in streams were closely related to the phosphorus content of the soils and rocks through which the water flowed. The amounts of phosphorus in the soil water largely determine the concentrations in nearby groundwaters and surface waters unless animal or human pollution is a factor.

When soils are devoid of oxygen, such as in bottom muds of streams or swamps or when land is under standing water, changes in phosphate solubility may occur. Where the phosphate is immobilized as insoluble iron phosphate, the above conditions can result in the ferric iron being reduced, a rise in pH and the production of more soluble ferrous phosphate. Similar conditions will occur in lakes, reservoirs, and ponds, increasing the available phosphorus for aquatic growth. Where soils contain aluminum or calcium phosphates, lack of oxygen does not necessarily result in increased soluble phosphorus in the soil solution.

Other Minerals

Many minerals can be found in agricultural wastes and waste waters and have their source in the feed to the animals, the crop being processed, and in the drinking and process water. Many of these minerals are adsorbed onto the soil particles and do not represent an environmental quality problem. Each soil varies in permeability and its ion removal capacity. Each soil should be evaluated to establish proper waste discharge rates and soil loading rates.

The insoluble hydrous oxides of Mn and Fe, which are ubiquitous in soils, provide one of the principal controls on the fixation of heavy metals, such as Ni, Cu, Co, and Zn, in soils. The principal factors that control the sorption and desorption of the heavy metals are redox, pH, and the concentrations of the metal of interest, competing metals, other ions capable of forming inorganic complexes, and organic chelates.

Certain agricultural wastes, such as those from the lye peeling of fruits and vegetables and where salt may be used for food processing and pickling, may contain high concentrations of sodium ions which can produce deleterious changes in soil structure. The application of high concentrations of animal urine to soils has the potential for similar problems. High concentrations of sodium cause dispersion of the soil particles, change of the effective pore size, and influence the permeability of the soil. The reduction in permeability decreases the value of the soil for waste water irrigation and disposal and for crop growth. Permeability becomes low when the exchangeable sodium exceeds about 15% of the total cations in the soil exchange complex. The effect of sodium can be countered by adding a soluble calcium source to the liquid being used for irrigation. No convenient method exists to regenerate the ion exchange capacity of soils in the field.

The effect of sodium depends not only on the amount of sodium applied but also on the amount of other cations present. From many studies carried out on the quality of irrigation water on western soils (6), Eq. (10.1) has been developed for use in determining the suitability of water for irrigation purposes.

$$\mathrm{Na}(\%) = \frac{\mathrm{Na} \times 100}{\mathrm{Ca} + \mathrm{Mg} + \mathrm{K} + \mathrm{Na}} \tag{10.1}$$

The units of the cations are in gram equivalents per million.

To be suitable for irrigation of western soils, the percentage of sodium, expressed by the above formula, should not exceed 80 and the total concentration of cations in equivalents per million should not exceed 25. In more humid regions, sodium may be a lesser problem due to the greater amount of leaching of soils in these regions.

The soil adsorption ratio (SAR) also can be useful in estimating potential problems due to excess sodium in irrigation liquid. The units of the cations in the SAR equation are milliequivalents per liter.

$$\mathrm{SAR} = \frac{\mathrm{Na}}{\left[\dfrac{\mathrm{Ca} + \mathrm{Mg}}{2}\right]^{1/2}} \tag{10.2}$$

Irrigation waters should have an SAR not more than 8–10. Where high sodium concentrations are expected in agricultural waste waters, the advice of competent soil scientists should be obtained to avoid adverse changes in soil structure.

Trace elements such as copper, zinc, molybdenum, chromium, nickel, iron, cadmium, and others can exist in wastes and waste waters. Sources include industrial processes that are discharged to municipal sewers, metal corrosion, and ingredients of animal feeds. When these wastes are disposed of on land, waste loading rates should not result in concentrations of heavy metals that will be toxic to plants.

No clear evidence of toxicities and yield reductions to crops have been reported, perhaps because of the relatively short experience of waste water disposal to land and the low trace element application rates. The effects of these elements on soil properties can be based on knowledge of soil composition and the chemical, physical, and biochemical relationships in the soils. In general, heavy metals are complexed by clay and organic compounds and are unavailable to the plants and unavailable for leaching.

Microorganisms

The removal of microorganisms from wastes or waste waters as they contact the soil is an important consideration in the use of land for waste disposal. Information from field situations indicates that bacteria and viruses are removed from waste water as it percolates through the soil. Unless fissures exist in the soil, percolation through even the coarsest soil will remove bacteria and viruses in a short distance. Most of the organisms in the applied waste are removed, generally greater than 90% removal, in a thin layer of soil at the surface. Several feet or more of soil are necessary for near complete removal of the applied microorganisms.

When organisms in a waste enter soil, they face competition for food supply, antibiotic materials from other microorganisms, and predation by other soil organisms. Competition and predation is greatest in the surface soil layer since oxygen is more abundant and rates of decomposition are greater. Human and animal pathogens are not adapted to the rigors of such an existence and are not expected to survive for lengthy periods. The persistence of microorganisms once they are removed by the soil depends on biological controls regulated by environmental factors such as temperature, organic matter content, and whether the soil is aerobic. Pathogenic microorganism survival can be expected to be the greater when normal biological activity is the least, such as under low temperatures and anaerobic conditions.

Soils that are suitable for waste disposal from the hydraulic, nutrient control, and adsorptive standpoint can provide satisfactory removal of pathogenic microorganisms. Microorganism removal is unlikely to limit waste or waste water application rates on soils.

Erosion

The amount of nutrients and pollutants moving to surface waters increases significantly when the surface runoff includes wastes not incorporated in the soil or includes eroded soil. Sediment can be an important pollutant due to the inert material, oxygen demand organics, and nutrients that are part of the sediment load. Soil erosion is greatest on steep slopes, sparsely covered land, in late winter when rains occur on land with little or no snow cover, and on cultivated land when heavy rains occur before adequate crop growth. The practice of spreading animal manures on frozen snow covered land as well as on land subject to runoff and erosion can lead to contamination of surface waters if the runoff reaches the waters.

The concentration of nutrients and pollutants to surface waters increases significantly when the surface runoff includes wastes not incorporated in the soil or includes eroded soil. Sediment can be an important pollutant due to the inert material, oxygen demand organics, and nutrients that are part of the sediment load. Soil erosion is greatest on steep slopes, sparsely covered land, in late winter when rains occur on land with little or no snow cover, and on cultivated land when heavy rains occur before adequate crop growth. The practice of spreading animal manures on frozen snow covered land as well as on land subject to runoff and erosion can lead to contamination of surface waters if the runoff reaches the waters.

Good soil conservation practices, incorporation of wastes in the soil as soon as possible, and avoidance of conditions likely to lead to erosion will minimize the quantity of pollutants from the soil it is used for waste disposal.

Fertilizer Applications

Except for manure applied to the soil as a fertilizer and/or soil conditioner, application of inorganic fertilizers for crop growth is not a land disposal problem in the strict use of the term. However, the excessive use of fertilizers can contribute to nitrogen problems in groundwaters and surface waters. It may be useful to attempt to put the environment quality problems caused by fertilizers in perspective with those due to waste disposal.

Chemical fertilizers are compounded from relatively simple inorganic compounds such as ammonia, ammonium nitrate, sulfates, polyphosphates, urea, mono- and dicalcium phosphates, and potassium chloride and sulfate. Calcium sulfate is present when a low analysis phosphate such as ordinary superphosphate is used (7). These materials are applied to the land singly or combined in various ways to provide the needed nutrients. Nitrogen is applied as ammonium, nitrate, or some combination of these. The form in which the nitrogen is applied is not of great importance since the ammonium is converted to nitrate by the soil organisms.

About 90% of the fertilizer nitrogen used for crops is in the form of ammonia or ammonium salts. This includes urea which is hydrolyzed to ammonium nitrogen shortly after being applied to the soil. Liquid ammonia is the largest single source of nitrogen fertilizer used. Depending on the rate and method of application, type of crop, and soil and climatic conditions, from 50 to 60% of the applied fertilizer nitrogen usually is recovered in the crop. Another 10–20% may volatilize or be denitrified. Any remaining nitrogen may move with water percolating through the soil. Fertilizer phosphorus is held by the soil and can accumulate in the soil.

Deep rooted crops offer an opportunity to manage a crop-waste disposal system to avoid excess nitrogen in subsurface waters. Detection of nitrates below the root zone does not imply that all of the nitrates eventually will appear in the groundwaters. Little is known of the fate of the nitrates between the root zone and the groundwaters. Where considerable distances exist between these boundaries, the passage of the nitrates will take time and depend upon the rate of water movement through the soil. If the nitrates are not removed either by crops or denitrification, and if the soil covers coarse material or fractured rock, the nitrates can reach groundwaters more quickly.

A study of nitrate movement in grass and cropland in the northern Great Plains indicated that the leaching of nitrates in that area is minimal since fertilizer nitrogen rarely is applied at a rate greatly in excess of crop utilization and since there is no evidence that water and nitrates penetrate below the rooting depth in fertilized grassland soils under natural moisture conditions (8). An evaluation of fertilizer applications to corn revealed that from 20 to 90% of the applied nitrogen was not utilized by the crop to which it was applied (9). Although immobilization of the nitrogen into the organic fraction of the soil was a possibility, the data did not indicate this as likely. A significant portion of the unutilized nitrogen was lost by denitrification and it was felt that losses by denitrification were equal to or greater than nitrate leaching losses.

An investigation on the use of dairy manure as a fertilizer for corn indicated that the nitrogen requirements of corn on a dairy farm in New York largely could be met by the adequate management of organic nitrogen (10). The soil organic nitrogen was a well-buffered reservoir that released nitrogen in proportion to the favorableness of the environment for corn production. The quantity of organic nitrogen in soil is a function of the annual losses and gains of organic nitrogen.

Organic nitrogen can supply nitrogen for crop growth and should be included in determining the quantity of supplemental inorganic fertilizer nitrogen. Proper management and application of the available nitrogen is important to avoid accumulation of nitrate in soil since once in contact with the soil, nitrogen from either an organic or inorganic source acts in a similar manner and can pose the same pollution concern.

Proper timing of nitrogen application offers the opportunity of both better nitrogen use by the crop and less excess nitrogen loss to the environment. When the nitrogen can be applied to approximately match the demand rates of the crops, a greater percentage will be incorporated into the crop. Controlled field experiments have indicated that the use of fertilizer nitrogen as a summer side-dressing to corn increased the yield and produced a greater residual effect on subsequent crops than did either spring or fall applications of fertilizer (9).

Fertilizer applications, either inorganic fertilizers or manure used for this purpose, should be adjusted to maximize crop return and to minimize loss to the environment. The yearly amounts of nitrate nitrogen lost from fertilized crop land will vary. The variability is related to the quantity of water movement and the relationship between the time of surplus water to the time of fertilization. Timing fertilizer applications to avoid excessive quantities of fertilizer nitrate nitrogen in the soil during periods with a high probability of surplus water will reduce the loss of nitrates to surface water and groundwater. Therefore, in the humid parts of the United States, fall and early spring applications of fertilizer should not exceed the immediate crop needs.

One approach to minimize soluble nitrogen loss is to keep the applied nitrogen in the form of ammonium nitrogen which is strongly absorbed by soil particles. Microbial nitrification can be inhibited chemically. Various methods and approaches have been studied (11). Such inhibition may permit more effective use of fall applied nitrogen. Application of ammonium nitrogen fertilizers in the fall often is more convenient because of availability of equipment, time, and generally better weather conditions.

The following approaches can act as guidelines to avoid potential nitrogen problems caused by excessive applications of agricultural wastes or fertilizers. Excess use of the latter not only adds to potential environmental quality problems but also adds unnecessarily to the cost of the crop. The rate of fertilizer application should be adjusted to more nearly approximate the crop need. The value of soil organic nitrogen should be included when determining fertilizer or waste application rates where crop response is important. Crop rotation should include at least one rotation of a deep-rooted crop to remove excess nitrates in the soil where heavy rates of manures and other organic agricultural waste are applied. Localized nitrogen problems can occur in and around manure storage and silo drainage areas. Excessive accumulation of nitrogen in a small area should be controlled especially if wells are located nearby.

Manure Disposal

In spite of the fact that the practice of distributing manure on land may be questionable from the profit standpoint because of alternative sources of inorganic fertilizers, the land remains the most appropriate point of disposal for both animal waste and wastes from many food processing operations.

These wastes may be applied to the land as a source of nutrients and organic matter or as maximum feasible loadings for disposal under conditions that do not create a pollution problem. The challenge is to properly incorporate the disposal of these wastes in a controlled land management program so that the applied wastes do not contribute to problems such as contamination of groundwater, pollution caused by excessive runoff, odors, or insect generation.

Application of animal wastes to land may have both beneficial and adverse effects. The chemical, physical, and biological characteristics of the soil may be altered. Under poor soil and water management practices, these wastes can create a pollution hazard especially from the nutrients lost in runoff and soil–water percolation. Sound soil and water conservation practices including erosion control, and possibly pretreatment of the wastes can minimize these problems. Land disposal of agricultural wastes may not be profitable and will have to be absorbed as part of the cost of production.

The nutrient content of animal wastes is a determining factor in the quantity of manure applied for either crop production or solely for land disposal. The nutrient content of animal wastes will vary depending upon many factors including the type of feed, age of the animal, and type of animal. Data on the nutrient content of specific wastes should be used in calculating the nutrient loads for land application of manures. Estimates of the nutrient content of animal manures are noted in Chapter 4. Consider-

Fig. 10.3. Land disposal of slurry animal wastes.

Fig. 10.4. Land disposal of semisolid animal wastes

ably more information is necessary to establish feasible loading rates for liquid and solid animal wastes that can be applied to the soil. The rates must be related to soil type, climate, rainfall patterns, type of crop, type of waste, and frequency of application. Manure can be spread on land either as a slurry (Fig. 10.3) or as a semisolid (Fig. 10.4). If the manure has been stored under anaerobic conditions, the odorous volatile compounds will be released by such spreading. An aerated liquid treatment system can be used in conjunction with a slurry waste disposal system to minimize odor problems and facilitate handling of the wastes.

Where soil fertility has not been maintained, manure applications can result in sizable crop yield increases. Application of both fertilizers and manure to soils should avoid over application. When manure application rates are excessive, a decrease in crop yield may occur due to inhibitory amounts of ammonia or nitrite nitrogen or salts in the soil.

Intermittent waste applications preserve the effectiveness of the soil to assimilate the wastes. Too frequent or too heavy a waste application can decrease the rate of infiltration and the oxygen input to the soil. For best results, concentrated organic wastes such as manures should be injected directly into the soil or applied to the surface and incorporated with the soil as soon as possible by discing or plowing. The problem of agricultural waste disposal is greater during winter months when lands cannot be plowed and under conditions when the soil is at or near saturation.

Infiltration tests in which both mineral fertilizers and dairy manure were applied to corn indicated manure application increased infiltration and heavy mineral fertilization decreased infiltration significantly as compared to manure with moderate fertilization (12). Where manure was compared to no manure on four crop rotations in conjunction with moderate fertilization, the manure did not significantly increase the rate of infiltration. The proper application of manures to land also can reduce erosion rates.

Thin spreading of fluidized manures on land will accelerate drying and will help avoid fly breeding. Average thicknesses of $\frac{1}{5}$ to $\frac{1}{25}$ inch were suggested, depending upon the season (13). Cumulative layering of thin layers of manure is possible, if drying rates are favorable, reducing the land area needed. Less than 200 ft^2/dairy cow and 1 ft^2/chicken was suggested as being sufficient for drying operations in areas where open air drying and soil characteristics for water removal were feasible.

Wastes spread upon the land for disposal should be incorporated with the soil soon after spreading to avoid pollution caused by runoff. Direct subsoil injection of liquid manures has shown promise on soils suitable for such equipment (14). The maximum land application that can be disposed of in this manner will depend upon the type of soil, crop utilization of the available nutrients, rate and frequency of waste application, and hydrologic conditions.

Poultry manure and broiler litter have been applied to land under conditions that have caused soil pollution and related problems. Where manure litter has been returned to the soil at rates exceeding 10 tons/acre annually, problems of excess salts and a chemical imbalance have occurred in the soil and problems of nitrate toxicity and grass tetany have occurred in the grass pasture (15). As a consequence of these potential health problems, broiler litter disposal rates greater than 4 tons/acre/year are not recommended on fescue pasture systems (16).

The plow-furrow-cover method was used to compare the disposal of poultry manure at rates of 0 to 45 tons of dry poultry manure per acre (14). The tests were on a loamy soil with a clay content of 15 to 33%. Soil and soil water analyses indicated that potassium, magnesium, and sodium were mobile and moved downward through the soil. The soil did not have a cover crop in these experiments. Ammonia nitrogen existed in the soil during the winter but generally was oxidized to nitrate nitrogen in the warmer months. Nitrate nitrogen in the soil was in low concentration shortly after manure application but increased to high concentrations in the manure application zone. Nitrate nitrogen concentrations as high as 700 mg/l were found in the soil water. With time and rainfall, the nitrate nitrogen was distributed throughout the soil profile and by the end of the experiment had leached below the lowest sampling point. Only very small amounts of phosphate were found in the soil water. Chloride and sulfate ions were found in the soil water at all depths. It was concluded that the disposal of poultry manure in these soils should be less than 15 tons/acre of dry material because of nutrient contamination of the soil water.

Farmers frequently are advised to apply dairy manure to crop land at a rate of not more than 20 tons/acre. Rates above this level generally have not provided a comparable yield increase. With the increasing herd size on dairy farms, some farms may not have adequate land to spread at this rate and greater amounts of manure per acre may have to be spread. Rates from 15 to 625 tons of dairy manure were applied per acre of corn to evaluate the effect of heavy loadings (17). These rates corresponded to loadings of 165 to 6800 lb of nitrogen per acre. Growth on all plots was good and no toxic symptoms were observed. At the highest loading rate, the thickness of the applied manure interfered with plowing. Nitrates, chlorides, and other salts moved through the well-drained sandy loam soils.

The application of beef cattle feedlot waste to the land is a common and accepted method of disposal. The solid wastes are usually stockpiled within the feedlot area or in a central storage area near the feedlot. Drying and/or microbial degradation of the wastes may occur during the stockpiling. These wastes are spread on fields as crop and weather conditions permit. The wastes generally are incorporated in the soil as soon as possible to make maximum use of the nutrients in the waste and to reduce surface water pollution.

Manure accumulations on and around beef feedlots can be a cause for concern. Many management factors can affect the amount of nitrate in these soils, such as type of feed, length of livestock feeding, length of lot occupation, and frequency of manure removal. Investigations of the nutrient accumulation under feedlots revealed that soil texture did not affect the nitrate distribution pattern beneath feedlots where the soil was of fairly constant composition (18). When the soil profile varied, both moisture and nitrate concentration increased as the clay content increased. Soil phosphorus analyses indicated little if any movement of phosphorus. The phosphorus was concentrated in the top layer of the soil or manure layer. Slight phosphorus movement was detected in soils with an extremely low exchange capacity. Nitrate concentrations up to 78 mg/l as nitrogen were found in groundwaters under the feedlots. The longer the feedlot existence, the greater the amount of nitrate in the soil profile.

The application of feedlot runoff to permeable loams and clay loams drastically reduced the permeability of these soils (19). This change in characteristics can be an advantage in sealing the land in a feedlot runoff lagoon to avoid groundwater infiltration. It can be a detriment when the runoff is applied to crop land for irrigation or recovery of nutrients. The detrimental role of feedlot runoff to crop land will depend on the rate and amount of runoff application and the land management methods.

When feedlot manure was applied to grain sorghum, grain yield was depressed when the manure was applied at annual rates of 120 and 240 tons/acre for two years. Early growth depression was observed on plots

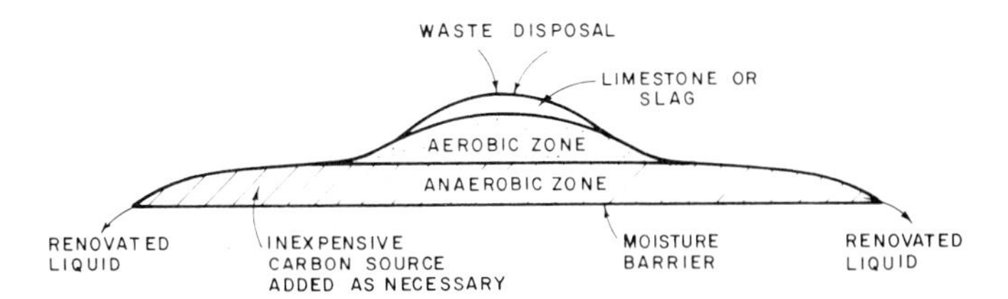

Fig. 10.5. Schematic cross section of the barriered landscape water renovation system (21).

receiving 30 and 60 tons/acre. This depression was attributed to high ammonium and salt concentrations in the seed zone. The depression could be relieved by applying irrigation water prior to seeding to decrease the ammonium and salt concentrations (19). For most conditions, it was suggested that an appropriate rate of beef cattle feedlot manure disposal on irrigated grain sorghum was 10 tons/acre every three years in the Texas high plains area.

In Kansas, maximum yields of irrigated corn silage occurred at beef cattle manure application rates of between 100 and 130 tons/acre (20). Larger applications depressed yields due to accumulations of soluble salts in the soil. Potentially toxic accumulations of ammonium ion were found in the surface 12 inches of the soil.

The amount of nitrogen that can be utilized by a crop places a limit on the quantity of wastes that can be applied to a given soil. The quantity of wastes that could be applied can be increased if the excess oxidized nitrogen that can be produced were denitrified in the soil. An engineered soil system to use the soil to remove the excess nitrogen in this manner, as well as for the removal of phosphorus and decomposition of organic matter, has been developed (21). The system consists of a mound of soil underlaid by a barrier impervious to water (Fig. 10.5). The barrier extends beyond the mound under the level soil at the edges. A thin bed of limestone or slag is placed on the top of the mound and the wastes are spread over it. As the water percolates through the mound, the organic particles are filtered out and decompose. The filter bed and the soil remove the major portion of the phosphate. The soluble organics and inorganics move into the aerobic soil zone where the ammonium ions are held in the exchange complex of the soil until they are nitrified. Most of the soluble organic matter is oxidized in the aerobic soil. The downward movement of the nitrates is stopped by the barrier and forced to move laterally through the anaerobic soil perched on the barrier. In the anaerobic zone, denitrification occurs. Carbon sources can be added to the denitrification zone, if needed, in the form of inexpensive organic matter such as grains. The renovated water moves from the edges of the barrier into the adjacent soil.

This barriered landscape water renovation system has been used with swine wastes. The renovated water has been shown to contain less than 2 mg/l of nitrogen and 0.07 mg/l of phosphate which amounted to 99.5% nitrogen removal and 99.8% phosphate removal (21). The long-term evaluation of this system will determine its maximum potential as part of an animal waste management system.

Farm livestock can harbor bacteria pathogenic to man and animals. The use of the land as a disposal media offers the opportunity to transmit these bacteria to susceptible hosts. The survival of salmonella organisms in soils was studied to determine the public health aspects of this possibility. *Salmonella typhimurium* was found to survive in soils up to 110 to 160 days depending upon soil characteristics. Soils with a low pH (5.3–5.5) and of an inorganic nature had the greatest rate of Salmonella die-off. Winter climatic conditions prevailed (34°–53°F mean temperature range) in the study (22). *Salmonella dublin* and strains of hemolytic *Escherichia coli* have been isolated from dairy cattle manure slurries (23). Although none of these pathogenic bacteria appeared to multiply in the slurry, they survived for up to 12 weeks. Movement of pathogenic bacteria would occur primarily with soil particle movement.

Liquid Wastes

Methods to dispose of waste water on the land can have the advantages of being economical, requiring minimum maintenance, and providing satisfactory results. Irrigation systems also can accommodate the seasonal wastes of typical food processing operations thereby avoiding more complicated treatment systems.

The application of waste waters to land is not a new approach to waste water treatment or management. The land disposal of municipal waste water can be traced back several centuries. Most of the past and current applications of waste water to land have been developed as convenient or economical methods for waste water disposal. The effects of these methods on the ecosystem, i.e., plant and animal life, soil characteristics, and groundwater quality, are only beginning to receive the necessary attention.

The number of communities that dispose of waste waters on land is increasing (Table 10.2), however, this number remains a small percentage of the total communities in the United States that have waste water treatment facilities. Over 55% of the land application systems noted in 1972 were identified as crop irrigation systems located in 13 western states (24). The remaining systems were either of the infiltration type or were unidentified as to method of application and were located in 32 states. In addition to

Table 10.2

Municipalities Using Land Application of Waste Water[a]

Year	Number of systems	Population served (millions)
1940	304	0.9
1945	422	1.3
1957	461	2.0
1962	401	2.7
1968	512	4.9
1972	571	6.6

[a] From reference 24.

the direct application of municipal waste waters to land, millions of gallons of other domestic waste waters are applied to the land by the combined use of the septic tanks and soil adsorption. The food industry has made effective use of the land for disposal. Fruit and vegetable canning waste, meat-packing waste, and dairy product wastes have been disposed of on the land in a controlled manner. The seasonal operation of some of these plants and their location in rural areas have made this approach attractive.

Ridge and furrow irrigation, spray irrigation, and overland flow irrigation have been used for the disposal of agricultural wastes. With the ridge and furrow method, wastes are discharged from a main ditch or gated pipe, and flow into the furrows and through the soil. Spray irrigation is the controlled spraying of liquid on the land. The liquid infiltrates and percolates through the soil. Lightweight metal or plastic pipes are used to transfer the waste from the source to the disposal field. Overland flow irrigation is the controlled discharge of liquid on the land with the liquid flowing downslope over the surface.

Spray irrigation systems are the more common system now in use. The land used for ridge and furrow systems is difficult to crop and the grasses frequently are left uncut. Spray irrigation systems require little land preparation prior to use. For both systems, the type of acceptable land is not as critical as the soil characteristics. Woodland, brush areas, pastures, and cropland have been used. Moderate slopes are preferred. Actual application sites have varied from flat to steeply sloping. The major constraint on slopes is the prevention of erosion. Where steep slopes have been used, natural vegetation was satisfactory to control erosion. Vegetation that has been used successfully with spray irrigation includes spruce, pine, oak, and larch in forests and wheat, corn, oats, clover, alfalfa and Bermuda, reed canary, fescue, and Sudan grasses on agricultural sites. The actual crop will depend on location and use of the harvested crop.

If designed correctly, the rate of waste application in a spray irrigation system will produce minimal damage to vegetative growth and will avoid surface runoff and erosion. The principal processes involved in spray irrigation are biological action and filtration. The organic matter is biologically removed in the upper soil layer and the particulate matter is filtered by the soil particles and surface vegetation. Adsorption of soluble inorganics can occur in the soil. Spray irrigation is restricted to soils that can take the necessary application rates and that are within reasonable pumping distances from the waste source.

Because of the variability of soil characteristics, effluent constituents, and use of a spray irrigation site, waste water application rates are determined on a site-by-site basis. Acceptable application rates have ranged up to 2 inches/week/acre for municipal waste water. The important considerations in determining a spray irrigation site are the ability of the soil to properly treat the wastes. Features such as the infiltration rate and the exchange capacity of the soil, soil texture, depth and drainage, and physical discontinuities require evaluation. Because the applied water eventually will enter the groundwater, the depth to the water table, groundwater rate and direction of movement, and groundwater quality are significant factors in site selection.

Deep, well-drained, loam soils are preferable for spray irrigation. Soils that are predominantly sand, silt, clay, or stony are not preferred. A high clay content reduces the permeability of a soil and coarse soils may permit waste water passage too quickly to provide for adequate treatment. When installing spray irrigation equipment, disruption of the established soil profile should be kept to a minimum.

The most widely used system involves spray irrigating on land that is flat or almost flat. The field is spray irrigated for a short time period and then allowed to rest for an equal or greater period of time before again being used for irrigation. Reed canary grass or other grass cover is used since when a cover crop is used, the infiltration rate is greatly increased. This method is used on soils with a good infiltration rate and where groundwater conditions can be managed. The amount of applied nitrogen can be a critical parameter with this method. The rate of application would be related to the amount of nitrogen utilization by the vegetation and to the amount lost by volatilization, denitrification, and immobilization. Drainage tiles can be installed when it is desirable to recover and reuse the treated waste water.

With overland flow irrigation, the liquid is irrigated on a slope and on a soil with poor infiltration capacity. Organic and inorganic waste removal is accomplished by the soil and plant surfaces. The waste water is renovated as it flows in a thin layer over the surface of the soil and through the living vegetation and the decaying material. A continuous vegetative cover is necessary to achieve the desired degree of filtration and physical removal of BOD and suspended solids. The overland runoff system requires trenches spaced approximately 100 ft apart per 1% of slope to achieve a desired retention and contact time with the soil. Slopes may vary between 2 and 6%. Waste water is applied intermittently with frequent 1–5-day drying periods. This system has been successful in improving the quality of canning waste water and treated municipal waste water.

A part of the irrigated waste waters is evaporated, some is transpired, and the remainder filters into the ground. The rate of evaporation depends upon temperature, humidity, and liquid surface area. The rate of transpiration depends upon plant leaf area, humidity at plant level, available moisture, and amount of solar energy. Under field conditions, it is difficult to separate water use by evaporation and transpiration. The two are combined as evapotranspiration or consumptive use. Normal evapotranspiration will range from 300 to 7000 gal per acre per day.

A portion of the mineral content of the waste water will be taken up by vegetation. The remainder will accumulate in the soil. Vegetation has an important role in the success of an irrigation system by keeping the soil loose and aiding the transmission of water. The type of vegetation will depend upon annual temperature variations. Reed canary grass has been successful in northern climates since it endures water and ice coverage. Brome grass and Kentucky blue grass can be suitable where water and ice problems are less severe. A mixture of blue grass, alfalfa, fescue, brome, timothy, and reed canary grass can be used to establish a vegetative cover. Local soil scientists can advise which type of vegetation is best suited to anticipated irrigation rates, soil types, and weather conditions.

Aeration in the root zone is essential for most vegetation. Occasional flooding can be tolerated especially when the plant is dormant. Sulfides and nitrite can be toxic to plant roots. These compounds can be present in specific wastes and can be formed in the soil when sufficient oxygen is not present. Liquid application rates can be increased where the soil is protected by cover crops or mulch from surface compaction by rain and can be increased as the temperature increases. The amount of water spread on the soil should be controlled to avoid long periods of flooding and subsequent anaerobic conditions. Odor problems can result from untreated or partially treated waste water accumulating in stagnant pools.

Every irrigation field should be designed on an individual basis. Data on percolation tests, soil logs, and type of vegetation need evaluation. The land can be drained to assist the movement of water through the soil if it can be shown that such drainage water will not contaminate surface streams or that the drainage water does not reach surface waters.

Waste water irrigation areas can be designed on the basis of maximizing combinations of infiltration, evapotranspiration, or nutrient uptake by the crop. By proper design, the runoff from such irrigation areas can contain very little organic material and mineral nutrients. The choice of the most appropriate system will depend upon climate, soil types, available land, and quantity of waste to be irrigated.

To utilize waste water irrigation as a long-term viable alternative, the concentration of salts in the root zone of the crops must be kept below detrimental levels. A balance must be achieved between the salts removed by deep percolation or drainage and the salts added by the irrigation water. With relatively large amounts of natural rainfall and waste waters having low salt concentrations, detrimental levels rarely are reached. Irrigation with waste waters having relatively high salt concentrations in regions of low natural rainfall can produce detrimental salt levels in the root zone and forage or cover crop inhibition. The latter conditions should be avoided when waste water irrigation systems are designed.

The volumetric loading and the cation loading rate are important design parameters for the distribution of waste waters on the land. The BOD loading rate is of less significance due to the effective removal action in the soil. Nitrogen loading rates should receive consideration. Reasonable predictions of loading rates at a given site can be made if soil conditions, type of cover crop, depth to water table, frequency and intensity of rainfall, and similar pertinent information are known. Final design application rates can be determined best by observing field conditions during irrigation.

Spray irrigation systems will be suitable for the disposal of liquid wastes when the hydraulic loading is maintained at a rate less than the infiltration rate of the soil, the soil is maintained in an aerobic condition and ponding of the liquid on the soil is avoided, a crop tolerant of wet soils is used, the SAR is kept below about 10, a resting period occurs between irrigation applications to avoid prolonged and excessively wet soils, the organic loading rate is kept below the rate at which the organic degradation occurs, and the nitrogen loading rate is not greatly in excess of the rate at which the nitrogen is taken up by the growing crop.

The advice of a competent soil scientist will prove useful when deciding on application rates. Adequate preliminary investigation is essential. Test wells for monitoring the groundwater and soil waters are desirable and warranted to protect the organizations using irrigation to dispose of waste waters. It can be dangerous and costly to assume that because the wastes are out of sight, they can be placed out of mind.

Application to Agricultural Wastes

Agricultural wastes are well suited for land disposal since most agricultural operations are in rural areas where land is readily available. Irrigation has been used to dispose of waste waters from dairy and milk processing operations as well as from food processing and canning operations. One of the most publicized spray irrigation systems has been that used at Seabrook Farms in New Jersey. A flow of 5–10 mgd of process water was sprayed over an 84-acre area of loamy sand. Each portion of the land was sprayed for 8 hr followed by a 24-hr recovery period (25). Canning wastes of 0.6–0.7 mgd have been sprayed on 90 acreas of grass and woodland pasture underlaid with clay. Application rates were about 0.5 inches/hr over an 8–12-hr period (26).

Packinghouse wastes have been applied at the rate of 1.3 mgd per 16-hr day to 240 acres of Kentucky fescue. The application rate was about 0.44 inches per day (27). Fruit and vegetable canning wastes were sprayed on a sandy area every two days (28). Suspended solids pumped with the waste water inhibited the wheat growing on the area. It was found necessary to disc the spray area every two weeks to prevent crusting of the topsoil.

Experience with dairy plant wastes (29) have noted many situations where milk plant and cheese plant wastes have been successfully treated by spray irrigation. Whey wastes have been applied to 30 acres at a rate of 5000 gal/day (30). Application rates of from 0.13 to 0.23 inches/hr have been used for the irrigation of milk plant wastes in Wisconsin (31). BOD loading rates were from 140 to 322 lb BOD/acre/day. An application rate of 2.3 inches/hr resulted in considerable runoff to a cover area.

In spray irrigation systems, only a small portion of the applied waste should appear as runoff. When fruit and vegetable wastes were disposed of by spray irrigation, the average runoff was about 37% of the raw waste flow and contained 60 lb BOD/day, a reduction of over 97% (32). Soil erosion was minimized by contour plowing the steepest slopes. The system was essentially an overland flow system on clay underlain by shale. A 24-hr spray irrigation program was used with at least two days rest between applications to specific land areas. The liquid loading was about 4600 gal/acre/day. Icing was experienced during the winter although BOD reductions were acceptable.

The results of a 12-month study on the spray irrigation of cannery wastes indicated that evaporation losses accounted for 18% of the total liquid applied, deep soil percolation accounted for 21%, and runoff accounted for 61%. Runoff from clay loam soil was greater than from sandy loam soil (33). The system removed 92–99% of the volatile solids and oxygen-demanding materials, 86–93% of the nitrogen, and 50–65% of the phosphorus. Treatment efficiency was improved by spreading the waste water load over a larger land area and by reducing the frequency of application. Phosphorus removal increased from 40 to 88% by decreasing the spray schedule. Evaluation of surface and subsoil samples and soil water samples indicated an increase in salinity with time. Both total dissolved solids and sodium increased in the older watersheds. Nitrogen and phosphorus remained in low concentrations. A substantial fraction of the phosphorus removed from the waste water was accounted for in the surface layer of soil.

The soil in this area was a heavy clay having a rate of surface infiltration estimated at 0.05 inches/hr with subsurface flow zero for practical purposes (34). This overland flow irrigation system was on slopes that were contoured to assure a uniform flow of water. Slopes ranged from 1 to 12% with the results of the study indicating that slopes between 2 and 6% were preferable. The screened liquid wastes were sprayed intermittently from a series of sprinkler lines. The liquid application rate in the test area was about 0.6 inches/day for a 5-day week. The design rate ranged from 0.25 inches/day in winter to 0.50 inches/day in summer. The desirable downhill distances between sprinkler lines were approximately 150–175 ft beyond the perimeter of the sprinklers.

The microbial action during the overland flow and the type of vegetation are important in the success of such studies. In the Paris, Texas studies (34), stands of both native and northern species of grasses initially developed. Continuous use of the overland flow system caused reed canary grass to predominate when readily available nutrients were supplied in quantity by the waste water. The hay harvested from the disposal site ranged up to 23% crude protein and contained about double the mineral content of good quality hay. Between one end and two harvests were feasible and essential to the continuous utilization of this site for nutrient removal.

The microbial populations in the disposal fields were similar to those present in all agricultural soils. The populations varied from field to field depending on food supply, temperature, and soil types. Although biological activity decreased during the winter, an increase in microbial population in cold weather compensated for the decrease in activity per unit population. Little loss of disposal efficiency was observed. No freezing or icing problems were identified.

Although the waste water quality varied with the cannery production schedule, characteristics of the runoff leaving the overland flow system were reasonably consistent. The runoff ranged from 40 to 80% of the applied waste water and had a BOD consistently less than 10 mg/l. Operating costs for the system were about $0.052/1000 gal of waste water.

Effluents from potato starch and potato processing plants have been disposed of by spray irrigation successfully. An annual dose rate of 8 to

20 inches/year was observed to be optimum on grasslands in Poland. At higher dose rates, 32 to 40 inches/year, the native grasses disappeared and a predominance of plants that could tolerate high nitrogen concentrations appeared (35). Potato processing plant wastes have been spray irrigated at the rate of 14,000 gal/acre. A weekly rotation program was involved so that each area was irrigated every 7 days.

The quantity of wastes that can be applied depends largely upon the infiltration rate of the soil since the systems may be expected to operate during months when crop growth is dormant and evapotransportation is small. Cannery waste application rates have ranged from 6000 to 100,000 gal/acre/day. Data for handling milk wastes indicate application rates of from 400 to 3000 gal/acre/day (36). Tables 10.3 and 10.4 (37) summarize data on application rates and waste volumes for a number of agricultural waste water irrigation systems.

Sewage and Sewage Sludges

Disposal of domestic sewage on land by irrigation has not been widely practiced in the United States primarily because of the distance between

Table 10.3

Rates of Application of Agricultural Waste Waters to Land

Flow rate (mgd)	Application rate (inches/day)	Area (acres)	Continuous or intermittent application	Soil type	Ref.
Canning					
0.19–0.44	1.4–3.4	200	—	—	31
—	3.0	—	I	Silt–loam	31
0.6–0.7	4–6	90	I	Heavy loam	26
0.14	—	7	I	Silt loam	31
0.3	—	8	I	Sandy loam	31
1.3	7	240	I	—	31
0.13	—	7	I	—	31
5–10	6.4	84	I	Loamy sand	31
0.65	—	43	I	—	32
3.8	—	400	I	Sandy–clay loam	33
Dairy					
0.012	—	2	—	Clay loam	31
0.075	—	48	—	Sandy loam	31
0.08	—	100	—	Clay loam over gravel	31
0.15	0.0008	—	—	—	31
0.01	0.08	2.5	—	—	31
0.001–0.005	3.1–5.3	0.2–1.1	I	Silt loam and sandy loam	31

Table 10.4

Spray Irrigation of Food Processing Wastes[a]

Type of processing	Applied BOD (lb/acre/day)	Liquid application rate (inches/day)	Crop
Cannery	210	6.4	Wooded, 8-month year
Cherries	810	3.6	—
Corn	860	3.4	—
Lima beans	65	0.4	—
Peas	1200	0.16	Cut grasses, waste includes both solid and liquids
Tomato	410	3.0	
Tomato	160	0.7	—
Vegetables	40	0.4	—

[a] From reference 37.

the collection and treatment facilities of a community and available land areas and soil types and because of a philosophy of using surface waters as a disposal medium. A number of investigations have indicated the feasibility of using irrigation to dispose of domestic wastes and to recharge groundwaters. A competent bibliography discussing sewage effluent as an agricultural resource, land disposal of liquid wastes, agricultural value of sewage sludge, and water reuse applications has been prepared (38). Based on the sanitary aspects, the available literature leads to the conclusion that untreated sewage should not be used for irrigation regardless of the crop grown, sewage effluent receiving at least primary treatment may be used for the irrigation of crops not to be used for human consumption, that the use of primary treated and secondary treated sewage effluent on feed and pasture crops for animal consumption is considered safe, and that clarified sewage effluent may be rendered bacteriologically safe for use on any irrigated crop by chlorination before irrigation (38).

If the pumping and distribution costs are not excessive and if the proper fundamentals of the use of land as a disposal media are utilized, disposal of treated sewage by land irrigation should be an acceptable method. The lack of sufficient large-scale investigations on the disposal of sewage by land irrigation has been a deterrent to wider consideration and application of this method of treatment and disposal. In recent years, several studies have attempted to provide detailed information on the utilization of this method.

Beginning in 1962, studies were conducted by an interdisciplinary team of engineers and scientists at Pennsylvania State University on the feasibility of a terrestial means of treated sewage disposal (39). The objectives were to find methods of waste disposal other than stream discharge, to conserve nutrients by growing useful vegetation, and to replenish the groundwater supply by natural waste water recharge. The project consisted of spraying treated sewage on croplands and forested areas. A solid set irrigation system was used.

Seven years of evaluation of this approach have shown that the soil and the harvested crops together have removed from 50 to 100% of the applied phosphorus and nitrogen. The nutrients in the soil water at the 4-ft depth averaged less than 0.05 mg/l of phosphorus and less than 9.0 mg/l of nitrate nitrogen. The use of corn, hay, trees, and soil to purify the effluent has been described as a "living filter." In years having little rain, the irrigated corn land resulted in higher yields. Hardwood trees averaged an annual diameter growth of 0.21 inch in the waste water treated plot but only 0.12 inch in the unfertilized control plot. White spruce irrigated with the waste water grew 13.5 ft over 7 years while trees on the control plot grew only 6.5 ft. Red pine growth was depressed when 2 inches of waste water were applied weekly.

The irrigation distribution systems were engineered to permit operation even at subzero temperatures. At subfreezing temperatures, much of the waste water was stored temporarily as ice which melted slowly, seeping into the soil as the temperature warmed. The soil in the forested areas was not frozen, although that in the grassy and cropland areas was frozen for part of the winter. The disposal load was shifted from the frozen cropland to the forestland in the winter and from forestland to cropland in the summer.

Over the 7 years, the results indicated that a satisfactory biological and hydrologic balance can be maintained if 2 inches of waste water were applied weekly to the land in the area of the study. The water table lay 10–75 ft below the surface in the valley areas and 50 to 325 ft in the adjacent uplands. The upland surfaces were covered by 5–161 ft of unconsolidated residues consisting of clay, silt, and sand mixtures that have accumulated from the limestone and dolomite bedrock. The Pennsylvania State University studies indicated that about 129 acres of this type of soil would be adequate for the treated wastes from a community of 10,000 population. The municipal wastes had received secondary treatment prior to the land disposal by irrigation.

BOD loading rates, soil types, and temperature were evaluated in the treatment of septic tank effluent and trickling filter effluent by soil disposal (40). About 68% of the BOD was removed by percolation through silica sand at BOD loading rates of 140 and 560 lb/acre/week. At a BOD loading of 140 lb/acre/week, 89% BOD removal occurred in a silt-loam soil. A wide temperature range (18°–35°C) and a wide variation in dosage period had no apparent effect on BOD removal efficiency. Treatment of waste water by soil disposal is limited more by the hydraulic capacity of the soil than by its ability to remove BOD.

Sewage sludge has been applied to land as a method of final disposal, as a method of land reclamation, and as a source of water and nutrients for grasses and other crops. Liquid sludge disposal is regulated by state and local health authorities because of possible odor and insect nuisance conditions and because of possible public health hazards from microorganisms. The disposal of liquid digested sludge on land is common for small waste treatment plants and is used by large cities where adequate land is available. Aerobic or anaerobic digestion normally is required before disposal of the sludge. Primary sludge contains the organic and inorganic matter removed by gravity in a primary or first-stage settling tank. These untreated solids have a high oxygen demand, may contain pathogenic organisms, rarely is disposed of on the land, and should receive further treatment in the treatment plant before disposal.

Pathogenic organisms may survive waste treatment processes and it is not advisable to use digested sludge on crops that may be eaten raw. Heat-drying of activated and digested sludge generally kills the pathogenic organisms. With use of suitable land disposal practices, the application of sewage sludges to soil, either as a soil conditioner or as a disposal method, can be practical.

The disposal of liquid digested sewage sludge on open land is a commonly used approach in small treatment plants. This approach eliminates a solid–liquid separation step. The disposal site should be within reasonable distance of the treatment plant. Transportation costs to acceptable areas can be expensive because large quantities of water are associated with the sludge solids. Depending upon the solids handling processes used, liquid sewage sludges can contain from 3 to 10% solids. If the disposal area is not owned by the organization discharging the sludge, the success of this approach depends upon the continued acceptance by the land owner. Digested sludge can emit noticeable odors when spread on land.

Land disposal of liquid sewage sludge is attractive because the process represents the final disposal of the material with someone else generally assuming responsibility for it after disposal, little capital investment is required, and complex mechanical operation and the use of chemicals is avoided. The approach is intellectually satisfying because it offers a way of recycling and reusing the organic and inorganic matter in the wastes. Economically, land disposal of these solids can be attractive especially if the sludge producer does not have to pay for the material to be hauled away.

Disposal of liquid sewage sludge on the land will remain popular at small treatment facilities because of the above advantages. It has been less satisfactory for large plants because of the costs of transporting the material to suitable disposal sites, the large volumes of material to be disposed of in this manner, and the need to have control over the land used for disposal. Large communities need to be assured that sufficient land is available for a long period. If a crop is to be grown, harvesting and marketing problems must be considered. Large communities are not in the crop production and managing business.

Sludge is a liability at all waste treatment plants. For small treatment plants, hauling and land disposal costs may be high on a per ton basis but the total volume of sludge is small and the total budget may be reasonable. It also avoids the need for a more qualified operator to handle alternative sludge treatment and disposal methods.

Most liquid sludge disposal is done by tank truck or pipelines although application methods vary. Application rates have ranged from 0.5 to 500 tons dry solids/acre/year. Application rates in the range of 0.5 to 50 tons dry solids/acre/year generally were where the disposal was in conjunction with field crops. Higher application rates have been for infrequent loadings to fill and sandy soil for reclamation and development of an organic soil (41). Sludge has been used for land disposal under many diverse climatic and cropping conditions. The rates of application are dependent upon whether it is being integrated into a crop production system or primarily for disposal. No sound guides to soil loading rates for sludges exist. Past and existing application rates appear to be those that have caused no observable adverse effect.

Little information exists on the long term effects of sludge application to soils. Besides the possible excess nitrogen concentrations in sludge applications, the heavy metal ion concentrations are of concern. Metals that have been found in digested sewage sludge include zinc, copper, chromium, cadmium, nickel, and lead. The accumulation of these metals in soil due to continued sludge applications may have an undesirable effect on crop yields. The quantity of metals in sewage sludge will be a function of the commercial and industrial activity of the municipality and can vary among communities. Because metals are complexed by organic matter, their greatest accumulation will be within the plow layer, the area that will have the greatest effect on growing plants. Alkaline soils or the addition of lime will tend to correct for metal toxicities by precipitating the metals in forms not available to plants. When sludges are applied to land, the long range effect of these inorganic constituents should receive close examination and monitoring.

The feasibility of land disposal of liquid sewage sludge is under investigation in a large scale experiment using Chicago sewage sludge (41). The relative costs of sludge disposal, $50–60/dry ton for drying, digestion, dewatering, and/or wet oxidation methods compared to about $15–20/dry ton for digestion and land disposal, indicates that considerable transportation costs can occur before the overall costs of land disposal approach those of current, conventional sludge disposal methods. This investigation will evaluate the practical amount, frequency, and economical method and time for applying digested sludge on crop land and the crops and cropping systems that will provide maximum absorption of certain elements supplied to the soil by digested sludge applications. Land disposal of sludges provides reclamation of strip mine areas, sandy areas, and other unproductive areas.

References

1. McGauhey, P. H., and Krone, R. B., "Soil Mantle as a Wastewater Treatment System," SERL Rep. 67-11. University of California, Berkeley, 1967.
2. Alexander, M., "Introduction to Soil Microbiology." Wiley, New York, 1961.
3. Kolliker, J. K., and Miner, J. R., Use of the soil to treat anaerobic lagoon effluent—renovation as a function of depth and application rates. *Annu. Meet., Amer. Soc. Agr. Eng*. Paper 69-460 (1969).
4. Schwartz, W. A., and Bendixen, T. W., Soil systems for liquid waste treatment and disposal: Environmental factors. *J. Water Pollut. Contr. Fed.* **42,** 624–630 (1970).
5. Webber, L. R., Lane, T. H., and Nodwell, J. H., Guidelines to land requirements for disposal of liquid manure. *Proc. Texas Ind. Water Waste Conf., 8th,* pp. 20–34 (1968).
6. Wilcox, L. V., The quality of water for irrigation use. *U.S., Dep. Agr., Tech. Bull.* **962,** (1948).
7. Nelson, L. B., Agricultural chemicals in relation to environmental quality: Chemical fertilizers, present and future. *J. Environ. Qual.* **1,** 2–6 (1972).
8. Power, S. F., "Leaching of Nitrate-Nitrogen Under Dryland Agriculture in the Northern Great Plains," Proc. Agr. Waste Manage. Conf., pp. 111–122. Cornell University, Ithaca, New York, 1970.
9. Lathwell, D. G., Bouldin, D. R., and Reed, W. S., "Effects of Nitrogen Fertilizer Applications in Agriculture," Proc. Waste Manage. Conf., pp. 192–206. Cornell University, Ithaca, New York, 1970.
10. Bouldin, D. R., and Lathwell, D. J., Behavior of soil organic nitrogen. *N. Y., Agr. Exp. Sta., Ithaca, Bull.* **1023,** (1968).
11. Prasad, R., Rajale, G. B., and Lakhdive, B. A., Nitrification retarders and slow release nitrogen fertilizers. *Advan. Agron.* **23,** 337–383 (1971).
12. Zwerman, P. J., Drielsma, A. B., Jones, G. D., Klausner, S. D., and Ellis, D., "Rates of Water Infiltration Resulting from Applications of Dairy Manure," Proc. Agr. Waste Manage. Conf., pp. 263–269. Cornell University, Ithaca, New York, 1970.
13. Hart, S. A., Thin spreading of slurried manures. *Trans. ASAE* (*Amer. Soc. Agr. Eng.*) **7,** 22–25 (1964).
14. Anonymous, "Poultry Manure Disposal by Plow-Furrow-Cover," Final Rep. Grant EC-00254. Submitted to the Office of Research and Monitoring, Environmental Protection Agency by the College of Agriculture and Environmental Science, Rutgers University, New Brunswick, New Jersey, 1972.
15. Hileman, L. H., "Pollution Factors Associated with Excessive Poultry Litter (Manure) Applications in Arkansas," Proc. Agr. Waste Manage. Conf., pp. 41–47. Cornell University, Ithaca, New York, 1970.
16. Wilkinson, S. R., Stuedemann, J. A., Williams, D. J., Jones, J. B., Dawson, R. N., and Jackson, W. A., Recycling broiler house litter on tall fescue pastures at disposal rates and evidence of beef cow health problems. *In* "Livestock Waste Management and Pollution Abatement," Publ. PROC-271, pp. 321–324. Amer. Soc. Agr. Eng., 1971.

17. Weeks, M. E., Hill, M. E., Karczmarczyk, S., and Blackmer, D., "Heavy Manure Applications: Benefit or Waste?" Proc. Anim. Waste Manage. Conf., pp. 441–447. Cornell University, Ithaca, New York, 1972.
18. Murphy, L. S., and Gosch, J. W., "Nitrate Accumulation in Kansas Ground Water," Proj. Completion Rep. Proj. A-016-Kan. Kansas Water Res. Inst., Manhattan, Kansas, 1970.
19. Stewart, B. A., and Mathers, A. C., Soil conditions under feedlots and on land treated with large amounts of animal wastes. *Int. Symp. Ident. Measure. Environ. Pollutants,* pp. 81–83. National Research Council of Canada, Campbell Publ. Co., Ottawa, Canada, 1971.
20. Murphy, L. S., Wallingford, G. W., Powers, W. L., and Manges, H. L., "Effects of Solid Beef Feedlot Wastes on Soil Conditions and Plant Growth," Proc. Agr. Waste Manage. Conf., pp. 449–464. Cornell University, Ithaca, New York, 1972.
21. Erickson, A. E., Tiedje, J. M., Ellis, B. G., and Hansen, C. M., "Initial Observations of a Barriered Landscape Water Renovation System for Animal Wastes," Proc. Agr. Waste Manage. Conf., pp. 405–410. Cornell University, Ithaca, New York, 1972.
22. Stewart, D. J., "The Survival of *Salmonella typhimurium* in Soils Under Natural Climatic Conditions," Res. and Exp. Record, Vol. 11, pp. 53–56. Ministry of Agriculture, Northern Ireland, 1962.
23. Rankin, J. O., and Taylor, R. J., A study of some disease hazards which could be associated with the system of applying cattle manure to pasture. *Vet. Rec.* **85,** 578–581 (1969).
24. Thomas, R. E., Land treatment of wastewater—an overview of methods. *J. Water Pollut. Contr. Fed.* **45,** 1476–1484, 1973.
25. Mather, J. R., The disposal of industrial effluent by woods irrigation. *Proc. Purdue Ind. Waste Conf.* **8,** 434–448 (1953).
26. Drake, J. A., and Bien, P. K., Disposal of liquid wastes by the irrigation method at vegetable cannery plants in Minnesota. *Proc. Purdue Ind. Waste Conf.* **6,** 70–75 (1951).
27. Miller, P. E., Spray irrigation at Morgan Packing Company. *Proc. Purdue Ind. Waste Conf.* **8,** 284–290 (1953).
28. Hicks, W. M., Disposal of fruit canning wastes by spray irrigation. *Proc. Purdue Ind. Waste Conf.,* **7,** 130–133 (1952).
29. Sanborn, N. H., Industrial waste disposal by irrigation. *Sewage Ind. Wastes* **25,** 1034–1040 (1953).
30. McKee, F. J., Dairy waste disposal by spray irrigation. *Sewage Ind. Wastes* **29,** 157–164 (1957).
31. Lawton, G. W., Engelbert, L. E., Rohlich, G. A., and Porges, N., Effectiveness of spray irrigation as a method for the disposal of dairy plant wastes. *Wis., Univ., Eng. Exp. Sta., Res. Rep.* **15,** (1960).
32. Luley, H. G., Spray irrigation of vegetable and fruit processing wastes. *J. Water Pollut. Contr. Fed.* **35,** 1252–1260 (1963).
33. Law, J. P., Thomas, R. E., and Myers, L. H., "Nutrient Removal from Cannery Wastes by Spray Irrigation of Grasslands," Water Pollut. Contr. Res. Ser. 16080. Robert S. Kerr Research Center, Federal Water Pollution Control Administration, Dept. of Interior, Ada, Oklahoma, 1969.
34. Gilde, L. C., Kester, A. S., Law, J. P., Neeley, C. H., and Parmelee, D. M., A spray irrigation system for treatment of cannery wastes. *J. Water Pollut. Contr. Fed.* **43,** 2011–2025 (1971).
35. Guttormsen, K., and Carlson, D. A., "Potato Processing Waste Treatment—Current Practice," Water Pollut. Contr. Res. Ser., DAST-14. Federal Water Pollution Control Administration, Dept. of Interior, 1969.
36. Schraufragel, F. H., Dairy wastes disposal by ridge and furrow irrigation. *Proc. Purdue Ind. Waste Conf.* **12,** 28–49 (1957).

37. National Canners Association, "Liquid Wastes from Cannery and Freezing Fruits and Vegetables," Final Rep., Proj. 12060 EDK. Environmental Protection Agency, 1971.
38. Law, J. P., "Agricultural Utilization of Sewage Effluent and Sludge—An Annotated Bibliography," CWR-2. Robert Kerr Water Research Center, Federal Water Pollution Control Administration, Dept. of Interior, Ada, Oklahoma, 1968.
39. Parizek, R. R., Kardos, L. T., Sopper, W. E., Myers, E. A., Davis, D. E., Farrell, M. A., and Nesbitt, J. B., Wastewater renovation and conservation, *Pa. State Univ. Stud.* **23,** (1967).
40. Thomas, R. E., and Bendixen, T. W., Degradation of wastewater organics in soil. *J. Water Pollut. Contr. Fed.* **41,** 808–813 (1969).
41. Harza Engineering Company, "Land Reclamation Project—An Interim Report." Solid Wastes Program, Environmental Control Administration, Public Health Service, U.S. Dept. of Health, Education and Welfare, 1968.

11

Nitrogen Control

Introduction

General

Major emphasis in waste treatment remains on the reduction of the carbonaceous oxygen-demanding substances released to surface waters. Early research and field experience with waste treatment plants established that ammonia oxidation also exerted an oxygen demand when present in the waste or the effluent. In the earlier part of this century, a large amount of nitrate in the effluents of a waste water treatment facility was a measure of a high level of treatment and operation and treatment plants were designed to accomplish oxidation of both carbonaceous and nitrogenous material in wastes.

The development of the standardized BOD test established that the nitrogenous oxygen demand (NOD) of untreated waste occurred a number of days, frequently greater than five, after the start of the test. The second stage of oxygen demand (NOD) of untreated wastes was considered of little consequence in receiving waters since the oxygen demand would occur considerably downstream from the discharge of the waste. Such a lag would not occur when adequately treated wastes were discharged, since nitrifying organisms could have had an opportunity to grow in the secondary facilities. With secondary effluents, the NOD would occur in the receiving waters closer to the point of discharge of the treated wastes. If adequate nitrification were to occur in the treatment plant, the NOD could be satisfied and not occur in the receiving waters.

Reliance on the 5-day BOD test, which does not measure NOD unless an adequate population of nitrifying organisms is in the BOD bottles, and the realization that treatment facilities could be designed and operated more economically without complete nitrification, led to an emphasis on removal of carbonaceous rather than nitrogenous oxygen demand. The lack of emphasis on meeting the nitrogenous oxygen demand in treatment plant design and operation occurred from about 1930 to 1960 in the United States.

Need for Control

Environmental concerns have caused a reexamination of this position with the result that interest is again being given to methods to control both the carbon and nitrogen content of a waste or treated effluent. The reasons nitrogen compounds are of concern include: the nitrogenous oxygen demand in receiving waters, ammonia toxicity to fish, increased chlorine demand due to ammonia if the water is treated as a potable source, the role of nitrogen as a controlling nutrient in eutrophication, and health problems in humans and animals. Regulatory authorities are beginning to set standards for nitrogenous materials in streams and effluents and treatment facilities are being required to produce effluents low in both unoxidized and total nitrogen (1). Control and/or management of nitrogen in municipal, industrial, and agricultural wastes is receiving greater emphasis.

The oxygen demand of reduced nitrogen compounds in receiving waters frequently has been neglected or assumed of minimum consequence. However, from the point of view of the oxygen demand on a stream, it is not insignificant. Municipal waste waters contain about 15 to 20 mg/l of nitrogen compounds and certain agricultural wastes such as animal and fertilizer production waste waters can have nitrogen concentrations in the 100–1000 mg/l range. A treatment plant with 20 mg/l of ammonia (NH_4—N) and 20 mg/l of carbonaceous BOD in the effluent and a flow of 1 mgd would place a carbonaceous oxygen demand of about 250 lb/day and a nitrogenous oxygen demand of about 750 lb/day on the receiving stream. The concern with NOD is becoming more important as the load of oxygen-consuming waste waters increases and as the desire for a less polluted environment continues. The nitrogenous oxygen demand load must be considered in any program of stream quality improvement.

The NOD or nitrification is a biological phenomena whose rate is a function of many factors such as the concentration of nitrifying organisms, the temperature, pH, dissolved oxygen concentration of the liquid, the type of available nitrogen compounds, and the concentration of any inhibiting compounds. The ammonia to be oxidized is that which is in excess of the nitrogen required for growth by the heterotrophic organisms in a waste treatment facility or a stream.

In untreated wastes, nitrogen will be in the form of organic nitrogen and ammonium compounds. The organic nitrogen is converted by microbial action to the ammonium ion. If adequate environmental conditions exist, nitrifying organisms can oxidize the ammonia. These organisms are autotrophs, obtaining their energy through the oxidation of ammonium ion or nitrite, as appropriate:

$$NH_4^+ + 1.5\,O_2 \xrightarrow{\text{bacteria}} 2\,H^+ + NO_2^- + H_2O \quad (11.1)$$

This reaction requires 3.43 gm of oxygen for 1 gm of ammonia nitrogen oxidized to nitrite. The nitrite can be oxidized to nitrate:

$$NO_2^- + 0.5\,O_2 \xrightarrow{\text{bacteria}} NO_3^- \quad (11.2)$$

This reaction requires 1.14 gm of oxygen for every gram of nitrite nitrogen oxidized to nitrate nitrogen. If all of the ammonium ion were oxidized, 4.57 gm of oxygen would be required per gram of ammonia nitrogen oxidized to nitrate. However, about 0.8% of the nitrogen is converted into cell material as carbon dioxide is fixed by the bacteria. Field experiments have indicated ratios of 3.22 mg oxygen utilized per milligram NH_4—N oxidized to NO_2—N and 1.11 mg oxygen utilized per milligram NO_2—N oxidized to NO_3—N (2), confirming the theoretical ratios.

High concentrations of ammonia in surface waters can be lethal to fish. The pH of the water has been found to be important in determining the concentrations of ammonia that are toxic. The phenomenon of the pH effect on the toxicity of ammonia appears ubiquitous in nature. Ammonia at low pH levels is usually toxic only in large quantities while at high pH levels, smaller amounts may be lethal.

The toxicity of ammonia depends upon the quantity of ammonia that enters the organism or plant cell. Cell membranes are relatively impermeable to ionized ammonia (NH_4^+) whereas un-ionized ammonia passes through cell membranes (3). Ammonia toxicity to fish is increased markedly at low dissolved oxygen concentrations (4).

In liquids, ammonia exists in two forms, the free or un-ionized form (NH_3) and the ionized form (NH_4^+). These forms exist in equilibrium which is affected by the pH of the liquid:

$$NH_3 + H_3O^+ \rightleftharpoons NH_4^+ + H_2O \quad (11.3)$$

Prevailing methods of ammonia measurement determine only the total ammonia concentration which is both the ionized and un-ionized forms.

The proportion of the total ammonia concentration that is in the ionized form at any pH, (F), is determined by the dissociation constants of aqueous ammonia and water and the law of mass action:

$$F = \frac{[NH_3]}{[NH_3] + [NH_4^+]} = \frac{10^{pH}}{10^{pH} + K_b/K_w} \tag{11.4}$$

K_b and K_w are the ionization constants of aqueous ammonia and water, respectively. The values of K_b and K_w increase with an increase in temperature. Mathematically, the relation between K_b/K_w and temperature is

$$K_b/K_w = [-3.4 \log_e (0.024\,\theta)] \times 10^9 \tag{11.5}$$

where θ is temperature in °C (5). The relationship between F, pH, and temperature is noted in Fig. 11.1. At a pH of about 9.0, about 50% of the total ammonia is in the un-ionized form. The proportion decreases markedly at lower pH levels.

Water quality standards utilize the total ammonia concentration to describe permissible levels rather than the un-ionized ammonia concentration. Such standards generally restrict the total ammonia in receiving waters to a maximum of 2.0 mg/l at a pH of 8.0 or greater (1). Under these conditions, the un-ionized ammonia nitrogen concentration at 20°C in stream water is 0.074 mg/l. If this un-ionized ammonia concentration does represent a critical threshold level to fish at other temperatures, then the total ammonia concentrations that may be permissible at other pH values and temperatures can be calculated using Eq. (11.4). Permissible total ammonia nitrogen concentrations to have an un-ionized ammonia concentration of 0.074 mg/l are shown in Fig. 11.2.

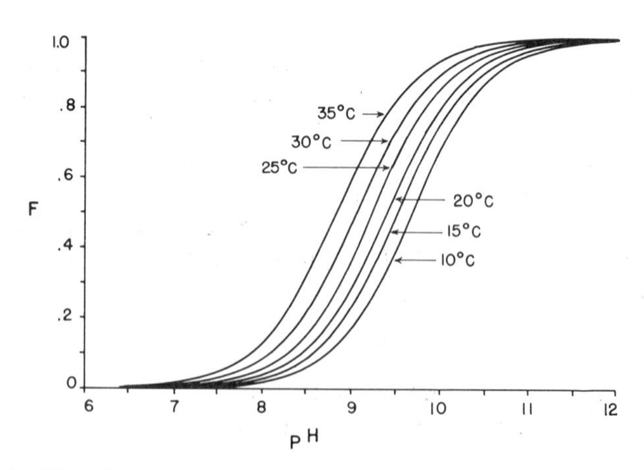

FIG. 11.1. Effect of temperature and pH on the fraction (F) of un-ionized ammonia.

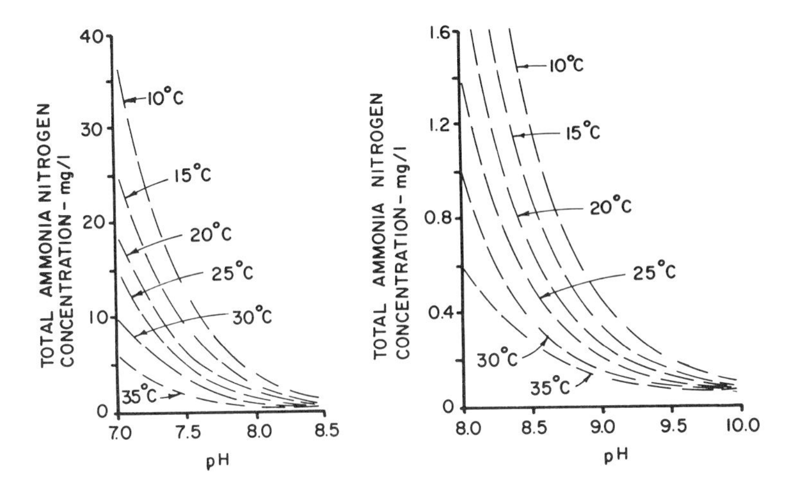

FIG. 11.2. Total ammonia nitrogen concentrations to obtain an un-ionized ammonia nitrogen concentration of 0.074 mg/l at the noted temperature and pH values.

The values in Fig. 11.2 permit total ammonia standards to be set as a function of other water quality parameters such as pH and temperature. This problem can be of particular concern where photosynthetic activity in a water increases the pH to above 8.0 during daylight hours. Water quality standards should more appropriately be set in terms of the un-ionized ammonia concentration, since it is the factor causing the toxic effect. The critical un-ionized ammonia nitrogen concentration limits the concentration of total ammonium nitrogen that should be in receiving waters to avoid fish toxicity.

Ammonia in waters can increase the chlorine demand of the waters if breakpoint chlorination is practiced or can reduce the relative effectiveness of the disinfection process (Chapter 12). Both free and combined available chlorine are capable of disinfection. However, the combined available chlorine compounds will require greater concentrations or longer contact periods than free available chlorine to accomplish comparable levels of disinfection. In disinfecting potable waters or treated effluents, free available chlorine will not predominate until the chlorine demand of ammonia is met, i.e., after the breakpoint. This point generally is not achieved until approximately 9.5 mg/l of Cl_2 is added per 1.0 mg/l of NH_4—N. If other reduced compounds or organic matter are present, the total chlorine demand will be increased proportionately.

Both oxidized and reduced nitrogen compounds can serve to stimulate the growth of aquatic life in surface waters. The type of limiting nutrients can be different in different bodies of water. Because smaller quantities of phosphorus are necessary per unit of aquatic growth, phosphorus has received the major emphasis in eutrophication control. Where phosphorus is naturally in excess in waters, nitrogen or even carbon can be the limiting nutrient.

In the United States, a nitrate concentration of 45 mg/l as NO_3 has been established as the maximum concentration that should be in potable waters. Maximum acceptable concentrations of nitrate reflect public health concerns and are based upon the judgement of public health officials. Acceptable concentrations vary among nations. High nitrate concentrations have caused methemoglobinemia in infants and have caused problems with animal health. In each area the problem is the same, difficulties in transport of oxygen in the blood. Nitrate-reducing bacteria in the intestinal tract of humans and animals reduce nitrate to nitrite. The nitrite can oxidize hemoglobin in the blood to methemoglobin which is unable to transport oxygen. Although nitrite is the cause of the problem, nitrites are rarely found in foods and water. The standards are placed on nitrate since it can be present in foods and water and is the precursor of nitrite. High nitrates in leafy foods such as spinach and in forage crops also can be a cause of the problem. Once high nitrate concentrations exist in surface waters and groundwaters, few practical processes exist for routine nitrate reduction or elimination. Generally the water supply is abandoned and other potable supplies are sought.

Opportunities for Nitrogen Control

Where excess nitrogen has the potential to cause the above problems, management systems must be utilized. A summary of the possible methods is noted in Table 11.1. The principal methods for removing nitrogen from municipal waste waters include biological nitrification followed by denitrification, ammonia stripping, algae harvesting, and ion exchange. The state of art of nutrient removal from agricultural wastes is in its infancy although some of the methods are being tested on the laboratory and pilot plant scale.

Traditionally the land has been the ultimate disposal medium for agricultural wastes. Where the waste applications to the land are excessive in terms of nitrogen, removal of nutrients prior to land disposal may be needed. The most feasible approaches for nitrogen control from agricultural waste waters appear to lie in the use of nitrification and denitrification processes, either prior to land disposal or in conjunction with the land. Ammonia stripping is also of interest in nitrogen control primarily to understand the losses that can occur in treatment processes, storage tanks, and feedlots. These losses represent a measure of nitrogen control with a specific waste but may have undesirable effects such as increased ammonia

Table 11.1

Methods for Reducing the Nitrogen Content of Waste Water

Method	Nitrogen compounds removed
Physical and chemical	
Land application	NH_3, NH_4^+, organic N
Electrochemical	NH_4^+
Ammonia stripping	NH_3
Ion exchange	NO_3^-, NH_4^+
Electrodialysis	NO_3^-, NH_4^+
Reverse osmosis	NO_3^-, NH_4^+
Breakpoint chlorination	NH_4^+, organic N
Biological	
Algal utilization	All forms
Microbial denitrification	NO_3, NO_2
Land application	All forms

concentrations in buildings and storage units and the release of ammonia to the environment.

Nitrogen control with all wastes is in a dynamic state. Utilization of process fundamentals and modifications of existing processes for nitrogen control are being investigated and demonstrated throughout the world. The fundamentals affecting nitrification, denitrification, and ammonia stripping as they are applied to wastes in general and to agricultural wastes in specific are discussed in this chapter. By the proper utilization of these processes and of land disposal (Chapter 10), agricultural waste management systems can be developed to avoid the problems associated with excess nitrogen that have been discussed previously.

Nitrification

Present-day knowledge on the transformation of nitrogen and the physiology of microorganisms involved in these transformations stems primarily from the research of agronomists and soil scientists. Nitrification can be defined basically as the biological conversion of nitrogen in inorganic or organic compounds from a reduced to a more oxidized state. In the field of water pollution control nitrification usually is referred to as a biological process in which ammonium ions are oxidized to nitrite and nitrate sequentially.

Pasteur in 1862 suggested that the oxidation of ammonia was microbiological. This suggestion was verified in classic studies with sewage and soil which showed that oxygen was essential and that alkaline conditions favored nitrification. The ubiquity of biological nitrification in soils has been well demonstrated. The autotrophic nature of the bacteria responsible for nitrification and the unavailability of special culture media to isolate and study them made it difficult for pioneer microbiologists to obtain them in pure culture. The first successful attempt to isolate them was made by Winogradsky (6) who showed that they would grow strictly on inorganic media.

For a long time it was considered that only autotrophic bacteria were responsible for nitrification. It is now known that heterotrophic bacteria, actinomycetes, and other fungi also can bring about oxidation of nitrogen to nitrite and nitrate.

The Nitrifying Organisms

Several genera of nitrifying organisms have been reported (7). *Nitrosomonas, Nitrosospira, Nitrosococcus,* and *Nitrosocystis* oxidize ammonia to nitrite. *Nitrosogloea, Nitrobacter,* and *Nitrocystis* oxidize nitrite to nitrate. Of these genera, only *Nitrosomonas* and *Nitrobacter* are generally encountered in aquatic and soil ecosystems and are the nitrifying autotrophs of importance. The other genera are rarely reported. Two new genera of obligate autotrophic nitrite-oxidizing bacteria, *Nitrospira* and *Nitrococcus* species, have been reported (8).

Although the nitrification process is largely due to autotrophic organisms, certain heterotrophic bacteria, actinomycetes, and fungi are recognized as nitrifying organisms. The genera involved include *Mycobacterium, Nocardia, Streptomyces, Agrobacterium, Bacillus,* and *Pseudomonas* (9). The first fungus to be recognized as a heterotrophic nitrifier was *Aspergillus flavus* which forms nitrate from amino nitrogen (10). Heterotrophic nitrification occurs only when nitrogen is present in excess of cellular needs and when an energy source other than the oxidation of nitrogen is available (9). Many of these heterotrophic nitrifiers are common in lakes, streams, and soils.

The nitrifying organisms of significance in waste management are autotrophic. Organic compounds have been reported to be inhibitory to the growth of nitrifying organisms but the claim that the organic compounds in general are inhibitors is over emphasized. The behavior of the nitrifying population in a pure culture can differ significantly from that occurring in the presence of organic matter in an ecosystem. The occurrence of nitrification in soils containing organic matter, trickling filters, activated sludge tanks, and in compost piles testifies that the process takes place freely in natural ecosystems containing varied degrees of organic matter.

Solids Retention Time

The concentration of nitrifying organisms is a major factor in the rate of nitrification. The quantity of nitrifiers is determined by the generation time of the organisms which in turn is related to the amount of energy obtained in the oxidation of ammonia or nitrite. Nitrification is an exothermic reaction. The free energy change in the oxidation of ammonium to nitrite and nitrite to nitrate is reported to be in the range of −65.5 to −84 kcal per mole of ammonium and −17.5 to −20 kcal per mole of nitrite oxidized, respectively (11). The energy is utilized for the assimilation of carbon in the form of carbon dioxide or bicarbonate ion. The ammonia oxidizing organisms obtain more energy than the nitrite oxidizing organisms. Assuming the efficiency of cell synthesis is the same, more ammonia oxidizing bacteria are formed than nitrite-oxidizing bacteria per unit of ammonium undergoing nitrification. In other words the nitrite-oxidizing bacteria utilize about three times more substrate than the ammonia-oxidizing bacteria while synthesizing the same amount of cell mass. This indicates why little significant accumulation of nitrite occurs in an ecosystem when there is no inhibition.

The energy obtained by nitrifying organisms is much lower (66 kcal/mole of substrate) than that obtained by heterotrophs in an aerobic system (686 kcal/mole of glucose as a substrate). Because of this energy difference there are low concentrations of nitrifying organisms, approximately one-tenth that of the heterotrophs, in a nitrifying biological treatment plant. The nitrifiers will be an even smaller proportion of the mixed liquor solids in a nitrifying unit. *Nitrosomonas* concentrations have been estimated to range from 5 to 50 mg/l when the MLSS of an actively nitrifying system ranged from 1900 to 2000 mg/l and from 50 to 100 mg/l when the MLSS in a nitrifying activated sludge system ranged from 5800 to 6000 mg/l (12).

Since the process of nitrification depends on the metabolism of a certain group of highly specialized aerobic organisms, it is imperative that these organisms be present in adequate numbers. High rates of aeration alone have not achieved nitrification in the absence of nitrifying organisms. It is necessary to provide a certain minimum detention time in a treatment system for the multiplication of these organisms. At very low solids detention times (SRT), the nitrifying organisms in an aerobic treatment process are removed from the system before they can multiply.

The generation time for the nitrifying organisms is comparatively long, on the order of 10 or more hours depending upon environmental conditions. The minimum SRT of a nitrification treatment unit must be greater than the growth rate of the organisms under the imposed environmental conditions. The growth rate, and therefore operating SRT of the nitrification unit is related to the temperature and the concentration of any inhibitor. Continuous flow studies of nitrification of poultry manure waste waters indicated that an SRT of 2 days was needed to accomplish maximum nitrification (Fig. 11.3) at 20°C. The minimum SRT for nitrification increases as the temperature decreases because of the decreased growth rate of the microorganisms.

The maintenance of a suitable SRT for nitrification is difficult in a trickling filter since there is poor control over the hydraulic and solids retention time. Better control can be achieved in a biological unit, such as an activated sludge unit, in which sludge recycle permits a separation of solids and hydraulic retention times. By proper sludge wasting, the SRT can be kept above the critical value. The development of a submerged aerobic filter (13) also permits efficient nitrification by maintaining long SRT values of tens of days with hydraulic retention times of less than 1 hr. The long SRT and stable characteristics of the submerged filter permitted nitrification at temperatures as low as 1°C and under variable loading conditions. If a biological system is to consistently obtain a required degree of nitrification, close control of the system SRT is needed to compensate for any change in environmental conditions.

In studies with domestic sewage, the growth rate of *Nitrosomonas* has been found to be a limiting factor and nitrification was achieved by maintaining a SRT greater than the *Nitrosomonas* growth rate. In a full-scale study, the maximum net specific growth rate constant (K_m) of *Nitrosomonas* was estimated at 0.25/day and was related to temperature in the following manner (14):

$$K_m = (0.18)^{0.12(T-15)} \tag{11.6}$$

The units of T are °C.

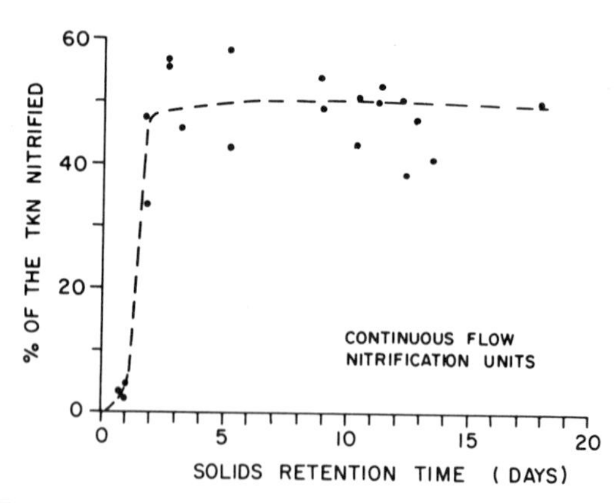

FIG. 11.3. Nitrification related to solids retention time (5).

Nitrification growth constants that have been observed for a number of activated sludge systems are (12)

Nitrosomonas	9°–17°C:	0.086–0.19
	15°–18°C:	0.18 –0.24
	6°–23°C:	0.02 –0.13
Nitrobacter	6°–23°C:	0.03 –0.14

The values varied with the type of the waste as well as with the temperature.

An empirical relationship has been developed (12) to estimate the minimum aeration time (t_m) that is required in a nitrifying activated sludge system:

$$t_m = \frac{s}{K_m S} = \frac{L[0.1 + 0.9(t_m)^{-0.5}]}{(1 + p)(K_m)S} \qquad (11.7)$$

where

s = increase in sludge solids, mg/l
S = mixed liquor suspended solids, mg/l
L = BOD of influent wastes, mg/l
p = rate of returned sludge, sludge flow rate/influent sewage flow rate
t_m = aeration period, volume of the aeration tank/flow of mixed liquor; i.e., hydraulic detention time

Equation (11.7) was developed using data from plug flow systems treating domestic wastes. Design criteria for nitrification systems should be based on relationships for solids retention time rather than hydraulic retention time. Equation (11.7) includes a term for the returned sludge and thereby incorporates the SRT concept.

Minimum SRT values are important in biological treatment units where there is the possibility of removing the nitrifying organisms faster than they can grow. Minimum SRT values are of little importance when considering nitrification in a soil. Soils contain ample quantities of nitrifying bacteria and the likelihood of removing them from the soil faster than they can reproduce is remote. Other environmental conditions have a larger effect on soil nitrification.

Examination of results obtained at a number of full-scale plants indicated that the rate of ammonia removal by nitrification varied little, from 0.6 to 1.0 mg N oxidized/hr/gm MLSS, from plant to plant (15). These rates provide an estimate of the time and solids concentration necessary to accomplish a necessary degree of nitrification.

Dissolved Oxygen

Since the nitrifying organisms are aerobic, adequate dissolved oxygen (DO) is necessary to support nitrification assuming other environmental conditions are satisfactory. Early studies on the effect of oxygen on nitrifying organisms indicated (16) that at low dissolved oxygen concentrations the growth of *Nitrosomonas* was suppressed measurably and *Nitrobacter* was effected to an even greater degree. The sensitivity of the *Nitrobacter* to low dissolved oxygen concentrations is one of the reasons complete nitrification is difflcult to accomplish in heavily loaded systems where the oxygen demand is significant and where adequate oxygen does not exist in the microbial floc.

A critical dissolved oxygen concentration exists below which nitrification does not occur. The critical DO concentration has not been precisely determined but it appears to be around 0.5 mg/l (Fig. 11.4) (17). The actual limit is more dependent on the rate of oxygen diffusion to the microorganisms than on the oxygen concentration in the mixed liquor. To assure that the dissolved oxygen concentration in a treatment unit is not limiting nitrification, the dissolved oxygen in a treatment unit generally is kept above about 1.0 mg/l. Nitrification proceeds at a rate independent of the dissolved oxygen above the critical concentration.

The recovery of nitrifying organisms from anaerobic conditions was found to be good. When a nitrifying sludge was kept anaerobic for 4 hr and then reaerated, its nitrifying ability increased to its original level after 20 min (12). Similar results have been obtained when nitrified poultry manure was stored without measurable dissolved oxygen. In one case the nitrified manure was stored for over 20 days yet when aeration was again applied, rapid rates of nitrification occurred within 48 hr. In a submerged

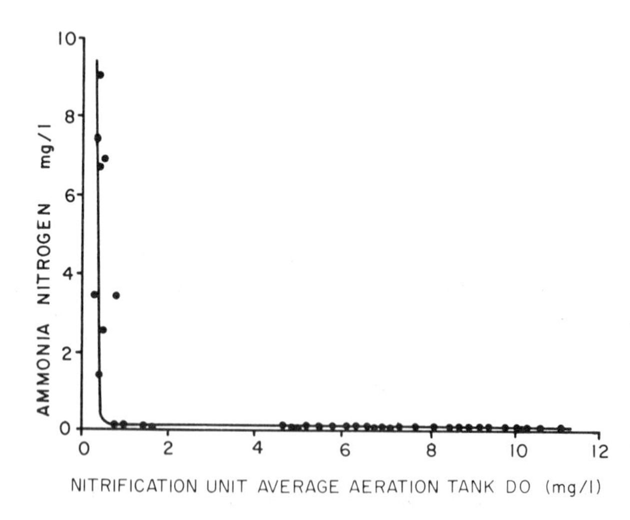

FIG. 11.4. Relationship of residual ammonia to dissolved oxygen in a nitrification system (17).

anaerobic filter, nitrification returned to normal conditions within four days of aeration following seven days of anaerobic conditions (13). Obviously the recovery rate of nitrification is related to the numbers of nitrifying bacteria in the system. While nitrification does not take place until adequate dissolved oxygen is in the biological system, when sufficient numbers of nitrifying bacteria are present, anoxic conditions of the above duration did not have a detrimental effect on the nitrifying ability of the organisms.

Temperature

Nitrification is affected by the temperature of the medium. Pure culture studies indicated that the optimum growth of nitrifiers occurred between 30° and 36°C (9). An exposure for 10 min at 53°–55°C and at 56°–58°C killed *Nitrosomonas* and *Nitrobacter*, respectively (18).

A theta (θ) factor of 1.070 has been suggested for relating the rates of nitrification at different temperatures (19) indicating that the growth rate constant roughly doubles for each 10°C increase in the range of 6°–25°C. Laboratory activated sludge studies indicated that the rate of nitrification increased throughout the range of 5°–35°C in reasonable agreement with the above concept (17). Such temperature relationships indicate why high rates of nitrification may be difflcult to obtain in the winter.

Information on nitrification at low temperatures indicates conflicting data. Different studies have indicated that nitrification did not develop below 10°C (20), that it was possible to maintain nitrification at 8°C (21), that little nitrification was achieved at temperatures below 6°C (12), and that nitrification could be accomplished at temperatures as low as 1°C (13). These studies provided little information to compare their conclusions regarding the temperature effect with the prevailing SRT values in the nitrification units. It is quite probable that when nitrification is reported as not being accomplished at low temperatures, the SRT of the unit was not exceeding the slower growth rate of the organisms at the lower temperatures. When longer SRT values are maintained at lower temperatures to compensate for the slower nitrifying organism growth rate, nitrification should occur even at low temperatures.

pH

The optimum pH for the growth of the nitrifiers is not sharply defined but in pure cultures it has been shown to be generally on the alkaline side (Fig. 11.5).

During nitrification, hydrogen ions are produced [Eq. (11.1)] and as a

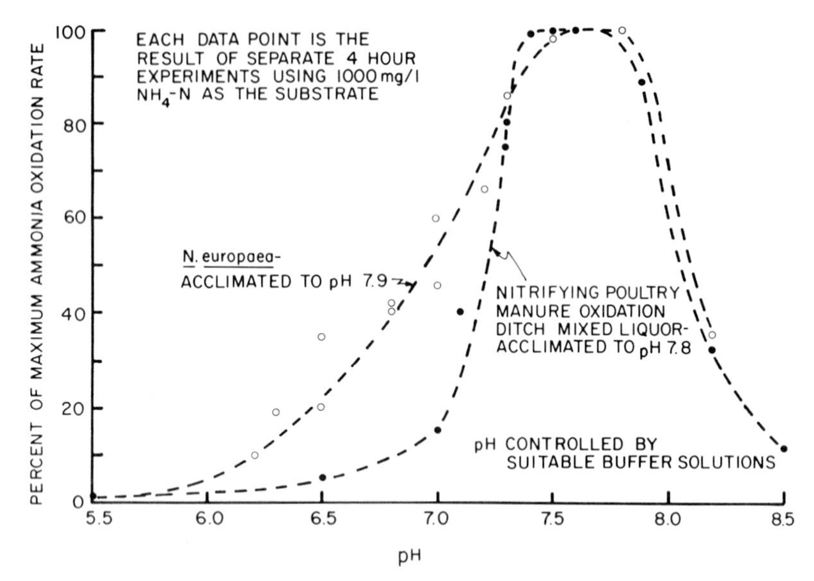

FIG. 11.5. Ammonia oxidation as affected by pH of the medium.

result the pH drops. Theoretically, a decrease of 7.1 gm of alkalinity as $CaCO_3$ should occur for every gram of nitrate nitrogen produced. If the system has adequate buffer capacity and/or if the ammonia concentration is low, the pH will decrease only slightly. If inadequate buffer capacity exists the pH will decrease noticeably. A decrease in pH can be a practical measure of the onset of nitrification.

The optimum range of pH for nitrification of municipal sewage has been indicated to be between 7.5 and 8.5. In a detailed study of nitrification, optimum pH was found to be 8.4. Ninety percent of the maximum rate occurred in the range of 7.8–8.9 and outside the ranges of 7.0 to 9.8 less than 50% of the optimum rate occurred (17). Nitrification rates varied from a maximum of 0.185 NH_3—N nitrified/day/gm MLSS at pH 8.4 to a rate of 0.020 at a pH of 6.0.

Nitrification can proceed at low pH levels. In a study of the aerobic digestion of sewage sludge, the pH of the mixed liquor was in the 5.0–5.5 range during active nitrification (22). The alkalinity decreased from 700 to 100 mg/l. Nitrification has occurred in both batch and continuous systems treating poultry manure waste water at pH levels in the 5.0–5.5 and 6.3–6.8 ranges (5). These ranges occurred naturally because of the oxidation of the high concentrations of nitrogen in this waste water. Low pH values resulting from nitrification also have been noted in compost piles and in soils.

Nitrifying organisms can adapt to low pH levels and achieve adequate nitrification. When the pH was adjusted to a range of 5.5 to 6.0 in a submerged aerobic filter (13) the nitrifiers adapted to the lower pH levels and the rate of ammonia oxidation reached that comparable to what had been achieved at a pH of 7.0.

Ammonia and Nitrite Concentrations

Although NH_4^+ is the energy source for the nitrifiers, excessive amounts can inhibit the growth of these bacteria. Ammonia is more inhibitory to *Nitrobacter* than to *Nitrosomonas* and the inhibition of *Nitrobacter* is caused by un-ionized (free) ammonia rather than the total ammonia concentration. The fundamentals of ammonia toxicity to living matter have been described earlier in this chapter.

High nitrite concentrations can reduce the activity of the nitrifying organisms at low pH levels. The nitrite toxicity is due to the undissociated nitrous acid (HNO_2) rather than the nitrite ion. These compounds exist in equilibrium in a liquid:

$$NO_2^- + H_3O^+ \rightleftharpoons HNO_2 + H_2O \qquad (11.8)$$

The equilibrium is a function of the pH with the concentration of nitrous acid increasing as the pH decreases. The quantity of nitrous acid at a specific pH can be determined using the law of mass action and the ionization constants of water and nitrous acid. Mathematically the nitrous acid concentrations can be obtained by Eq. (11.9) (23):

$$HNO_2\,(\text{mg/l}) = \frac{46}{14} \times 10^{(3.4-\text{pH})} \times NO_2 - N \qquad (11.9)$$

Pure culture studies have indicated that free ammonia (16, 24–26) and undissociated HNO_2 (23) were more inhibitory to nitrite oxidation than the NH_4^+ or NO_2^- concentration per se. The dissociation equations for ammonium hydroxide and HNO_2 indicate that the concentration of free ammonia increases with an increase in the pH, and the concentration of undissociated nitrous acid increases with a decrease in pH.

Studies on the nitrification of poultry manure (5) have shown that the free ammonia concentration in the solution affected the oxidized nitrogen species that predominated. These relationships were observed at SRT values long enough to keep an actively nitrifying population in the biological unit and the dissolved oxygen concentrations were well above critical levels for nitrification.

Because the MLVSS is an imperfect measure of the active nitrifying organisms, the actual free NH_3 concentrations in the continuous nitrification units were compared to the percent nitrite and nitrate production (Fig. 11.6). The ratio used for comparison was the ratio of % nitratification to the % nitrification (% NO_3–N formed/% NO_2–N + NO_3–N formed). This ratio represents nitrate formation as a fraction of the overall nitrite and nitrate formation that occurred under a specific set of conditions. This approach was taken since free NH_3 is toxic to the nitrate formers and therefore should be directly related to this ratio.

Nitrate formation predominated below a free ammonia concentration of 0.02 mg/l. When the free ammonia concentration was greater than 0.02 mg/l, nitrate formation rapidly decreased and nitrite formation increased. A continuous nitrification unit with a solids retention time of 216 days sustained nitrate formation at a free ammonia concentration of 0.033 mg/l. The long SRT might have provided an opportunity for the nitritifying population to adapt to the slightly higher concentration of free ammonia in the unit.

Ammonia is both a substrate for and an inhibitor of nitrification. Because nitrification induces an error in the comparison of the BOD of an untreated waste water and the effluent from a secondary treatment unit, modified procedures for inhibiting the nitrification in the BOD test have been recommended (27–29). The fact that ammonia can cause inhibition of nitrification has been utilized in some of these procedures. A 0.1 *M* ammonium chloride solution was shown to inhibit nitrification in sewage effluents without affecting the carbonaceous demand (28).

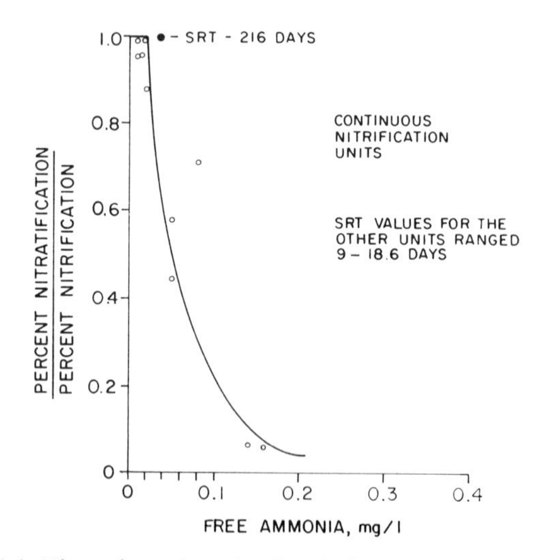

FIG. 11.6. Nitrate formation related to the free ammonia concentration (5).

The use of 0.1 *M* NH_4Cl to inhibit nitrification is not universal. With an actively nitrifying mixed liquor, this concentration of ammonia was used as a substrate and oxidized (5). Some nitrification occurred even when 0.5 *M* ammonium chloride was added as a potential inhibitor. The tolerance to high ammonia concentrations was related to the previous ammonia concentrations to which the nitrifying populations had been exposed, to the SRT of the units, to the physiological state of the organisms, and to the numbers of nitrifying organisms in the system. While 0.1 *M* ammonium chloride may inhibit nitrification in effluents with small concentrations of nitrifiers adapted to low ammonia concentrations, it is unlikely to be an inhibitor where high concentrations of nitrifiers have been adapted to high ammonia concentrations. Care should be taken in the use of ammonium compounds as inhibitors of nitrification. Other inhibitors would be more appropriate.

Loading Rates

Design loading rates should be based upon the nitrification rates per quantity of nitrifying organisms per unit time (milligrams NH_4-H oxidized/milligrams organisms/time) under variable environmental conditions. Parameters of this nature are only beginning to become available. Empirical relationships utilizing common parameters for substrate (COD, BOD, TKN) and nitrifying organisms (MLSS, MLVS, MLVSS) have been employed. In addition to actual nitrification rates, a key parameter is the SRT of the biological unit. The majority of nitrification studies have been done with municipal wastes. With wastes that vary little in quality or quantity, the SRT and liquid detention time also vary little. Under such circumstances, gross loading parameters such as pounds of BOD/day/pound MLVS may suffice.

Using available gross parameters, a loading factor of 0.3–0.4 lb BOD/day/MLSS is a biological treatment unit such as an activated sludge aeration tank has been suggested as an upper limit for good nitrification (30). In nitrification studies with poultry manure waste water, maximum nitrification occurred at a loading factor less than 0.8 lb COD/day/lb MLVSS (5). Because of inhibition, nitrite formation resulted at loadings ranging from about 0.15 to 0.8 lb COD/day/lb MLSS and nitrate formation occurred at loadings less than 0.15 lb COD/day/lb MLSS. Figure 11.7 summarizes some of the available data relating process loading rates and degree of nitrification.

It should be recognized that parameters such as these are empirical and have only a vague relationship to the fundamental factors affecting nitri-

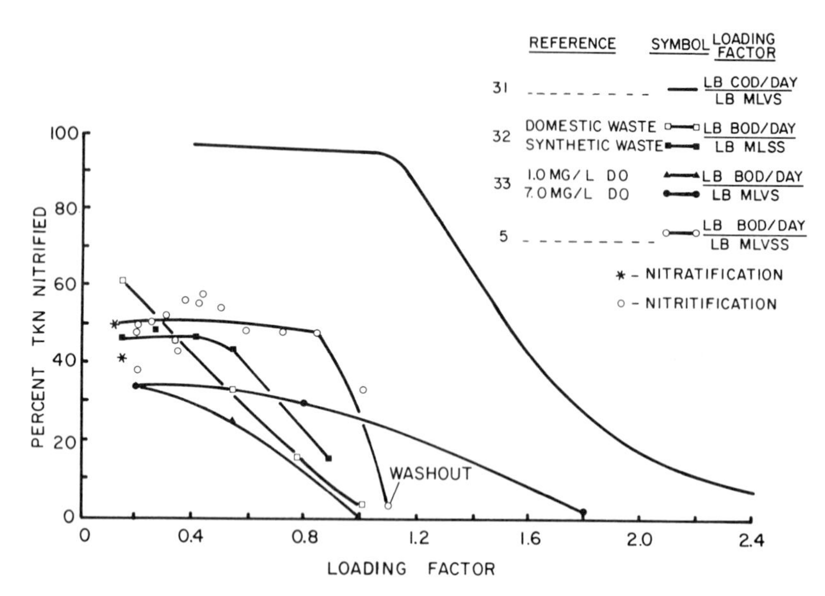

FIG. 11.7. Nitrification as affected by the loading factor; comparison of data from various studies.

fication. Experimental results and design parameters should be described in terms of

$$\frac{\text{Pounds of oxidizable nitrogen applied}}{\text{Pound of nitrifying organisms/day}}$$

to achieve a specific degree of nitrification.

Denitrification

General

Microbial denitrification takes place under anoxic conditions where nitrites and nitrates are used as terminal electron acceptors in place of molecular oxygen. Fundamental knowledge on this subject primarily stems from the observations reported on denitrification in soils. Denitrification rates are a function of the respiration rate of the organisms in the waste mixture or the soil, the dissolved oxygen in the immediate vicinity of the organisms, and the availability of nitrite and nitrate in the liquid surrounding the organisms.

The nitrates and nitrites are reduced to gaseous nitrogen resulting in a reduced nitrogen content of the waste water as the gas escapes from the liquid. The composition of the gas produced is a function of the environmental conditions. Nitrogen gas (N_2) is the primary gaseous end product. Other gases produced include nitrous oxide (N_2O), nitric oxide (NO), CO_2, and H_2. These gases are innocuous and have no undesirable effect in the environment. Denitrification therefore offers a mechanism for removal of nitrogen from waste waters that causes few additional problems. Uncontrolled denitrification in secondary clarifiers can hinder sedimentation.

Denitrification is brought about by facultative heterotrophic bacteria. Most of the active denitrifying organisms belong to the genera of *Pseudomonas, Achromobacter, Bacillus,* and *Micrococcus.*

True dissimilatory nitrate reduction occurs when both nitrite and nitrate are used as electron acceptors without the accumulation of potentially toxic concentrations of end products such as nitrite and ammonia. This reaction is different from the process of nitrate reduction for the purpose of cell protein synthesis which is known as assimilatory nitrate reduction. The primary end product of this latter process is ammonium which enters the biochemical pathways leading to protein synthesis.

The primary reactions involved in denitrification are

$$\text{Reduced organic matter} + NO_3^- \xrightarrow{\text{bacteria}} NO_2^- + CO_2 + H_2O + \text{oxidized organic matter} \quad (11.10)$$

which illustrates the reduction of nitrate to nitrite and

$$\text{Reduced organic matter} + NO_2^- \xrightarrow{\text{bacteria}} N_2 + CO_2 + H_2O + \text{oxidized organic matter}$$

which illustrates the reduction of nitrite to nitrogen gas. During denitrification, the pH will increase as hydroxide ions are produced. The degree of pH change is related to the amount of denitrification and the buffer capacity of the liquid.

To achieve high nitrogen removals by biochemical denitrification in a waste water treatment facility, a high degree of nitrification is a prerequisite. Denitrification as a method of nitrogen control is useful where a nitrified liquid exists, such as from irrigation drains, where a nitrified waste water can be produced economically, and where it can be shown that nitrogen removal will have a beneficial effect. Denitrification is an important factor in soils since it reduces the amount of nitrate that can be leached to surface waters and groundwaters and because it reduces the amount of nitrogen available for plant growth.

There are many environmental factors that affect the rate of denitrification in soils and controlled biological treatment systems. The important factors governing denitrification in an ecosystem include: (a) organic matter, (b) oxygen tension, (c) pH and, (d) temperature.

Organic Matter

Substrates that can act as hydrogen donors are necessary for the denitrification of oxidized nitrogen. These are oxidizable organic compounds and act as energy sources for the denitrifying population. The compounds must be available in concentrations to meet the combined requirements of organism growth and nitrite and nitrate reduction. The rate of denitrification is a function of the rate of carbon utilization and resultant electron acceptor demand. Both endogenous and exogenous carbon sources have been explored to achieve denitrification.

Denitrification in an activated sludge system can be accomplished by using the endogenous electron acceptor demand for the microorganisms contained in the activated sludge without the addition of any exogenous electron donors (33–36). The reaction can be illustrated as:

$$C_5H_7O_2N + 4\,NO_3^- \longrightarrow 5\,CO_2 + NH_3 + 2\,N_2 + 4\,OH^- \quad (11.12)$$

Based upon Eq. (11.12), approximately 1.9 mg of microbial solids are required to reduce 1.0 mg of nitrate as N. Since not all of the microbial cell is oxidized, a greater amount of microbial cells per unit nitrate will be needed than the theoretical amount.

The endogenous electron acceptor demand of the microorganisms occurs at a low rate and exogenous hydrogen donors are added to increase the denitrification rates. Untreated sewage and other wastes have been added for this purpose but have been undesirable since these sources contain nitrogen in forms other than nitrate and can add unwanted amounts of organic and ammonia nitrogen to the effluent from the treatment facility. Where high nitrogen removals may not be necessary, such as where the wastes are to be applied to the land, untreated wastes can be an acceptable source of hydrogen donors.

To achieve high nitrogen removals a simple carbon source, such as sugar or alcohol, is preferred.

$$5\,CH_3OH + 6\,NO_3^- \longrightarrow 5\,CO_2 + 3\,N_2 + 7\,H_2O + 6\,OH^- \quad (11.13)$$

$$5\,C_6H_{12}O_6 + 24\,NO_3^- \longrightarrow 30\,CO_2 + 12\,N_2 + 18\,H_2O + 24\,OH^- \quad (11.14)$$

The theoretical requirement of these compounds is 1.9 mg methanol per mg NO_3-N reduced and 2.6 mg glucose per mg NO_3-N reduced. The quantity of methanol (C_m) required for denitrification is a function of the nitrate, nitrite, and dissolved oxygen concentrations. The requirement has been expressed as (37)

$$C_m = 2.47\,NO_3\text{-}N + 1.53\,NO_2\text{-}N + 0.87\,DO \quad (11.15)$$

All the terms are in milligrams/liter.

A number of organic compounds can be used with the decision usually made on the basis of the chemical cost. Methanol has been the least cost chemical. When methanol is used to obtain rapid denitrification rates in waste water treatment facilities, care should be taken to avoid gross over additions. Denitrification will go to completion if enough methanol is added but too much methanol can cause excess COD levels in the effluent from a treatment plant.

Nitrate also can be denitrified with the concurrent oxidation of sulfur. Autotrophic sulfur oxidizing bacteria are able to use nitrate as their terminal electron acceptor when the supply of oxygen is limited.

Dissolved Oxygen

Denitrification does not take place until the medium is essentially oxygen free. This has been taken to be zero dissolved oxygen. Some studies suggest that denitrification may take place up to about 0.5 mg/l dissolved oxygen (38) while other reports have indicated that nitrate was not reduced at dissolved oxygen concentrations of about 0.2 to 0.4 mg/l (39, 40).

The role of dissolved oxygen in denitrification is related to the approximately 16% difference in energy yield when dissolved oxygen and nitrate are used as the terminal electron acceptors. The energy yield is about 570 kcal/mole of glucose metabolized when nitrate is the terminal electron acceptor and about 686 kcal/mole of glucose when oxygen is the acceptor. The facultative organisms will utilize nitrate only when adequate oxygen is not available. The difference in energy yield indicates that fewer organisms will exist in a denitrification unit than in an aerobic unit with the same organic loading.

The dissolved oxygen concentration surrounding the microorganisms is the important factor governing denitrification. Even in aerobic treatment systems with measurable dissolved oxygen in the liquid, it is possible to have some denitrification occur if the microenvironment adjacent to the organisms is devoid of oxygen. This can occur if the oxygen demand of the microorganisms is greater than the rate at which the oxygen is transferred to the organisms through the liquid. Similar conditions have occurred in soil denitrification studies and can occur wherever localized anoxic pockets occur in aerobic systems.

In pure culture studies, there have been conflicting reports in which denitrification was observed in environments with adequate aeration (41–43). The term "aerobic denitrification" has been used to explain such results. The explanation is that the microenvironment of the organism is depleted of dissolved oxygen because of the rapid metabolism of organic matter. Under these conditions, the cells may utilize nitrate or nitrite present in its immediate vicinity as an electron acceptor, although there may have been a considerable amount of residual dissolved oxygen in the culture medium.

Aerobic denitrification can have importance in systems with high, mixed liquor suspended solids concentrations, such as aerated systems treating agricultural wastes. In oxidation ditches used for the treatment of poultry wastes, nitrogen losses attributed to denitrification occurred even though bacterial nitrification took place simultaneously and high dissolved oxygen concentrations were in the mixed liquor. Unaccounted for losses of nitrogen also have been reported to occur in actively nitrifying activated sludge and trickling filter plants (33, 44–46).

pH

No correlation between pH and other denitrification parameters has been observed (47) although it is believed that active denitrification occurs under neutral or slightly alkaline conditions (48, 49). Inconsistencies between specific studies are likely caused by the differences in the behavior of the various microbes and by the complex microbial interactions in studies. The microorganisms responsible for denitrification are ubiquitous and should be able to adapt to pH levels within the broad range of microbial activity of about 5 to 9.5. A study of denitrification of nitrified poultry manure waste water indicated that the pH values that resulted in the nitrification step (5.0–6.5) were not detrimental (5) to denitrifying organisms.

Temperature

As with any biochemical process, denitrification also shows a temperature dependency. Little temperature effect on denitrification was observed in the range of 20°–30°. The microbial activity decreased significantly as the temperature was decreased to 10°C (50). Denitrification rates varied from 0.18 mg/l NO_3-N reduced/hr/mg/l denitrifying organism at 27°C to less than 0.01 at 3°C (51). The variation of denitrification rate with temperature can be represented by the Arrhenius Law.

Redox Potential

An interdependence of redox potential and denitrification was demonstrated by several investigators. Above 350 m*v* nitrates accumulated and below 320 m*v*, they disappeared (52). Nitrate was reduced at 338 m*v* (53). Rapid losses of nitrogen via denitrification have occurred when the redox potential was dropped to 300 m*v* or below (54). Redox potential was not indicated as a limiting factor in the reduction of nitrite (55).

Solid Retention Time

Microorganisms obtain more energy when oxygen is used as a hydrogen acceptor than when nitrate is used. Therefore, the generation time of bacteria will be greater and the cell mass yield will be less under denitrification conditions than under aerobic conditions. The longer generation time will require somewhat longer design solids retention times in a biological denitrification reactor.

Based upon a study of suspended growth denitrification, the minimum SRT was found to be 0.5 days at 20°C and 30°C and 2 days at 10°C (Fig. 11.8). For practical applications a SRT of at least 3–4 days at 20°C and 30°C and 8 days at 10°C was recommended (50). Nitrate reduction efficiency in denitrification can be controlled by adjusting the SRT of the process to assure adequate numbers of denitrifying organisms and denitrification rates as environmental conditions change.

Practical Application

A variety of systems have been proposed for denitrification of wastes and waste waters. These include: an activated sludge unit followed by a mixed basin without aeration for denitrification, bypassing a portion of

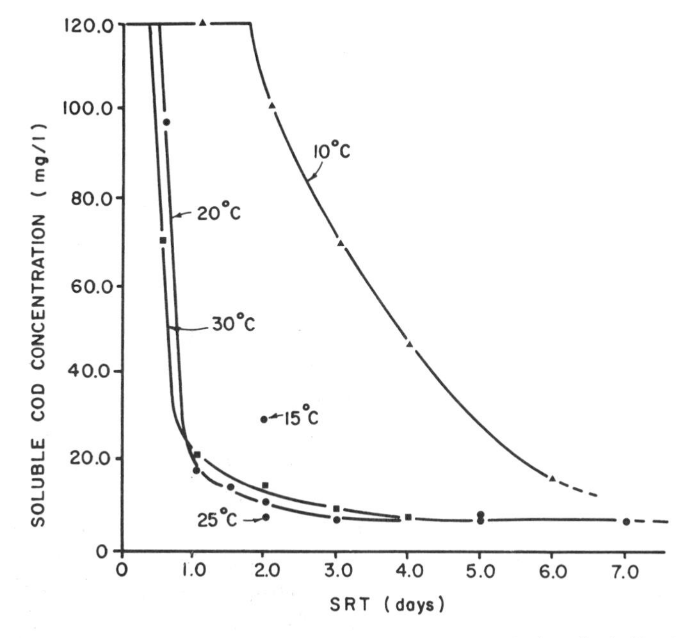

FIG. 11.8. Average steady state effluent COD concentration during denitrification (50).

untreated waste to a final denitrification basin to increase the respiration rate, three-stage nitrification–denitrification process with solids recycle around each stage in which the first stage is used to remove the organic matter in the waste water, the second stage is to nitrify, and the third stage is to denitrify using an exogenous carbon source, anaerobic filters, denitrification using oxidation ditches, anaerobic ponds, and modified land disposal systems.

The practice of denitrification following nitrification has been tried in laboratories, in pilot plants, and in large-scale plants to remove the oxidized forms of nitrogen from the sewage effluents. Two different sets of conditions are necessary to control the processes of nitrification and denitrification and frequently the two processes are separated. The majority of denitrification processes have been applied to municipal waste waters. Some information is available on the denitrification of waste waters from agricultural operations

In studying the feasibility of denitrifying irrigation drainage water containing relatively little organic matter, methanol was used as a hydrogen donor (37). Three possible processes were used: (a) an anaerobic pond with detention times of several days and recycling of the accumulated seed, (b) an anaerobic activated sludge system, and (c) an anaerobic filter. Depending on the temperature, efficient nitrogen removals from the agricultural subsurface drainage water were accomplished with an anaerobic filter having a HRT of 0.5–2 hr. Backwashing of the filter over a long period of operation was not found to be necessary (56).

Laboratory denitrification studies conducted in small anaerobic lagoons with fertilizer plant effluent containing 1000 mg/l of nitrate nitrogen indicated that a ratio of about 3.19 mg COD/mg of nitrate nitrogen accomplished both carbon and nitrogen removal at 10 days detention time. The percentage removal of nitrate nitrogen increased with longer detention times (57).

The use of aerated units for storage and treatment of animal waste slurries and waste waters will minimize the odor problem normally present with anaerobic storage of the wastes and can provide some degree of treatment prior to ultimate disposal. Nitrogen losses caused by denitrification can be obtained in these units under proper operating conditions.

With long liquid detention times and adequate oxygen, the opportunity exists for nitrogen removal by a nitrification–denitrification cycle. Nitrogen losses have been determined by nitrogen balances on the units. In the majority of cases, losses due to volatilization have been included in the balances or have been determined to be negligible. The losses noted in Table 11.2 (58–61) represent those presumably due only to denitrification. Nitrogen losses in excess of 50% have been recorded in aerated poultry waste units when the redox potential remained below 350 m*v* (62). Un-

Table 11.2

Nitrogen Losses Attributed to Denitrification—Aerated Poultry Waste Units

Liquid temperature (°C)	Dissolved oxygen (mg/l)	Nitrogen loss (%)	Ref.
12–23	2–7	36	58
5–21	2–6	66	59
12–23	0–2	50–60	60
11–18	0–6	70–80	61

controlled nitrification–denitrification patterns have been observed in oxidation ditches for swine and aeration units for cattle feedlot wastes. Aerated units treating animal wastes will contain ample unoxidized organic matter. Soluble BOD levels range from 200 to 3000 mg/l and additional oxygen demand occurs because of the particulate matter. No exogenous carbon compounds are needed to achieve denitrification in these units.

Because of variations in waste water characteristics, close operational control is necessary to obtain satisfactory nitrification and denitrification in any biological system that is contemplated for nitrogen control.

Suitable design parameters to accomplish biological denitrification with agricultural wastes are not yet available. The basic parameters needed for design are the required denitrification loss, the nitrate removal rate per denitrifying organism (milligrams NO_3-N removed/milligram MLVSS/hour), the solids retention time, and the type of biodegradable organic material. Nitrate removal rates should be related to operating temperature and pH.

Algal Systems

Algal systems, in which excess nutrients are incorporated into algal cell mass that is harvested from the liquid, can be used as a method of nitrogen removal. The fundamentals of oxidation and aerobic ponds are discussed in Chapter 6, and can be applied to these systems.

An algal system consisting of algal growth, harvesting, and disposal has been evaluated as a possible means of removing nitrate nitrogen from agricultural drainage water in California (62). About 75 to 90% of the 20 mg/l influent nitrogen was assimilated by shallow algal cultures. The system was augmented by inorganic nutrients and carbon dioxide. Theoretical hydraulic retention times varied from 5 to 16 days depending on the time of the year. The most economical and effective algal harvesting system tested was flocculation and sedimentation followed by filtration of the settled solids. The algal cake contained about 20% solids and could be flash-dried to about 90% solids. Utilization of the algal cake as an animal food supplement or soil conditioner was suggested.

Algal systems for nutrient control can be satisfactory where there is adequate land area, sunlight, reasonably uniform warm temperatures and when the algal cells can be adequately harvested and utilized or disposed of.

Ammonia Stripping

General

Ammonia removal by desorption from a liquid waste has received close scrutiny as a method for nitrogen control. Studies with domestic sewage (63–65), and industrial waste waters (66) have indicated that nitrogen removal by ammonia desorption was technically feasible. The process has been tried on a large scale with municipal waste waters. Two practical limitations have been noted: (a) inability to operate the process at ambient air temperatures near freezing and (b) the buildup of calcium carbonate scale on a stripping tower. Detailed discussion of ammonia stripping as applied to municipal waste waters is available (67).

Ammonia stripping is not an ultimate disposal process but simply transfers the nitrogen from a liquid to a gaseous phase. As such, the process is not a total solution to the excess nitrogen problem. If the stripped ammonia nitrogen falls in adjacent water bodies, it can contribute to nitrogen enrichment in these waters. Examples of such enrichment have occurred near cattle feedlots (Chapter 3). Atmospheric ammonia has been indicated as the most important input of nitrogen into the surface waters of semiarid northeastern Colorado.

If the stripped ammonia falls on crop land, there are minimal environmental problems and the ammonia can be a benefit. Plants can utilize the ammonia in rain, and plant leaves act as a natural sink for atmospheric ammonia. A field crop growing in air containing ammonia at normal atmospheric concentrations might satisfy as much as 10% of its nitrogen requirement by direct absorption of ammonia from the air. Annual ammonia absorption by plant canopies could be about 20 kg/hectare (68). The mechanisms involved in such absorption were not explained but were speculated to be combinations of physical absorption, chemical exchange, and metabolic assimilation.

Ammonia stripping can have a positive role in nitrogen control if the stripped nitrogen does not cause additional environmental problems. The principles of ammonia desorption are common for the various methods involved in such desorption. The fundamentals of ammonia desorption, controlling parameters for diffused aeration systems, and possible application to agricultural waste waters are discussed in this section.

Ammonia is very soluble in water (Table 11.3) (69) with the solubility increasing as pressure and temperature increase. All of the ammonia nitrogen in a liquid is not in a form capable of desorption. Only un-ionized ammonia (NH_3) is available for desorption [Eqs. (11.1) and (11.2)].

The desorption of ammonia from waste waters involves the contacting of a liquid and a gas phase in units such as spray towers, packed columns, aeration towers, or diffused air systems. Whatever the mode of contacting, the two phases are brought together to transfer the ammonia. One phase usually flows counter-current to the other. Ammonia being transferred from the liquid to the gas phase must pass through the interface (Fig. 11.9) and the gas and liquid film layers. During the removal of ammonia from an aqueous solution by air stripping, the greatest resistance to mass transfer occurs in the transfer from the liquid to the gas phase because of the high solubility of ammonia. Under equilibrium conditions a steady transfer of ammonia to air from the liquid takes place and the rate of transfer is dependent upon the concentration profiles of ammonia in both transfer resistance layers.

The amount of ammonia desorbed from its solution into air is directly proportional to the concentration of ammonia in the liquid, interfacial area of exposure, duration of desorption, temperature, and atmospheric pressure. If the total area of interfacial surface of a volume of liquid is A_i, then the change in the ammonia concentration in the liquid (ΔC) because of desorption from the entire surface in the duration Δt is

$$\Delta C = -k' \cdot A_i \cdot F \cdot C \cdot \Delta t \qquad (11.16)$$

Only the desorbable ammonia concentration can be removed by stripping. The desorbable concentration is equal to the total ammonia concentration times the fraction available to be desorbed, F, [Eq. (11.2)], i.e., $F \cdot C$. Equa-

Table 11.3

Solubility of Ammonia in Water[a]

Pressure (mm Hg)	Solubility at: 0°C gm/gm	0°C cm^3/cm^3	20°C gm/gm	20°C cm^3/cm^3	40°C gm/gm	40°C cm^3/cm^3
800	—	—	0.544	714	0.329	429
1000	1.094	1440	0.629	826	0.386	504
1800	1.847	2430	0.906	1190	0.543	709

[a] From reference (69).

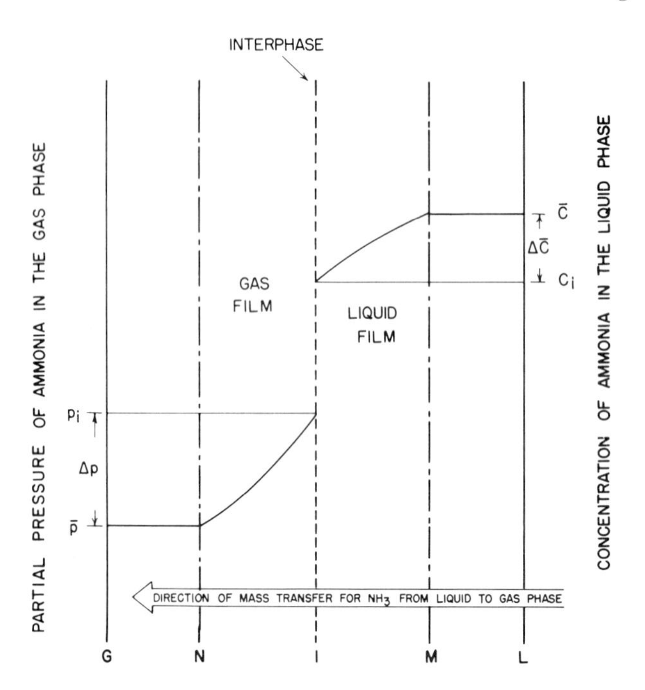

FIG. 11.9. Schematic of the transfer of ammonia from a liquid to a gas phase.

tion (11.16) indicates that greater quantities of ammonia can be desorbed by increasing the time of exposure, area of exposure, and the concentration of ammonia in the liquid. k' is a desorption constant at a given temperature and pressure in units of area^{-1} time^{-1}.

In ammonia desorption systems such as aeration towers or diffused aeration systems, the interfacial gas–liquid area is difficult to estimate and the terms k' and A can be grouped into an overall desorption coefficient (K_D) for the system being utilized. K_D has the units of time^{-1}. The numerical values of K_D are system specific and depend upon the nature of the liquid, temperature, aeration device, rate of aeration, air rate to liquid volume or liquid flow rate ratios, and system turbulence.

Ammonia removal in batch desorption units can be expressed as:

$$\log_e \frac{C_1}{C_2} = K_D F(t_2 - t_1) \tag{11.17}$$

where C_1 and C_2 are the concentrations of total ammonia at times t_1 and t_2, respectively. Batch ammonia desorption occurs as a first-order reaction.

Comparable equations can be written for continuous flow desorption systems. If V is the volume of the liquid in the unit in which the desorption of ammonia is carried out, and Q is the rate of flow of the liquid into the unit, Q is the rate of outflow, assuming no losses due to evaporation. If C_1 and C_2 are the concentrations of ammonia in the influent and effluent, respectively, then a mass balance on the unit at equilibrium will show:

$$Q \cdot (C_1 - C_2) = K_D \cdot F \cdot C_2 \cdot V \tag{11.18}$$

which can be written as

$$(C_1 - C_2)/C_2 = K_D \cdot F \cdot t_H \tag{11.19}$$

where t_H is the average liquid detention time in the unit and the average time of desorption in a continuous flow desorption system.

For whatever decrease in ammonia that may be required, the time of liquid detention or desorption for which a batch or continuous flow unit should be designed may be determined if K_D and F are known. F is dependent upon pH and temperature (Fig. 11.1), and can easily be determined. K_D is independent of pH but is a function of the factors noted earlier.

Equations (11.16–19) are valid in systems where the pH and temperature, and hence F, are constant. When these parameters are not constant, these equations must be modified. Such modifications are available and permit the evaluation of ammonia desorption systems as pH varies during desorption (5).

The important variables affecting ammonia desorption during diffused aeration are concentration of un-ionized ammonia in the liquid, rate of air flow, volume of liquid in batch units and rate of liquid flow in continuous units, desorption coefficient, and duration of desorption.

Temperature

An increase in temperature will increase the desorption coefficient K_D and the removal of ammonia from a solution. Ammonia desorption from poultry and dairy manure waste waters have provided information on the manner in which K_D changes with temperature (Fig. 11.10). The intercepts of the temperature-K_D lines have a positive intercept on the Y axis indicating that the desorption of ammonia can occur only at temperatures greater than 3°–5°C. The slope of these lines provides an estimate of the effect of temperature on the efficiency of ammonia desorption and the ability to predict the effect of changes in ambient conditions on the process. The temperature effect is greater at higher aeration rates. The quantity of ammonia removed by a unit volume of air is directly proportional to the concentration in the liquid (Fig. 11.11) and is related to the temperature of the liquid.

Data on ammonia desorption from poultry and dairy manure waste

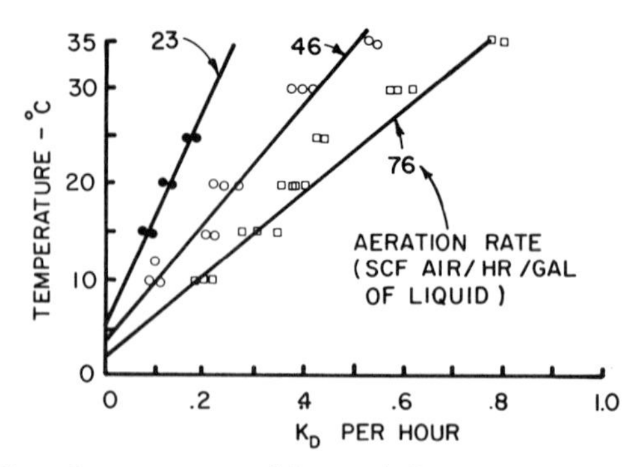

FIG. 11.10. Effect of temperature at different air flow rates on the ammonia desorption rate constant; batch units (5).

waters indicated that K_D values at different temperatures can be related by (5)

$$K_{D_2} = K_{D_1} \cdot 1.063^{(\theta_2 - \theta_1)} \tag{11.20}$$

where K_{D_1} and K_{D_2} are the desorption rates at the same rate of air flow and at temperatures θ_1 and θ_2 in °C.

Aeration Rate

At any given temperature, higher values of K_D can be obtained by increasing the rate of aeration. The relationship of K_D to air flow rate (A) per liter of liquid (L_Q) was linear (Fig. 11.12) over a temperature range of 10° to 35°C. The K_D values when the diffused air flow rate is zero provide an estimate of the rate of ammonia desorption when there is no desorption, i.e., under ambient quiescent conditions.

It is possible to develop an empirical relationship to relate K_D, air flow rate, and temperature. The following relationship was based on data collected at air flow rates 6 to 20 SCFH/l and temperatures from 5° to 35°C (5) during the diffused air desorption of ammonia from poultry and dairy manure waste waters:

$$K_D = 0.021 \exp\left[0.091 \frac{A}{L_Q} + 0.062(\theta - 5)\right] \tag{11.21}$$

The parameters in Eq. (11.21) have the following units: K_D, hour^{-1}, A, standard cubic feet of air flow per hour (SCFH); L_Q, liquid volume in liters; θ, temperature in °C. When air flow is zero, Eq. (11.21) reduces to

$$K_D = 0.021 \exp[0.062(\theta - 5)] \tag{11.22}$$

Equation (11.22) has been shown to be a reasonable estimate of ammonia desorption under quiescent conditions (5).

Viscosity

Physical properties, such as viscosity, can influence the mass-transfer coefficient of gas transfer systems. Viscosity was found to affect the desorption coefficient (K_D). Straight and parallel lines were obtained in logarithmic plots of viscosity and K_D at different air flow rates (Fig. 11.13). The data were obtained over a temperature range of 10° to 30°C, and air flow rates varied from 3–20 SCFH/l (11–76 SCFH/gal of liquid). With an increase in the viscosity there was a proportionate decrease in K_D. This relationship was expressed as:

$$K_D = a(\mu)^{-b} \tag{11.23}$$

where μ is the viscosity of the liquid, b is the slope of the lines in Fig. 11.13, and a is the value of K_D when μ is one centipoise. A comparison of the slopes,

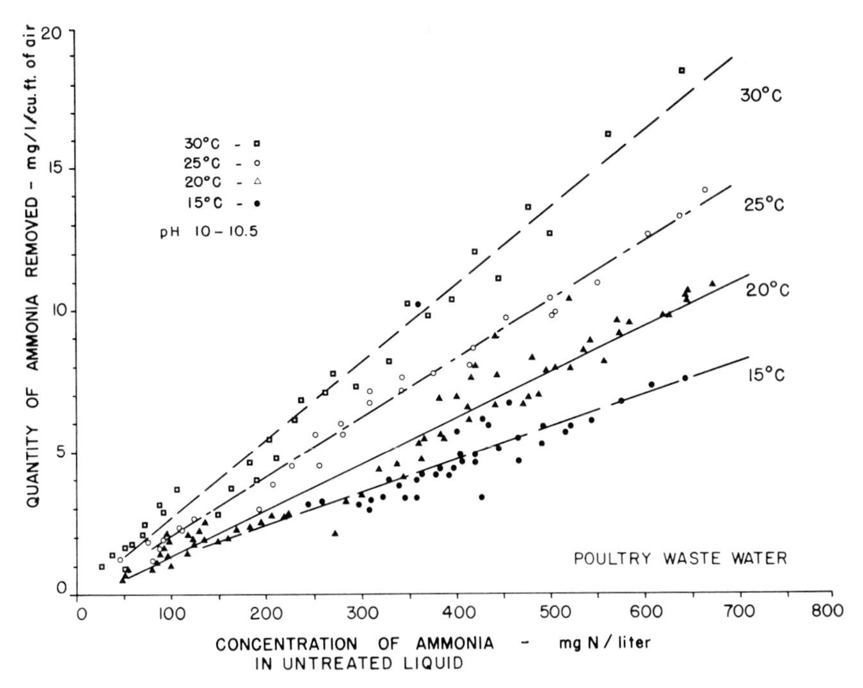

FIG. 11.11. Quantity of ammonia removed per quantity of air related to temperature and ammonia concentration (5).

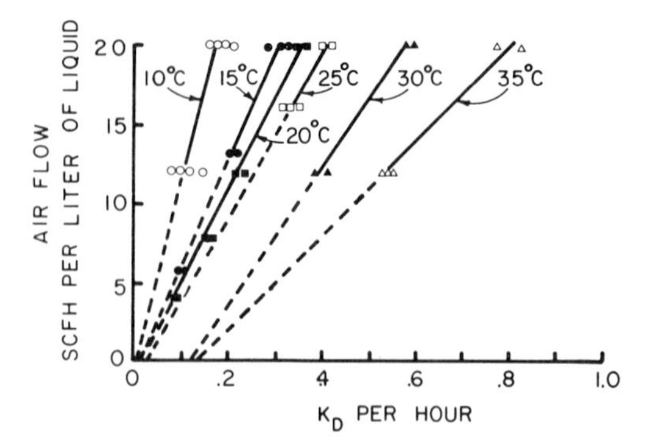

FIG. 11.12. Effect of air flow rates at different temperatures on the ammonia desorption constant; batch studies (5).

b, indicated that neither the type of liquid nor the air flow rate within the range of 6–20 SCFH/l of liquid had a significant effect. A slight difference in this relationship was noted at an air flow rate of 3 SCFH/l. The value of b ranged from 2.9 to 3.9 with an average of 3.5. The value of a represents an estimate of the maximum K_D value that can be obtained at the noted air flow rates. The fact that Eq. (11.23) was consistent over many air flows with different wastes and over a broad temperature range suggests its usefulness for general design purposes with wastes of different characteristics.

By knowing the viscosities of the suspensions of wastes, and using Eq.

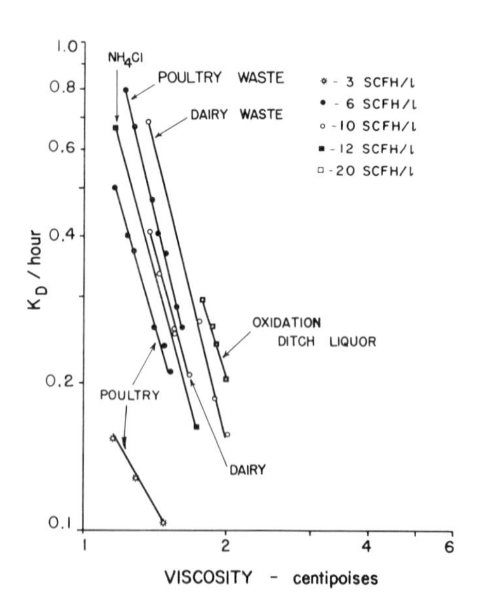

FIG. 11.13. Effect of viscosity on the desorption of ammonia from animal waste solutions (5).

(11.23), it is possible to predict the value of K_D of a liquid (K_{D_2}) if the K_D of a similar liquid (K_{D_1}) and the viscosities of both liquids (μ_1 and μ_2) are known.

$$\frac{K_{D_1}}{K_{D_2}} = \left(\frac{\mu_1}{\mu_2}\right)^{-b} = \left(\frac{\mu_2}{\mu_1}\right)^{b} \tag{11.24}$$

This relationship holds only when the air flows to be used for the desorption of the two liquids are the same. Equation (11.24) permits the extrapolation of available data. Desorption coefficients obtained with one waste can be used to estimate desorption coefficients of different wastes or of wastes modified by in-plant changes. The equation also permits better use of laboratory time in obtaining K_D values since the equation may be used to relate data from different wastes and only a small amount of confirmatory data may be needed.

Practical Application

The equations presented for ammonia desorption using diffused aeration can be used to estimate the magnitude of the design parameters to remove ammonia from a specific waste or to calculate the amount of ammonia desorption that occurred in a specific treatment unit or body of water. The predictive relationships for K_D determined in both laboratory and pilot plant studies have shown close correlation (5).

There are many factors that influence ammonia desorption and therefore many tradeoffs that can be made to obtain a specific ammonia removal, each requiring different desorption times to accomplish the desired results. The basic equations [Eqs. (11.17–19)] can be used to indicate the relative importance of some of the factors affecting ammonia desorption (Fig. 11.14). With a specific waste, little difference in desorption time occurs at pH values of 10–11 at temperatures from 10°C to 25°C. Below a pH of 10, the time to accomplish a specific removal increases rapidly. Higher removal efficiencies require increased desorption times. The time to accomplish 75% removal is double that to accomplish 50% removal at 20°C. In the same manner, it takes twice as much time to accomplish 99% removal as to accomplish 90%.

A change in temperature affects the desorption time. At a pH of 10 to 11, a 10°C temperature drop doubles the time necessary to obtain a specific removal. At a pH of 8–9, a 5°C temperature decrease doubles the desorption time to achieve the desired removal. The larger temperature effect at the lower pH values occurs because of the cumulative effect of pH on the un-ionized ammonia concentration and the effect of temperature on K_D.

To achieve high ammonia removals, any or all of the following must be increased to optimum values: pH, temperature, air flow, and time. If lime

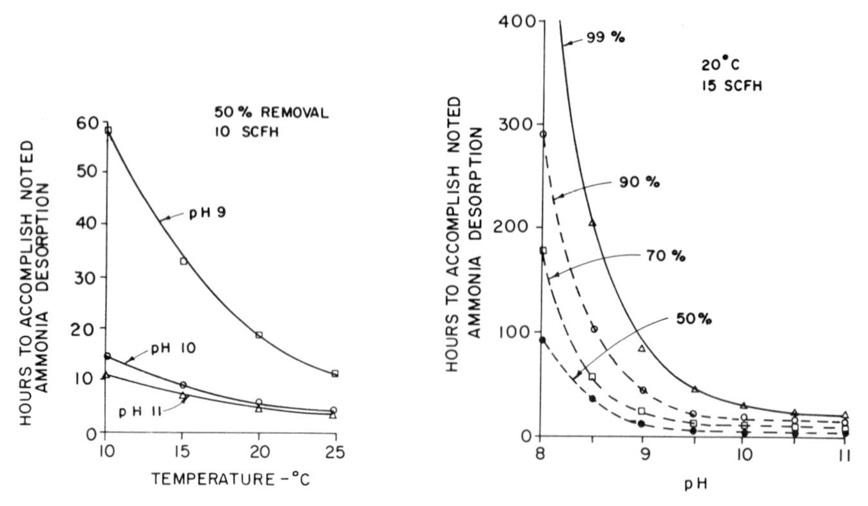

FIG. 11.14. Desorption time to obtain specific ammonia removals as a function of temperature and pH.

or magnesium hydroxide are used to adjust the pH value of waste water, the quantities of resultant sludge can be large. While higher pH values may afford ease of ammonia removal, the disposal of the resultant solids may create another problem. Although greater removals of ammonia can be obtained by raising the temperature of the liquid, it usually is impractical to heat waste waters. With low strength, high volume wastes such as domestic waste water, it may not be practical to increase the liquid detention time to achieve better ammonia removal since this will increase tank size. Longer detention times can be a lesser problem with high strength, low volume wastes which may require extended aeration times to reduce the BOD of the wastes.

The quantity of air required to accomplish a specific ammonia removal by desorption is an important design parameter. Quantities of air needed achieve ammonia desorption of 90% have ranged from 300 to 750 ft^3 of air/gal of waste water. Results of specific studies are shown in Table 11.4 (5, 65, 67, 70). The quantity of air needed for desorption is hundreds of times larger than that needed for activated sludge units which is in the range of 1–2 ft^3 of air/gal of sewage.

As noticed from Fig. 11.14, achievement of high ammonia removal percentages requires high air flow rates. At Lake Tahoe, 80% ammonia nitrogen removal was obtained at 200 ft^3 of air/gal of sewage effluent while to achieve 98% removal, 800 ft^3/gal was required in a 24-ft aeration tower (67). At this full-scale desorption facility, ammonia removal efficiency was reduced substantially as the temperature fell below 20°C. The lower limit of removal efficiency was reached as the temperature reached 0°C. The lower limit of removal efficiency was reached as the temperature reached 0°C. The Tahoe temperature results are compatible with the information shown in Fig. 11.10.

Greater amounts of ammonia nitrogen are removed per unit volume of air when the ammonia concentration in the liquid is high. This indicates a greater efficiency of air usage for desorption with wastes having large ammonia concentrations and/or not requiring high removal efficiencies.

Agricultural waste waters, such as fertilizer plant waste water or animal waste waters, have high ammonia concentrations which favor greater efficiency of air utilization. Animal waste waters can be held in holding tanks or lagoons for long periods of time prior to disposal. Unlike when ammonia is removed from municipal wastes, short liquid detention times are not an important factor when treating animal waste waters. These longer detention times permit using lower air flow rates and pH levels to accomplish desired ammonia removals. The ammonia desorption data from units treating municipal effluents are not directly applicable to agricultural waste waters.

The relationships presented in this section permit a better understanding of ammonia desorption as a nitrogen removal mechanism in both natural

Table 11.4

Desorption of Ammonia from Liquid Wastes

Type of waste	Temperature (°C)	Quantity of air used (ft^3/gal)	pH	Removal (%)	Ref.
Tower aeration					
Petroleum waste water	—	300–480	8.8–10.5	34–95	66
Secondary sewage effluent	23	430–750	10.8–11.0	63–98	65
Secondary sewage effluent	—	200–800	11.5	80–98	67
Anaerobic digester supernatant	—	440	11.2	82	70
Poultry manure waste water	21	280	9.3–11.2	44–77	5
Diffused aeration					
Poultry manure waste water	20	78	10.5	25	5
Dairy manure waste water	20	78	10.5	21	5
Poultry manure oxidation ditch mixed liquor	20	230	10.5	33	5
Tap water	20	78	10.2	23	5

and treatment systems. The relationships can be used to evaluate existing facilities for their efficiency of ammonia desorption and to aid the design of equipment suitable for ammonia desorption.

References

1. Mt. Pleasant, R. C., and Schlickenrieder, W., Implications of nitrogenous BOD in treatment plant design. *J. Sanit. Eng. Div., Amer. Soc. Civil Eng.* **97,** SA5, 709–719 (1971).
2. Wezernak, C. T., and Gannon, J. J., Evaluation of nitrification in streams. *J. Sanit. Eng. Div., Amer. Soc. Civil Eng.* **94,** SA5, 883–895 (1968).
3. Warren, K. S., Ammonia toxicity and pH. *Nature (London)* **195,** 47–49 (1962).
4. McKee, J. E., and Wolf, H. W., "Water Quality Criteria," Publ. 3-A. State Water Quality Control Board, Sacramento, California, 1963.
5. Loehr, R. C., Prakasam, T. B. S., Srinath, E. G., and Joo, Y. D., "Development and Demonstration of Nutrient Removal from Animal Wastes," Rep. EPA-R2-73-095, Office of Research and Monitoring, Environmental Protection Agency, Washington, D.C., 1973.
6. Winogradsky, S., Recherches sur les organismes de la nitrification. *Ann. Inst. Pasteur, Paris* **4,** 213–231, 257–275, and 760–771 (1890).
7. Breed, R. S., Murray, R. G., and Smith, N. R., eds., "Bergey's Manual of Determinative Bacteriology," 7th ed. Williams & Wilkins, Baltimore, Maryland, 1957.
8. Waterbury, J. B., Remsen, C. C., and Watson, S. W., *Nitrospira* sp. and *Nitrococcus* sp. —two new genera of obligate autotrophic nitrite oxidizing bacteria. *Bacteriol. Proc.* **45,** (1968).
9. Alexander, M., Nitrification. *In* "Soil Nitrogen" (W. V. Bartholomew and F. E. Clark, eds.). Publ. No. 10, pp. 307–343. Agr. Soc. Agron. Madison, Wisconsin, 1965.
10. Marshall, K. C., and Alexander, M., Nitrification by *Aspergillus flavus. J. Bacteriol.* **83,** 572–575 (1962).
11. Gibbs, M., and Schiff, J. A., Chemosynthesis: The energy relationships of chemoautotrophic organisms. *In* "Plant Physiology" (F. C. Steward, ed.), Vol. 18, pp. 279–319. Academic Press, New York, 1960.
12. Downing, A. L., Painter, H. A., and Knowles, G., Nitrification in the activated sludge process. *Inst. Sewage Purif., J. Proc.,* 130–158 (1964).
13. Haug, R. T., and McCarty, P. L., Nitrification with thc submcrgcd filtcr. *J. Water Pollut. Contr. Fed.* **44,** 2086–2102 (1972).
14. Hopwood, A. P., and Downing, A. L., Factors affecting the rate of production and properties of activated sludge in plants treating domestic sludge. *Inst. Sewage Purif., J. Proc.* 435–452 (1965).
15. Jenkins, S. H., Nitrification. *J. Inst. Water Pollut. Contr.* **68,** 610–618 (1969).
16. Meyerhof, D., Untersuchengen uber den atmungsvorgang nitrification de bakterien. *Pfluegers Arch. Gesamte Physiol. Menschen Tiere* **164,** 352; **165,** 229; **166,** 240 (1916).
17. Wild, H. E., Sawyer, C. N., and McMahon, T. C., Factors affecting nitrification kinetics. *J. Water Pollut. Contr. Fed.* **43,** 1845–1854 (1971).
18. Gibbs, W. M., The isolation and study of nitrifying bacteria. *Soil Sci. Proc.* **8,** 427–436 (1919).
19. Balakrishnan, S., and Eckenfelder, W. W., Discussion—recent approaches for trickling filter design. *J. Sanit. Eng. Div., Amer. Soc. Civil Eng.* **95,** SA1, 185–187 (1969).
20. Rimer, A., and Woodward, R. L., Two stage activated sludge pilot plant operations—Fitchburg, Massachusetts. *J. Water Pollut. Contr. Fed.* **44,** 101–116 (1972).

21. Mulbarger, M. C., Nitrification and denitrification in activated sludge systems. *J. Water Pollut. Contr. Fed.* **43,** 2054–2070 (1971).
22. Jaworski, N., Lawton, G. W., and Rohlich, G. A., Aerobic sludge digestion. *Int. J. Air Water Pollut.* **5,** 93–102 (1963).
23. Boon, B., and Laudelout, H., Kinetics of nitrate oxidation by *Nitrobacter winogradsky*. *Biochem. J.* **85,** 440–447 (1962).
24. Winogradsky, S., and Omeliansky, V., Uber der influss der organischen substangen aus die arbeit der nitrifizierenden bakterien. *Zentrabl. Bakteriol., Parasitenk. Infektionskr.* **5,** 329–343, 377–387, and 429–440 (1899).
25. Aleem, M. J. H., The physiology and chemoautotrophic metabolism of *Nitrobacter agilis*. Ph.D. Thesis, Cornell University, Ithaca, New York, 1959.
26. Boulanger, E., and Massol, L., Etudes sur les microbes nitrificateurs. *Ann. Inst. Pasteur, Paris* **18,** 181–196 (1904).
27. Montgomery, H. A. G., and Borne, B. J., The inhibition of nitrification in the BOD test. *Inst. Sewage Purif., J. Proc.* 357–368 (1966).
28. Siddiqui, R. H., Speece, R. E., Engelbrecht, R. S., and Schmidt, J. W., Elimination of nitrification in the BOD determination with 0.1 M ammonia nitrogen. *J. Water Pollut. Contr. Fed.* **39,** 579–589 (1967).
29. Young, J. C., Chemical methods for nitrification control. *Proc. Purdue Ind. Waste Conf.* **24,** 1090–1102 (1969).
30. Balakrishnan, S., and Eckenfelder, W. W., Nitrogen removal by modified activated sludge process. *J. Sanit. Eng. Div., Amer. Soc. Civil Eng.* **96,** SA2, 501–512 (1970).
31. Mechalas, B. J., Allen, P. M., and Matyskiela, W. W., "A Study of Nitrification and Denitrification," Report from Contract 14-12-498. Federal Water Quality Administration, Dept. of Interior, 1970.
32. Balakrishman, S., and Eckenfelder, W. W., Nitrogen relationships in biological treatment processes. I. Nitrification in the activated sludge process. *Water Res.* **3,** 73–81 (1969).
33. Wuhrmann, K., Objectives, technology, and results of nitrogen and phosphorus removal processes. *In* "Advances in Water Quality Improvement" (E. F. Gloyna and W. W. Eckenfelder, eds.). pp. 21–48. Univ. of Texas Press, Austin, 1968.
34. Nilsson, E. S., and Westberg, N., Bacterial denitrification of nitrate containing waste waters. *Vattenstands Forutsagelser* **23,** 35–40 (1967).
35. Wuhrmann, K., Stickstoff—und phosphorelimination ergebnisse von versuchen in technischen masstab. *Schwiz. Z. Hydrol.* **26,** 520–558 (1964).
36. Balakrishnan, S., and Eckenfelder, W. W., Nitrogen relationships in biological treatment process. III. Denitrification in the modified activated sludge process. *Water Res.* **3,** 177–178 (1969).
37. McCarty, P. L., Beck, L., and St. Amant, P., Biological denitrification of waste water by addition of organic chemicals. *Proc. Purdue Ind. Waste Conf.* **24,** 1271–1285 (1969).
38. Wheatland, A. B., Barrett, M. J., and Bruce, A. M., Some observations on denitrification in rivers and estuaries. *Inst. Sewage Purif., J. Proc.* **2,** 258–271 (1959).
39. Skerman, V. B. D., and MacRae, I. C., Influence of oxygen availability on the degree of nitrate reduction by *Pseudomonas denitrificans*. *Can. J. Microbiol.* **3,** 505–530 (1957).
40. Skerman, V. B. D., and MacRae, I. C., Influence of oxygen on reduction of nitrate by adapted cells of *Pseudomonas denitrificans*. *Can. J. Microbiol.* **3,** 215–230 (1957).
41. Meikeljohn, J., Aerobic denitrification. *Ann Appl. Biol.* **27,** 558–573 (1940).
42. Marshall, R. O., Dishburger, H. J., MacVicar, R., and Hallmark, G. D., Studies on the effect of aeration on nitrate reduction by *Pseudomonas* species using N^{15}. *J. Bacteriol.* **66,** 254–258 (1953).
43. Sacks, L. E., and Barker, H. A., Influence of oxygen on nitrate and nitrite reduction. *J. Bacteriol.* **58,** 11–22 (1949).

44. Truesdale, G. A., Wilkinson, R., and Jones, K., A comparison of the behavior of various media in percolating filters. *Eng. J.* **60,** 273–287 (1961).
45. Barth, E. F., Mulbarger, M., Salotto, B. V., and Ettinger, M. B., Removal of nitrogen by municipal waste water treatment plants. *J. Water Pollut. Contr. Fed.* **38,** 1208–1219 (1966).
46. Johnson, W. K., Removal of nitrogen by biological treatment. *In* "Advances in Water Quality Improvement" (E. F. Gloyna and W. W. Eckenfelder, eds.), pp. 178–189. Univ. of Texas Press, Austin, 1968.
47. Khan, M. F. A., and Moore, A. W., Denitrifying capacity of some Alberta Soils. *Can. J. Soil Sci.* **48,** 89–91 (1968).
48. Valera, C. L., and Alexander, M., Nutrition and physiology of denitrifying bacteria. *Plant Soil* **15,** 268–280 (1961).
49. Nommik, H., Investigations on denitrification in soil. *Acta Agr. Scand.* **2,** 195–228 (1956).
50. Stensel, H. D., Loehr, R. C., and Lawrence, A. W., Biological kinetics of suspended growth denitrification. *J. Water Pollut. Contr. Fed.* **45,** 244–261 (1973).
51. Dawson, R. N., and Murphy, K. L., The temperature dependency of biological denitrification. *Water Res.* **6,** 71–83 (1972).
52. Pearsall, W. H., and Mortimer, C. H., Oxidation reduction potentials in water logged soils, natural waters, and muds. *J. Ecol.* **27,** 483–501 (1939).
53. Patrick, W. H., Nitrate reduction rates in submerged soil as affected by redox potential. *Trans. Int. Congr. Soil Sci. 7th, 1960,* pp. 494–500 (1961).
54. Meek, B. C., Grass, L. B., Willardson, L. S., and MacKenzie, A. J., Nitrate transformations in a column with controlled water table. *Soil Sci. Soc. Amer., Proc.* **34,** 235–239 (1970).
55. Kefauver, M., and Allison, F. E., Nitrite reduction by *Bacterium denitrificans* in relation to oxidation-reduction potential and oxygen tension. *J. Bacteriol.* **73,** 8–14 (1956).
56. Tamblyn, T. A., and Sword, B. A., The anaerobic filter for the denitrification of agricultural subsurface drainage. *Proc. Purdue Ind. Waste Conf.* **24,** 1135–1150 (1969).
57. Adams, C. E., Krenkel, P. A., Eckenfelder, W. W., and Bingham, E. C., Removal and recovery of high concentrations of nitrogen. Paper presented at the *Int. Congr. Ind. Waste Water,* Stockholm, Sweden, 1970.
58. Loehr, R. C., Anderson, D. F., and Anthonisen, A. C., An oxidation ditch for the handling and treatment of poultry wastes. *In* "Livestock Waste Management and Pollution Abatement," Publ. PROC-271, pp. 209–212. Amer. Soc. Agr. Eng., 1971.
59. Stewart, T. A., and McIlwain, R., Aerobic storage of poultry manure. *In* "Livestock Waste Pollution and abatement," Publ. PROC-271, pp. 261–263. Amer. Soc. Agr. Eng., 1971.
60. Hashimoto, A. G., "An Analysis of a Diffused Air Aeration System Under Caged Laying Hens," Paper NAR-71-428. Amer. Soc. Agr. Eng., 1971.
61. Dunn, G. G., and Robinson, J. B., "Nitrogen Losses Through Denitrification and Other Changes in Continuously Aerated Polutry Manure," Proc. Agr. Waste Manage. Conf., pp. 545–554. Cornell University, Ithaca, New York, 1972.
62. Brown, R. L., "Removal of Nitrate by an Algal System," Water Pollut. Contr. Res. Ser., Rep. 13030 ELY, REC-R2, DWR 174-10. Water Quality Office, Environmental Protection Agency, Washington, D.C., 1971.
63. Rudolfs, W., and Chamberlain, N., Loss of ammonia nitrogen from trickling filters. *Ind. Eng. Chem.* **23,** 828–830 (1931).
64. Nesselson, E. J., Removal of inorganic nitrogen from sewage effluent. Ph.D. Thesis, University of Wisconsin, Madison, 1953.
65. Culp, G., and Slechta, A., Nitrogen removal from waste effluents. *Pub. Works* **97,** 90–93 (1966).

66. Prather, B. V., Wastewater aeration may be key to more efficient removal of impurities. *Oil Gas J.* **57,** 78–89 (1959).
67. Culp, R. L., and Culp, G. L., "Advanced Wastewater Treatments." Van Nostrand-Reinhold, Princeton, New Jersey, 1971.
68. Hutchinson, G. L., Millington, R. J., and Peters, D. B., Atmospheric ammonia: Adsorption by plant leaves. *Science* **175,** 771–772 (1972).
69. Weast, R. L., ed., "Handbook of Chemistry and Physics," 50th ed. Chem. Rubber. Publ. Co., Cleveland, Ohio, 1969.
70. Bennet, G. E., "Development of a Pilot Plant to Demonstrate Removal of Carbonaceous, Nitrogenous and Phosphorus Materials from Anaerobic Supernatant and Related Process Streams," Rep. Proj. 17010 FKA. Federal Water Quality Administration, Dept. of Interior, 1970.

12

Physical and Chemical Treatment

Introduction

There are a variety of processes that can be classified as physical or chemical waste treatment processes. The need for higher quality effluents from municipal and industrial waste treatment facilities has required increasing use of these processes. Examples are chemical precipitation for phosphorus control, filtration to remove particulate matter in final effluents, adsorption for removal of soluble organics, and ion exchange and reverse osmosis for water reclamation and reuse. The wastes from agricultural industries have not yet had to be treated to achieve high levels of effluent quality. This is a result of the availability of alternatives other than disposal in surface waters. The high strength of agricultural wastes and the economics of most agricultural industries preclude the widespread utilization of processes capable of producing a highly treated waste effluent. Other opportunities for waste management such as in-plant waste control and partial treatment at the production facility followed by land disposal or municipal treatment will have higher immediate priorities.

The basic physical and chemical processes can be applied to agricultural wastes for either disposal or reuse purposes. Many industries are recognizing that waste waters are a resource that can be reclaimed for subsequent utilization of the treated waste water or utilization of the separated organic or inorganic fractions. The processes that have been used primarily to obtain high quality waste water for either discharge or reuse include: filtration, ion exchange, adsorption, reverse osmosis, and electrodialysis. Carbon adsorption has been used for the reconditioning and reuse of olive brines (1). A mixed-bed ion exchange unit has been investigated to reduce the salt content of food processing brines (2). The primary interest in these processes is the reclamation and reuse of the brines rather than treatment of the wastes. Similar reuse opportunities exist with other brine wastes such as from sauerkraut and pickle processing. Because of the availability of water and raw materials, waste water or by-product recovery by the above physical and chemical processes is not widely practiced in U.S. agriculture production operations. Land disposal of wastes remains the primary method used for solid and liquid agricultural waste recovery and reuse.

Physical and chemical processes have had application in the treatment of agricultural wastes. The physical and chemical processes that may have possible application to agricultural waste treatment include: disinfection, chemical precipitation, incineration, sedimentation, and flotation. The fundamentals of these processes and their application to agricultural wastes are discussed in this chapter. More discussion of these and other physical–chemical processes as applied to other wastes is available (3, 4).

Disinfection

The purpose of disinfection is to reduce the total bacterial concentration and to eliminate the pathogenic bacteria in water. The microbes of sanitary significance in water are either pathogenic (disease-producing) or nuisance-producing organisms. The production of a potable water supply requires that the bacterial concentration be zero or very low to avoid disease transmission. Disinfection has been practiced since the early 1900's for public water supplies.

There has been an increased emphasis on the disinfection of waste waters before they are released. This emphasis is caused by increased recreational use of surface waters and by continual use of surface waters for water supply and waste disposal. Many states require disinfection of the effluent from all waste treatment facilities discharging to surface waters.

Pathogenic organisms in human and animal wastes may survive for days or more in surface waters depending upon environmental conditions. Factors affecting microbial survival include pH, temperature, nutrient supply, competition with other organisms, ability to form spores, and resistance to inhibitors. The ability of pathogenic organisms to cause disease in man depends upon their concentration, virulence, and ingestion and resistance by their hosts. The pathogenic organisms of interest include bacteria, protozoa, worms, and viruses.

In the disinfection of waters and waste waters, the presence of specific pathogens is not determined. Rather the reduction of a group of indicator organisms, coliforms, to acceptable concentrations is used as a measure of disinfection efficiency and performance. The coliform organisms are ubiquitous organisms commonly found in the intestinal tract of man and animals. There are a number of strains of coliform organisms that are measured by the traditional coliform test (5) such as *Escherichia coli* which is a common coliform organism in the intestinal tract of man and animals. *Aerobacter aerogenes* is a coliform organism usually of nonfecal origin.

Generally, it is the entire group of coliform organisms that is used as an indicator of bacteriological quality, safety of potable water supplies, and efficiency of disinfection. The use of the coliform index is based on two premises: (a) that pathogenic organisms are equally or more susceptible to disinfection than coliform organisms and (b) the presence of coliform organisms suggests the possible presence of human or animal wastes and pathogenic organisms.

The direct count of coliform organisms in a water usually is not done. The most probable number (MPN) test is used to estimate the number of organisms in a sample of water. The MPN is a statistical value indicating the most probable concentration of organisms that were present in the initial sample. A number of different size water samples are examined for the presence of coliform organisms using specific fermentation tube methods. A membrane filter method which permits more direct counts may be preferred (5).

Disinfection of potable and waste waters is used to reduce the concentration of coliform organisms to a low level. With potable waters, the level is such that the possibility of pathogenic organisms being in the water is extremely remote. Drinking water standards in the United States limit the number of coliform organisms in potable water to less than 1 per 100 ml. The combined value of coliform organisms, as indicators of potential pathogens, as part of bacteriological standards for drinking water quality, and of disinfection to meet these standards, has been demonstrated by the decrease in waterborne bacterial diseases since the adoption of these concepts and controls.

Chlorination as a disinfection process for both potable water and untreated or treated waste waters was initiated prior to 1900 in both England and the United States. About 1907 the first plant scale applications were made of chlorine for sewage disinfection and drinking water supplies (6) in the United States. Rapid application of chlorination to municipal water and waste water facilities followed.

There are a number of chemicals and methods that can be used for disinfection such as chlorine, iodine, ozone, quaternary ammonium compounds, and ultraviolet light. Due to low cost, efficiency of disinfection, and ease of application, chlorine is the most common chemical used for disinfection. Both gaseous chlorine and solid chlorine compounds such as calcium or sodium hypochlorite can be used.

The chlorine demand of a water is a function of many factors. Chlorine is an oxidizing agent and will react with a number of compounds including organic waste matter, reduced inorganics, and living matter such as microorganisms. Survival of microorganisms to disinfection will depend upon the microbial species, the protection afforded by solids, and environmental conditions. The efficiency of chlorine disinfection is influenced by the amount and type of chlorine present in the solution, the concentration of other chlorine-demanding substances, time of contact, temperature, and the type and concentration of microbial life. The chlorine demand will be less in waters having low turbidity and suspended solids. The chlorine demand of filtered potable water is less than that of treated waste water effluent.

Chlorine combines with a wide variety of materials in a water, and a number of reactions occur. Many of them compete for the use of chlorine for disinfecting purposes. The chlorine demand of a water must be satisfied before chlorine residuals remain available for disinfection. The residual chlorine compounds are able to be disinfecting agents because they inactivate key enzyme systems in microorganisms after the chlorine species has penetrated the cell wall. Chlorine compounds capable of disinfection are either free available chlorine (FAC) or combined available chlorine (CAC) compounds. The free available chlorine compounds are hypochlorous acid (HOCl) and hypochlorite ion (OCl^-). These can be formed by the addition of either gaseous chlorine or hypochlorite salts to water, i.e.,

$$Cl_2 + H_2O \longrightarrow H^+ + Cl^- + HOCl \tag{12.1}$$

$$NaOCl + H_2O \longrightarrow Na^+ + OH^- + HOCl \tag{12.2}$$

Use of gaseous chlorine tends to decrease the pH of a water as hydrogen ions are produced while use of hypochlorites tends to increase the pH as hydroxyl ions are produced. The disinfecting ability of chlorine compounds increases as the temperature increases.

The distribution of hypochlorous acid and hypochlorite ion is a function of pH

$$HOCl \rightleftharpoons H^+ + OCl^- \tag{12.3}$$

and also is affected by the temperature of the solution. The disinfecting properties of the two FAC compounds is different with the hypochlorous acid being the better disinfecting agent. Free available chlorine exists as HOCl at low pH levels. Below a pH of 6.5, hypochlorous acid predominates while above a pH of about 8.5, hypochlorite ions predominate. The maximum concentration of hypochlorous acid occurs at a pH of about 7.5. Of the common forms of aqueous chlorine, only undissociated hypochlorous acid is regarded as an effective antiviral agent. Other chlorine disinfectants can inactivate viruses but only at concentrations considerably higher than that needed for hypochlorous acid. Both free available and combined available chlorine compounds will destroy bacteria effectively when proper concentrations and contact times are employed.

Chlorine also will react with other chemicals in the solution to produce additional disinfecting compounds. When ammonia is present, chloramines will be formed.

$$NH_4^+ + HOCl \longrightarrow NH_2Cl\,(\text{monochloramine}) + H_2O + H^+ \quad (12.4)$$

$$NH_2Cl + HOCl \longrightarrow NHCl_2\,(\text{dichloramine}) + H_2O \quad (12.5)$$

$$NHCl_2 + HOCl \longrightarrow NCl_3\,(\text{nitrogen trichloride}) + H_2O \quad (12.6)$$

Nitrogen trichloride is not observed at neutral or greater pH values. The relative concentrations of mono- and dichloramines are a function of pH and temperature. Above a pH of about 8.5 monochloramines predominate while below a pH of 6.5 dichloramines predominate. While the mono- and dichloramines are disinfecting agents, they require greater contact times to accomplish satisfactory disinfection. About 25 times as much chloramine concentration is needed to obtain the same bacterial reduction as free available chlorine during the same contact period.

Free available chlorine compounds will not exist in a water until after the ammonia has been oxidized. The point at which this occurs is called the breakpoint. In waters containing ammonia, chlorine compounds will exist as combined available chlorine before the breakpoint and as free available chlorine after the breakpoint (Fig. 12.1). This point generally is not reached until approximately 9.5 mg/l of Cl_2 is added per 1.0 mg/l of NH_4–N. With treated municipal effluents having ammonia concentrations of about 20–30 mg/l, the practice of breakpoint chlorination would require about 190–280 mg/l of chlorine. Agricultural waste waters can have much higher ammonia concentrations. As a result, breakpoint chlorination may be practiced in potable water production but is rarely practiced in disinfection of waste waters. In order for waste water effluent chlorination to be practical and effective, the waste water must be well treated.

Water contact recreation activities are an important concern when waste waters are discharged to surface waters. Bacteriological criteria for these activities govern the desired coliform concentration in discharged waste waters. The value of 1000 MPN per 100 ml of water has been used as a guideline for water contact recreation waters. Greater work is needed to define specific bacterial concentrations but this criteria has provided a workable

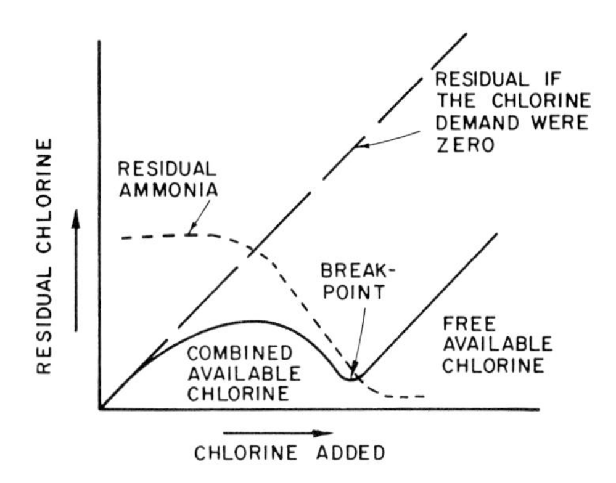

Fig. 12.1. Free and combined chlorine distribution in water.

guideline. As a consequence, chlorination of waste water effluents generally is designed to result in a coliform concentration in the effluent of from 1000 to 5000 MPN per 100 ml or less. Studies on the chlorination of a waste water effluent have indicated that a MPN of about 1000 per 100 ml frequently can be obtained if the product of the combined chlorine residual and the chlorine contact detection time in minutes is equal to or greater than 20 (7). The range of permissible values occurs due to the use and quantity of water into which the chlorinated effluent is discharged. An adverse effect of waste water effluent chlorination may be the toxicity of the resultant chlorine compounds to surface water microorganisms and fish.

Chlorination of waste waters or treated effluents also will reduce the BOD of the waters and oxidize reduced compounds. Soluble ferrous and manganous compounds can be oxidized to insoluble ferric and manganic compounds by chlorine. Hydrogen sulfide can be oxidized by chlorine. The reaction occurs at a pH about 9. Theoretically, the oxidation of 1 mg/l of H_2S requires about 8.5 mg/l of chlorine. Nitrites are oxidized to nitrates by chlorine. The oxidation of 1 mg/l of nitrite requires about 1.5 mg/l of chlorine. Any BOD reduction or oxidation of inorganic compounds is incidental to basic chlorination process since chlorination is not practiced solely to reduce the oxygen demand of waste waters.

Available information on the application of chlorination to agricultural wastes is not extensive. Chlorination of effluents from duck farm waste water facilities on Long Island is required with the requirement being that a minimum chlorine residual of 0.5 mg/l should persist after 15 min contact time. When settling lagoons were used at the duck farms, from 0.85 to 4.6 lb of chlorine/1000 ducks/day were required to reduce the coliform density in the effluents to 2300/100 ml or less (8) and 24 lb of chlorine/1000 ducks/day were needed to reduce the coliform densities to less than 100/100 ml. The economics became more favorable when secondary treatment of the duck wastes was accomplished. Chlorination of treated meat-packing wastes has achieved a greater than 99% bacterial kill with an average of about 1500 MPN/100 ml resulting in the effluent (9).

Lime and chlorine have been used to suppress odors in hog wastes (10). The chlorine demand of anaerobic swine wastes collected from below slotted floors was about 0.1 lb of chlorine/100 lb hog/day. About one-half this concentration would suppress odors. Chlorination of diluted swine waste eliminated some of the odor and improved floculation and dewatering (11). There are more appropriate methods to minimize odors and to treat wastes. The most appropriate role for chlorination with agricultural wastes is to reduce the bacterial concentration of any wastes discharged to surface waters.

Chlorination is an effective process for waste water disinfection. Capital and operating costs are relatively modest when compared with some of the other processes required in waste water treatment. Control of the process is straightforward and relatively simple. When properly applied and controlled, chlorination of waste waters for disinfection is an effective measure for improving the bacteriological quality of the waste water and protecting humans and animals against transmission of enteric diseases by the water route.

Chemical Precipitation

The addition of chemicals to waste waters offers an opportunity to precipitate particulate and colloidal material thereby reducing the oxygen demand of the waste water. Soluble organic compounds are poorly removed by this process. Soluble inorganic compounds, such as phosphates, can be removed if insoluble precipitates can be formed. The amount of material precipitated from a waste water is a function of amount of chemical added, the pH of the solution, and the type of constituents in a waste water. Chemical precipitation has obtained suspended solids removals of up to 90+% and BOD removals of up to 50–70% from municipal waste waters. Comparable removals can be achieved with other wastes.

Chemical precipitation of waste water as a treatment process can be considered to be intermediate between a primary and secondary process. Chemical precipitation is useful to achieve partial treatment prior to discharge of industrial wastes to a municipal system, to achieve lower loadings on subsequent treatment units, and remove inorganic compounds such as phosphates. The quantity of chemical sludge that is generated requires careful evaluation since it increases sludge handling and disposal costs.

The coagulants commonly used to cause precipitation in waste waters are alum [aluminum sulfate—$Al_2(SO_4)_3$], ferric salts such as ferric sulfate [$Fe_2(SO_4)_3$], or ferric chloride ($Fe\ Cl_3$), and lime. Alum reacts with the natural alkalinity of the waste waters, or added alkalinity if necessary, to form an insoluble aluminum hydroxide precipitate which coagulates colloids and hastens the precipitation of other particulate matter. Lime reacts with the bicarbonate alkalinity of waste water to form calcium carbonate which will precipitate and enmesh other particulate matter. Insoluble calcium carbonate occurs above a pH of 9.5. Ferric salts are used to develop insoluble ferric hydroxides which can precipitate colloids and increase the sedimentation rate of other particulates in the waste water. Both alum and the ferric salts have the ability to precipitate negatively charged colloids in a water. Chemical precipitation of a waste water is combined with sedimentation to remove the particulate matter.

Anionic, cationic, and nonionic organic polyelectrolytes also can be used to precipitate colloidal matter in waste water either separately or in combination with inorganic coagulants. A wide variety of polyelectrolytes are available from many manufacturers. Except for previous situations where specific polyelectrolytes have been proved successful, there is little to guide the choice of a polyelectrolyte to be used with a particular waste water. Evaluation of suggested polyelectrolytes on samples of the waste water normally is done using jar tests in the laboratory to determine proper types and dosages.

The quantity of chemical to achieve specific removals depends upon characteristics of the waste water such as pH, alkalinity, solids content, phosphate concentration, and related factors that affect the coagulant demand. These factors vary from waste water to waste water with the result that empirical relationships are used to estimate the necessary chemical dose. Laboratory jar tests with a representative sample of the specific waste water enable an estimation of the feasible type and quantity of chemical. General predictive relationships for waste waters remain to be developed.

The results of the jar test experiments provide a point of departure for proper chemical requirements which must be refined under actual operating conditions. The jar test procedure represents controlled conditions such as waste water characteristics, degree of mixing, quiescent settling conditions, and time of reaction, each of which may vary in a treatment facility. Removal in practice can be poorer than estimated from jar test experiments especially since clarification characteristics of the actual suspensions can be different from those observed in the jar tests.

While chemical precipitation can be utilized with agricultural wastes, there are few recorded cases of it being used as a major treatment process. Cow and dairy shed washings were treated with 500 mg/l of alum and followed by sedimentation for 1 hr. The BOD and suspended solids content of these wastes were reduced by 30 and 70%, respectively (12). Other types of treatment methods are preferred for animal wastes and waste waters.

One place where chemical precipitation may be warranted is for the removal of phosphorus from agricultural waste waters. The phosphorus problem is minimized when waste waters are discharged to land, a common method for disposal of agricultural wastes. However, for highly dilute wastes discharged to surface waters, the application of phosphorus removal methods to agricultural waste waters becomes more critical.

The chemicals used to precipitate phosphorus are lime, alum, and ferric salts. Lime reacts with orthophosphates in solution to precipitate hydroxylapatite. The apatite precipitate, represented by $Ca_5(OH)(PO_4)_3$, is a crystalline precipitate of variable composition. pH control is important with optimum precipitation occurring at pH levels above 9.0 when lime is used. When alum is used to remove phosphates in solutions with pH values above 6.3, the phosphate removal mechanism is by incorporation in a complex with aluminum or by adsorption on aluminum hydroxide floc. Ferric ions and phosphate react at pH values above 7.0 to form insoluble ferric phosphate precipitates.

Experiments have been conducted with poultry manure, dairy manure, and duck farm waste waters to elucidate the type and concentration of chemicals that can best remove phosphates from these wastes (13). The results indicated a number of relationships useful for prediction of chemical dosages to achieve specific removals of phosphates. A comparison of the predictive relationships that were examined indicated that the most sensitive parameters were: chemical dosage per initial total phosphate concentration versus total phosphate remaining for alum with dairy manure and poultry waste water (mg/l alum/mg/l initial PO_4 versus mg/l PO_4 remaining), chemical dosage per remaining total or orthophosphate concentration versus percent total or orthophosphate removal for alum, lime, and ferric chloride with poultry, dairy manure, and duck farm waste waters (mg/l of chemical/mg/l PO_4 remaining versus % PO_4 removal), chemical dosage per initial calcium and total hardness versus percent total and orthophosphate removal for poultry and dairy manure waste waters (mg/l chemical/mg/l hardness versus % PO_4 removal), and lime dosage per initial alkalinity versus the pH after lime addition for poultry and dairy manure waste water (mg/l lime/mg/l initial alkalinity versus pH).

In this study, the percent orthophosphate removal was actually a measure of the orthophosphate disappearance as measured by the procedure for orthophosphate analysis. The soluble orthophosphates are converted to the insoluble state by the chemicals and may still be in suspension as particulate phosphate compounds. To achieve low total phosphate concentrations in the effluent from a chemical precipitation process, an efficient solids removal unit, such as a sand bed or well-operated clarifier, is needed.

Many of the above relationships are not unexpected since they relate to the fundamentals of chemical precipitation reactions. The most useful relationships occurred when the chemical dosage per remaining phosphate concentration was plotted versus the percent phosphate removed. A typical relationship is shown in Fig. 12.2. Both ortho and total phosphate removals and the amount remaining could be related in this manner. With poultry and dairy manure waste waters, the data points using alum fit the relationships better than did the data obtained with lime. Similar relationships were obtained with duck waste water for alum, lime, and ferric chloride. If this relationship is valid for other waste waters it offers an approach to design and operation for the removal of phosphates from waste waters by chemical precipitation.

Calcium and magnesium ions are important in phosphorus precipitation reactions and can be related to chemical dosage and phosphate removal (Fig. 12.3). Similar relationships were obtained for total hardness and for

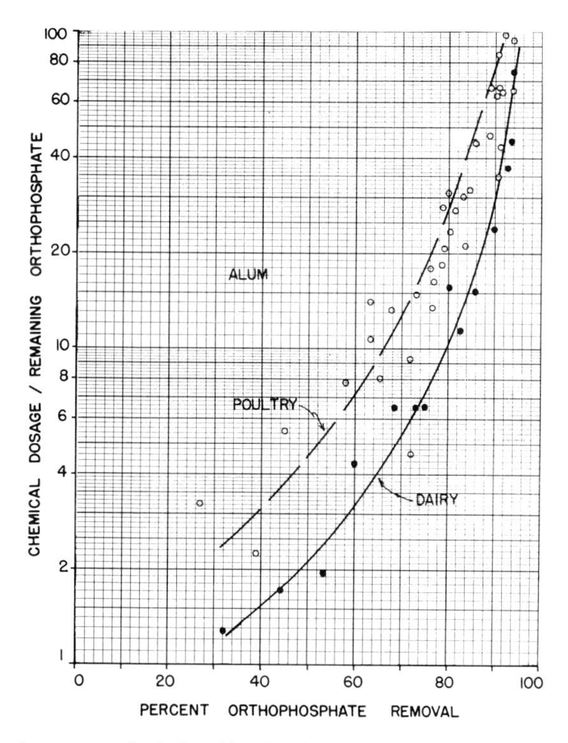

Fig. 12.2. Phosphate removal relationships for dairy and poultry manure waste waters (13).

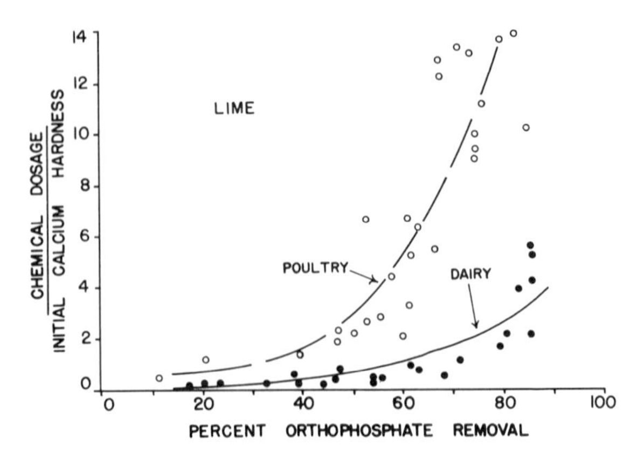

Fig. 12.3. Chemical dosage for phosphate removal related to the initial calcium hardness of the agricultural waste waters (13).

alum in both poultry and dairy manure waste waters. This relationship also offers an opportunity for control of phosphorus precipitation processes.

Decisions on the most appropriate chemical for phosphorus removal from animal waste waters are difficult. The chemical of choice will depend upon the required dosage and chemical cost and the costs of ultimate solids disposal. The alum requirements were less than those of lime for most poultry manure waste waters. For duck waste water, lime requirements were less.

Although this study (13) was directed toward chemical means of removing phosphates from animal waste waters, it should not be inferred that this is the most effective method of phosphate control with these wastes. The results indicated that required chemical concentrations were in proportion to the characteristics of the waste water, i.e., alkalinity, hardness, or phosphate. Ratios of chemical dosages per initial orthophosphate concentration ranged up to 8–10 for alum and lime at low, residual orthophosphate concentrations (less than 5–10 mg/l), and high orthophosphate removals (greater than 90%). Sludge production averaged between 0.5 and 1.0 mg/l suspended solids increase per mg/l chemical used.

To achieve low residual phosphate concentrations, a waste water containing 100 mg/l of orthophosphate may require about 800–1000 mg/l of chemicals which may produce an additional 400–1000 mg/l of suspended solids for ultimate disposal. The large chemical demand and sludge production are decided disadvantages to this method of phosphate control for concentrated animal waste waters.

The general characteristics of agricultural wastes and waste waters are such that a high degree of treatment also will be necessary to remove BOD, suspended solids, and other constituents. Chemical precipitation of phosphates will add to costs and operational problems. Approaches other than conventional liquid waste treatment methods are needed for phosphorus removal from agricultural wastes. Except in unique situations, controlled land disposal should be considered as a high priority method for phosphorus control from agricultural waste waters because it is more amenable to normal agricultural production operations, avoids the need for chemical control and treatment plant operation, and eliminates additional problems of chemical costs and sludge production, handling, and disposal.

Sedimentation

Sedimentation is the process most commonly used to remove settleable solids from sewage and industrial wastes in waste water treatment and from surface waters in water treatment. It is one of the most widely used processes in water and waste water treatment.

Waste waters contain a wide variety of solids having different densities and settling characteristics and ranging from discrete particles to flocculant solids. While recognizing that sedimentation of waste water solids is not an "ideal" situation, the factors involved in sedimentation can be observed by considering the gravity settling of discrete particles in an ideal sedimentation basin. Figure 12.4a represents an ideal basin of length L and

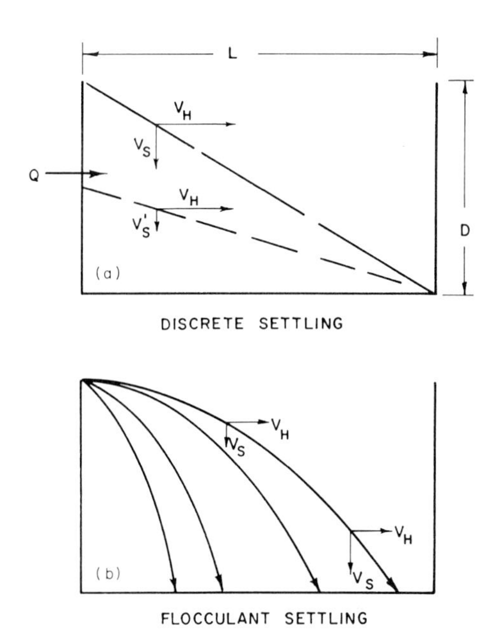

Fig. 12.4. Schematic illustrations of discrete and flocculant particle sedimentation.

width W. Each discrete particle will have a horizontal and vertical velocity. The horizontal velocity is determined by the vertical cross-sectional area and the flow, i.e., $V_S = Q/D \cdot W$. The vertical velocity is a function of the settling characteristics of the particle. In an ideal basin, it is assumed that all particles are uniformly distributed at the influent end of the tank, that there is no effect due to density or velocity gradients throughout the tank, and that all solids remain out of the flow after they reach the bottom of the tank. Therefore in Fig. 12.4a all particles having a vertical velocity of V_S or greater would be removed from the basin. Particles having a velocity less than V_S would be removed in proportion to their settling velocity, V_S', and to the distance they were above the bottom of the tank at the influent.

The critical settling velocity of a discrete particle such that it will settle out, i.e., V_S, establishes the design of a settling tank. Engineers have chosen to use gallons per square foot-day (surface overflow rate) rather than feet per hour as the design criteria. The two items can be related by noting that when a particle has a settling velocity of V_S, the settling time of the particle is equal to the time it takes to move in the horizontal direction through the basin, i.e.,

$$t_v = t_h = \frac{D}{V_S} = \frac{L}{V_H} \tag{12.7}$$

$$V_S = \frac{V_H D}{L} = \frac{QD}{D \cdot W \cdot L} = \frac{Q}{W \cdot L} \text{ (flow/day/surface area)} \tag{12.8}$$

In an ideal sedimentation tank, depth is not an important factor controlling gravity settling and efficiency of removal is a function of overflow rate rather than depth. With sewage and industrial wastes, idealized sedimentation does not occur with the result that settling patterns of the type noted in Fig. 12.4b occur, especially with flocculant particles. Under these conditions, efficiency of removal is related to both overflow rate and to detention time since particle size increases with time.

Consideration of the horizontal velocity, V_H, is important since it can affect the removal efficiency. As the quantity of water flowing through the sedimentation tank increases, the efficiency of removal will decrease and there will be greater solids carried over the effluent weirs.

In addition to the effective settling zone in a sedimentation basin, the tank also contains other important components such as an inlet zone, an outlet zone, and a solids removal zone. The inlet zone serves to reduce the velocity from the order of feet/second in the entering pipes to feet/hour in the tank and to distribute the influent solids as uniformly as possible. Baffled inlet zones are used to reduce the velocity and to provide both horizontal and vertical distribution of the solids.

The function of the outlet zone is to remove the clarified liquid from the settling zone. Low rates of flow are desirable to avoid carryover of solids. Weirs are used to maintain low overflow velocities. The length of overflow weir is a function of the solids in the sedimentation basin and the liquid flow. Overflow rates of 10,000–15,000 gal/day/linear foot of weir are common for sewage sedimentation tanks. Figure 12.5 indicates the overflow weirs on a sedimentation tank of a water treatment plant.

The design of a sedimentation tank is determined by the surface overflow rate needed to achieve a required solids removal. Rates from 600 to 1000 gal/ft^2/day have been used in conjunction with detention times of 1.5–2 hr for municipal sewage. For specific industrial or agricultural wastes, pilot plant or full-scale tests may be needed to obtain the necessary criteria for the desired solids removal. The efficiency of solids removal is affected by factors such as variable flow rates, short circuiting, turbulence, sludge density currents, and inlet and outlet conditions.

Attention must be given to the handling of the settled solids. When the settled sludge represents a significant portion of the total basin flow, such as in final clarifiers for activated sludge units, the solids loading and withdrawal methods are important. The solids loading is important in determining the area requirements of a tank if thickening is important. Settled solids may be removed by mechanical or hydraulic methods. Mechanical methods include rakes or scrapers that are driven by chains and motors. Such scrapers move the settled solids slowly along the bottom of the tank

Fig. 12.5. Effluent overflow weirs on a water treatment sedimentation tank.

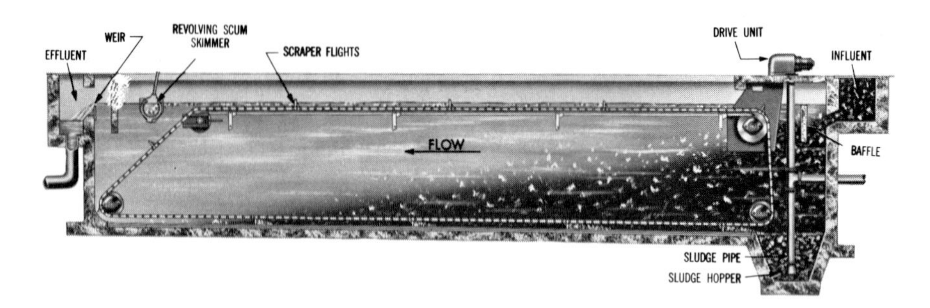

Fig. 12.6. Rectangular primary settling basin with combination chain and scraper surface skimmer and sludge collector (courtesy Rexnord Envirex, Inc.).

to an exit at an end of a rectangular tank (Fig. 12.6) or the center of a circular tank. Hydraulic methods involve the removal of the sludge at the point of deposition (Fig. 12.7). The selection and design of the sludge removal equipment will depend upon the type of sludge to be handled. Hydraulic sludge removal methods are applied most commonly to low density solids such as settled activated sludge. Because of the biological activity of microbial solids, the time they may be retained in a final clarifier is limited to avoid gas production and solids resuspension.

Sedimentation units can be rectangular, circular, or square. They can be classified as primary units, i.e., used before secondary units such as activated sludge or trickling filter units or as secondary or final units which are used to clarify the effluent from the secondary biological units. Many older secondary clarifiers are not equipped with scum scraping and removal

Fig. 12.7. Peripheral feed circular clarifier with hydraulic sludge removal system (courtesy Rexnord Envirex, Inc.).

devices. Because of the possibility of floating solids in a final clarifier, a number of states are requiring scum removal units on all final clarifiers. Without scum removal, any solids that do float in a final clarifier will go over the effluent weirs unless physically removed by the operator. Since the effluent BOD from a treatment plant is a function of the oxygen demand of any solids in the effluent, it is prudent to include scum removal units on all secondary clarifiers. Figure 12.8 illustrates the floating scum, scum removal mechanism, and effluent weirs of a primary sedimentation unit treating duck farm wastes.

As early as 1904, investigators suggested that a shallow sedimentation basin with large surface area would be superior to conventional basins (14).

Fig. 12.8. Primary sedimentation unit treating duck farm wastes.

Horizontal trays have been installed in sedimentation basins to increase the net surface area. However, difficulties with sludge removal and floating material resulted in limited utilization of the tray settling concept. Low Reynolds numbers and laminar flow conditions also favor better sedimentation, Practical application of minimal sedimentation depth, increased horizontal surface area, and laminar flow conditions has led to the development of the tube settler (15) which permits shorter liquid detention times and better sedimentation. Tube settlers are being used at both water and waste water treatment plants.

Because of the need for high quality waste water effluents, sedimentation tanks no longer are used as a sole waste water treatment method. They are incorporated as part of a total treatment facility as a primary and/or secondary clarifier. Surface overflow rates for secondary clarifiers usually are less than that for primary units because of the settling characteristics of biological sludges. Standard design overflow rates are difficult to quote since they are related to the characteristics of the solids. As an example, in the extended aeration treatment of citrus wastes, the secondary clarifier was able to function effectively at about 350–390 gal/ft^2/day, half the design rate, because of light microbial floc and solids carryover at higher rates (16). Sedimentation tanks have been used as an integral part of many waste systems treating fruit and vegetable processing wastes, meat-packing wastes, and liquid animal wastes.

Flotation

Dissolved air flotation is a process which increases the rate of removal of suspended matter from liquid wastes. The process achieves solids–liquid separation by attachment of gas bubbles to suspended particles, reducing the effective specific gravity of the particles to less than that of water. The gas is dissolved in the liquid as it flows through a pressure tank at a pressure of about 30–50 psig (Fig. 12.9). The pressurized liquid is released in the influent end of a flotation tank where the dissolved gas comes out of solution and attaches to the suspended particles in the waste flow. Although air usually is used for treatment of wastes, any gas can be used. Greater quantities of carbon dioxide will go into solution than will air at a given pressure. However, unless carbon dioxide is used for other purposes or is readily available, it usually is more expensive.

Within limits, the number of air bubbles produced will be directly proportional to the product of the absolute pressure and the rate of flow of the pressurized flow. Air solubility in water increases as the temperature decreases and the temperature of the waste water is a factor in the effectiveness of the dissolved air flotation process. The removal rate of both floatables and settleables is a function of the size of the particle. Coagulants such as alum, ferric chloride, and clays can be used to increase the particle size. When chemicals are advantageous, chemical addition, coagulation, and floculation units are utilized prior to the flotation unit. Bench scale tests can be used to evaluate chemical type and dosage as well as detention times, pressurized flow rate, air to solid or liquid ratios, and other design criteria for various wastes.

Pressurization of both influent and effluent liquid can be used (Fig. 12.9). Generally effluent pressurization is preferred since it avoids problems of emulsification and floc destruction that can occur when raw or flocculated wastes pass through the pressurization pump. Effluent pressurization readily

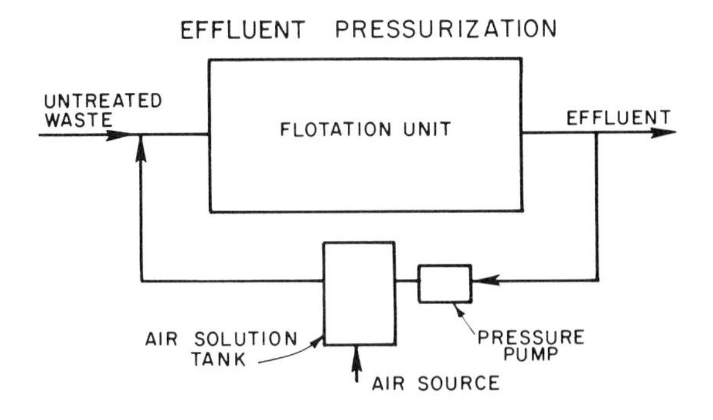

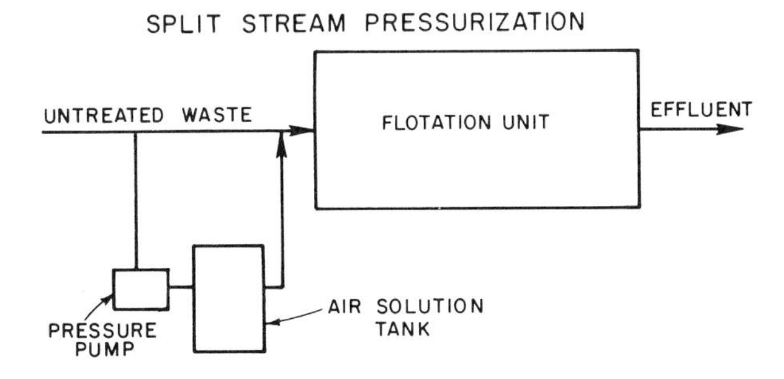

Fig. 12.9. Schematic of dissolved air flotation systems.

allows flocculation to precede dissolved air flotation. Pressurized flow volumes are dependent on the solids concentration in the waste and generally range from 20 to 30% of the raw waste flow. Figure 12.10 illustrates a dissolved air flotation unit to remove both settleables and floatables. Both scum and settleable solids removal equipment is needed.

Dissolved air flotation has been widely applied to the ore industry, to oil refinery wastes, and for thickening of activated sludge. With agricultural wastes, it has been used with food processing waste waters, meat-packing wastes, and edible oil refinery waste waters. Solids removal of 60 to 93% were accomplished by dissolved air flotation of peach rinse water and tomato processing waste water (17). The volume of the float material generally increased with an increase in total hydraulic load while the concentration of floated suspended solids generally decreased as the hydraulic rate increased. Dissolved air flotation was used to pretreat the raw waste from a hog processing facility. The grease from the flotation unit was recovered, rendered, and sold. Removals in the flotation unit averaged 62% for grease, 32% for suspended solids, and 33% for BOD. Oils in the effluent

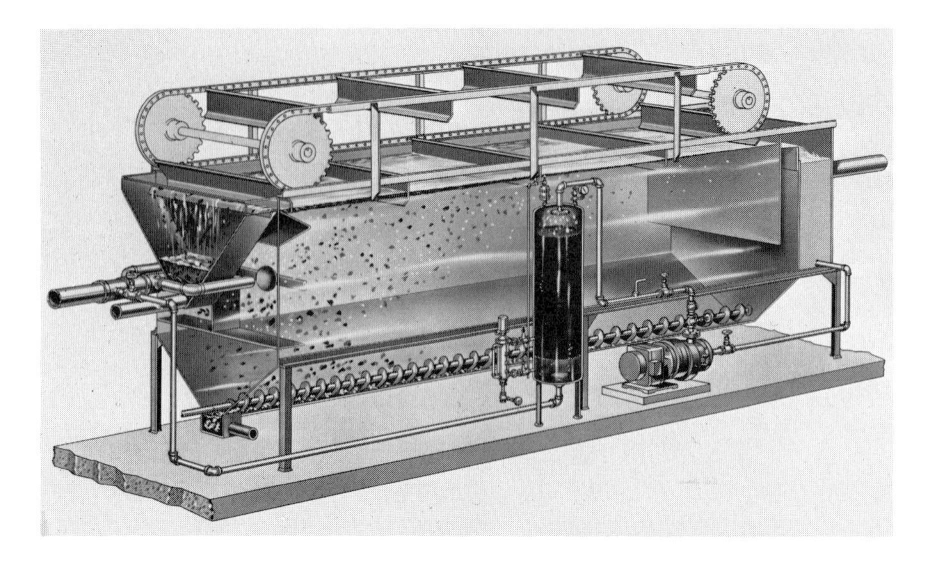

Fig. 12.10. Dissolved air flotation unit for removal of floatable and settleable solids from a waste flow—effluent pressurization (courtesy Rexnord Envirex, Inc.).

from a fatty oil refinery have been recovered using both cationic and anionic flocculants in a dissolved air flotation unit (18). Dissolved air flotation can be applied to any wastes having significant amounts of particles whose density is close to that of water.

Incineration

The purposes of incineration are reduction in volume and sterilization of the end products of the process. It is not a total disposal method since depending upon the characteristics of the waste about 10–30% of the initial waste weight will remain as ash. The volume of the waste will be reduced to a much larger degree. The basic concepts and applications of incineration have been summarized (19) and only the important factors affecting incineration of agricultural wastes will be covered here.

The principal characteristics that affect the suitability of incineration for a waste are moisture, volatility, inert material, and calorific value. Since the water must be removed to achieve incineration, the lower the initial water content, the less energy and added fuel that will be needed for the process. Dehydration, natural drying, and sludge thickening and dewatering are valuable processes to be used prior to incineration. Moisture control is a process over which the designer and operator has some control.

The inert material represents the amount of ash that will result from the process requiring final disposal. Efforts should be made to reduce the quantity of inerts in a waste. Virtually all of the combustible material exists in the waste as volatiles and with proper operation, all of the volatiles will be combusted. Not all of the heat released by the combustion of the waste may be recovered to be used in a drying process since the off-gases will contain considerable heat energy. The calorific value of the waste will determine the quantity of heat released. If the moisture of the waste is sufficiently low, the combustion process can be self-supporting. The BTU value of wastes on a moisture-free basis is about 8000–9000/lb for fresh sewage sludge, 5000–6000/lb for digested sewage sludge, and 5100–5900/lb for fresh or air-dried chicken manure (20).

The design of an incinerator is based upon combustion and heat calculations. The principal sources of heat in the oxidation of the wastes are carbon and hydrogen. The heat released is approximately 14,100 BTU/lb of carbon oxidized and 51,600 BTU/lb of hydrogen oxidized on a moisture-free basis. In addition to fuel and air for combustion, adequate time, temperature, and turbulence must be designed into an incinerator. Temperatures in an incinerator should be above 1200°F since below that temperature, incompletely oxidized particulate, volatile, and odorous compounds may be released in the exit gases. Incinerator operating temperatures normally range between 1200° and 1800°F. If the incinerator is consistently operated at lower temperatures, incomplete combustion and greater air pollution nuisances may result.

Air pollution regulations require adequate pollution controls on an incinerator. For small incinerators, afterburners may be sufficient while with larger ones settling chambers, cyclones, fabric filters, spray systems, or electrostatic precipitators may be necessary. Air pollution control equipment is chosen to meet a particular pollution control objective.

Incineration is one of the more expensive processes that can be used with agricultural wastes, especially since the incinerators will require suitable air pollution control methods to handle any odors and particulate matter that can be generated. It may become more feasible for very large agricultural production operations as available land for disposal decreases and environmental quality constraints increase. It is not likely to be one of the more important processes for agricultural waste disposal in the near future.

References

1. Mercer, W. A., Maagdenberg, H. J., and Ralls, J. W., "Reconditioning and Reuse of Olive Processing Brines," Proc. 1st Nat. Symp. Food Process. Wastes, pp. 281–293. Pacific Northwest Water Lab., Federal Water Quality Administration, 1970.

2. Ralls, J. W., Mercer, W. A., and Yacoub, N. L., "Reduction of Salt Content of Food Processing Liquid Waste Effluent," Proc. 2nd Nat. Symp. Food Process. Wastes, pp. 85–108. Pacific Northwest Water Lab., Environmental Protection Agency, 1971.
3. Culp, R. L., and Culp, G. L., "Advanced Wastewater Treatment." Van Nostrand-Reinhold, Princeton, New Jersey, 1971.
4. Weber, W. J., "Physical and Chemical Treatment Processes." Wiley, New York, 1972.
5. "Standard Methods for the Examination of Water and Wastewater," 13th ed. Amer. Pub. Health Ass., Chicago, Illinois, 1971.
6. Laubusch, E. J., "Chlorine: Its Development, Characteristics and Utility for Disinfection and Oxidation," Proc. 3rd Sanit. Eng. Conf., pp. 6–17. University of Illinois, Urbana, 1961.
7. Classen, H. W., Chlorination of wastewater effluents. *Pub. Works,* 63–66, Jan. (1969).
8. Gates, C. D., "Treatment of Long Island Duck Farm Wastes," Res. Rep. No. 4. New York State Water Pollution Control Board, Albany, New York, 1959.
9. Baker, D. A., and White, J. E., "Treatment of Meat Packing Wastes Using PVC Trickling Filters," Proc. 2nd Nat. Symp. Food Process. Wastes, pp. 287–312. Pacific Northwest Water Lab., Environmental Protection Agency, 1971.
10. Hammond, W. C., Day, D. L., and Hansen, E. L., Can lime and chlorine suppress odors in liquid hog manure? *Agric. Engr.* **49,** 340–343 (1968).
11. Irgens, R. L., and Day, D. L., Laboratory studies of aerobic stabilization of swine wastes. *J. Agr. Eng. Res.* **11,** 1–10 (1966).
12. Painter, H. A., Treatment of wastewaters from farm premises. *Water Waste Treat. J.* 352–355, March/April (1957).
13. Loehr, R. C., Prakasam, T. B. S., Srinath, E. G., and Joo, Y. D., "Development and Demonstration of Nutrient Removal from Animal Wastes," Rep. EPA-R2-73-095, Office of Research and Monitoring, Environmental Protection Agency, Washington, D.C., 1973.
14. Hazen, A., On sedimentation. *Trans. Amer. Soc. Civil Eng.* **53,** 45–62 (1904).
15. Culp, G. L., Hsiung, K. Y., and Conley, W. R., Tube clarification process—operating experience. *J. Sanit. Eng. Div., Amer. Soc. Civil Eng.* **95,** SA5, 829–837 (1969).
16. Eidsness, F. A., Goodson, J. B., and Smith, J. J., "Biological Treatment of Citrus Processing Wastes," Proc. 2nd Nat. Symp. Food Process. Wastes, pp. 271–286. Pacific Northwest Water Lab., Environmental Protection Agency, 1971.
17. National Canners Association, "Waste Reduction in Food Canning Operations," Final Rep., Proj. WPRD 151-01-68. Federal Water Quality Administration, Dept. of Interior, 1970.
18. Seng, W. C., "Removal and Recovery of Fatty Materials from Edible Oil and Fat Refinery Effluents," Proc. 2nd Nat. Symp. Food Process. Wastes, pp. 337–366. Pacific Northwest Water Lab., Environmental Protection Agency, 1971.
19. Corey, R. C., "Principles and Practices of Incineration." Wiley, New York, 1969.
20. Sobel, A. T., and Ludington, D. C., "Destruction of Chicken Manure by Incineration," Proc. Nat. Symp. Anim. Wastes Manage., Publ. SP-0366, pp. 95–98. Amer. Soc. Agr. Eng., 1966.

MANAGEMENT APPROACHES

13

Management

Introduction

The theme of this book has been one of agricultural waste management, i.e., continued production of food while minimizing adverse environmental quality effects that may be a result of such production. A balance must result between agricultural production, profit, and environmental quality objectives. Maximum profit frequently implies no concern with environmental quality or pollution control, and maximum environmental quality control can result in severe economic constraints on agricultural production. The balance between these extremes must be the goal of both agricultural producers, environmental control organizations, and the public.

The increased efficiency of modern agricultural methods is a result of the need to economically meet the increasing demand for food and fiber. These efficiencies have generated a variety of potential and real environmental quality problems. Proposals to return to less efficient agricultural production methods are not consistent with national and international food needs and labor realities. There are, however, available waste management methods that should be utilized in a coordinated manner at each agricultural operation to reduce the excesses that may cause degradation of the environment. Feasible waste management methods will incorporate the best use of science and technology to accomplish needed agricultural production with the least undesirable impact on the environment.

Adequate information is available to avoid gross pollution from agricultural production. Good practice environmental quality guidelines are being developed by both agricultural organizations and governmental agencies. Two approaches are necessary to assure both short and long range environmental quality consistent with satisfactory agricultural production: (a) a program of education for the agricultural producer, the public, and their elected representatives to assure adequate knowledge of environmental quality problems, alternative solutions, and the costs associated with the solutions, and (b) a program of research and development of technically and economically feasible agricultural waste management methods to develop better solutions and to keep abreast of continuing needs. These programs must consider comprehensive waste management approaches that are integrated with agricultural production operations. There are no simple or single solutions to these problems and it is no longer possible for a component of society to dispose of its waste by export. There is no longer any "away" into which wastes can be disposed. Acceptable solutions will result where approaches are developed that will provide the nation with the desired level of environmental quality and the public is convinced of the need of paying for it.

Little is known about the additional food production costs that would be associated with various levels of improvement in environmental quality. When a substantial proportion of producers incur additional production costs, these costs will be reflected in food prices. A major unknown is the increase in food costs that the public would have to bear in order to achieve desired levels of environmental quality. Only with such information can the public decide whether the benefits of environmental quality are enough to justify the costs.

A good environment is a national asset but one that is difficult to measure by direct economic methods. The concept that availability of fresh air, pure water, and good natural surroundings should be part of the criteria for our standard of living is widely accepted. Humans and their activities are part of nature and its cycles. Every measure that affects these cycles must be judged with reference to the total consequences to the environment and to humans. Land planning and site selections for certain types of agricultural production are necessary considerations when decisions are made on which regions and environments should be preserved for the future, where and how an exploitation of the natural resources may be allowed without the environment suffering undue damage, and which environmental quality management methods are necessary and appropriate for agricultural production.

An important agricultural management goal is to use approaches that will avoid the creation of agricultural residues or wastes that cannot be readily reused or recycled or that will result in environmental problems. Minimum pollution should be an important agricultural production goal. Agricultural production philosophy, as well as that of other industrial activity, requires a change from that of maximum yield to that of adequate yield consistent with adequate environmental quality. The practical aspects of the latter philosophy can be difficult to achieve.

The ultimate disposal of agricultural wastes is of prime importance as decisions are made regarding the type and degree of waste management that is needed. The appropriate degree of management will represent the minimum level of control or treatment that will permit discharge of the wastes without subsequent nuisance or pollution. The type of management will be a function of the minimum degree of treatment and the type and concentration of wastes to be treated.

The land remains the most logical point of ultimate disposal for most agricultural wastes. Other possible points of ultimate disposal include: treated liquid wastes to surface waters, solids to the land and separate treatment of liquid wastes, solids destruction and residual inerts to the land, partial treatment of the wastes followed by reuse, and joint industrial-municipal cooperation on waste treatment.

Management includes proper analysis of planning, economic, legal, labor, and technical constraints on a production system. The concept of management should not be our concern. Rather the concern should be the development of sufficient information to be used in management decisions for agriculture that maintain and enhance the environment while permitting an adequate profit. The challenge for agricultural producers, and the engineers and scientists who work with and assist them, is to develop the most economical and equitable combination of alternatives. Technology and science must be employed to reduce the number of mistakes in environmental management and improve our ability to estimate future needs and problems. A number of engineering and scientific alternatives have been presented in previous chapters. An attempt to integrate these alternatives into feasible management approaches is made in this chapter.

Governmental Action

The development and application of regulations to control pollution from nonpoint sources, such as general agricultural runoff or percolation, are difficult. Few states or federal regulations have attempted this type of control. The 1972 Amendments to the Federal Water Pollution Control Act contained the first specific concern about agriculturally related nonpoint sources of pollution in federal water pollution control legislation. The Amendments indicated that each state waste treatment management plan was to identify agricultural nonpoint pollution sources, including runoff from manure disposal areas, and set forth feasible procedures to control such sources. Effluent limitations and the use of the best practicable and best available technology were applied to control the point waste discharges from agricultural operations.

Many states have water quality regulations or administrative codes or guidelines that are applicable to agricultural wastes. Agricultural wastes may not be explicitly cited as potential pollution sources in these regulations although the regulations are sufficiently broad to be applied to agricultural pollution problems, especially those that are point sources. A number of states in which livestock production is an important factor have established additional legislation or guidelines that are specific for livestock waste management.

States generally have air pollution statutes or guidelines that deal with air pollution problems that can be associated with agricultural production facilities. The majority of the air pollution control provisions identify permissible ambient air qualities in regions of the state. Odors and particulate matter arising from feedlot dust and feed processing activities are examples of items that can be subject to control.

Agricultural producers need to be aware of the various state regulations and guidelines and any local zoning or health ordinances that may apply to their operations. The producers also should know who to contact about specific problems. Generally state or county extension agents and employees of the state pollution control agencies can provide information about the applicable regulations or guidelines. Advice on appropriate pollution control measures should be obtained before a problem occurs and especially before an operation is enlarged or a new operation developed.

There are many approaches utilized to control potential pollution from agricultural operations. Many states have developed registration and permit procedures for control of runoff from animal feedlots. While the procedures vary throughout the nation, they usually contain the provision that a feedlot operator shall obtain a permit from the appropriate state agency, correct any pollution hazard that exists, and assure that the operation conforms to all applicable federal, state, and local laws. Specific minimum pollution control methods such as runoff retention ponds, dikes, and distance from dwellings may be included. A map of the areas indicating land use, streams, wells, houses, roads, topography, and other salient features frequently is requested.

While control of potential pollution from feedlots is a recent item, pollution from food processing operations has been included in the pollution control regulations of many states for decades. Food processing wastes are point source wastes frequently discharged to surface waters and their discharge is controlled by the prevailing pollution control regulations.

Guidelines, rather than regulations, are common approaches to minimize pollution problems from agricultural operations. A reasonable approach has been established by the Province of Ontario, Canada which has developed a suggested code of practice for livestock buildings and disposal of animal wastes (1). Good practice guidelines, such as this one, permit the better approaches to be elucidated without placing them in the status of firm rules or regulations. Such an approach is advantageous while better treatment, disposal, and management methods are being developed.

The intent of the Ontario Code of Practice is to: (a) assist farmers in avoiding unnecessary and undesirable situations which could lead to disputes concerning pollution, (b) serve as guidelines for anyone concerned with the establishment of new livestock or poultry production units, (c) serve as guidelines for anyone concerned with making a major renovation or expansion of existing livestock or poultry enterprises, (d) be flexible enough in interpretation and application to cover special cases which may exist or develop from time to time, and (e) serve as a basis for a sound, considered plan of farm operation, giving due regard to waste disposal management, without being specific in design requirements. Key measures in the Code include the provisions of: enough land area on which to dispose of the wastes, sufficient waste storage capacity, and sufficient distance between livestock and poultry buildings and neighboring human dwellings.

Farmers can apply for a certificate of approval which indicates that he is following accepted guidelines for environmental quality control. Recommended land areas for disposal were based on the number of equivalent animal units contributing the wastes. The animal units were based upon the pollution potential as determined by the age, size, and feed of the animal and the period of confinement of the animals. The suggested animal unit relationship is presented in Table 13.1.

Minimum tillable land requirements for disposal of the animal wastes were suggested for both loam and sandy soils for varying size of the animal units (Table 13.2). The recommended acreages were not necessarily the most economical from the standpoint of efficient crop production. When efficient crop production is considered, the necessary acreage may be doubled. The minimum acreage noted in Table 13.2 was considered to be that required to avoid the risk of groundwater pollution by nitrogen compounds.

Suggestions on adequate distance between livestock buildings and human dwellings were (a) at least 2000 ft from land presently zoned for residential use; (b) at least 1000 ft from dwellings on adjacent property; (c) at least 300 ft from the center line of any public road; and (d) at least 200 ft from the lot lines of the site on which the production unit is situated. The Code noted that the first two distance requirements may be amended depending upon the type of waste treatment or disposal method that is used.

Table 13.1

Equivalent Animal Units Based on Pollution Potential[a]

Animal	No. of animals equivalent to one animal unit	Basis of production
Dairy cow (plus calf)	1	Annual
Beef cow (plus calf)	1	Annual
Beef steer (400–1100 lb gain)	2	Annual
Bull	1	Annual
Market hog (40–200 lb gain)	15	As marketed
Dry sows (plus litter)	4	Annual
Laying hens	125	Annual
Chicken broilers (4–5 lb)	1000	As marketed
Pullets	300	As marketed
Turkey broilers (11–12 lb)	300	As marketed
Horse	1	Annual
Mature sheep (plus lambs)	4	Annual

[a]From reference 1.

Manure handling and storage suggestions described in the Code were (a) sufficient storage capacity for six months accumulation of wastes should be provided where a liquid manure handling system is to be used, (b) weather, cropping programs and other local conditions may permit satisfactory operations with a shorter storage period, (c) where liquid manure is spread on land within 1000 ft of neighboring dwellings, it should be incorporated into the soil as soon as possible, and preferably within 24 hr of application, (d) where solid manure is spread on land within 600 ft of neighboring dwellings, it should be incorporated into the soil as soon as possible and preferably within 24 hr of application, and (e) solid and liquid manure storage units should be well managed to keep odor problems to a minimum and to prevent runoff into streams, wells, and other bodies of water.

An increasing number of states are developing broad guidelines to control wastes from agricultural production operations. The majority focus on animal production operations. These guidelines include provisions comparable to those in the Ontario, Canada guidelines as modified by situations and concerns within the state. The guidelines of Washington, Maine, Maryland, and New York offer examples of the various approaches and concerns.

The purpose of the guidelines of the state of Washington (2) is to aid animal producers in meeting air and water pollution regulations at minimum expense. Common sense approaches, such as do not apply animal wastes closer than 50 ft to an open supply of water such as a creek, river, lake, or pond; do not place a heavy application of animal waste closer than 100 ft to a domestic water supply; and fence waterways so that they cannot be contaminated by direct access of cattle are included. The guidelines also suggest minimum land areas which should be available for dairy waste disposal if the disposal area is utilized only as a pollution control device. These minimum land areas are based upon the estimated fertilizer content of dairy wastes. A minimum land area of 0.73 acres/dairy cow or 500 dairy animal manure-days/acre/year was recommended for western Washington. Other good practice guidelines are available to animal producers in Washington (3).

The Maine guidelines (4) outline rates for the spreading of manure on the land. The maximum rates were developed based on the physical and chemical characteristics of Maine soils and knowledge of the movement of manure liquids and residues on and through each soil type. The limiting factor in determining application rates was the pounds of nitrogen to be ap-

Table 13.2

Minimum Land Requirements for Disposal Wastes from Livestock Production Operations[a]

Number of animal units housed or marketed per year[b]	Minimum acreage for livestock or poultry production (acres)	
	Loam to clay soil	Sandy soil
40 or less	20	30
41–60	30	45
61–80	40	60
81–100	50	75
101–120	60	90
121–140	70	105
141–160	80	120
161–180	90	135
181–200	100	150
201–220	110	165
221–240	120	180
241–260	130	195
261–280	140	210
281–300	150	225
301–320	160	240
321–340	170	255
341–360	180	270
361–380	190	285
381–400	200	300

[a] From reference 1.
[b] Whichever is greater.

plied per acre. Tables are provided to indicate manure application rates on specific Maine soil types. Suggestions related to spreading on slopes, snow or frozen ground, distances from surface waters, and type of crop are included in the guidelines.

The Maryland guidelines (5) emphasize control of odors from confined poultry operations. Factors affecting odor production and detailed techniques for odor control are discussed.

The New York guidelines (6) discuss farm sites and planning, handling and storage of manure, manure disposal, and disposal of dead animals. Criteria for the frequency of manure cleaning, handling, and management and for the ventilation of confined animal operations are discussed for beef and dairy cattle, veal calf, horse, laying hen, duck, turkey, and swine operations. Specific New York pollution control regulations applicable to agriculture were identified.

The state of Pennsylvania developed a manual (7) to guide engineers, geologists, and soil scientists in locating and evaluating sites for spray irrigation of waste waters and in designing systems to distribute and apply the waste water to the land. The information suggests spray irrigating only one day a week permitting the soil to dry and reaerate on the other days and suggests the use of fixed irrigation lines where spraying operations are expected under freezing conditions. A topographic and soils map for the spray field is to be furnished as part of the application for a permit to utilize spray irrigation for waste water disposal and renovation. Discussion of operating conditions, runoff control, choice of vegetative cover, erosion control, waste water treatment before irrigation, and other important design parameters are included in the manual.

Not all of these suggestions are applicable in other parts of North America. The suggestions would have to be adapted to local soil types, temperatures, and crop production. Codes or guidelines of this type offer a valuable management approach to the potential pollution problems from livestock production and represent a useful model for other governmental organizations. Codes or guidelines are preferable to governmental regulations since there is the danger that regulations will be applied uniformly to producers despite the wide variety of measures needed to properly protect the environment and the wide differences between agricultural production operations. Codes or guidelines can be changed more easily than regulations when better practices become available.

Specific state or national regulations for controlling general agricultural runoff do not seem appropriate. With the exception of food processing wastes and feedlot runoff, agricultural runoff is ubiquitous and not amenable to regulatory control. Adherence to concepts included in the above guidelines will minimize the contaminants from agricultural operations. Point sources of agricultural waste discharges can be controlled by existing state and federal regulations.

A national permit system has been established for discharge of wastes to the surface waters of the United States. The Environmental Protection Agency administrator is authorized, after opportunities for public hearings, to issue permits for waste discharges under certain conditions if the discharges meet the requirements of the 1972 Amendments to the Federal Water Pollution Control Act. States are authorized to conduct their own discharge permit programs when approved by the EPA.

The 1972 Amendments require that every company that discharges waste into a waterway must apply for a permit. Companies must disclose fully the amount and nature of their pollutants and assure EPA that they meet existing state water quality standards. By 1977 all companies must employ the "best practicable" control technology and by 1983 they must install the "best available" technology. The goal is the elimination of water pollution by 1985. The term "best practicable" technology means a level of control achieved by the least polluting treatment facilities in a given industry. "Best available" technology infers that where it is economical, the most advanced waste treatment processes are to be installed. The best practicable technology includes in-plant changes to reduce the amount of pollution requiring treatment.

Effluent limitations will be established for each industry. The limitations will be based on the applicable technology for the waste and the industry. The limitations will be established in terms of pollutional units, such as BOD or suspended solids, per unit of production, such as per pound of live weight killed for the meat product industry or per 1000 lb of milk or milk equivalent for the dairy products industry. Weight, rather than concentration (milligrams/liter), units are utilized to avoid the possibility of diluting a waste to meet the effluent requirements. The effluent limitations will be based primarily upon the oxygen demanding (BOD, COD, or TOC), solids (suspended or total solids), and bacterial (coliform) contaminants in the waste effluent to be discharged. The limitations can be expanded to nutrients (nitrogen and/or phosphates), salts, and other pollutional parameters as necessary.

The permit system was developed as a method for controlling point source industrial waste discharges. Although agricultural pollution is in many respects a different problem than industrial waste and difficult to control through the permit system, the permit approach does apply to some agricultural pollution sources. The following agricultural industries are those for which effluent limitations are to be developed: meat product and rendering processing, dairy product processing, canned and preserved fruits and vegetable processing, canned and preserved seafood processing, feedlots, and fertilizer manufacturing. New treatment facilities that meet the performance standards of the 1972 Amendments will not be subject to more stringent standards for at least 10 years.

The permit program does not apply to general agricultural runoff or to irrigation return flow. The permit system is not an effective approach to solving these problems. Administratively, the permit program is oriented toward large industries with point sources of discharges.

By development of the effluent limitations for portions of the agricultural industry, the government has taken the leadership from the industries and associated agricultural organizations. Agricultural producers and agricultural organizations should exert the leadership necessary to develop, demonstrate, and thereby indicate the acceptable and appropriate pollution control methods for their industry. These producers and organizations should finance technical and economic studies by private organizations and undertake these studies themselves. Past development of feasible agricultural waste management systems has been largely on an individual operation basis. There has not been the development of industry-wide alternatives that have been given industry support and demonstrated throughout the country to produce factual cost and operating information.

The mutual nonprofit co-op approach should be explored as a potential arrangement for the development of waste management approaches. A possible arrangement could be an organization consisting of individual producers, processors, or farmers with the common goal of defining environmental pollution problems resulting from their activities, seeing that research having an objective of solving those problems is conducted, and developing a coordinated approach for implementing feasible systems for handling their wastes.

Agriculture in many parts of the country has moved from a small family enterprise to a business type operation. Governmental agencies view large agricultural operations as they do other industries and require that operators of these enterprises manage their facilities so that they do not pollute the environment. These operators are in the best position to define their own problems and to develop solutions to the problem.

The pollution of the soil, water, and air resources within a region cannot be attributed solely to agriculture. It is the summation of contributions from all activities. Modern agricultural practices have contributed to new water and air pollution problems and will be required to control these problems as will other sources in the region. While management economics may compel food producers to use production practices that will sustain or even enhance the present rate of waste production, it is incumbent upon these producers to take all necessary steps to assure that the wastes do not adversely affect the environment.

Decision Making

General

A satisfactory management system for environmental quality control in agriculture consists of a combination of alternatives which will provide a suitable level of environmental quality and which will be economically feasible. With the broad spectrum of environmental quality effects, of management interests and capabilities, of legal constraints, and of level of technology, it is apparent that a separate management system must be developed for each agricultural operation or for agricultural operations that have common problems, operational methods, and environmental quality needs.

The legal, administrative, financial, and political constraints will narrow the range of alternatives for a successful agricultural waste management systems. Any costs assigned to an alternative management component should be real so that an estimate can be made of the benefits lost by using a particular alternative. Sacrifices in production efficiency to achieve a desired level of environmental quality need to be estimated.

In obtaining agricultural waste management decisions on a national level, there will be tradeoffs between environmental quality objectives and food availability and food price objectives. The effect of environmental constraints on food production and cost require close analysis when environmental decisions are being made. At the national level and the actual production level, the satisfactory alternative will be the one that achieves the highest level of the co-objectives of adequate profit and adequate environmental quality with the resources that are available.

Focusing on profit and environmental quality objectives may be satisfactory in arriving at production management, including waste management alternatives at a specific production facility. However, other societal goals such as economic growth, alleviation of poverty and hunger, and national self-sufficiency are important in arriving at national policies concerning agricultural waste management. The thrust of this discussion will be on decisions regarding waste management systems at individual agricultural production operations.

Environmental quality requirements will differ from location to location. As with other industries, agricultural production operations should be located where minimum cost treatment facilities are possible, other factors being equal. The cost of adequate waste management facilities will be an increasing factor for agriculture production facilities and in certain cases will determine the economic success or failure of such facilities.

The total agricultural operation should be considered in developing suitable waste management systems. The agricultural waste management problem generally is viewed as a waste treatment and/or disposal problem. Superficially, this appears logical since the problem is one of disposing of the accumulated wastes without contaminating the environment. Waste treatment and disposal are, however, only parts of an adequate waste management scheme. Consideration of waste management begins with the type and quantity of input and continues with the effect of the production environment through methods of waste handling, treatment, and disposal, through the potential of waste or wastewater reuse, and through a consideration of the overall economics of the agricultural production operation as it is affected by waste management (Fig. 13.1).

When methods to handle, treat, and dispose of agricultural wastes are contemplated to avoid water pollution, it is equally important that other types of pollution be avoided. Although the ideal solution to the agricultural waste problem is to dispose of the waste onto the land, more information is required as to the quantity of nutrients, solids, and water the land can take before the crops or soil structures are adversely affected or before water pollution problems occur. Air pollution can result from drying and incineration, and odor nuisances can be generated by uncontrolled anaerobic operations, holding tanks, improperly timed land disposal operations, and within large scale animal production and agricultural processing operations. Transferral of pollution from one sphere to another will not be a satisfactory waste management alternative.

No new facility or expansion of existing facilities for agricultural production should be considered without prior planning which should include the probable environmental effects of the disposal of wastes from the facility. Consideration should be given to air movement patterns, rainfall and runoff relationships, ultimate waste disposal site and methods, soil characteristics, environmental quality criteria, the effects of waste treatment, and possible changes in production techniques and in environmental

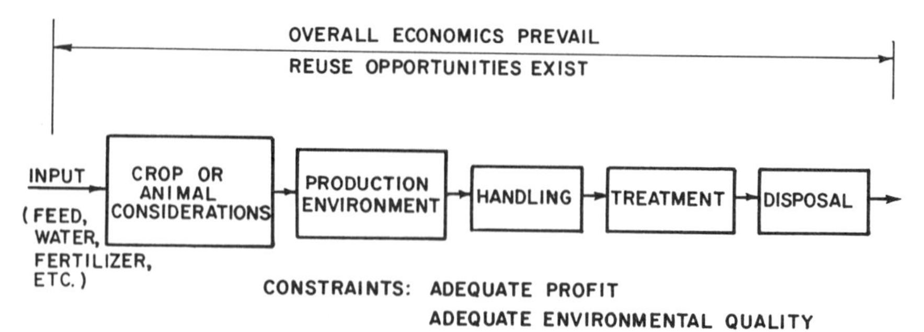

Fig. 13.1. Components of a comprehensive agricultural waste management system.

quality restrictions. An evaluation also should be made of the probable risks of civil suits and governmental action in response to valid complaints of inadequate nuisance and pollution abatement.

An acceptable waste management scheme will not only be technically feasible but also will be economically competitive with other approaches, require a minimum of maintenance and may be operated on an automatic or semiautomatic schedule, not create environmental conditions objectionable to plant workers, housed animals, and the public, and allow for expansion and new technology.

Feasible waste management systems should be produced in cooperation with individuals responsible for animal environment, housing, and feeding and with individuals knowledgeable in waste treatment, land use, and agricultural economics. A degree of coordination must exist on a broader and higher level than currently is observable.

Opportunities

Both internal and external opportunities exist in agricultural waste management systems. A number of these are outlined in Fig. 13.2. Considerable improvement can be made by "in-plant" changes. These include changes in process equipment, production processes, elimination of leaks, reduction in waste volume, separation of wastes for possible reuse or treatment and disposal according to waste strength, recovery of raw materials, change in the physical characteristics of the waste, i.e., liquid slurry or solid to accommodate waste treatment and disposal, and even a judicious choice of location for the facility.

Other opportunities exist after the wastes exit or are removed from the facility. These include the use of proper waste treatment design criteria, alternative treatment methods to obtain specified effluent requirements, treatment of wastes for acceptance on the land rather than for discharge to surface waters, runoff control, reuse of cooling water or partially treated water elsewhere in the facility such as for flushing or raw product rinsing, and a rccovery of portions of the wastes for by-product utilization.

The final disposal point offers additional alternatives to be evaluated. The final residuals can be disposed of into combinations of the air, water, or land. Such disposal can be done directly or indirectly via discharge to municipal waste disposal systems. The latter approach requires that the industry pay the municipality for the discharge of its wastes but it relieves the industry of the burden of final disposal. The proximity of concentrated animal production as well as food processing operations to the urban areas suggests the possibility of an integrated urban–industrial–agricultural waste management concept, however, little experience has been reported in this area.

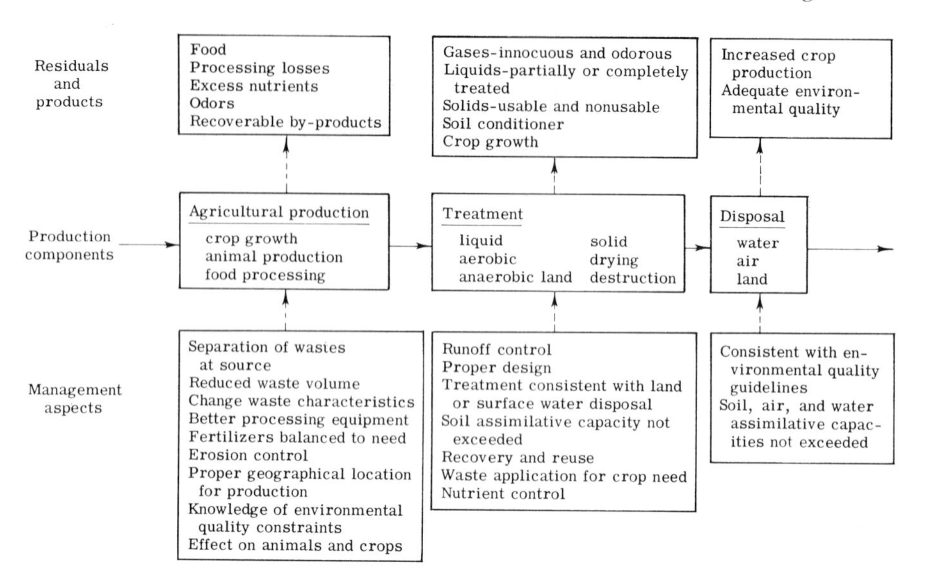

Fig. 13.2. Conceptual agricultural waste management systems.

A materials balance of the production facility, including as appropriate flow, nutrient, solids, energy, and labor balances, should be prepared as an initial step in evaluating the proper components of the combined production-waste management system. The balances will help identify if and where process changes are needed and the types of treatment and disposal that are appropriate.

Selection of Treatment Alternatives

A waste management system consists of a number of individual components. After "in-plant" processing changes have been evaluated, the unit processes for treatment can be considered. A primary consideration in evaluating unit treatment processes is the point of ultimate disposal of the treated or untreated waste. Agriculture may consider both land and water as primary points of disposal. Another equally important consideration is the degree of treatment for the disposal that is required at present and that may be required in the future. For each situation, there are a number of combinations of unit processes which will satisfy the conditions. Known and possible future environmental quality constraints must be kept in mind so that unit processes selected initially can be expanded or readily modified to meet future needs. Details on specific unit processes applicable to animal production and food processing operations are described in earlier chapters.

The selection of a suitable waste management system will be different for the small farm feedlots and food processing operations than for the larger feedlots and food processing operations. The farmer-feeder and small processing operation may be better able to return their wastes to land. The large operations may not have the cropland available and may have to acquire it as an added cost or operate more expensive waste management systems.

A major consideration for possible unit processes is their simplicity and reliability. Animal and food processers are in the business of producing food. Waste management considerations are not among their highest priority items. Treatment and disposal processes for these operations should be as foolproof as possible, able to be operated and maintained by individuals not trained in waste management, should be capable of handling varying waste loads, and should be consistent with equipment normally found in these operations. The simpler a process is to operate and control, the more likely it is to operate successfully and continuously. Regardless of how efficient a process may be when operating under rigidly controlled conditions, if it is sensitive to minor changes in waste characteristics, or if it requires constant expert control and supervision, it is not likely to operate as intended under field conditions.

While there are many unit processes that may be used with agricultural wastes, a number of these have been developed and utilized with municipal and industrial wastes. Extensive experience using some of these processes with agricultural wastes is minimal. However, the fundamentals of the processes permit their adaptation to a variety of agricultural wastes. It is important to properly evaluate the characteristics of agricultural wastes when deciding on feasible unit processes. Lack of such consideration will result in unit processes that fail to perform as expected. The literature holds examples of a number of such situations.

One of the more difficult factors to determine is the cost of specific unit processes. Many potentially feasible agricultural waste treatment processes are in the laboratory and development stage. Laboratory results have the disadvantage that they may not reflect actual conditions and may not duplicate the practical field scale environment. Pilot plants can demonstrate the feasibility of specific processes but accurate costs can be obtained only from full-scale treatment studies.

Local conditions and circumstances will determine the process or processes that will be selected for a given operation. Full-scale plants and processes will have to be constructed and operated for a period of time before sound operational feasibility and cost estimates can be obtained. In the absence of such data, the limitations of feasibility and cost projections must be recognized.

It is at the full-scale level that real engineering and operational problems are found. The talents and knowledge of the waste management engineer are very important in devising the proper waste management system. The transformation of prior applications of the processes and of laboratory and pilot plant data into workable engineering designs is difficult and should not be taken lightly. The success of the system depends on how the engineer selects and combines the unit processes, how he selects and specifies the performance of equipment, and how well the engineering design details are completed. In the waste management area, the development of adequate engineering design knowledge is a real gap in the development and application of new and better methods for treatment and disposal of agricultural wastes. The combined talents of a sanitary engineer, a soil scientist, an animal scientist, and an economist are useful in developing optimum waste management systems.

The many unit processes available, the range of efficiencies over which they operate, the different ways they can be combined, and the need to include capital and operating cost factors make it apparent that exhaustive search of all possible alternatives to meet the treatment and disposal objectives at a minimum cost is extremely difficult by normal design procedures. Engineers use their experience and subjective judgement to reduce the number of potential alternatives for detailed consideration. By doing so, especially with agricultural wastes where practical results are not plentiful and many nonconventional options are available, equally acceptable or better solutions may be overlooked.

Within recent decades, the techniques of operations research and systems analysis have been applied to waste treatment problems. These techniques were initially used to solve military transportation and logistics problems and have been refined for many other problems. Since the early 1960's the techniques have been applied to water resources, water quality, and waste management problems. These techniques enhance the designers ability to objectively examine a larger range of alternative treatment systems while reducing the cost and duration of the preliminary selection procedure. Another advantage of this approach is that it permits evaluation of tradeoffs of cost and effectiveness of process changes as well as choices of unit processes in arriving at alternative solutions.

A typical methodology for at least initial evaluation of feasible unit process combinations involves development of a computer program which includes the desired unit processes and which yields the cost of treatment alternatives for given influent and desired effluent qualities. In essence the program simulates the alternatives that would be open to a design engineer in selecting a treatment system utilizing a variety of unit processes. Each treatment process has its own operating characteristics and economic relationship which are or have been estimated by laboratory and field investigations.

This approach, while not having been used extensively for agricultural wastes, has been used to make preliminary selections of waste treatment systems for cannery wastes (8), for phosphorus removal alternatives for duck waste water (9), and for the storing, handling, and spreading of hog manure (10). Components of these techniques are linear programming, dynamic programming, and sensitivity analysis. A series of computational design schemes are available to assist in the development of appropriate liquid waste management systems (11). While these schemes were developed for municipal waste waters, and the basic assumptions in the schemes will have to be altered for use with agricultural wastes, these schemes may be useful in delineating the estimated costs and effectiveness of proposed unit process systems.

For agricultural wastes, many typical management systems may be considered. Depending upon the desired complexity of the preliminary systems design and the type of waste to be treated, a number of basic components can be incorporated as in Fig. 13.3. Aspects of each of these components are outlined in Fig. 13.2. The preliminary program to select feasible management alternatives is concerned with minimizing the costs of treating and disposing of quantities of BOD, solids, and nutrients by several treatment and disposal stages. After the more feasible treatment and disposal combinations are delineated, more detailed investigation and computational programs can be developed for final design decisions.

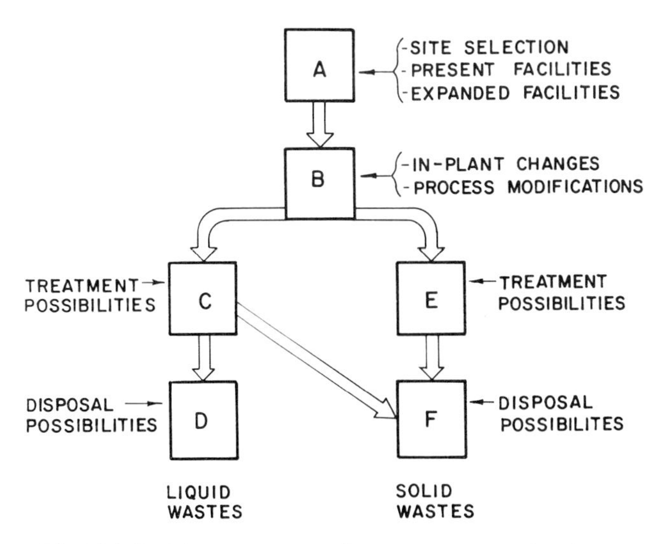

Fig. 13.3. Decision components of a waste management system.

Knowledge of many interactions are important to the success of the above approach. A number of these, such as rates of solids production and destruction in various unit processes and operating and capital costs of specific processes, are either not known or are known only as general terms. The lack of precise interactions does not invalidate the approach but rather directs attention to the types of interactions that do occur and the type of research and investigation which are necessary to more rapidly advance our knowledge and solutions. One of the strengths of the approach is that even if the precise interactions are not known, assumptions that span the range of reality can be made and their effect on the results can be observed. In this manner, factors to which the results are sensitive can be delineated and investigations initiated to more clearly define these factors. More effective use of research and developmental monies could result.

Residual solids and sludge handling is an important factor in the selection of unit processes. Almost all treatment processes are solids concentration processes in which either the clarified or purified liquid is released or the gases from a drying or dewatering step are released. Sludge handling and disposal frequently is an inadequately considered factor in many unit processes. The ease and cost of solids disposal are factors that may have the greatest effect on the cost of waste management systems. Treatment and handling processes for residual solids must be selected on the basis of the character of the sludges they produce.

In locations where adequate land is available, almost any kind of sludge and solids can be disposed of by hauling or pumping to these sites. Land loading rates should be consistent with good practice guidelines and the soil assimilative capacity. All of the other methods for solids disposal involve a dewatering or further concentrating step. While there are many ways to treat the liquid wastes to the desired effluent quality, there are few ways to satisfactorily and economically dewater sludge and dispose of residual solids. Adequate consideration should be given to the quantity and quality of solids when selecting unit processes and waste management systems.

Joint Industrial-Municipal Cooperation

Discharge of untreated or partially treated wastes to a municipal treatment systems can be a possibility for both municipalities and a number of agricultural operations. The advantages include assurance that the wastes are adequately treated or disposed of, the community retains the economic base of the industry, and the industry avoids increased capital investment in waste treatment facilities and avoids the need for increased waste treatment personnal.

The details and charges regarding the use of a municipal sewerage system by an agricultural or industrial operation generally are covered by the sewer ordinances and sewer service charges of the municipality. Most municipalities have at least a general provision in their waste ordinances stipulating that harmful and objectionable materials may not be discharged to their sewerage systems. A number of items can be specifically prohibited. Examples of prohibited wastes are liquids above a certain temperature, liquids having a pH value outside a specific range, improperly shredded garbage, oils, fats, and greases above a specific concentration, any material that can cause obstructions in the sewers such as ashes, feathers, sawdust, food processing bulk solids, paunch manure, etc., waste toxic to the biological processes used to treat the wastes, and radioactive wastes. The logic is that an industrial operation should not place any waste into the sewers that will cause problems in the sewers, with subsequent treatment processes, or will not be removed by these processes. The burden of proof that the waste will not be detrimental to the system and can be removed by conventional treatment processes rests with the industry. In addition, many municipalities will not allow, or allow only with special permission and surcharges, the discharge of wastes that have abnormal concentrations of BOD or suspended solids. Guidelines for the preparation of specific ordinances are available (12,13).

Although some municipalities have detailed restrictions and enforcement on what can be discharged to their sewerage systems, the majority accept industrial wastes without examination, pretreatment, or charge. The cost of the industrial waste treatment is then passed to all the users. The future will see many more municipalities charging industry for accepting their wastes.

The need of complete treatment of wastes to meet more rigid effluent standards, rising costs, and the public demand for better service has led many communities to levy special charges on industrial wastes discharged to their sewerage systems. The logic behind these surcharges is that the cost of providing sewerage service should be paid directly by the users as the most equitable means of charging for the service. A number of typical service charges and examples of their use are available (14).

Communities will utilize surcharges for industrial wastes in an increasingly larger number of situations. The Presidential 1971 Environmental Message proposed that "... municipalities receiving Federal assistance in constructing treatment facilities be required to recover from industrial users the portion of project costs allocatable to treatment of their wastes." This proposal was implemented in the 1972 Amendments to the Federal Water Pollution Control Act (PL 92-500) which indicated that federal grants for construction of municipal treatment facilities shall not be approved until the applicant adopts a system of charges to assure that each recipient of waste treatment services will pay its proportionate share of the costs of operation, maintenance, and replacement of the waste treatment services. Industrial users of the treatment facilities are to pay proportionate costs based upon the characteristics of the wastes discharged to municipal systems.

A properly designed service charge will distribute costs more closely related to the service provided than any other way of raising revenue for treatment facilities. Effectively administered user charges can improve the management of industrial wastes. Charges related to the cost of treating wastes, based at least on volume and strength, can create an incentive for industry to pretreat, change processes, and manage their wastes more effectively. When the cost of the service is directly borne by the user, he has an interest in seeing that the treatment system is effectively planned and managed. This is useful in bringing local pressure to minimize costs through a reduction in unnecessary excess capacity, substitution of operation costs for construction costs where they are lower, efficient plant operation, staging of construction where possible, and use of lower cost technologies if applicable. User charges provide a relatively stable source of revenue to meet waste treatment costs which allow for a businesslike management of the sewerage system and provide for an orderly operation, maintenance, replacement, and expansion of the system.

The type and magnitude of the industrial waste surcharge will affect the decision of an agricultural operation to discharge its liquid wastes to a municipal system. There are several formulas in existence for industrial waste surcharges. The constant rate formula is a common one and involves charging on the basis of a production unit such as water use, number of employees, or quantity of product produced. These formulas are simple to administer but are only a crude way of taking differences in the waste strength into account. Because testing of the effluent from an operation usually is not done with this formula, the charge is not related directly to the quality of the waste sent to the sewerage system and therefore is not likely to influence the industry to reduce its waste quantity or quality. For small municipalities where the gain in accuracy from waste testing and surveillance is small compared to the expense involved, this type of charge may be satisfactory.

A quantity-quality formula can be used where the expense of more detailed waste testing and enforcement can be justified. There are over 150 municipalities using this type of a surcharge and the number is increasing. The formula indicates the extra charge which is to be made when the formula is applied to wastes of above average characteristics. Surcharges based on this formula take into account both the volume and pollutional quality of the waste.

A typical surcharge of this type would be

$$C = V[Y_1 + Y_2(B - B_n) + Y_3(S - S_n) + \cdots Y_n(N - N_n)] \qquad (13.1)$$

Where C = charge; V = volume; Y_1, Y_2, Y_3, Y_n = surcharge rates for specific items; B, S, N = actual concentrations of pollutants causing increased costs, and B_n, S_n, N_n = normal concentrations of pollutants. The waste characteristics usually included in surcharge calculations are volume, BOD, and suspended solids. Other characteristics such as nutrients or chlorine demand can be included as appropriate. The definition of "normal" concentrations varies widely among municipalities.

The surcharge rates also will vary among communities, reflecting the type of treatment and the effect a characteristic will have on a treatment process. The liquid volume affects both collection and treatment costs, BOD affects primary, secondary, and tertiary treatment costs as well as sludge disposal costs, and suspended solids affect solids handling, treatment, and disposal costs. Both capital and operating costs should be reflected in the surcharge rates.

The advantage of the quantity-quality formula is that it is an equitable charge since it takes the characteristics of the industrial effluent and the costs associated with treating the waste directly into account. This can be important when one considers that in a small community, a continuous or seasonal agricultural operation can contribute a large percentage of the volumetric or pollutional load to the community treatment plant. Another advantage is that it provides an inducement for the agricultural or industrial operation to reduce the quantity and quality of its waste. The chief disadvantage is that the administration of the surcharge requires continued monitoring and sampling.

Decisions to discharge industrial wastes to a municipal system are determined by the proximity to the system, the quality and quantity of the wastes, and the advantages of no, partial, or complete treatment by the industry. Because most agricultural operations are located in rural areas, the discharge of agricultural wastes to a municipal system is constrained to those operations that can exist in urban or suburban areas. These are limited to food processing facilities such as meat-packing, canning, freezing, and milk processing operations. Animal production facilities such as confined feeding operations produce low volume, high strength wastes that exist as slurries, semisolids, or solid matter. Wastes of this nature are unlikely to be accepted by a municipality.

Some opportunities also exist for agricultural–municipal cooperation in the disposal of solid wastes. Examples are land fills used for dead poultry and animals and the use of animal and food processing wastes in municipal composting operations. Relationships generally are less formal than with acceptance of liquid wastes. Acceptance or denial of agricultural solid wastes by a municipality generally is based on whether the wastes can be handled with little difficulty. The agricultural producer has the responsibility of transporting the solid wastes to the disposal sites. The feasibility of such transport is a function of the proximity of the disposal site and the quality and quantity of the solid wastes.

Animal Wastes

General

Animal wastes are defecated in semisolid form and there is considerable logic in treating and disposing of these wastes as a semisolid or solid material rather than adding water for liquification before treatment and disposal. The latter approach requires the treatment of both the animal wastes and the contaminated water. However, because of inadequacies with present handling equipment and difficulties in obtaining the necessary labor to handle these wastes as a solid material, liquid waste handling, treatment, and disposal methods are common at many animal production facilities. In such cases treated or untreated animal waste slurries or solids separated from the liquid animal wastes will represent the animal solid waste media requiring disposal.

One of the first animal waste management decisions is to determine whether a liquid or solid waste system is to be used. No clear-cut advantages occur with either system and the decision will be related to the ability of each system to fit into an existing or new facility, labor costs, and ultimate disposal possibilities and constraints. When considering animal solid waste disposal, it is important to obtain the initial wastes in as dry a condition as possible.

Any proposed manure management system must accomplish more than simply relocating a nuisance to await development of further problems and complaints. Year-round adequacy must be provided. Nearly all present means of manure disposal are satisfactory during parts of the year but may be troublesome at other times. The management system must satisfactorily treat and dispose of the wastes without jeopardizing the health or well-being of the producer, the animals, or the public.

More emphasis must be given to the planning of animal waste management systems before new animal production projects are started. It is necessary to adequately estimate the quantities produced and to link these to capital and operating costs of the waste system and to the overall costs of running the facility. To view either the animal wastes or the production unit in isolation is self-deceptive.

In view of the constraints placed upon the discharge of wastes to surface waters, it does not seem logical to approach animal waste management by attempting to produce an effluent suitable for such discharge. In addition to the use of land as a disposal site, manure can have a positive effect on certain soils. Where subsoils have been exposed by erosion or farming, manure is useful to improve soil aeration, increase water penetration, and reduce erosion. Opportunities also should be explored for possible reclamation, processing, and recycling of liquid and solid components for reuse in production operation and in other situations.

Animal production units fall into two main groups, (a) the normal unit where disposal on land will be the prime objective with or without the use of some waste treatment and (b) a smaller number of operations which are single purpose, large scale enterprises. Many of the latter are maintained on land that is inadequate for waste disposal purposes. These units will require special treatment and disposal methods and larger capital commitments for waste management. The large livestock producer is in the animal rather than the crop farming business. He relies on commercially available feeds for a considerable portion if not all of his feeding requirements and may have a minimum interest in utilization of the resultant manure in crop production.

The four general methods of manure management are dry handling—excreta handled dry or mixed with bedding, distributed on land either daily or after storage, and may be used as part of a recovery and reuse program; semidry—a semisolid slurry disposed of on the land by spreading; semiliquid–mixed with water for distribution on the land by tankers; and liquid manure—mixed with water for liquid irrigation, for treatment and disposal, or for reuse of portions of the liquid and/or solid fraction. Odors can be common with the last three methods.

Components of animal waste management systems include modification at point of production, handling and transport, storage, processing, treatment, recovery and utilization, and disposal. Some components may be combined or omitted depending upon the purpose of the system. It is also important that the waste management system fit into existing production systems, be consistent with available labor, and be as mechanized as possible.

Feedlots

Where animals are housed in the open, such as feedlots, runoff from these areas has the potential for serious water pollution problems. The characteristics of feedlot runoff and resultant water quality problems have been discussed in earlier chapters. The pollution from open feedlots will be related to the rainfall–runoff relationships that exist in a specific area. Pollution caused by runoff is reduced when animals are completely housed as is the case in many hog and dairy, and in most poultry operations.

The major considerations affecting the magnitude of the feedlot runoff problem and the possibilities for abatement include the fact that the wastes are more diffuse, generally nonpoint in origin, and not amenable to conventional waste treatment technology used for defined and controlled point source discharges, that feedlot runoff is related to rainfall and snow melt, thus the predictability of waste flow in terms of quantity, quality, and duration is difficult, and that the characteristics of feedlot runoff is variable due to precipitation and to differences in the wastes from different animal classes. These and other geographical, cultural, and socioeconomic factors illustrate that feedlot waste management considerations are complex.

Many factors can be incorporated into suitable waste management systems for feedlots such as site selection and location, lot management, runoff management, waste disposal, and the type of feedlot. At least three types of feedlots can be used for beef cattle production: unpaved lots, paved lots, and completely housed lots. Each has its advantages and cost relationships. Paved lots permit greater numbers of animals per acre and greater mechanization of feed distribution but have higher labor costs to maintain clean cattle (15). Complete confinement reduces runoff problems to a minimum and permits a maximum number of animals per acre. Partially enclosed units, open to the atmosphere on at least one side, are preferred at this time. Environmentally controlled enclosed units are useful only under extreme climatic conditions which affect cattle performance. A review of the alternatives for cattle feedlot construction (15) concluded that confined, housed feeding will become more popular with beef producers as labor shortages and pollution problems increase, that there is little difference between the overall materials cost of paved and nonpaved feedlots, and that the cost of a structure is about 50% of the total materials cost of a confined housed feeding unit.

Site selection and feedlot location can reduce the potential of water pollution from feedlots materially. The number of animals and surface of the lot will determine the area of the lot that is required. A site should be chosen that is away from a stream or waterway. Animals in confined feedlots should not have free access to a stream since this will permit a natural stream to run through the lot and be the receptor of feedlot runoff. Drainage of feedlots through adjacent property, road ditches, or directly into natural streams or lakes should not be permitted. Land area should be available so that drainage from the lots can be retained on property owned by the feedlot.

Drainage from the feedlots will be affected by the slope of the lots. Slopes of 4–6% will keep the lots drier. Feedlot runoff contains greater pollutants when the lot has been permitted to remain wet. Where inadequate slopes are available, internal drainage and manure mounds should be provided. New lots should be properly drained during construction. Snow buildup should be avoided since it can increase runoff and pollution potential. Animal shelters should be placed where snow buildup can be minimized.

Odor nuisances can be minimized by locating feedlots downwind from nearby towns and residences. Buffer zones of trees between the feedlot and adjacent homes will help decrease odor problems. In the drier climates, such as in the southwest, dust problems occur. Natural and man-made windbreaks may help this problem. Many of the gross water and air pollution problems from feedlots can be minimized by judicious selection of feedlot sites.

Lot management includes items such as number of animals per lot area, frequency of manure cleaning, and diversion of excess water. More animals per lot will increase the manure accumulation and the need of manure cleaning. On hard surfaced lots, there is less opportunity for precipitation to percolate to the soil. Suitable internal feedlot drainage is important in lot management.

The quantity of manure is increased when high roughage rations are used or if bedding is added. Manure on the feedlot surface will protect the soil from erosion and can reduce the overland flow velocity thus decreasing the particulates in the runoff. In most areas, manure removal once or twice a year or whenever a pen of cattle is marketed is satisfactory.

Increasing the feedlot area will increase the quantity of runoff generally in direct relation to the increase in area since runoff is a function of the precipitation falling on the lot or of the water flow through the lot. A critical feedlot management item is to divert all water outside the feedlot from flowing through the lot. All water from adjacent land and roof drains should be diverted to avoid contact with the feedlot wastes. Diversion of extraneous water provides for a drier environment for the animals, possibly reduced odor, and a marked decrease in the volume of feedlot runoff to be managed. Even with proper site selection and lot management some runoff will result and will require suitable management. Runoff control from feedlots is more difficult in the humid climates of the Midwest than in the drier ones of the Southwest.

For the foreseeable future, the expense of using conventional waste treatment techniques for feedlot runoff and wastes will be unnecessary because land is generally available for runoff control and waste disposal. The most practical approach for feedlot runoff control appears to be a combination of protective dikes and levees to prevent the entry of rainfall from outside the feedlot area, plus collecting dikes, levees, and holding ponds to collect the rainwater falling directly on the lot (Fig. 13.4). After collection and retention, some acceptable means of disposing of the liquid and solid material such as fertilization and irrigation on available pasture and crop land should be employed. Overflows from the retention ponds may be allowed only during times of rainfall that exceed the design criteria. The size of retention ponds is based on criteria of a specific maximum probable rainfall frequency for the local area. The rainfall and frequency will vary with geography and rainfall patterns. Criteria such as the maximum 7-day, 10-year rainfall or the maximum probable 24-hour, 25-year rainfall frequency have been used.

Retention ponds will minimize the pollution caused by runoff from animal production units housed in the open. In dry areas, they may be the only treatment needed. Where retention ponds are not satisfactory to protect the receiving waters, a combination of lagoons, oxidation ponds, and aerated systems may be possible. With proper operation, runoff retention ponds are a very feasible waste management technique for confined, unenclosed feedlots.

Kansas was one of the first states in the country which adopted regulations for the control of water pollution from animal feedlot operations. The regulations stipulated that the operation of any newly proposed confined feeding operation using waste ponds or lagoons must register with Kansas State Department of Health prior to construction and operation of the facilities. If in the judgement of the Department a confined feeding operation did not constitute a water pollution problem, provision for water pollution control facilities was not required. If such a confined feeding operation did constitute a water pollution potential or if water pollution occurred as a result of the operation, the operator must provide water pollution control facilities constructed in accordance with plans and specifications approved by the Department. A permit is required for water pollu-

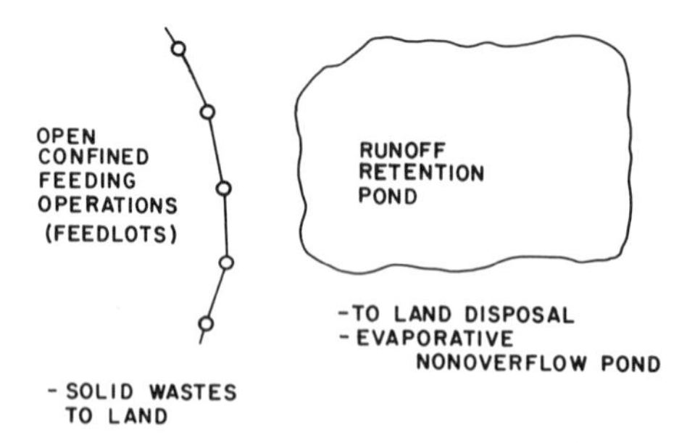

Fig. 13.4. Runoff detention ponds and liquid and solid disposal for beef feedlots.

tion control facilities and is granted subject to continual and satisfactory operation of the pollution control facilities.

The minimum water pollution control facilities for confined feeding of cattle, swine, and sheep were stated as retention ponds capable of containing three inches of surface runoff from the confined feeding area and all other waste contributing areas, The liquid and solids from the ponds are expected to be distributed on nearby land in a nonpollutional manner. Each confined feeding operation was to be reviewed and evaluated on its own merits. Modifications to these regulations have been considered or adopted in other states. The judicious application of these regulations has not hampered the growth of the feedlot industry and data are available to suggest that well-constructed and well-administered regulations can be a help to the industry and may enhance economic growth (16). Obsolete facilities or facilities in improper locations may not be able to survive but new and better facilities will replace them.

Grass terraces offer another opportunity for runoff control and can be incorporated with runoff retention ponds. Runoff flowing over grass or agricultural land prior to retention ponds can be expected to be reduced in volume by soil percolation and in pollutional characteristics by solids removal on the land and by oxidation of some of the organic matter in the stubble or grass.

Runoff control systems using a settling channel, porous dams, and a detention pond for holding the liquid portion of the runoff have been investigated (17). The objective of these systems was to remove settleable solids from the runoff. Disposal of the settled solids was on adjacent fields using suitable equipment and the liquids in the retention pond were pumped onto adjacent crop land. The settling channels were successful with the heavier particles such as undigested grain settling out rapidly. The porous dams consisted of crushed rock supported by wood planking. Reducing the velocity of runoff flow was a low maintenance method of controlling the settleable solids content of runoff reaching a liquid detention pond. In a continuous flow system, up to 50% of the total solids in feedlot runoff were removed by a series of three porous dams. This approach is feasible for separating discrete particulate material from animal waste runoff. It may be less successful with wastes that contain fine particulate matter such as dairy or poultry manure slurries. Such fine matter may be difficult to remove from the settling areas and may plug the porous dams.

Disposal of animal wastes on land is the method of choice. This approach should not be considered a panacea since surface runoff from lands receiving the animal wastes may contain a higher pollutional load than that of land that did not receive the wastes. The disposal of accumulated feedlot wastes on land should be done with care. Incorporation of the solid waste after field spreading will reduce subsequent runoff and odors. Wastes should not be placed on snow or frozen ground if runoff and thawing may cause pollution. Waste spreading areas should not be adjacent to streams or other surface waters. Crop and grassland irrigation systems can be used for the accumulated liquid wastes with subsequent runoff being controlled if applications are excessive. A portable irrigation system has the advantages of moderate cost and flexibility. A large area can be irrigated with a small system if labor is available. Changing of disposal areas can be accomplished if necessary for expansion or changes in land use. Land requirements will vary with soil types and climatic conditions.

The various alternatives for feedlot runoff management have been evaluated with regard to both daily operating costs per animal and initial investment costs for each component in the system (18). Year-round occupancy of the feedlot was assumed. The results indicated that open feedlots have lower investment and operating costs than confinement building systems although they have a higher potential for air and water pollution. Housing and feedlot facility costs were about 80% of the total waste management costs. A manure irrigation system for handling slurry and flushed wastes cost about one-half as much as other waste handling methods such as solids handling equipment. The report concluded that a cold confinement building with a deep pit and a slurry handling system without soil injection had the lowest daily operating costs for confinement building systems.

The potential for impairment of groundwater quality around animal production operations should receive close scrutiny since once the groundwater becomes contaminated, it is difficult to return it to its original condition or to use it for potable purposes. Standards for maintenance of groundwater quality have not been set as yet but most states have provisions in their statutes to protect the groundwater and maintain its quality. These provisions become of concern when the land disposal of agricultural wastes is practiced.

Pollution of groundwater can also result from the direct movement of runoff into inadequately cased or poorly protected wells located in the immediate vicinity of a feedlot. Feedlot drainage should be away from all wells and wells should not be drilled in geological formations likely to intercept feedlot runoff. Contamination of groundwater under feedlots is determined by the number of animals per surface area, soil texture and structure, depth to water table, and frequency of manure removal. Coarse-textured sandy soils permit greater movement of pollutants to groundwater than do fine-textured clay soils. Infrequent manure removal retards the movement of contaminants to the groundwater. Factors that increase the quantity of water moving through the soil will increase the movement of pollutants to groundwater.

Enclosed Confined Operations

In contrast to beef feedlots, most other animal production utilize enclosed confined facilities. Examples are hogs, dairy cattle, broilers, and laying hens. The facilities represent point sources of wastes which are less subject to the probabilistic nature of hydrologic conditions. A greater variety of treatment processes are possible. With few exceptions, the ultimate disposal of the untreated or treated wastes will be on the land, although it may be possible to produce an acceptable effluent for discharge to surface waters with dilute wastes from some animal production facilities. Less operational flexibility of the treatment facility is possible when discharge to surface waters is practiced since specific effluent requirements are established by the regulatory agencies. Greater operational flexibility is possible when discharge to the land is practiced since release of the wastes can be matched to crop needs, to the soil assimilative capacity, and to the work and seasonal schedule of the producer.

While many treatment processes are possible only a few are in general use. For liquid wastes these are the aerobic oxidation pond, the aerated lagoon, the oxidation ditch, the rotating biological contactor, and the anaerobic lagoon. For concentrated waste slurries it is storage and intermittent cleaning with disposal on the land, and for drier wastes it is land disposal, refeeding, composting, and drying for use as an agricultural or home garden soil conditioner.

A waste management system for enclosed confined facilities will deal with wastes having a consistency of "as defecated," as a liquid, or as a "dry" material. The most feasible approach will depend upon the nature of the waste, the ultimate disposal method, and labor requirements and availability. Justification of a liquid manure system frequently is done on the basis of reduced labor demand, and the possible return of the nutrients in the manure to the land. To obtain the benefits of improved labor distribution, reasonably long storage periods and a planned disposal schedule must be available and maintained. The long storage can be under both anaerobic and aerobic conditions.

Farm waste storage is not a controlled anaerobic digestion system since the microbial reactions take place in an uncontrolled manner. Such storage is storage under anaerobic conditions that cause partial decomposition and odor. The odor nuisance is caused more by the movement of the stored slurry, i.e., pumping and disposal, than by the fact of its storage.

Daily manure cleaning and spreading on the land or daily manure cleaning, manure storage, and intermittent disposal on the land are common manure handling operations at dairy farms. The simplest form of waste handling from a free-stall operation consists of a tractor with a rear or front mounted blade or bucket and a ramp for pushing the manure into a spreader. This system requires daily spreading. Difficulties in disposing of the wastes on the land result when the land is wet or frozen or when crops are being grown on the land. Labor is distributed throughout the year. A holding tank permits flexibility during periods when the wastes cannot be disposed of on the land. In northern climates, holding capacity of from three to five months is suggested. These holding tanks are not aerated and odorous gases and compounds will result because of the anaerobic conditions in the tanks.

Another development has been the use of manure stackers—barn cleaner elevators up to 60 ft in length, running on a semicircular track—which are used to stack the dairy manure in a concrete storage area during the winter months. This system does not require a large capital input and is flexible. The investment in this system is intermediate between daily hauling and liquid systems. Advantages include storage during inclement weather and when fields are unavailable. A tractor with a manure loader is required. The stacking area should be fenced off and drainage and runoff from the area kept from nearby streams. The area can be a source of nuisance insects and may not be the best system for summer storage.

The primary wastes associated with concentrated poultry operations consist of three categories, (a) manure, (b) litter, and (c) other wastes such as dead birds. The most common methods of disposing of the other wastes are landfill, incineration, and rendering. The latter is feasible in highly concentrated poultry production areas. Almost all of the manure from broiler production is dropped on litter which serves to absorb moisture from the manure and assist in drying. Many materials are used as litter such as sawdust, wood shavings, peanut hulls, chopped straw, and bark and chips. Litter is one of the most significant wastes from commercial turkey operations since the first eight weeks or so of the turkeys' lives are spent on litter. The commercial egg production in the United States contains no litter. Consequently the wastes from egg laying operations and the remainder of a turkey production operation consists almost entirely of manure.

Various methods have been developed to handle, treat, and dispose of the wastes from commercial poultry operations. In most egg laying facilities the wastes are permitted to accumulate, in either a slurry or a drier form, in pits under the birds for periods of one or more years before cleaning. The odors from these buildings and from cleaning operations have been a source of nuisance complaints when the facilities are near residential areas. The major method of treatment and disposal of poultry wastes has been land spreading. Lagooning, composting, and drying have been used with various degrees of success. Manure with litter generally is disposed of on the land.

Swine wastes are characterized by having a fluid nature, a noticeable odor even when fresh, and a rapid decomposition rate. The pollution potential of swine waste is associated with surface runoff, odors, and land disposal. Daily removal of swine manure is not common in open and partially confined lots. The manure is removed as convenient for the operator and disposed of on nearby land. In total confinement systems either a solid floor, a slotted floor, or a gutter arrangement is used to accommodate manure removal.

In a solid floor operation, the wastes are scraped and removed to a storage unit or a lagoon. The slotted floor system eliminates manure handling in the production areas. The manure, spilled food, and water fall through a slotted floor into a storage unit. Manure stored under the slotted floor will become anaerobic unless the unit is aerated such as with an in-house oxidation ditch. The gutter system collects the wastes by scraping or flushing across the floor and into a gutter. The wastes then are moved hydraulically to a lagoon, storage tanks, or aerated system. All swine waste handling systems require regular removal and/or careful storage of manure to avoid odors, flies, and surface water and groundwater pollution. The ultimate disposal of the stored or aerated manure should be on the land. Diversion of runoff from all surface areas containing manure is desirable.

Odors have caused problems and when the contents of anaerobic storage units have been agitated. The released gases have caused the death of some hogs and the unconsciousness of a few humans. Air movement in these facilities should be directed from the animal area to minimize these problems. Aeration systems to prevent the generation of odors and toxic gases can eliminate these situations.

Land disposal of stored wastes should be done by incorporating the wastes in the soil as soon as possible. Anaerobic, stored waste should not be disposed of in close proximity to residences, motels, or other places frequented by humans especially during seasons of outdoor recreation. All liquid or slurry wastes should be transported in liquid tight containers maintained in a clean and neat condition. The advantage of the manure storage and land disposal system is its simplicity and low cost. It is also one of the traditional methods of waste handling and disposal and as such has been accepted by the agricultural community.

A system that incorporates an aerobic unit for a storage tank or following an anaerobic lagoon can minimize odor problems and obtain higher degrees of effluent quality. Two feasible aerobic units that can follow an anaerobic lagoon or can be used separately to treat liquid manures are an oxidation pond and an aerated lagoon. The bottom of oxidation ponds are sealed to prevent groundwater pollution and to maintain proper water levels. Some seepage does occur. The amount permitted by various states range from 0.1 to 0.25 inches/day (36 to 90 inches/year). The waste inflow from the confinement operations may not be adequate to overcome permitted seepage losses. An additional source of water would be needed to maintain proper water depth.

Oxidation ponds for most animal manures could be nonoverflowing ponds which would have distinct advantages for pollution control. The large land area needed and the need for make-up water can be disadvantageous. For more liquid wastes such as milking parlor, poultry processing, milk processing, and duck wastes, the ponds would have an overflow which would be required to meet local or state discharge requirements.

The three types of aeration systems that have been used with animal wastes are the aerated lagoon, the oxidation ditch, and the rotating biological contactor. The design parameters of these systems for animal wastes are not as well understood as for other wastes. Knowledge is available to adequately estimate the size of the aerobic units. Experience with aerobic treatment units for liquid animal wastes has shown that these processes can have high BOD removal efficiencies. The BOD in the effluent from the units is almost entirely due to the oxygen demand of the solids in the effluent and to nitrification. Because of the high concentration of original waste material, the resultant effluent will still have a significant residual total BOD. As such the effluent is more suitable for land disposal than for disposal to a receiving stream.

One of the more interesting challenges is to design these aerobic systems to meet a specific effluent standard which is governed by land disposal requirements rather than by stream criteria. Because land disposal is an integral part of animal waste systems, conventional solids removal systems need not follow these aerobic systems.

Certain animal wastes, such as duck production waste waters, are a liquid rather than a slurry and can be treated by conventional liquid waste treatment methods. As with many industrial wastes, the conventional methods have to be adapted to the characteristics of the specific waste. Duck production waste water has the characteristics of strong domestic sewage and has been successfully treated by a combination of aerated lagoons and settling ponds followed by chlorination (19).

Separation at the source is both a feasible approach and a waste management philosophy. The treatment of any waste is a direct function of both the strength and volume of the wastes. Since animal wastes are generated as a semisolid, it is logical to separate these wastes from wash or clean-up waters at the source, and to treat the semisolid wastes and the liquids in separate units. The liquids can be treated in liquid systems while the solids can be disposed of directly on the land or by systems to handle semisolid such as drying, composting, and incineration.

Separation of specific wastes at the source offers the opportunity for treatment and disposal systems that are the best for each type of waste. Such separation is done in modern milking parlors. Modifications of this system include the use of water flushing of animal wastes with separation of the settleable fraction prior to treating the liquid fraction. The separated solids would be disposed of on the land.

Drying and composting are feasible with animal wastes. Poultry wastes, because of its lower initial moisture content, requires less energy to dry than will wastes containing a higher water content. In-house drying of poultry wastes by greater air circulation, by the addition of heat, or by both offers the possibility of least cost drying systems for these wastes. The success of drying and composting requires a market or ultimate disposal for the product.

One of the alternatives for animal producers who wish to practice adequate waste management is to handle and dispose of the animal wastes in a dry or semisolid manner rather than with a water carriage system that will require a high degree of treatment. This alternative will require greater investigation and knowledge about methods of handling and transporting of concentrated animal wastes. Animal wastes are defecated as a semisolid having a moisture content ranging from 75% for poultry to about 80–85% for hog and cattle manure. Liquid manure handling and disposal techniques have been used because of the costs and difficulties involved in dry waste handling. However, drier systems may become more feasible as animal production operations increase in size, environmental quality constraints increase, and as markets and uses for the dried and composted products develop.

Liquid and slurry systems are available to minimize the gross pollution that can be caused when untreated animal wastes are released to the environment. Only a small fraction of the total solids will be reduced in a biological unit treating animal wastes. Unless a high evaporation rate occurs, these units will not reduce the volume of material to be handled. An aerated biological treatment unit capable of maintaining a dissolved oxygen residual will eliminate the odor problem and reduce the oxygen demand of the effluent. The residual solids and soluble salts represent the final disposal problem from such units.

Current systems may remove the organic oxygen-demanding material but will not remove large amounts of inorganic nutrients. These latter items, especially nitrogen and phosphorus can affect the nutrient and oxygen balance of a stream. Nitrates in animal waste waters are unlikely to occur except from a long detention time aerobic unit. The majority of the nitrogen in animal waste waters is as ammonia or organic nitrogen. The current approaches to nitrogen and phosphorus removal for animal waste waters have been discussed in Chapters 10, 11, and 12.

Odor Control

Odor nuisance from animal production units can be caused by odors from fans of buildings used for confinement animal production, wastes which are spread on land, emptying of waste storage tanks, manure heaps near roads and houses, uncontrolled anaerobic lagoons either for storage or for treatment, inefficient or improperly maintained incineration or drying plants, and silage operations. Management considerations for odor control with animal wastes include moving and disposing of the manure when it is as fresh as possible, use of aeration methods, and drying the wastes either in-place or with heat before odors are generated.

The odors from animal waste systems primarily are caused by anaerobic conditions occurring during waste storage. Odors also originate from the animals but if the odors resulting from the handling, storage, and disposal of anaerobic wastes are eliminated, few odor nuisances will result.

A number of commercial, proprietary chemicals have been reported to be effective in preventing and/or suppressing odors. An impartial evaluation of 44 commercial masking agents, counteractants, and deodorants found that masking agents and counteractants were the most effective odor control products, that deodorants were moderately effective, and that digestive deodorants were least effective (20). In laboratory and field tests, powdered digestive enzyme preparations and powdered yeast or bacteria starter cultures were not effective in controlling odors, especially where liquid manure handling was used. Such preparations have to compete with the enzymes and bacteria present in the wastes and massive dosages were found to be necessary before reasonable odor control could be obtained (20, 21).

Not all commercial chemicals are equally effective for odor control and care should be taken when making decisions to use these compounds. The use of an effective chemical may not be necessary each time accumulated anaerobic wastes are handled and spread on the land. Odor control additives during times of unfavorable atmospheric conditions and other emergency situations may be warranted.

Soil filtration of odor carrying gases has been shown to be successful for the removal of up to 200 mg/l of ammonia, 20 mg/l of hydrogen sulfide (21), and 775 mg/l of methyl mercaptan (22). The removal rates are dependent upon soil characteristics, temperature, moisture, and acclimation. Soil depths from 6 inches to 4 ft have been shown to be effective.

Many of the obvious cases of odor nuisances would have been minimized if the production facilities had been in locations less susceptible to air movement that can carry the odors to residential areas. When waste management systems that may produce odors are selected, proper site location will help minimize odor problems. Knowledge of wind movement, dispersal of the odors in the atmosphere, and the general meteorology of an area greatly can assist in planning for odor control at a site. Consideration of air drainage basins in developing land use plans can be helpful in minimizing odor nuisances.

Because the odors are produced microbially under anaerobic conditions, chemicals to inhibit the bacteria and/or oxidize the organic matter have been tried. Hydrated lime and chlorine have been used successfully to reduce hydrogen sulfide, ammonia, and methane production in hog and poultry wastes (23, 24). Potassium permanganate was found satisfactory to control odors in a cattle feedlot (25). Generally, large quantities of these chemicals must be used due to the chemical demand of the concentrated animal wastes. Little commercial use of this type of chemical control is practiced.

A positive method of odor control is to prevent their generation. Control methods of this nature involve modifying the environmental conditions under which the wastes are stored. Two processes are effective and can be incorporated in many animal waste systems: drying and aeration. In each case, anaerobic conditions are prevented.

Fresh animal wastes are rarely offensive. Drying of these wastes to conditions that prevent microbial action has been shown to be an effective odor control method. Natural air movement without added heat, increased air movement using fans or blowers, and the addition of heat and air movement have effectively controlled odors in poultry houses. Commercial dryers also have been used with animal wastes. Storage of the dried material under conditions that do not produce odors is essential. Poultry wastes which have been dried to a moisture content of approximately 30% or lower are less odorous than those not dried. The odor emitted and its intensity depends to a considerable extent on the temperature at which the material is stored and the temperature and humidity conditions at the time they are spread. Similar relationships can be expected for other animal wastes.

Aeration can be utilized with liquid systems to prevent odors by maintaining aerobic conditions. In addition, the liquid animal wastes are partially oxidized and the pollution potential of the resultant slurry is reduced. Nitrogen control by nitrification–denitrification can be possible in these systems. Mechanical surface aeration, using surface rotors as in an oxidation ditch or floating aerators as in an aerated lagoon can be used. Diffused aeration has been used in a few cases although the oxygen transfer relationships for these systems in concentrated animal waste slurries appear less favorable than for surface aerators.

The key to a successful aeration system for odor control is to supply oxygen at a rate equal to or greater than the oxygen demand. If the oxygen supplied is considerably less than the oxygen demand, odors may still result. The oxygen transfer relationships of possible aeration equipment is affected by the characteristics, solids content, and temperature of the wastes. Care should be taken in designing aeration systems for animal waste systems to insure adequate oxygen supply.

There is little information to indicate the minimum quantity of aeration per unit of animal waste that will prevent odor production. Traditional concepts of aerobic waste treatment emphasize the maintenance of residual dissolved oxygen concentrations of 1 to 2 mg/l. These concepts were established to produce a liquid effluent acceptable for discharge to receiving surface waters and may not be applicable to concentrated wastes where land disposal is practiced and odor control is the prime consideration. Ammonia can be released from aerated systems when nitrification does not occur. The discharge of the ammonia in a confined animal unit can be undesirable. An adequate oxygen supply and bacteria to maintain nitrifying conditions in the liquid will eliminate the ammonia release.

The greatest benefit of aeration as a method of odor control occurs when it is used from the time storage is first utilized. Sufficient aeration under these conditions will prevent odor production. Aeration of stored manure after it has been permitted to accumulate under anaerobic conditions will strip the existing odorous compounds from the wastes, with possible severe odor problems.

When odorous animal wastes result, proper application to the land can minimize odor problems. Spreading thinly creates less problems because the wastes dry much more rapidly than when spread in a thick layer. With proper management and suitable soil type, discing thinly spread waste into plowed or worked soil has alleviated odor problems. Soil injection of wastes or the use of the plow-furrow-cover method will minimize odor release during disposal.

Food Processing Wastes

The type of waste management suitable for a food processing operation will depend upon the characteristics of the wastes, the location of the operation, the degree of waste treatment required, and available points of discharge for the treated liquid wastes and for the collected solid wastes. As with almost all industrial waste situations, a unique waste management solution may exist for each operation.

One of the first approaches is to investigate the opportunities for in-field and in-plant modifications to reduce waste quality. Techniques to reduce the pollutional load from fruit and vegetable processing operations include: harvesting equipment that leaves more of the stems, leaves, and culls in the field; field washing and/or processing of the crop with the residue returned to the crop land; air and liquid separators to remove extraneous material from the crop; modification of peeling and pitting operations to use less water and waste less crop; recovery and reuse of the process water throughout the plant; and separation of waste process streams at the source for potential by-product utilization and separate waste treatment and disposal of wastes with different characteristics.

Two major waste streams contribute to waste water flows in a food processing plant: water used in food product processing and water used for cooling. Water conservation can be practiced by the use of low volume, high pressure sprays for plant cleanup, elimination of excess overflow from water supply tanks, use of automatic shut-off valves on water lines, use of mechanical conveyors to replace water flumes, and by investigations into the proper quantity of required water to use only the amount of water needed for a specific operation. The reuse of cooling water as product wash water is an example of a separation and reuse possibility.

Other modifications to food processing operations for waste product separation become apparent when materials and waste load balances are conducted at a plant and when by-product utilization opportunities are available. The development of dry caustic potato and fruit peeling methods has reduced both the volume and strength of wastes from these operations. Separation of blood for use as animal feed, fat and grease for rendering, and paunch manure for land disposal can be practiced at a meat slaughtering and meat-packing installation to reduce the pollutional characteristics and treatment costs of the resultant flow. Other examples of potential by-product recovery include protein recovery from shellfish wastes, potato starch separation and evaporation recovery of protein-containing wastes, fungi and microbial solids production in biological waste treatment, and whey solids separation. A waste survey of a food processing facility, including a water balance and mass balances of product, BOD, solids, and nutrients can reveal other opportunities for waste management by in-plant modifications.

Use of waste monitoring equipment such as temperature recorders, conductivity measurements, flow measurement devices, and turbidity measurements can be used to make the materials balances at a processing plant and to pinpoint potential improvements in processing operations. Such waste production indices can be used to measure the performance of both management and production workers. By using these techniques, waste handling costs can be reduced, water conservation and reuse can be enhanced, and savings of raw materials may result.

After in-plant waste management opportunities have been explored, decisions can be made on the appropriate methods of treatment and disposal of the resultant wastes. The discharge of untreated or screened food processing wastes directly to a municipal waste treatment plant may be a desirable method if satisfactory waste surcharge arrangements can be made. The seasonal nature of many food processing operations, such as canneries, and the strength of these wastes can cause serious problems at municipal plants. The strength of typical food processing wastes are stronger than municipal wastes. In addition, the food processing plants often are adjacent to small communities whose treatment plants may not have the capacity to absorb the food processing wastes without difficulty. The result is large slug loads on the municipal treatment plant caused by seasonal, diurnal, and clean-up conditions. It is difficult to operate or design an efficient municipal waste treatment plant under such conditions.

For the above reasons, either pretreatment of the food processing waste before discharge to the municipal system or complete treatment of the waste by the processing plant are common. In either case, a waste equalization unit is valuable to reduce the variability of waste volume and strength during operation. More uniform waste characteristics will result from the equalization unit which make its effluent more amenable to treatment plant design and operation.

Biological methods of many types can be used with food processing wastes since they are organic in nature. Certain wastes may need nutrient supplementation for optimum biological treatment. Feasible processes include aerated lagoons, oxidation ponds, anaerobic ponds, trickling filters, activated sludge modifications, rotating biological contactors, chemical precipitation, and spray irrigation. The methods can be used separately or in combination. The advantages and disadvantages of each process and examples of their use with food processing wastes have been noted earlier. Reduction of the solids content by screening, sedimentation, or flotation may be advisable with certain wastes before biological treatment. The degree of treatment may be that required for discharge to surface waters, to land disposal, or to municipal facilities.

While all of the above methods have been used successfully with food processing wastes, the most common methods of treatment and disposal have been oxidation ponds, aerated lagoons, and spray irrigation. These are used because of the availability of land near most processing plants and the minimum maintenance and control that is needed. For cannery wastes, the oxidation pond may also act as a surge equalization unit for the wastes which are seasonally produced. Where odors occur due to temporary heavy loadings and aeration equipment does not appear justified, the addition of sodium nitrate in quantities usually sufficient to satisfy 20–30% of the BOD can be added for odor control.

Solids from food processing operations can be used for animal feed, for compost, and disposed of by landfill. Landfilling is the most common method for solids disposal.

Nutrient control, on one hand to obtain satisfactory biological treatment efficiency and on the other to avoid excess nutrients discharged to surface waters, is important with food processing wastes. Because some wastes from fruit and vegetable processing operations may be nitrogen and phosphorus deficient for adequate biological treatment, supplementary nutrients can be added. The common approach of adding nitrogen and phosphorus to nutrient deficient wastes is to attempt to obtain a BOD:N:P ratio of 100:5:1 in the waste. This ratio is satisfactory if one wishes to assure no nutrient deficiency but is of little use if the purpose is to have the nitrogen and phosphorus levels be low in the effluent. The above ratio was designed to assure adequate nutrients in high rate biological treatment. However, for a stationary or declining growth biological treatment system, such as are most treatment systems, the above ratio will result in excess nutrients in the effluent. With the long solids retention time such as a matter of days in the common treatment systems, endogenous respiration of the microbial cells will release nutrients to the system. These nutrients will be used in the synthesis of new microbial cells. To avoid an excess of nutrients in the effluent, nutrients should be added to the system in proportion to the difference in the amount of nutrients lost in the microbial solids and effluent leaving the system and in the nutrients entering the system.

Cropping Patterns and Soil Management

Although these areas do not fall solely into the waste management area, they are an important part of agricultural production. Proper crop and soil management can go far to minimize the release of sediment, organic matter, and nutrients to surface waters.

Historically, cropping patterns have been exploitive of native plant nutrients. On a national basis more nutrients are being removed by crops than are being replaced by fertilizers. This should not indicate that fertilizers are not contributory to excess nutrients in surface waters and groundwaters. In specific areas of the country, farmers add more nutrients than their crops can recover. The remaining fertilizer nutrients have the potential of entering surface waters and groundwaters depending upon soil management methods.

Erosion can be a major contributor of organics and nutrients to surface waters. The physical removal of organics and nutrients by erosion is non-selective since the compounds can be removed in any chemical form. However, the organic and finer particles of soil are more vulnerable to erosion than are the coarser soil fractions. Organic matter will be among the first constituents to be moved by runoff because of its low density and high concentration in surface soils. Significant quantities of nitrogen and phosphorus may be removed in the organic phase.

Practices recommended for reduction of runoff losses include use of crop residues on the soil, application of animal manure in conjunction with crops, minimum tillage on slopes, terracing, strip cropping, contouring, and diversions, early growth of crops, sod crops in rotation, and avoidance of bare land surfaces. The above practices will also reduce sediment loss since generally the higher the runoff rate, the greater the sediment load.

Reduction of runoff and reduction of nutrients and pollutional matter from agricultural land normally are complimentary but there are also specific practices that can reduce the availability of nutrients lost to surface streams by runoff and to groundwater by percolation. Feasible practices are (a) avoid manure application on frozen soils, (b) limit the amounts of fertilizer to that needed by the crops, (c) minimize fall application of fertilizers, (d) apply fertilizer consistent with crop need, (e) spread manure on growing crops or stubble rather than on bare fields, (f) keep feedlots and barnyards out of waterways, (g) divert excess water around areas where wastes accumulate, (h) include crops which have a high nutrient demand in a crop rotation to withdraw excess nutrients from previous crops and from decaying plant residues, (i) avoid aerial application of fertilizers, and (j) avoid application of manures on slopes capable of rapid runoff.

The role of cropping systems on loss of nitrogen from agricultural lands has been summarized concisely (26). The role is complex and is a function of the quantity of organic matter mobilized from the soil, the amount of nitrogen taken up in crop growth, and the amount and nature of crop residues and manure returned to the soil. Many good farming practices are possible to minimize nutrient loss from crop land. These include consideration of (26) type of nitrogen compounds added, time of application in relation to weather and crop utilization, soil type, crop species, method of fertilizer placement, i.e., on or in the soil, rate of application, and supporting crop production practices.

There are a number of ways to minimize nutrient losses without sacrificing the benefits of fertilizers in agriculture. This can be accomplished by controlling rates and times of nitrogen application to fit the needs of the growing crops thereby limiting the buildup of nitrates during periods when leaching is likely, using smaller and more frequent fertilizer applications, avoiding overirrigation to minimize leaching, development of slow release fertilizers that would release nutrients to the soil at a rate close to the need of the crop, and use of appropriate soil conservation and water control practices such as contour cropping, maintenance of adequate plant cover, return of crop residues and manure to the soil, and growth of ground cover on steep lands. The development of slow release fertilizers will depend upon their economic feasibility and the applicability of other management practices.

Many agricultural soils have a high nutrient content accumulated from decaying plants over centuries. Normal water movement through these soils would contain a nitrate concentration whether fertilizers were used or not. Nitrate in groundwaters and surface waters has many sources. Its movement and residence time in the soil depends upon the hydrology of the area, application rate, crop growth, and environmental factors. It is simplistic to relate nitrogen concentrations in tile drain flow or in groundwater base flow only to fertilizer applications. Use of the previous guidelines will avoid excess loss of fertilizer nutrients but may not eliminate the loss of natural nutrients from the soil.

Education and Research

The breadth of agricultural waste management requires that individuals competent in this area have an understanding of many engineering and scientific disciplines. While each individual must have depth in a specific discipline, he must also have the knowledge of how other disciplines interface with his, how to talk with and understand workers in other disciplines, and what data are needed by other disciplines. An individual trained or having interests solely within one discipline is not likely to develop successful agricultural waste management alternatives since he is unlikely to understand the interactions that must take place in an agricultural operation.

The education of an individual competetent in agricultural waste management, whether it is informal or formal education, must be similar in manner to that of a tree, i.e., strength in one or more disciplines (the trunk) to support and interact with other disciplines (the branches). Examples of the discipline that should be involved in such education are noted in Fig. 13.5.

Core disciplines in such education are sanitary engineering, agricultural engineering, and soil science. Knowledge of the fundamentals and practices of sanitary engineering are important because they provide an understanding of feasible treatment and disposal processes, of mass and energy balances for waste separation, reduction, and reuse possibilities, of public health protection, and of rules and regulations governing environmental control and waste discharge. The fundamentals of soil science are important because the agricultural products originate from the soil and the soil is the logical place for disposal and possible reuse of the wastes from agricultural production. Proper methods to manage crop and noncrop lands for production, waste disposal, and public recreation depend upon a knowledge of soil physics, chemistry, and microbiology. Knowledge of agricultural practices,

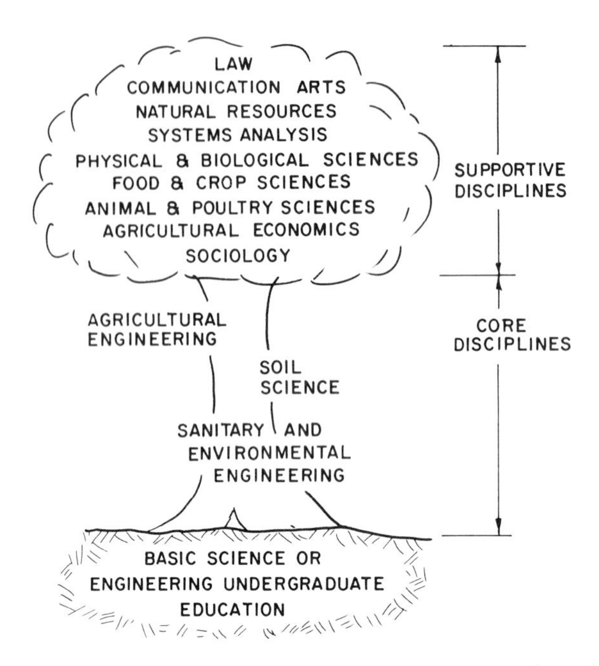

Fig. 13.5. Components of an educational system for competence in agricultural waste management.

construction and operational procedures at a production operation, general soil and water relationships, available equipment, and of the economics of production permits an agricultural engineer to have a valuable input into agricultural waste management.

Knowledge of the fundamentals of sanitary engineering, agricultural engineering, and soil science will be inadequate to solve agricultural waste management problems unless the individual understands agricultures' unique situation and problems and unless the specifics of agriculture are taken into account when the fundamentals are utilized. The understanding can be developed by courses in the supportive disciplines noted, especially agricultural economics, food and crop sciences, and animal and poultry sciences. As part of any educational process in agricultural waste management, the individual should be required to utilize his acquired knowledge on a specific agricultural waste management problem to demonstrate his ability to integrate and use the information from many disciplines on a real problem. The demonstration can be accomplished by imaginative theses, class problems, term reports, and site visits. Solutions to agricultural waste problems will occur from a new breed of individuals educated in an interdisciplinary manner using the fundamentals and practices from each discipline as appropriate.

Specialization at the graduate level will be required to develop the competence necessary for success in this professional area. The home of educational efforts in agricultural waste management logically belong in an agricultural engineering department because of the large engineering component that is required in the educational program. In some institutions the program may have its headquarters in another engineering department. The key to the success of the program will be the enthusiasm, dedication, and commitment of both the staff members of the program and the administrative officials of the institution. The education program must be broad and flexible to incorporate courses and concepts from many disciplines.

Educational and training activities in agricultural waste management also should be geared to educating the general public and the agricultural community that there is a problem, the magnitude of the problem, and the costs of its solution. The education of the general public, the general agricultural community, consulting engineers, and regulatory agencies can be done through the activities of the cooperative agricultural extension program now existing in all states. Education of these extension agents in proper waste management processes and systems also is an important part of the current educational needs.

In a similar manner, solutions to agricultural waste problems and the development of feasible agricultural waste management systems will result from research that cuts across disciplinary lines. Considerable progress in finding solutions to problems can be obtained by in-depth research within single disciplines. Examples might be the development of better liquid or solid waste treatment systems, better materials handling procedures, more feasible reuse possibilities, or proper waste application rates on the soil. However, the better approaches must be integrated into feasible systems, with the overall costs and environmental effects properly evaluated. The development of the total management systems requires input from many of the disciplines noted in Fig. 13.5. Integration of the feasible components into possible systems and their overall effect should be demonstrated, first conceptually and then at the laboratory and field level, as rapidly as possible.

There is a need to initiate studies that will investigate the interrelationships of agricultural production, waste management, and environmental quality so that unnecessary difficulties and problems can be anticipated and avoided. The environmental problems discussed earlier in this book have arisen because of rapid progress in efficiency of agricultural production without adequate anticipation or understanding of the difficulties that would occur with the associated waste treatment and disposal. There is a need to develop predictive and decision-making tools that will permit a better understanding of the impact of changes in agricultural production operations on environmental quality constraints and vice versa.

Management and systems studies are needed to integrate the relevant technical, economic, and legal aspects of agricultural waste management. Because of the current state of knowledge, some of the data and knowledge necessary to delineate "best" approaches or alternatives are either not available or are known in very general terms. An important feature of such studies would be estimates of the type of additional activities that are likely to significantly advance the development of approaches and processes to meet the desired environmental and cost objectives of the production operation and of the public. There are few simple solutions to agricultural waste management problems.

There is an almost complete lack of cost information relating to agricultural waste management processes and system. It is imperative that all studies collect and report information that can be used to estimate the cost of the various processes. Laboratory studies will be necessary to delineate feasible processes. Field studies on as large a scale as possible should follow as quickly as possible to determine the performance of the processes when exposed to the vagaries of environmental conditions and actual waste management practices. All such projects should include the collection of detailed cost as well as performance data so that adequate evaluation of the process can be made.

More specific economic studies should be conducted to evaluate: (a) the effect of the costs of waste control and abatement on the costs of agricultural production, (b) the costs that will ultimately be borne by the consumer, and (c) the need for possible subsidies to insure adequate agricultural waste control and abatement. Large-scale increases in agricultural production costs may only be met by a form of subsidy or by increasing the cost of the product to the consumer directly.

New technology will be needed to involve waste management as an integral and conscious effort in every phase of animal production. Solutions to agricultural waste management problems will require innovative and imaginative approaches using the fundamentals from many disciplines. It is perhaps fortunate that this area is only currently being developed. Traditions and inertia to change have not been established. It is hoped that resistance to traditional and outmoded concepts will be maintained in this field as new and better waste management systems are developed.

There are few research efforts or educational programs being developed to train the necessary individuals or to integrate individual treatment and disposal components in an overall, interdisciplinary manner. It is no longer proper for results from individual research projects to be presented to professions for random utilization nor is it proper to attempt to educate people in agricultural waste management without developing an integrated, interdisciplinary program. Better organization is required in both cases.

Research on components of waste management systems will benefit by the application of the concepts of systems analysis, and overall waste management systems will benefit by a program of integrated research by individuals in many disciplines. Better educational and research programs will result from the thought of individuals from a number of disciplines as the programs are formulated and continually reevaluated.

Summary

The agricultural producer has both immediate and long term waste management problems. The immediate problems have arisen because of changes in production, because of the national emphasis on enhancement of environmental quality, and because a lack of knowledge of how to adequately deal with the waste produced from these operations. Current solutions to many of the problems are only stop-gap approaches until better methods become available. The increasing numbers of individuals active on these problems and the increasing funds being spent on research and development projects relating to agricultural pollution problems will, in time, evolve satisfactory solutions to the problems. In the interim, there are a number of good practice or management guidelines that can minimize some of the environmental problems caused by agriculture. A variety of these guidelines have been outlined in this chapter with details on specific processes described elsewhere in the book.

The long term approach for agriculture production must be based upon not only optimal production of the product such as egg production or animal weight gain or maximum fruit or vegetable production but also on the management of the entire production scheme such that it is consistent with the maintenance of acceptable environmental quality not only to the animals, and to the producers, but to society as a whole. No longer can wastes be discharged indiscriminately. Such discharge invariably intrudes upon someone else's living space.

There is no longer any way an individual sector of society can dispose of wastes by simple export. Ultimate disposal of most of the stabilized material will have to be on agricultural lands. Care must be taken in using the land for waste disposal. While the traditional approach has been to consider the land as a disposal "sink" for residual waste solids and liquids, there is uncertainty that this can be sustained on a long term basis. Greater information is needed on the waste assimilative capacity of specific soils to avoid secondary pollution problems or the development of soil conditions that may lead to reduced crop production. A major effort should be placed upon reuse and recycle when developing feasible waste management systems.

Indiscriminate discharge of untreated or partially treated wastes to receiving waters is to be avoided. The advice of state water pollution control agencies, consulting engineers, and other competent to deal with the waste problems from agriculture should be obtained before decisions are made on waste treatment and disposal methods for a particular operation. Agriculture in general and each crop or animal producer in particular should recognize that the changing practices of agriculture can and are having an adverse effect on the quality of the environment. The producers should seek out the most qualified individuals to provide solutions that will avoid these effects rather than rely on inadequate information or hearsay.

The agricultural waste problem usually is viewed as a treatment and disposal problem. While this is the most obvious aspect of the problem, waste management systems should consider all aspects of an agricultural production operation rather than concentrate only on the treatment and disposal aspects. While the technical and scientific ability to manage the environmental quality problems associated with agriculture is far from complete, sufficient knowledge is available to minimize the gross adverse environmental effects that occurred in the past through ignorance.

References

1. Anonymous, "Agricultural Code of Practice for Ontario," The Ministry of the Environment and The Ministry of Agriculture and Food, Ottawa, Canada, April 1973.
2. Anonymous, "Guidelines for Handling Animal Wastes as Related to Water and Air Pollution Control," Coop. Ext. Serv., Bull. EM 3107. Washington State University, Pullman, 1969.
3. Anonymous, "Livestock Waste Management Guidelines," Coop. Ext. Serv., Bull. EM 3479. Washington State University, Pullman, 1971.
4. Anonymous, "Maine Guidelines for Manure and Manure Sludge Disposal on Land," Misc. Rep. No. 142. University of Maine, Life Sciences and Agricultural Experiment Station, and the Maine Soil and Water Conservation Commission, Orono, Maine, 1972.
5. Anonymous, "Guidelines for Control of Poultry Manure Odors from Caged Layer Houses." Bureau of Air Quality Control, Dept. of Health and Mental Hygiene, Environmental Health Administration, Baltimore, Maryland, 1972.
6. Anonymous, "Guidelines for Controlling Contaminant Emissions from Animal Production Farms," New York State Dept. of Environmental Conservation, Albany, New York, 1972.
7. Anonymous, "Spray Irrigation Manual," Publ. No. 31. Bureau of Water Quality Management, Pennsylvania Dept. of Environmental Resources, Harrisburg, Pennsylvania, 1972.
8. State Water Resources Control Board, "Cannery Waste Treatment, Utilization, and Disposal," Publ. No. 39. State Water Resour. Contr. Bd., Sacramento, California, 1968.
9. Schulte, D. D., and Loehr, R. C., Analysis of duck farm waste treatment systems. *In* "Livestock Waste Management and Pollution Abatement," Publ. PROC 271, pp. 73–76. Amer. Soc. Agr. Eng., 1971.

10. McKenna, M. F., and Clark, J. H., "The Economics of Storing, Handling, and Spreading of Liquid Hog Manure for Confined Feeder Hog Enterprises," Proc. Agr. Waste Manage. Conf., pp. 98–110. Cornell University, Ithaca, New York, 1970.
11. Smith, R., Preliminary design of wastewater treatment systems. *J. Sanit. Eng. Div., Amer. Soc. Civil Eng.* **95,** SA1, 117–145 (1969).
12. Water Pollution Control Federation, "Regulation of Sewer Use," MOP-3. Water Pollut. Contr. Fed., Washington, D.C. 1963.
13. American Public Works Association, "Guidelines for Drafting a Municipal Ordinance on Industrial Waste Regulations and Surcharges," Spec. Rep. No. 23. Amer. Pub. Works Ass., Chicago, Illinois, 1959.
14. Maystre, Y., and Geyer, J. C., Charges for treating industrial wastewater in municipal plants. *J. Water Pollut. Contr. Fed.* **42,** 1277–1291 (1970).
15. Gilbertson, C. B., Beef cattle feedlots—production alternatives. *Winter Meet., Amer. Soc. Agr. Eng.* Paper 70-908 (1970).
16. Doll, R., "Economic Impact of Agricultural Pollution Control Programs," Proc. Agr. Waste Manage. Conf., pp. 9–16. Cornell University, Ithaca, New York, 1972.
17. Gilbertson, C. B., McCalla, T. M., Ellis, J. R., and Woods, W. R., Methods of removing settleable solids from outdoor beef cattle feedlot runoff. *Annu. Meet., Amer. Soc. Agr. Eng.* Paper 70-420 (1970).
18. Butchbaker, A. F., Garton, J., Mahoney, G. W. A., and Paine, M., "Evaluation of Beef Feedlot Waste Management Alternatives," Final Rep., Proj. 13040 FXG. Environmental Protection Agency, 1971.
19. Loehr, R. C., and Schulte, D. D., "Aerated Lagoon Treatment of Long Island Duck Wastes," 2nd Int. Symp. Waste Treatment Lagoons, pp. 249–258. Federal Water Quality Administration, Dept. of Interior, Kansas City, Missouri, 1970.
20. Burnett, W. E., and Dondero, N. C., The control of air pollution (odors) from animal wastes—evaluation of commercial odor control products by an organoleptic test. *Winter Meet., Amer. Soc. Agr. Eng.* Paper 68-609 (1968).
21. Burnett, W. E., and Sobel, A. T., "Odors, Gases, and Particulate Matter from High Density Poultry Management Systems as they Relate to Air Pollution," Proj. Rep. No. 2, N.Y. State Contract No. 1101. Cornell University, Ithaca, New York, 1968.
22. Gumerman, R. C., and Carlson, D. A., Hydrogen sulfide and methyl mercaptan removals with soil columns. *Proc. Purdue Ind. Waste Conf.* **21,** 172–181 (1969).
23. Hammond, W. C., Day, D. L., and Hansen, E. L., Can lime and chlorine suppress odors in liquid hog manure? *Agr. Eng.* **49,** 340–343 (1968).
24. Deible, R. H., Biological aspects of the animal waste disposal problem. *In* "Agriculture and the Quality of Our Environment," Publ. No. 85, pp. 395–399. Amer. Ass. Advan. Sci., Washington, D.C., 1967.
25. Faith, W. L., Odor control in cattle feed yards. *J. Air Pollut. Contr. Ass.* **14,** 459–460 (1964).
26. Aldrich, S. R., "The Influence of Cropping Patterns, Soil Management, and Fertilizer on Nitrates," Proc. 12th Sanit. Eng. Conf., pp. 153–176. University of Illinois, Urbana, 1970.

APPENDIX: CHARACTERISTICS OF AGRICULTURAL WASTES

The references noted in this Appendix can be located in the list of references included with Chapter 4

Table | Page

Table A-1

Waste Characteristics from Vegetables, Fruit, and Wine Processing[a,b]

Product	COD (mg/l)	BOD (mg/l)	BOD/COD (mean)	pH
Beets	1,800–13,200	1,200– 6,400	0.51	5.6–11.9
Beans, green	100– 2,200	40– 1,360	0.55	6.3– 8.6
Beans, wax	200– 600	60– 320	0.58	6.5– 8.2
Carrots	1,750– 2,900	800– 1,900	0.52	7.4–10.6
Corn	3,400–10,100	1,600– 4,700	0.50	4.8– 7.6
Peas	700– 2,200	300– 1,350	0.61	4.9– 9.0
Sauerkraut	500–65,000	300– 4,000	0.66	3.6– 6.8
Tomatoes	650– 2,300	450– 1,600	0.72	5.6–10.8
Apples	400–37,000	240–19,000	0.55	4.1– 7.7
Cherries	1,200– 3,800	660– 1,900	0.53	5.0– 7.9
Grape juice	550– 3,250	330– 1,700	0.59	6.5– 8.2
Wine	30–12,000	30– 7,600	0.60	3.1– 9.2

[a] From reference 3.
[b] Represents samples from processing steps; minimum inclusion of wash and cooling waters.

Table A-2

Waste Characteristics from Fruit and Tomato Processing[a]

	Percent of values equal to or less than the indicated value		
	10%	50%	90%
Peaches (flow)[b]	27	37	50
Suspended solids[c]	0.013	0.047	0.17
Cherries (flow)	11.5	16.0	21.0
Suspended solids	0.01	0.018	0.028
Apples (flow)	13	27	52
Suspended solids	0.018	0.04	0.09
Tomatoes (flow)	56	78	103
BOD[c]	0.18	0.42	0.65
Suspended solids	0.07	0.35	0.63

[a] From reference 4.
[b] Gallons/case of #303 cans.
[c] Pounds/case of #303 cans.

Table A-3

Fruit Waste[a]

Product	Flow (gal/ton raw product)	BOD (lb/ton raw product)	COD (lb/ton raw product)	P/BOD	N/BOD
Pears	5800	78	105	0.0012	0.006
Peaches	7500	66	98	0.0027	0.011
Purple plums	7000	64	95	—	—
Apples	8200	53	71	0.0010	0.0023

[a] From reference 5.

Table A-4

Waste Characteristics from Fruit Canning[a]

Product	Volume (gal/case)	BOD (mg/l)	Suspended solids (mg/l)
Apples	25–40	1680–5530	300–600
Apricots	57–80	200–1020	200–400
Cherries	12–40	700–2100	200–600
Cranberries	10–20	500–2250	100–250

[a] From reference 6.

Table A-5

Waste Characteristics from Fruit and Baby Food Products[a]

	Average (mg/l)	Range (mg/l)
BOD	820	370–3000
COD	2000	780–4000
Suspended solids	180	55– 500

[a] From reference 7.

Table A-6

Waste Characteristics from Pickle Processing and Packing[a]

	pH	BOD (mg/l)	Chlorides (mg/l as Cl^-)	Acidity (mg/l as $CaCO_3$)
Packing plant effluent	3.6–4.4	800– 5,400	2,500–14,000	400– 2,300
Sweet liquor	2.9–3.1	200,000–400,000	10,000–17,000	11,000–17,000
Cook room waste water	2.8–6.6	25– 6,000	750– 4,400	300– 2,500
Cleanup wash water	3.9–7.0	0– 800	25– 1,000	25– 500

[a] From reference 8 [Barnes, G. E., and Weinberger, L. W., *Wastes Eng*. Jan. (1958)].

Table A-7

Frozen Pea Processing[a]

	Per 1000 lb peas processed	
	Average	Range
Flow (gal)	3500	2900–5300
Suspended solids (lb)	10	7–17.5
BOD (lb)	24	18.5–31
COD (lb)	41	33–48

[a]From reference 9.

Table A-8

Waste Characteristics from Potato Processing[a]

	Waste production per ton of potatoes processed
Process water (gal)	4200
BOD (lb)	50– 90
COD (lb)	210
Suspended solids (lb)	60–110
Total phosphate as PO_4 (lb)	0–6
Total nitrogen as N (lb)	3.5

[a]From reference 10.

Table A-9

Characteristics of Corn and Pea Processing Waste[a]

	Average values (mg/l)	
Parameter	Corn	Pea
COD	2536	1650
BOD	1564	772
TOC	1632	962
Total solids	2520	6815
Total volatile solids	1490	917
Suspended solids	210	264
Ash	920	5898
Ammonia nitrogen	35	45
Phosphate as P	8.5	13
pH	6.9	—

[a]From reference 11.

Table A-10

Waste Characteristics from Processing of Various Vegetables[a]

Vegetable	Type of process	Volume of water (gal/ton)	BOD		Suspended solids	
			mg/l	lb/ton	mg/l	lb/ton
Carrots	Washing	600	243	1.4	4120	24.7
		1600	78	1.2	2190	35.0
	Canning	1250	1400	17.5	2000	25.0
		1350	1100	14.8	1830	24.7
	Dehydration	4000	1220	48.8	703	28.1
			(860)[b]	(34.4)	(157)	(6.3)
Potatoes	Canning	2000	1100	22	1250	25.0
		3500	220	7.7	990	34.6
		> 1000	3000–12,000	30–120	—	—
	Lye peeling	3700	2570	95	1005	37.2
	Dehydration	6820	310	21.1	1217	83
			(330)	(22.5)	(297)	(20.3)
Broad beans	Canning	1500	333–666	5–10	—	—
Beetroot	Canning	1000	4000	40	1250	—
			1580–7600	—	740–2220	—
Spinach	Canning	7200	280	20.2	580	41.7
		6750	730	49.3	—	—
		6200	600	37.2	—	—
Peas	Freezing	3000–4000	1000	35–40	—	—
		350	8000–9000	28–31.5	—	—
	Canning	1500	1130–3730	17–56	—	—

[a] From reference 12. British Crown Copyright, reproduced by permission of the Controller, Her Majesty's Stationery Office.

[b] Values in parentheses relate to settled waste waters.

Table A-11

Characteristics of Wine Processing Wastes[a]

Parameter[b]	Pressing season		Nonpressing season	
	Range	Avg.	Range	Avg.
Flow (gpd)	30,000–205,000	125,000	30,000–120,000	85,000
Temperature (°F)	59–70	65	38–73	54
pH	5.3–11.3	8.6	4.9–9.5	8.0
Dissolved oxygen	1.6–7.8	5.3	0–10	5.5
Suspended solids	5–460	200	5–448	170
Chlorides	10–60	20	20–5000	600
Total phosphate (PO_4)	22.1–40.0	27.8	7.7–25.7	14.6
Orthophosphate (PO_4)	10–30	16.8	0.8–4.8	2.4
Total Kjeldahl nitrogen	1.0–5.7	2.4	1.2–12.4	4.4
Nitrates	0.9–2.3	1.4	0.09–0.5	0.3
BOD	131–5000	2300	130–5000	2000

[a] From reference 13.
[b] Except as noted, the parameters are in terms of mg/l.

Table A-12

Typical Winery Process Waste Streams[a,b]

Waste stream	BOD_5	COD	Total solids	Total volatile solids	pH
Baker wash	3240	3865	1300	506	5.9
Press wash	2000	3900	1635	955	7.2
Bottle wash	75	138	600	260	9.7
Cooling water	15	40	120	—	8.0
Washwaters from blenders, filters, and ion-exchange processes	2470	3375	1450	765	7.0

[a] From reference 78.
[b] Parameters are given in terms of mg/l, except where noted.

Table A-13

Characteristics of Meat-Packing Wastes[a,b]

Item	Average	Range
BOD (mg/l)	1240	600–2720
COD (mg/l)	2940	960–8290
Organic nitrogen (mg/l)	85	22– 240
Grease (mg/l)	1010	250–3000
Suspended solids (mg/l)	1850	300–4200
Volatile suspended solids (%)	92	80– 97

[a] From reference 21.
[b] Composite samples from 16 plants.

Table A-14

Characteristics of Slaughterhouse Wastes

Item	Range	
BOD (mg/l)	360–1880	900–2600
Suspended solids (mg/l)	480–1450	600–1800
Organic nitrogen (mg/l)	36– 510	—
Grease (mg/l)	88– 440	200– 800
Chlorides (mg/l) as Cl^-	190–5690	—
pH	6.5–8.4	6.8
COD (mg/l)	—	2000–4900
Flow (gal/1000 lb live wt.)	—	770–3600
Reference	19	20

Table A-15

Characteristics of Wastes from Killing and Processing[a]

Animal	Hogs	Hogs	Broilers
BOD (lb/1000 lb live weight)	6.5–9.0	2.3	9–11
Flow (gal/1000 lb live weight)	435–455	155	825–1525
Type of plant	Killing and processing	Killing	Killing and processing

[a] From reference 15; *Wastes Engr.*, 135–138 (1961).

Table A-16

Characteristics of Packing and Slaughtering

Type of plant	lb/1000 lb live weight							Ref.
	BOD	Suspended solids	Grease	Organic nitrogen (as N)	Ammonia nitrogen (as N)	Total phosphorus (as P)	Soluble phosphorus (as P)	
Packing	5.2–23.5	2.4–22.1						22
Packing	14.6	12.0	1.6	← 1.7 →				22
Slaughtering	8.4–20	3.5–11						22
Meat processing	5.2–17.8	4.2–21.5	1.2–23.5	0.3–1.2	0.1–0.7	0.06–0.2	0.02–0.13	23
Hog and beef processing	14.8	11.5	5.3					20

Table A-17

Poultry Processing Plant Waste

Combined wastes[a] from 13 processing plants	
Item	Range (mg/l)
BOD	150–2400
COD	2–3200
Suspended solids	100–1500
Dissolved solids	200–2000
Volatile solids	250–2700
Total solids	350–3200
Total alkalinity	40– 350
Total nitrogen	15– 300
BOD—lb/1000 birds processed	25
Suspended solids—lb/1000 birds processed	13

Slaughtering and dressing[b]		
Item	Average (mg/l)	Range (mg/l)
BOD	630	400–1200
COD	2620	1600–3700
Suspended solids	350	200– 600
Total nitrogen	65	57– 92
Ammonia nitrogen	5.8	3.1–11.6
pH	—	6.4– 8.1
Flow (gal/bird)	5	—

Table A-17 (continued)

Broiler processing[c]

Item	Average	Range
BOD—mg/l	665	315–1000
lb/1000 birds processed	33	25 (blood recovery)
Suspended solids (mg/l)	400	276– 480
lb/1000 birds processed	20	—
Flow (gal/bird)	6	—

Broiler processing plant[d]

Source of waste	Characteristic (mg/l)			
	BOD	COD	Total solids	Fat
Scalder water	690	1,000	1,460	280
Oil bag	220	220	380	—
Eviscerating table	370	760	1,120	—
Feather line	1,070	1,260	1,370	—
Eviscerating line	630	1,520	1,250	720
Lung tank	100,000	106,000	138,500	—
Chiller water	860	1,100	1,140	620
Sump pump	14,700	13,900	—	—
Wash water from killing room	14,200	14,000	—	—

[a] From reference 25.
[b] From reference 26.
[c] From reference 27.
[d] From reference 76.

Table A-18

Average Composition of Milk and Milk Products[a] (weight/100 gm)

Product	Fat (gm)	Protein (gm)	Lactose (gm)	Total organic solids (gm)	Ca (mg)	P (mg)	Cl (mg)	S (mg)	Total ash (gm)	BOD (mg/l)
Skim milk	0.08	3.5	5.0	8.6	121	95	100	17	0.7	73,000
Milk	2.0	4.2	6.0	12.2	143	112	115	20	0.8	100,000
Whole milk	3.5	3.5	4.9	11.1	118	93	102	19	0.7	99,000
Half and half	11.7	3.2	4.6	19.5	108	85	90	16	0.6	167,000
Coffee cream	18.0	3.0	4.3	25.3	102	80	73	12	0.6	219,000
Heavy cream	40.0	2.2	3.1	45.3	75	59	38	9	0.4	399,000
Chocolate milk	3.5	3.4	5.0	18.5	111	94	100	19	0.7	145,000
Churned buttermilk	0.3	3.0	4.6	8.0	121	95	103	15	0.8	71,000
Cultured buttermilk	0.1	3.6	4.3	10.0	121	95	105	17	0.7	64,000
Sour cream	18.0	3.0	3.6	24.6	102	80	73	12	0.6	218,000
Yoghurt	3.0	3.5	4.0	10.8	143	112	105	19	0.7	91,000
Evaporated milk	8.0	7.0	9.7	27.0	757	205	210	39	1.6	206,000
Ice cream	10.0	4.5	6.8	41.3	146	115	104	20	0.9	290,000
Whey	0.3	0.9	4.9	6.3	51	53	195	8	0.6	45,000
Cottage cheese whey	0.08	0.9	4.4	6.1	96	16	95	8	0.8	42,000

[a] From reference 28.

Table A-19

Milk Plant Waste Water Coefficients[a]

Type of product	Waste volume (lb/lb milk processed)		BOD (lb/1000 lb milk processed)	
	Average	Range	Average	Range
Milk	3.25	0.1–5.4	4.2	0.2–7.8
Cheese	3.14	1.6–5.7	2.0	1.0–3.5
Ice cream	2.8	0.8–5.6	5.7	1.9–20.4
Condensed milk	2.1	1.0–3.3	7.6	0.2–13.3
Butter	0.8	—	0.85	—
Powdered milk	3.7	1.5–5.9	2.2	0.02–4.6
Cottage cheese	6.0	0.8–12.4	34.0	1.3–71.2[b]
Cottage cheese and milk	1.84	0.05–7.2	3.47	0.7–8.6[b]
Cottage cheese, ice cream, and milk	2.52	1.4–3.9	6.37	2.3–12.9
Mixed products	2.34	0.8–4.6	3.09	0.9–6.9
Overall	2.43	0.1–12.4	5.85	0.2–71.2

[a] From reference 28.
[b] Whey included, whey excluded from all other operations manufacturing cottage cheese.

Table A-20

Milk Plant Processing Waste Data from Five Plants (mg/l)[a]

	Range of averages	Total range of data
BOD	940–4790	400–9440
COD	1240–7800	360–15,300
Ammonia	7–36	1–76
Organic nitrogen	36–150	9–250
Alkalinity	81–505	0–1080
pH	4.8–6.8	4.2–9.5
Total solids	2280–6490	1210–11,990
Suspended solids	360–1040	270–1980
Volatile suspended solids	300–1000	200–1840

[a] From reference 29.

Table A-21

Ion Content of Milk Plant Wastes—Five Plants (mg/l)[a]

	Range of averages	Total range of data
Nitrogen	43–180	10–320
Phosphorus	17–132	6–194
Potash	37–160	4–450
Sodium	166–470	98–796
Calcium	58–78	32–107
Magnesium	35–50	13–140
Chloride	160–560	57–1000

[a] From reference 29.

Table A-22

Composition of Whey Products[a]

	Parameter (%)					
Product	Water	Total nitrogen	Fat	Lactose	Acid	Ash
Raw cheese whey	93–94	0.7–0.9	0.05–0.6	4.5–5	0.2–0.6	0.5–0.6
Condensed whey	50–60	7–8	1–2	28	1–3	5–6
Dried whey	2–6	12–14	0.3–5	65–70	2–8	8–12

[a] From reference 81.

Table A-23

General Characteristics of Fish Processing Waste Water[a]

Type of fish processing	Flow		BOD			COD		Total solids		Suspended solids		Grease	
	gpm	gal/ton of fish	mg/l	lb/ton of fish	lb/ton of product	mg/l	lb/ton of fish	mg/l	lb/ton of fish	mg/l	lb/ton of fish	mg/l	lb/ton of fish
General	—	465–9100	2700–3440	8–120	21–24	—	—	4200–21,800	—	2200–3020	—	—	—
Bottom fish	105–450	—	190–1726	—	74	—	—	—	—	300	—	—	—
Herring, menhaden, and anchovies	680–1440	240–1020	620–1005	—	—	—	—	—	—	—	—	—	—
Salmon	—	—	173–3900	—	3.2–80	5920	—	88–7400	—	40–4780	—	—	—
Sardine	40–1000	—	100–2000	—	—	170–1700	—	—	—	100–2100	—	60–1340	—
Tuna	—	—	890	48	—	2270	129	17,900	950	1090	58	287	15

[a] From reference 30.

Table A-24

Smelt and Perch Waste Water Characteristics[a]

Parameter (mg/l)	Smelt		Perch		Mixed smelt and perch		Combined perch and smelt[b]	
	Average	Standard deviation	Average	Standard deviation	Average	Standard deviation	Average	Standard deviation
BOD_5	1150	630	1850	1790	3040	1410	4.5	2.0
COD	1965	1220	3350	2890	4800	4340	8.0	3.6
TOC (filtered)	213	117	283	147	366	113	0.57	0.22
Suspended solids	600	490	935	745	1400	724	2.3	1.3
Volatile suspended solids								
(% suspended solids)	85%	13%	87%	16%	89%	13%		
Total solids	1310	685	1810	925	3070	2380		
Total phosphate								
Filtered	19	5	15	9	19	6		
Unfiltered	22	6	18	8	22	9		
TKN	120	42	122	63	136	50		
Nitrate	0.32	0.07	0.50	0.28	1.06	0.73		
Nitrite	0.01	0.004	0.03	0.02	0.03	0.02		
Oil and grease	37	5	24	12	46	28		
BOD_5:N:P (mean values)	60:6:1		100:6:1					

[a] From reference 77.

[b] Values are given in lb/1000 lb of fish processed (landed weight).

Table A-25

Poultry Manure Characteristics

Nutrients				
Total nitrogen	Ammonia nitrogen	P_2O_5	K_2O	Ref.
		lb/bird/day		
0.0062	0.0035			2
0.0012–0.0057		0.0010–0.0045	0.0005–0.0019	36
0.0030	0.0006	0.0026		37
0.0036	0.0023			38
0.002–0.0046				85
		mg/kg of manure		
14,860	838			2
		Percent dry solids		
1.8–5.9		1.0–6.6	0.8–3.3	39

Ca	Mg	S	Fe	Zn	Cu	Ref.
lb/1000 gal of manure						
300	24	26	3.9	0.75	0.12	36

Table A-26

Pollutional Characteristics of Poultry Manure

Dry total solids	Dry volatile solids	BOD	COD	BOD/COD	Pop. equiv.[a]	Ref.
		lb/day/bird[b]				
		0.018	0.08	0.23	9.2	2
	0.035		0.033			36
0.063	0.044	0.015	0.050	0.30	11.3	37
0.066	0.051	0.015	0.057	0.26		38
		0.015–0.032				40
0.067	0.048	0.012				41
0.062						83
0.048–0.058	0.039–0.044	0.007	0.032–0.043			85
		mg/kg of waste				
		44,160	192,000			2

[a] Chickens equivalent to per capita BOD in domestic sewage.

[b] 4–5 lb birds.

Table A-27

Duck Waste Water Characteristics

Parameter	lb/1000 ducks/day			mg/l		
	Average	Range	Range	Range	Range	Average
BOD	—	—	16–30	80–220	—	—
COD	82	18–260	—	—	140–7520	810
Total solids	96	26–320	—	—	330–6340	1010
Suspended solids	54	10–237	52–123	—	17–4630	340
Total nitrogen	—	—	4.7–6.4	—	—	—
Phosphates (as PO_4)						
Total	—	—	4.7–10.5	—	—	—
Soluble	5.5[a]	1.0–13.4	2.5–4.6	—	31–130[a]	42.5[a]
Water use (gal/duck/day)	15	5–27	10–34	—	—	—
Reference	43[b]	43	42	42	43	43

[a] Orthophosphate.

[b] Data from reference 43 originated from over 50 grab samples during the duck production season.

Table A-28

Hog Manure Nutrient Characteristics

Total nitrogen	P_2O_5	K_2O	Ref.
	Pounds per day		
0.042–0.060	0.029–0.032	0.034–0.062	36
0.032	0.025		38
0.064			48
0.07			49
0.043	0.037	0.016	50
	Percent dry solids		
0.2–0.9	0.14–0.83	0.18–0.52	39
0.7	0.4	0.4	54

Elements in Hog Manure

Element	lb/1000 gal manure	Feces[a] (mg/l)	Urine[a] (mg/l)	Feces (mg/gm solids)	Urine (mg/gm solids)
Ca	41	25,100	340	71.0	8.3
Mg	6.6	8,020	88	22.5	2.2
Zn	0.5	510	2.3	1.4	0.05
Cu	0.13	108	0.16	0.3	0.004
Fe	2.3	456	1.1	1.3	0.03
Mn	—	176	0.3	0.5	0.008
Na	—	2,630	1,300	7.4	31.0
K	—	10,200	2,300	29.0	56.0
P	—	16,700	178	47.0	4.4
S	12	1,040	1,100	2.9	27.1
N	—	34,600	5,000	97.7	122.6
Ref.	36	79	79	79	79

[a] Feces at 65% moisture, urine at 96% moisture.

Table A-29

Pollutional Characteristics of Hog Manure

Total solids	Volatile total solids	Suspended solids	BOD_5	COD	Ammonia nitrogen	Total nitrogen	Ref.
			lb/day/100 lb animal				
0.80	0.62		0.20	0.75	0.024		38
0.97	0.80		0.43	0.96			48
0.50	0.35			0.47			51
			0.25				40
			0.22				49
0.59	0.47	0.47	0.20[a]	0.52			52
		0.05–0.28	0.13–0.35				47
0.23			0.33	0.55			50
0.47–0.86			0.19–0.34				53
			0.32	1.4			54
			gm/l of slurry				
6–8			27–33	70–90	5.5–7.5		55
0.7–6.2			2–52	5–143	2.7		56
				30			49
			1.3–13				40
			20–34[a]	62–74			52
		7–16	18–20				47
			25				44
			16–22	47–75			75
	80–84%		29–47	61–90	1.9–4.6	5.6–9.2	79

Oxygen demand of feed consumed	Ref.
BOD–0.037 to 0.043 lb/lb feed	75
COD–0.11 to 0.18 lb/lb feed	75

Animal weight range (lb)	Waste production rate (lb feces and urine/day/100 lb animal)	Ref.
40– 80	7.0	79
80–120	5.0	79
120–160	4.2	79
160–200	3.4	79

[a] Ultimate BOD.

Table A-30

Dairy Cattle Manure Characteristics

Total solids	BOD_5	COD	Total nitrogen	Ammonia nitrogen	P_2O_5	Pop. equiv.[a]	Ref.
			lb/day/1000 lb animal				
	1.5						57
10.4	1.5	8.4	0.38		0.12		38
6.8	1.3	5.8	0.37	0.23		0.13	58
6.7–8.5	1.3–1.6						53
2.5	0.93	5.6	0.16		0.05		50
			Nutrient concentrations (lb/1000 gal)				
	Ca	Mg	S	Fe	Zn	Cu	
	17	8.7	5.8	0.3	0.12	0.04	36
Total nitrogen mean = 11.6 gm/l (range = 9–17 gm/l).							73

[a]Dairy cattle equivalent to per capita BOD in domestic sewage, i.e., 0.13 head of dairy cattle produces the BOD of one person in domestic sewage.

Table A-31

Dairy Cattle Manure from Stanchion Barns[a]

Fresh manure including bedding and urine[b]

Parameter	Average	Range
Total solids (%)	13.5	10.6–16.6
Volatile solids (% of TS)	35.5	81–90
COD (mg/l)	136,500	97,000–166,000
Total nitrogen (mg/l as N)	4,500	3,100–6,500
Ammonia (mg/l as N)	1,750	360–3,300
Total phosphorus (mg/l as P)	1,250	880–1,730
Potassium (mg/l as K)	3,250	2,050–4,900
pH	6.9	5.4–8.0

[a]From reference 72.

[b]Data collected from November through August during two trials in separate years: about 30% less bedding was used during trial 2; average values of COD, total nitrogen, ammonia nitrogen and % volatile solids were statistically different between the two trials.

Table A-31 (continued)

Manure, stacked and stored before land disposal[b]

Parameter	Average	Range
Total solids (%)	23	18–30
Volatile solids (% of TS)	83	75–89
COD (mg/l)	227,000	169,000–268,000
Total nitrogen (mg/l as N)	5,450	4,100–6,900
Ammonia (mg/l as N)	1,800	700–2,500
Total phosphorus (mg/l as P)	1,500	810–2,100
Potassium (mg/l as K)	5,000	3,800–6,900
pH	7.6	6.6–8.6

[b] Analysis of the stacked manure as spread on the land during spring through summer; manure stacked for winter storage.

Seepage from stacked stanchion barn manure and bedding

	Winter[c]		Summer	
Parameter	Average	Range	Average	Range
Total solids (%)	2.8	1.8–4.3	2.3	1.7–2.9
Volatile solids (% TS)	55	52–59	53	50–58
Suspended solids (%)	0.35	0.20–0.83	0.24	0.20–0.30
BOD (mg/l)	13,800	4,200–31,000	10,300	4,400–21,700
COD (mg/l)	31,500	21,000–41,000	25,900	16,400–33,300
Total nitrogen (mg/l as N)	2,350	1,500–2,900	1,800	1,200–2,770
Ammonia (mg/l as N)	1,600	980–1,980	1,330	780–2,200
Total phosphorus (mg/l as P)	280	64–560	190	90–340
Potassium (mg/l as K)	4,700	3,000–7,200	3,900	3,000–4,900
pH	7.1	6.4–7.8	7.6	6.8–8.1

	Winter	Summer
Total precipitation (inches)	15.0	9.4
Seepage volume (gal/cow/day)	3.0	1.2

[c] Winter data collected from November through May, summer data from June through August.

Table A-32

General Characteristics of Milking Center Wastes

Waste volume (gal/animal/day)	BOD		Suspended solids		Nitrogen total	Remarks	Ref.
	mg/l	lb/animal/day	mg/l	lb/animal/day	mg/l		
2.6–6.1	292–3620	0.02–0.09	3085–7190	0.18–0.19	172–1812	Cow shed	59
5.1–28.3	200–2170	0.06–0.13	700–3780	0.11–2.5	100–1220	Cow shed and dairy	59
7.4–9.5	300–1440	0.02–0.14	220–790	0.02–0.07	130–1202	Cow shed and dairy	60
4.4–9.4	455–2120	0.03–0.22	390–1660	0.02–0.16	620–4200	Milking parlor	59
18.9	472	0.09	817	0.17	100	Milking parlor	59
10–30	450–4330					Cow shed excluding dung	40
9.5	610–2000	0.08	1490–2380	0.15		Milking parlor	61

Table A-33

Characteristics of Milking Center Wastes (24 Dairy Farms)[a]

	Average	Range
Flow (gal/cow/day)	4.0	1.8–16.8
BOD (lb/cow/day)	0.127	0.01–0.24
Total dry solids (gm/l)	5.0	0.8–10.4
Phosphorus (mg/l)		
Soluble as P	57.6	6–183
Ammonia nitrogen, soluble (mg/l)	132.	5–625
Nitrate plus nitrate nitrogen, soluble (mg/l)	1.6	0.3–6.5

[a] From reference 62.

Table A-34

Beef Cattle Manure Characteristics

Pollutional characteristics (lb/animal/day)

Total dry solids	BOD_5	COD	Total nitrogen	Ref.
3.6	1.0	3.3	0.26	58
9	1–2	9	0.26	64
5	1.9	19.4	0.16	44

Slurry manure from confined beef (feces and urine)[a]

Item	% of Constituent (wet basis)
N	0.29
P_2O_5	0.18
K_2O	0.31
H_2O	85.0
Volatile solids	11.6
Total solids	15.3
Ash	3.7
NH	0.05
pH	7.3

Table A-34 (continued)

Particle size analysis of fresh manure[a] (oven dry weight basis)

Particle size	% of Total
Greater than 4 mm	2.45
4 mm to 2000 μm	25.09
2000 to 500 μm	36.69
500 to 210 μm	4.75
210 to 2 μm	1.01
Less than 2 μm	30.02

[a] From reference 65.

Table A-35

Characteristics of Waste Material Removed from Outdoor, Unpaved Beef Cattle Feedlots[a]

Parameter (tons/acre except where noted)	Animal stocking rate: 100 ft²/animal		200 ft²/animal	
	Mean	Range	Mean	Range
Moisture (% dry solids)	99	50–186	84	32–170
Total solids (dry solids)	379	209–815	263	66–635
Volatile solids (dry solids)	129	76–175	64	28–134
Total nitrogen	3.6	0.6–7.6	1.8	0.05–5.7
NH_4-N	0.6	0.35–0.70	0.2	0.01–0.46
Total P	0.44	0.18–0.69	0.2	0.02–0.55

Parameter (ppm dry weight)	Mean	Range
Na	1050	550–1660
K	4620	530–10,200
Ca	2560	300–4770
Mg	1700	315–3145
Zn	45	25–62
Cu	6	1.5–11.3
Fe	4090	195–9910
Mn	510	24–1630

[a] From reference 84.

Table A-36

Characteristics of Production Livestock[a,b]

Animal	Weight of animal (lb)		Quantity (gal[c])	Dry matter (lb)	BOD (lb)	COD (lb)	Organic carbon (lb)	Total nitrogen (lb)
Cow	1100	Urine	23.9	—	0.32	0.50	0.2	0.07
		Feces	11.7	—	0.76	17	3.92	0.42
			35.7	10.4	1.1	17.5	4.12	0.49
Calf	550	Urine	1.9	—	0.32	0.53	0.25	0.15
		Feces	5.4	—	0.44	6.07	1.15	0.08
			7.3	6.0	0.76	6.6	1.4	0.23
Pig	150	Urine	0.44	—	—	—	—	—
		Feces	0.72	—	—	—	—	—
			1.2	1.1	0.3	—	—	0.05
Sheep	—	Urine	0.46	—	0.12	0.19	0.07	0.018
		Feces	0.76	—	0.10	1.07	0.23	0.03
			1.2	1.2	0.22	1.26	0.30	0.048
Hen	4.4	Total	0.03	0.08	0.02	—	—	0.004

[a]From reference 12. (Crown Copyright, reproduced by permission of the Controller, Her Majesty's Stationery Office.)

[b]Volume and composition will vary considerable depending on type and weight of animal, feed.

[c]U.S. gallons.

Table A-37

Composition of Animal Wastes[a]

Animal	Weight/head (lb)	Feed (dry)/head/day (lb)	Hydraulic load (gpd)[b]	5-day BOD (lb/day)[c]	Total solids (lb/day)	Susp. solids (lb/day)	N (% by wt.)		Phosphate (% by wt.)[b]		Potassium oxide (% by wt.)[b]		Carbohydrates (% by wt.)[b]	
							Solid	Liquid	Solid	Liquid	Solid	Liquid	Solid	Liquid
Horse	950–1400	30–45	20.0	6.50	15.0	12.5	0.50	1.2	0.30	Trace	0.30	1.60	27.1	1.9
Cattle	900–1250	25–40	20.0	6.50	15.0	12.5	0.30	0.9	0.21	0.03	0.18	0.93	16.7	1.3
Calf	450–600	15–20	15.0	3.50	10.0	8.0	0.32	0.95	0.20	0.01	0.15	0.80	15.9	0.9
Pig	100–300	8–12	5.0	0.70	1.7	1.1	0.6	0.3	0.50	0.15	0.12	0.50	15.5	0.3
Small pig	20–90	3–8	3.0	0.40	1.1	0.6	0.3	0.4	0.13	0.13	0.10	0.44	12.2	0.7
Rabbit	4.0–4.5	0.25–0.33	0.4	0.08	1.2	0.9	0.4	0.2	0.10	Trace	0.07	0.50	30.0	0.3
Rat	1.5–2.0 oz	0.04–0.05	0.2	0.04	0.08	0.02	0.2	0.3	0.05	Trace	—	0.10	16.0	0.2
Mouse	0.5–1.2 oz	0.01–0.015	0.1	0.01	0.02	0.01	0.4	0.3	0.05	—	—	0.10	15.0	0.2
Sheep	100–150	4.0–6.0	6.0	1.20	2.50	2.0	0.65	1.7	0.51	0.02	0.03	0.25	30.7	0.9
Monkey	4.0–5.0	0.25–0.40	2.5	0.10	0.20	0.10	0.9	1.2	0.30	0.02	0.05	0.30	17.0	0.1
Dog	50–70	0.25–0.40	1.0	0.08	0.30	0.18	0.75	1.1	0.30	Trace	0.02	0.58	17.0	0.1
Cat	4.0–4.5	0.17–0.37	0.2	0.06	0.15	0.07	0.70	1.2	0.20	Trace	0.02	0.50	17.5	0.1
Turkey	20–25	0.75–0.80	0.4	0.12	0.35	0.20	1.2	0.5	1.00	—	—	0.60	31.0	0.1
Chicken	2.0–4.0	0.25–0.30	0.2	0.07	0.12	0.05	1.4	0.5	0.90	—	—	0.50	30.0	0.1
Duck	3.5–5.0	0.25–0.30	0.4	0.10	0.15	0.07	1.5	0.8	0.89	—	—	0.40	12.0	0.1
Goose	7.5–9.5	0.50–0.75	1.0	0.20	0.35	0.10	1.8	0.8	1.20	—	—	0.70	14.0	0.1

[a] From reference 69; *Water Wastes Engr*. **6,** A14–A18 (1969).

[b] Including spilled water.

[c] Including spilled food.

Notes: All observed animals were kept in cages or stalls. COD for all wastes was 1.30 to 1.50 times higher than 5-day BOD. Data obtained on confined animals at a scientific laboratory.

Table A-38

Moisture and Main Nutrients of Poultry Manures[a,b]

Material	Details		Moisture	N	P_2O_5	K_2O	Mg	Ca	No. of samples
CHICKEN									
1. Deep litter									
Straw	Sampled in house		35–60	1.4–2.9	1.8–3.7	1.1–1.5	0.18	—	3
		M	52	2.2	1.8	1.1			
							(1)		
Straw + peat	Sampled in house		35–42	2.5–3.1	1.9–2.5	1.4–1.5	—	4.9–5.0	3
		M	35	3.0	2.5	1.5		5.0	
Straw + sawdust	Sampled in house		61	1.3–1.8	2.0–2.3	0.6–1.4	—	—	3
		M		1.3	2.0	0.6			
			(2)						
Wood shavings	Sampled in house		6–71	1.0–3.5	0.4–5.2	0.9–2.3	—	—	10
		M	25	2.2	2.5	1.4			
Wood shavings	12–14 Months build up in house		31–32	1.7–2.6	1.1–2.1	1.1–1.3	—	—	2
Wood shavings	5 Years build up in house		25	3.3	4.8	1.8	—	—	1
Wood shavings	Stored 2–3 months under cover		13–81	1.4–3.4	0.7–6.1	0.8–2.8	—	—	7
		M	23	2.0	1.7	1.0			
Wood shavings	Stored more than 12 months under cover		28–64	0.4–2.7	0.6–3.4	0.4–1.6	—	—	7
		M	39	2.0	2.8	0.9			
Wood shavings	Stored less than 6 months in open		61–67	0.9–1.8	1.7–2.4	0.4–1.0	—	—	5
		M	67	0.9	1.7	0.5			
Wood shavings	Stored more than 6 months in open		9–65	0.8–3.1	0.7–3.4	0.8–2.0	—	—	7
		M	51	2.1	2.5	1.8			
Wood shavings	Unknown storage time or conditions		13–36	1.5–3.3	1.4–3.8	0.9–2.7	—	—	3
		M	29	2.2	2.5	2.0			
Unknown type of litter	Stored 2 years under cover		9	2.4	2.9	2.0	—	—	1

Unknown type of litter	Unknown conditions		10–81	0.3–4.1	0.1–5.2	0.1–11.2	0.18–0.53	2.2–8.4	145
		M	31	1.6	1.7	1.2	0.38	2.6	
							(4)	(3)	
Dried	Unknown litter		6–56	1.0–5.1	2.3–3.0	1.2–1.9	—	—	9
		M	9	4.6	2.8	1.7			
Ash	Burnt deep litter		1–96	0.1–1.0	4.0–18.1	3.1–8.1	—	—	9
		M	4	0.3	8.2	3.7			
Range of deep litters (excluding dried and ash)			6–81	0.3–4.1	0.1–6.1	0.1–11.2	0.18–0.53	2.2–8.4	197
		M	31	1.7	1.8	1.3	0.40	5.0	
							(5)	(6)	
2. Broiler									
Wood shavings	8–12 Weeks in making		22–70	0.7–3.4	0.2–2.5	0.6–1.7	—	—	5
		M	43.2	2.2	1.7	1.1			
Wood shavings	8–22 Weeks in making		18–66	0.8–2.8	1.9–2.0	0.4–1.6	—	—	3
		M	24	1.6	1.9	1.4			
Wood shavings	10 Weeks in making, stored 3 weeks in open		26–30	2.6–3.6	2.7–3.3	1.7–1.8	—	—	2
Wood shavings	Stored 4–6 months under cover		24–36	2.0–3.3	2.2–2.8	1.6	—	—	2
Wood shavings	Stored less than 12 months in open		23–73	0.6–3.7	0.7–2.4	0.4–1.5	0.11–0.22	0.4–1.4	4
		M	66	1.8	1.4	0.5	0.15	0.5	
Wood shavings	Stored 3 years in open		50	1.3	2.3	1.3	—	—	1
Wood shavings	Stored 4 years in open		31	2.2	3.2	1.2	—	—	1
Wood shavings	Unknown time and conditions		12–69	0.8–3.6	0.2–4.0	0.3–2.4	0.22	1.4–4.0	40
		M	30	2.4	2.1	1.4		1.9	
							(1)	(11)	
Unknown type of litter	Stored 2 years under cover		90	2.4	2.9	2.0	—	—	1
Unknown type of litter	Stored 6–8 months in open		34–72	0.6–2.3	0.9–1.8	0.3–1.2	—	—	4
		M	61	1.2	1.0	0.8			
Unknown type of litter	Stored 1 year in open		25–33	1.5–2.4	2.1–2.3	1.6–1.9	—	—	2
Unknown type of litter	Stored 2 years in open		—	2.3–2.4	1.0–2.9	0.6–2.0	—	—	2
Unknown type of litter	Unknown conditions		4–81	0.4–5.4	0.2–4.2	0.1–3.2	0.28–0.58	1.4–3.2	62
		M	29	2.4	2.2	1.4	0.50	1.7	
							(3)	(7)	

Table A-38 (continued)

Moisture and Main Nutrients of Poultry Manures[a,b]

Material	Details		Moisture	N	P_2O_5	K_2O	Mg	Ca	No. of samples
Ash	Burnt broiler litter		16	0.2	38.3	5.8	—	—	1
Range of broiler litters (excluding ash)			4–90	0.4–5.4	0.2–4.2	0.1–3.2	0.11–0.58	0.4–1.0	129
		M	28	2.4	2.2	1.4	0.22	1.7	
							(8)	(22)	
Overall range and median of deep litter and broiler litter (excluding dried and ash)			4–90	0.3–5.4	0.1–6.1	0.1–11.2	0.11–0.58	0.4–8.4	326
		M	31	1.8	1.8	1.2	0.25	1.9	
							(13)	(28)	
3. Battery									
	Fresh droppings		18–86	0.3–5.7	0.3–4.5	0.2–4.3	0.04–0.36	0.4–2.8	73
		M	75	1.5	1.1	0.6	0.12	0.8	
							(19)	(8)	
	Stored 1 year under cover		34–74	0.8–2.7	1.2–3.5	0.6–1.8	0.24	—	5
		M	67	1.2	2.0	1.1			
							(1)		
	Stored 4 years under cover		16	4.5	2.0	1.8	—	—	1
	Stored 1 year in open		58–76	0.5–1.9	1.2–1.3	0.2–0.6	—	—	3
		M	66	1.3	1.2	0.6			
	Stored 1½ years in open		71	0.8	0.8	0.6	—	—	1
	Unknown conditions		4–95	0.1–6.8	0.1–7.8	0.04–3.4	0.10–0.80	0.5–8.0	151
		M	64	1.9	1.7	1.0	0.19	1.7	
							(10)	(15)	
Dried			4–82	2.1–7.9	0.5–6.1	1.3–5.7	0.27–0.65	2.4–10.2	72
		M	14	4.2	4.3	2.0	0.40	4.9	
							(20)	(32)	
Ash	Burnt droppings		8–14	0.5–0.7	4.8–23.0	2.2–20.4	—	—	5
		M	9	0.6	6.9	6.6			

Range of battery manures (excluding dried and ash)			4–95	0.1–6.8	0.1–7.8	0.04–4.3	0.04–0.80	0.4–8.0	234
		M	71	1.7	1.4	0.7	0.13	1.6	
							(30)	(23)	
4. Slurry									
	Droppings and washings stirred before sampling		77–99	0.3–1.8	0.1–0.8	0.1–0.6	0.04–0.20	—	17
		M	94	0.6	0.4	0.2	0.06		
							(6)		
	Droppings and washings unstirred before sampling		95	0.3	0.1	0.1	—	—	1
	Fresh slurry		89	0.8	0.6	0.3	0.1	—	1
	Unknown conditions		81–99	0.1–3.8	0.1–0.9	0.1–0.5	0.13	1.1–1.4	20
		M	88	0.5	0.5	0.2		1.3	
							(1)	(3)	
Range of slurry			77–99	0.1–3.8	0.1–0.9	0.1–0.6	0.04–0.20	1.1–1.4	39
		M	92	0.6	0.5	0.2	0.08	1.3	
							(8)	(3)	
DUCK									
1. Straw litter	2 Months in open		6	0.4	1.0	0.6	—	—	1
	2 Years in open		81	0.6	1.2	0.6	—	—	1
2. Droppings	Fresh		32–60	0.8–3.2	1.4–2.3	0.6–2.2	—	—	2
3. Slurry	Unknown conditions		—	0.5–0.6	0.5–0.6	0.1	—	—	2
GOOSE									
Droppings	Fresh		75	0.6	0.6	0.8	—	—	1
PIGEON									
Droppings	Unknown conditions		61	3.0–4.8	1.3–2.2	0.7–1.4	—	—	3
		M		3.4	1.4	1.2			
			(1)						
TURKEY									
1. Litter types									
Straw	Fresh		64–67	0.4–0.5	0.5–0.6	0.4	—	—	2
Peat	Fresh		10	3.8	2.5	1.4	—	—	1
Wood shavings	Fresh		30–56	1.4–3.3	1.2–3.0	0.7–1.8	—	—	7
		M	42	1.9	1.5	0.9			

Table A-38 (continued)

Moisture and Main Nutrients of Poultry Manures[a,b]

Material	Details		Moisture	N	P_2O_5	K_2O	Mg	Ca	No. of samples
Wood shavings	6 Months in open		56–70	0.7–1.8	1.1–1.9	0.8–1.2	—	—	4
		M	63	1.5	1.5	0.9			
Unknown type of litter	1 Month in open		81	2.2	1.3	0.6	—	—	1
Unknown type of litter	Well weathered in open		40	0.6	1.0	0.1	—	—	1
Unknown type of litter	Unknown conditions		34–69	0.5–6.0	1.1–4.3	0.1–1.7	—	—	9
		M	58	1.0	1.9	0.6			
2. Droppings									
	Fresh		55	1.8	1.4	0.9	—	—	1
	Stored unknown time and conditions		36–81	2.2–3.2	1.2–5.3	0.6–2.6	—	—	2
	Burnt			0.4	12.2	7.4	—	—	1

[a] Percent in material as received.

[b] The range in the composition and the median value M are given where possible; where the number of determinations differs from the total number of samples, the number is given in parentheses below the analysis.

Table A-39

Organic Matter, Heavy Metals, and Other Nutrients in Poultry Manures[a,b]

Material	Details	% Na	% Cl	NH_4-N (ppm)	NO_3-N (ppm)	% OM	pH	% Ash	% $CaCO_3$	Total Cu (ppm)	Avail. B (ppm)	Total As (ppm)	Avail. Cu (ppm)	Total Zn (ppm)
CHICKEN														
1. Deep litter														
Straw	Sampled in house	0.2				49								
		(1)				(1)								
Straw + peat	Sampled in house						7.1–	12.4–						
							7.4	14.0						
	M						7.3	13.4						
							(3)	(3)						
Straw + sawdust	Sampled in house					26								
						(1)								
Wood shavings	Stored more than 6 months in open			3476	513		8.6				6.1			
				(1)	(1)		(1)				(1)			
Wood shavings	Unknown storage time						6.8							
							(1)							
Unknown type of litter	Unknown conditions	0.6	1.0	196	252–	24–66	5.2–	8.4–	1.6–	23.0–	6.1–			
					5000		7.5	28.5	22.9	45.0	17.9			
	M				3200	41	6.8	21.0	9.2		9.1			
		(1)	(1)	(1)	(4)	(9)	(6)	(3)	(5)	(2)	(5)			
Dried unknown litter				1185–	133–	7.8	9.4–							
				4097	819		11.1							
	M			3280	169	7								
				(7)	(7)	(3)	(2)							
Ash	Burnt deep litter					9–16	8.0							
						(2)	(1)							
2. Broiler														
Wood shavings	Stored less than 12 months in open			875				7.2						
				(11)				(1)						
Wood shavings	Unknown time and conditions						7.2	4.5–						
								45.0						
	M							9.1						
							(1)	(12)						

Table A-39 (continued)

Organic Matter, Heavy Metals, and Other Nutrients in Poultry Manures[a,b]

Material	Details	% Na	% Cl	NH_4-N (ppm)	NO_3-N (ppm)	% OM	pH	% Ash	% Ca CO_3	Total Cu (ppm)	Avail B (ppm)	Total As (ppm)	Avail Cu (ppm)	Total Zn (ppm)
Unknown type of litter	Unknown conditions	0.32–	0.4–	0.3–	0.08–	20–65	8.4–	1.2–	0.7–	150				
		0.38	0.6	10,000	200		8.8	42.0	11.7					
	M		0.5	5,400	30	44		32.1	2.5					
		(2)	(5)	(7)	(7)	(7)	(2)	(8)	(6)	(1)				
3. Battery														
	Fresh droppings			220–	1.7–	20–32	7.5	3.9–	3.7–	648				
				10,000	500			86.2	4.4					
	M			3,300	36	21		10.2	4.3					
				(12)	(7)	(3)	(1)	(6)	(3)	(1)				
	Unknown conditions	0.42		0.3–	0.02–	21–56	6.2–	4.8–	1.0–	18	12.1–			
				10,000	533		8.4	48.0	12.8		15.1			
	M			4,800	200	39	7.7	28.2	3.1					
		(1)		(21)	(7)	(4)	(6)	(16)	(9)	(1)	(2)			
	Dried	0.29–	0.70–	526–	24–		7.5–	16.8–	1.1–	18–79	8–13	0.3	39	328–
		0.51	1.1	10,000	1,000		7.8	60.2	16.9					1,110
	M	0.38	1.0	5,040	140		7.6	23.0	14.8	48	9			416
		(3)	(10)	(19)	(10)		(6)	(37)	(12)	(12)	(3)	(1)	(1)	(16)
4. Slurry														
	Droppings and washings stirred	0.10–		2,000–										
	before sampling	0.20		5,140										
	M			4,670										
		(2)		(3)										
	Droppings and washings unstirred			3,000										
	before sampling			(1)										
	Unknown conditions							5.2						
								(1)						

DUCK			
1. Straw litter	2 Months in open	9	
2. Droppings	Fresh	25	
GOOSE			
Droppings	Fresh	15	
TURKEY			
Litter types			
Unknown type of litter	Unknown conditions	29–56 (2)	26.0 (1)

[a] Data refer to material as received.

[b] The range in the composition and the median value **M** are given where possible; where the number of determinations differs from the total number of samples, the number is given in parentheses below the analysis.

Table A-40

Moisture and Main Nutrients in Animal Manures[a,b]

Material	Details		Moisture	N	P_2O_5	K_2O	Mg	Ca	No. of samples
CATTLE									
1. FYM type									
Straw litter	Fresh from broiler beef		73–83	0.4–2.3	0.3–3.0	0.4–2.1	0.07	—	4
		M	78		0.4	0.6			
							(1)		
Straw litter	6 Months in open		64–86	0.4–0.9	0.2–1.0	0.2–1.3	0.05–0.13	0.2–0.3	12
		M	79	0.6	0.4	0.4	0.07		
							(5)	(2)	
Straw + wood shavings	2–3 Months in open		78–86	0.4–0.6	0.3–0.6	0.2–1.4	0.07–0.08	—	4
		M	81	0.5	0.4	0.2			
							(2)		
Wood shavings	1 Year under cover		66–71	0.4–1.6	0.1–0.2	0.5–0.7	—	—	5
		M	67	0.4	0.2	0.6			
Wood shavings	Unknown conditions		60–78	0.4–0.6	0.1–0.3	0.3–0.9	—	—	4
		M	74	0.5	0.2	0.6			
Unknown litter	Unknown conditions		8–89	0.2–3.5	0.1–2.7	0.1–4.4	0.02–0.11	0.1–0.4	135
		M	77	0.6	0.3	0.7	0.04	0.1	
							(10)	(3)	
Range of FYM types			8–89	0.2–3.5	0.1–3.0	0.1–4.4	0.02–0.13	0.1–0.4	164
		M	77	0.6	0.3	0.7	0.04	0.2	
							(18)	(5)	
2. Feces									
Milk cows	Fresh		83–89	0.2–0.6	0.1–0.4	0.04–0.4	0.02–0.11	0.2	14
		M	85	0.4	0.2	0.1	0.06		
							(7)	(2)	
Beef cattle	Fresh		86	0.3	0.2	0.1	0.05	—	1

Calves	Fresh		84	0.6	0.2	0.1	0.06	—	1
	With some straw		82	2.3	3.0	2.1	—	—	1
	Unknown conditions		78–86	0.3–0.5	0.1–0.4	0.04–0.6	—	—	7
		M	82	0.4	0.2	0.3			
Range of feces			78–89	0.2–2.3	0.1–3.0	0.04–0.6	0.02–0.11	0.2	24
		M	85	0.4	0.2	0.2	0.06		
							(9)	(2)	
3. Urine									
	Fresh undiluted		91–92	0.58–1.29	0.001–0.005	0.49–2.05	0.01–0.04	—	13
		M	92	0.79	0.002	1.6	0.02		
							(4)		
	Stored undiluted		—	0.4–0.8	0.1–1.0	0.01	—	—	3
		M		0.7	0.8				
						(1)			
4. Slurry									
Feces + urine			73–98	0.2–1.9	0.1–2.4	0.1–4.2	0.03–0.06	—	14
		M	90	0.5	0.2	0.6	0.05		
							(3)		
Feces, urine, + washings	2–3 Days in tank		86–99	0.1–2.7	0.03–0.4	0.06–0.5	0.01–0.07	0.04–0.4	15
		M	94	0.45	0.1	0.2	0.02	0.06	
							(12)	(8)	
Feces + urine + washings	Unknown conditions		40–99	0.005–4.8	0.0004–0.5	0.001–2.8	0.009–0.1	0.03–0.2	102
		M	96	0.3	0.2	0.3	0.04	0.1	
							(34)	(7)	
Range of slurry			40–99	0.005–4.8	0.0004–2.4	0.001–4.2	0.009–0.14	0.03–0.4	131
		M	96	0.3	0.1	0.3	0.03	0.1	
							(52)	(18)	
DOG									
Feces	From kennels		70–74	0.3–0.9	0.1–0.5	0.06–0.2	—	—	3
		M	72	0.4	0.2	0.1			

Table A-40 (continued)

Moisture and Main Nutrients in Animal Manures[a,b]

Material	Details		Moisture	N	P_2O_5	K_2O	Mg	Ca	No. of samples
HORSE									
1. FYM type									
	Stable manure		53–78	0.5–0.9	0.3–0.7	0.4–1.0	—	—	10
		M	68	0.7	0.5	0.6			
2. Feces									
	Unknown storage conditions		73–91	0.5–1.9	0.03–0.4	0.1–1.2	—	—	7
		M	75	0.6	0.3	0.6			
3. Urine									
	Unknown storage conditions		89–93	1.1–1.6	Trace	0.8–1.5	—	—	7
		M	91	1.4	Trace	1.2			
MINK									
Feces	Unknown conditions		68–70	1.8–2.1	2.5–3.2	0.2–0.3	0.11	—	3
		M	68	2.1	3.2	0.3			
PIG									
1. FYM type	Unknown type of litter and conditions		54–98	0.3–1.0	0.05–1.4	0.2–0.7	0.13–0.18	—	14
		M	79	0.6	0.6	0.4			
							(2)		
2. Feces	Unknown storage conditions		1–82	0.1–0.8	0.4–8.1	0.2–0.9	—	0.1	8
		M	76	0.6	0.8	0.4			
								(1)	
3. Urine	Unknown storage conditions		96–99	0.1–1.2	0.03–0.5	0.0006–1.0	0.002	—	9
		M	97	0.4	0.1	0.5			
							(1)		
4. Slurry									
Pork pigs	1–4 Weeks under cover		83–99	0.2–1.0	0.025–0.9	0.04–0.9	0.001–0.1	0.01–0.4	44
		M	92	0.6	0.2	0.2	0.03	0.36	
							(29)	(6)	

	4 Weeks in open		92–90	0.02–0.4	0.015–0.4	0.1–0.3	0.01–0.09	0.1	19
		M	98	0.2	0.2	0.2	0.02		
								(4)	
Fatteners	10 Weeks in open		96–99	0.1–0.2	0.1–0.2	0.1–0.2	0.03–0.1	—	16
		M	98	0.18	0.14	0.13	0.085		
							(4)		
Liquid from unstirred slurry	Under cover		96–99	0.1–0.5	0.02–0.3	0.1–0.2	0.04	—	6
		M	98	0.4	0.1	0.2			
							(1)		
Sediment from unstirred slurry	Under cover		90–96	0.1–0.6	0.03–0.36	0.1–0.3	0.05	0.1–0.2	8
		M	95	0.4	0.3	0.1		0.15	
							(3)	(4)	
	Unknown conditions		31–99	0.01–4.8	0.01–4.2	0.02–3.3	0.003–0.16	0.02–1.0	57
		M	95	0.4	0.2	0.2	0.03	0.078	
							(25)	(14)	
Range of pig slurry			31–99	0.01–4.8	0.01–4.2	0.02–3.3	0.001–0.16	0.01–1.0	150
		M	96	0.4	0.2	0.2	0.02	0.1	
							(81)	(28)	
RABBIT									
Feces	Unknown conditions		31–78	0.4–1.3	0.6–4.2	0.3–3.2	—	—	3
		M	58	0.5	1.2	0.5			
SHEEP									
Feces	Unknown conditions		56–87	0.7–1.7	0.01–1.0	0.2–2.1	—	—	10
		M	68	0.8	0.45	0.35			

[a] Percent in material as received.

[b] The range in the composition and the median M are given where possible; where the number of determinations differs from the total number of samples, the number is given in parentheses below the analysis.

Table A-41

Organic Matter, Heavy Metals, and Other Nutrients in Animal Manures[a,b]

Material	Details	% Na	% Cl	NH_4-N (ppm)	NO_3-N (ppm)	% OM	pH	% Ash	Total Cu (ppm)	Total B (ppm)	Total Pb (ppm)	Total Zn (ppm)	Avail. B (ppm)	Total As (ppm)	Avail. Cu (ppm)	% $CaCO_3$
CATTLE																
1. FYM type																
Straw litter	Fresh from broiler beef			770												
				(1)												
Straw litter	6 Months in open	0.06						2.8								
								(1)								
Wood shavings	1 Year under cover							2.6–								
								2.9								
	M							2.6								
								(4)								
Wood shavings	Unknown conditions				6.0			2.0–								
								2.4								
					(1)			(2)								
Unknown litter	Unknown conditions			14–	1,400	12–50		0.8–	65	47						
				8,600				36.5								
	M			2,000		15		6.25								
				(3)	(1)	(15)		(76)	(1)	(1)						
2. Feces																
	Unknown conditions					14–20		2.4								
	M					20										
						(3)		(1)								
3. Urine																
	Fresh undiluted			11–			7.1–									
				1,227			8.1									
	M			867			7.7									
				(4)			(6)									
	Stored undiluted	0.22–	0.3–	486–	44–63			7.0								
		0.26	0.6	7,163												
		(2)	(2)	(2)	(2)			(1)								

4. Slurry											
Feces + urine			0.01								
			(3)								
Feces + urine	2–3 Days in tank		0.004–						0.6–	0.6	2.0–
+ washings			0.01						2.0		6.0
		M	0.01						1.0		3.0
			(5)						(10)	(1)	(10)
Feces + urine	Unknown conditions		0.01–	0.18–	0.2–				16		1.0
+ washings			0.02	2,700	2.6						
		M	0.02	90							
			(4)	(12)	(2)				(1)		(1)
HORSE											
1. FYM type											
	Stable manure					18–26					
		M				21					
						(8)					
2. Feces											
	Unknown storage conditions				600	8–25		3.0			
		M				21					
					(1)	(3)		(2)			
PIG											
1. FYM type											
	Unknown type of litter and conditions		0.14			11	6.4	2.0–	236–		
								15.6	366		
		M				11			350		
			(2)			(4)	(1)	(2)	(3)		
2. Feces											
	Unknown storage conditions					17		3.0			
						(1)		(1)			
3. Urine											
	Unknown storage conditions				300	2	8.4	1.0			
					(1)	(1)	(1)	(1)			
4. Slurry											

Table A-41 (continued)

Organic Matter, Heavy Metals, and Other Nutrients in Animal Manures[a,b]

Material	Details		% Na	% Cl	NH_4-N (ppm)	NO_3-N (ppm)	% OM	pH	% Ash	Total Cu (ppm)	Total B (ppm)	Total Pb (ppm)	Total Zn (ppm)	Avail. B (ppm)	Total As (ppm)	Avail. Cu (ppm)	% Ca CO_3
Pork pigs	1–4 Weeks under cover		0.02–		2,000–				23.1	4–100			2–4		0.05		
			0.1		8,300												
		M	0.05		4,000					10			3				
			(6)		(4)				(1)	(9)			(5)		(1)		
	4 Weeks in open		0.04–		700–	2–20											
			0.08		1,000												
		M	0.08		850	8.5											
		M	(16)		(16)	(16)											
Fatteners	10 Weeks in open								28.3								
									(1)								
Sediment from unstirred slurry	Under cover		0.05						1.1–	83–							
									1.4	103							
		M							1.15	93							
			(1)						(4)	(4)							
	Unknown conditions				322–	25–84	25–57	5.8–		6–365	8.0		2–323	0.9–		2–26	5.2–
					3,017			7.7						1.0			5.5
		M			930	46	25	6.6		30			10			16	
					(5)	(4)	(3)	(4)		(10)	(1)		(9)	(2)		(3)	(2)
Range of pig slurry			0.02–		322–	2–84	25–57	5.8–	1.1–	4–365	8.0		2–323	0.9–	0.05	2–26	5.2–
			0.1		8,300			7.7	23.1					1.0			5.5
		M	0.06		900	10	25	6.6	1.3	27			6			16	
			(23)		(25)	(20)	(3)	(4)	(6)	(23)	(1)		(14)	(2)	(1)	(3)	(2)
SHEEP																	
Feces	Unknown conditions						10–36		3.6–								
									6.0								
		M					30										
							(3)		(2)								

[a] Data refer to material as received.
[b] The range in composition and the median value M are given where possible; where the number of determinations differs from the total number of samples, the number is given in parenthesis below the analysis.

Table A-42

Moisture and Main Nutrients in Organic Manures[a,b]

Material	Details		Moisture	N	P_2O_5	K_2O	Mg	Ca	No. of samples
Blood									
	Dried		—	9.6–13.0	0.8	—	—	—	3
		M		12.2					
					(1)				
	Liquid		—	0.65	—	—	—	—	1
Bones									
	Crushed		—	0.1–4.7	19.5–21.0	—	—	—	2
	Steam-heated		11	3.9	23.0	—	—	—	1
Bone meal			0.2–14	0.3–4.6	14.1–33.2	—	—	—	6
		M	11	3.6	23.6				
Bone and meat meal			6	3.9–12.3	0.9–19.0	—	—	—	3
		M		6.4	14.3				
			(1)						
Composts									
Miscellaneous									
	Cow manure and bracken		—	1.9	0.03	0.3	—	—	1
	Dano type		11–72	0.3–0.7	0.4–0.9	0.3–0.5	—	—	7
		M	33	0.5	0.5	0.4			
	Poultry manure and sawdust		60	0.8–1.2	0.05–1.4	0.3–0.8	—	—	2
			(1)						
	Sewage sludge and straw		7–39	0.9–1.4	0.5–1.0	0.1–0.2	—	—	4
		M	14	1.25	0.7				
						(2)			
	Strawy type		59–86	0.2–1.3	0.04–0.4	0.1–1.0	—	—	44
		M	75	0.5	0.1	0.4			
	Unknown type		5–76	0.6–1.1	0.4–0.8	0.2–0.9	—	—	3
			72	0.9	0.5	0.4			

Table A-42 (continued)

Moisture and Main Nutrients in Organic Manures[a,b]

Material	Details		Moisture	N	P_2O_5	K_2O	Mg	Ca	No. of samples
Mushroom compost	No information on sampling or activator		24–82	0.3–3.4	0.3–3.3	0.1–2.1	0.08–0.30	—	94
		M	66	0.7	0.6	0.8	0.28		
					(20)	(19)	(3)		
	Sampled before peak heat		61–79	0.1–2.0	—	—	—	—	
		M	75	0.5					
	Sampled after peak heat		57–77	0.5–1.1	—	—	—	—	
		M	69	0.6					
	Spent compost		15–76	0.1–2.9	0.1–3.7	0.3–3.3	—	—	66
		M	64	0.6	0.5	0.9			
Feathers			10–81	2.5–13.1	0.1–0.4	0.1–0.7	—	—	3
		M	67	4.9	0.2	0.2			
Fish meal			—	6.3–8.9	6.0–8.9	—	—	—	2
Offal			36–63	0.9–4.2	—	—	—	—	3
		M	49	1.5					
Hoof and horn			—	6.5–13.2	0.8	—	—	—	8
		M		12.2					
					(1)				
	Fused		—	4.1	2.5	—	—	—	1
Hoof meal			10	14.5	—	—	—	—	1
Leaf mould	Surface layer from wood		36–45	0.4	0.2	0.3	—	—	2
Sea fern			42	3.5	—	—	—	—	1
Seaweed			18–80	0.1–1.3	0.1–0.6	0.1–3.4	—	—	8
		M	68	0.6	0.3	1.0			
Sewage sludge									
	Digested sludge		2–96	0.2–3.6	0.2–3.6	0.03–0.7	0.09–0.21	10.8	20
		M	66	1.4	1.0	0.1			

							(2)	(1)	
	Unknown conditions		3–99	0.1–6.9	0.0046–4.8	0.009–2.8	0.01–0.18	0.3–0.9	275
		M	60	1.0	0.6	0.2	0.10	0.5	
							(7)	(5)	
	Burnt sludge		1–61	0.1–0.8	0.8–5.8	0.1–0.9	—	—	8
		M	25	0.4	1.6	0.3			
Shoddy									
	Clean uniform sample		7–17	3.2–14.9	0.3–10.0	0.1–12.0	0.09	1.3	33
		M	10	11.6	2.0	0.6			
			(9)		(6)	(7)	(1)	(1)	
	Very poor grades		7–62	0.4–3.0	—	—	—	—	7
		M	10	1.7					
			(3)						
	Oily wool waste, fresh from factory		3–54	0.4–3.0	—	0.6	—	—	10
		M	10	1.95					
			(4)			(1)			
	Pressed wool stored 2–3 weeks in open		24	1.3	—	—	—	—	1
	Wool contaminated with synthetic fiber		4–23	1.5–14.5	0.2–0.5	0.3–1.8	—	—	4
		M	8	3.9	0.3	0.9			
	No details		9–32	1.2–13.5	0.2–1.2	0.2–1.3	—	—	29
		M	11	4.6	0.3	0.4			
			(4)		(6)	(6)			
Slaughterhouse waste			82–99	0.2–0.9	0.03–0.2	0.05–0.06	—	—	2
	+ Shoddy		26	2.5	0.2–0.6	0.4–0.5	—	—	2
Soot			2–15	0.3–6.1	0.1	0.1–0.2	—	—	27
		M	14	3.6		0.1			
			(4)		(2)	(4)			

[a] Percent in material as received.

[b] The range in composition and the median value M are given where possible; where the number of determinations differs from the total number of samples, the number is given in parentheses below the analysis.

Table A-43

Organic Matter, Heavy Metals, and Other Nutrients in Organic Manures[a,b]

Material	Details	% Na	% Cl	NH_4-N (ppm)	NO_3-N (ppm)	% OM	pH	% Ash	% Ash free	Total ppm Cd	Cr	Cu	Pb	Ni	Zn	B	Available ppm Cd	Cr	Cu	Pb	Ni	Zn	B	% $CaCO_3$	No. of samples
Composts																									
Miscellaneous	Dano type							24.7–				268–	68	31	76–										
								78.0				850			850										
	M							48.1							260										
								(4)				(2)	(1)	(1)	(3)										
	Sewage, sludge and straw																								
								20.3–																	
								45.0																	
	M							29.3																	
								(4)																	
	Strawy type							1.3–																	
								9.3																	
	M							3.7																	
								(41)																	
	Unknown type					13–18																			
	M					13																			
						(3)																			
Mushroom compost	Before peak heat			194–	18–		6.9–																		
				264	88		7.8																		6
	M						7.0																		
				(2)	(2)																				
	After peak heat			66–	0–18																				26
				1,155																					
	M			186	4																				
	No sampling or activator details	0.09	0.1–	58–405	9–468	20–26	5.8–																		
			0.3				8.1																		
	M		0.2	208	50		7																		
		(1)	(3)	(9)	(7)	(2)	(75)																		
	Spent compost						5.0–	3.3–																	
							6.7	23.2																	
	M						5.9	12.9																	
							(10)	(46)																	
Leaf mould						9		33.8																	
						(1)		(1)																	
Seaweed						15																			
						(1)																			
Sewage sludge	Digested					17–74					213–	48–		11–	84–							13–	300–	1.0	
											218	630		290	6,300			1	22			305	330		
	M					31					215	336		259	856							78	320		
						(7)					(3)	(7)		(4)	(9)			(1)	(1)			(6)	(3)	(1)	
	Unknown	0.01–		20–	1–32	5–44	4.8–	7.5–		0.5–	2–	4–	1–	3–	19–	3.7–		0.3–	1–	2–	0.36–	9–	0.7–	3.2	
		0.73		1,515			12.6	78.8		19	5,000	8,306	319	2,290	15,876	117		1,220	1,932	25	941	22,890	39	17.8	
	M	0.5		56	3	18	6.5	37.2		10	31	220	35	38	405	9.2		2	42		7	84	5.1		
		(3)		(3)	(4)	(46)	(24)	(66)		(4)	(72)	(129)	(4)	(55)	(148)	(5)		(15)	(70)	(2)	(20)	(70)	(17)	(2)	

	Burnt sludge		1–2					226–	290–		20–	795–									
								9,000	2,600		150	5,300									
		M						605	486		22	1,805									
			(2)					(6)	(7)		(5)	(7)									
	Dry weight basis						< 1–	2–	5–	20–	5–	80–	7.8–	< 1–	0.13–	3–	2–	< 1–	10–	2.5–	20.5
							1,500	9,797	11,600	3,000	11,000	54,500	117	32	170	2,550	115	3,000	12,000	221	
		M					150	148	500	600	80	1,200		23	4	60	9	25	450	32	
							(47)	(123)	(190)	(51)	(109)	(207)	(2)	(3)	(62)	(106)	(52)	(66)	(116)	(12)	(1)
Shoddy																					
	Clean uniform sample		11–71		12.1–	81.7															
					23.3																
		M	64		18.9																
			(5)		(3)	(1)															
	Wool contaminated with synthetic fiber		65																		
			(1)																		
	Unknown type				12.1–				4,600												
					12.3																
					(2)				(1)												
Soot			37–45	3.4																	
			(2)	(1)																	

[a] Data refer to material as received.

[b] The range in composition and the median value M are given where possible; where the number of determinations differs from the total number of samples, the number is given in parentheses below the analysis.

Table A-44

Moisture and Main Nutrients in Farm Wastes[a,b]

Material	Details		Moisture	N	P_2O_5	K_2O	Mg	Ca	No. of samples
Barley									
	Strippings		—	4.7	1.7	2.4	—	—	1
Bracken			14	2.0	0.2	0.5	—	—	1
Duckweed			94	0.2	—	—	—	—	1
Mustard									
	Mainly husks		12	3.3	—	—	—	—	1
Poultry carcass ash			23	1.2	27.6	2.2	—	—	1
Rape dust			—	5.1–5.3	1.6–2.5	1.5	—	—	2
						(1)			
Silage									
	Pea haulm + cocksfoot		81–83	0.3–0.5	0.1	0.5–0.6	—	—	2
	Pea pods		82–86	0.4–0.5	0.1	0.2–0.3	—	—	2
Silage effluent									
	From grass		86–96	2.3	0.1–0.3	0.3–0.4	—	0.1	3
		M	94		0.1	0.4			
				(1)				(2)	
	From kale		—	0.1	0.09	0.6	—	—	1
	From pea haulm		95	0.1–0.4	0.04–0.1	0.3–0.5	0.03	—	3
		M		0.3	0.09	0.3			
			(1)				(1)		
Straw									
	Barley		7–69	0.2–1.3	0.1–0.4	0.4–1.7	0.01–0.04	0.2–0.4	10
		M	17	0.4	0.2	0.9		0.3	
							(2)		

Cocksfoot		15	0.9	—	—	—	—	1
Oat		8–17	0.3–0.8	0.2–0.7	0.8–1.0	—	0.3–0.5	5
	M	13	0.45	0.4			0.4	
					(2)			
Pea		16–23	2.3	0.2–0.3	—	—	0.9–1.8	2
Wheat		11–47	0.2–0.9	0.1–0.3	0.3–1.4	—	—	48
	M	18	0.5	0.2	0.8			

[a] Percent in material as received.

[b] The range in composition and the median value M are given where possible; where the number of determinations differs from the total number of samples, the number is given in parentheses below the analysis.

Table A-45

Organic Matter, Heavy Metals, and Other Nutrients in Farm Wastes[a,b]

Material	Details		% Na	NO_3-N (ppm)	pH	% OM	% Ash
Bracken						82	4.7
						(1)	(1)
Silage effluent							
	From pea haulm		0.004	0.1	5.7		
			(2)	(1)	(1)		
Straw							
	Barley		0.1–0.2			75	3.1–6.7
		M					4.6
			(2)			(1)	(8)
	Oat		0.57			78	
						(1)	
	Pea						8.4
	Wheat					76–79	3.3–18.2
		M					5.5
						(2)	(47)

[a] Data refer to material as received.

[b] The range in composition and the median value M are given where possible; where the number of determinations differs from the total number of samples, the number is given in parentheses below the analysis.

Table A-46

Moisture and Main Nutrients in Industrial and Miscellaneous Wastes[a,b]

Material	Details		Moisture	N	P_2O_5	K_2O	Mg	Ca	No. of samples
Brewers waste			6–94	0.9–3.2	0.1–0.4	1.0	—	0.02	2
						(1)		(1)	
Carbon (activated) waste			48	0.9	0.4	0.03	—	—	1
Castor meal			11	5.5	2.4	1.4	—	—	1
Chicory waste			—	1.5	—	—	—	—	1
Cockleshell			2	0.3	0.1	0.1	—	—	1
Cocoa waste			6–71	0.8–2.9	0.6–1.5	—	—	10.2	8
		M	48	1.8	1.0				
								(2)	
Coconut fiber/matting waste									
	Dust		18	0.2–0.9	0.1	0.4–0.7	—	—	3
		M		0.4		0.6			
			(1)		(1)				
	Fiber		—	0.1	—	—	—	—	1
Coffee waste			42–70	0.7–3.1	0.033–12.5	0.013–0.9	—	0.1	6
		M	62	1.5	0.2				
						(2)		(1)	
Coffee-chicory residue			66–67	0.8–1.7	0.6	0.015	0.04	—	2
					(1)	(1)	(1)		
Cotton cake			11	4.1–6.9	1.7	1.6–2.8	—	—	1
		M		4.8					
				(3)		(2)			
Cuttle fish			—	—	0.2	—	—	—	1
Felt waste			—	13.6	—	—	—	—	1
Flax									
	Cavings and cleanings		6	—	3.6	9.2	—	—	1

Table A-46 (continued)

Moisture and Main Nutrients in Industrial and Miscellaneous Wastes[a,b]

Material	Details	Moisture	N	P_2O_5	K_2O	Mg	Ca	No. of samples
	Residues from retting tanks (dried)	7	0.8	0.2	0.7	—	—	1
	Flax and jute dust	9	1.2	—	—	—	—	1
Flue dust		16	0.1	0.2	0.2–0.8	—	—	1
	M				0.5			
					(5)			
Fly ash		13–33	0.1–0.8	0.01–15.8	0.04–1.6	—	—	4
	M	26	0.1	0.1	0.3			
Ginger root		—	1.8	—	—	—	—	1
Glue factory waste		28–64	1.4–4.7	0.01–1.1	0.04–0.6	—	—	6
		41	2.1	0.2	0.25			
Hemp waste		66–68	0.7–1.5	—	—	—	—	4
	M		0.9					
		(2)						
Hops								
	Spent	9–87	0.6–5.7	0.2–3.4	0.006–2.6	0.12–0.16	—	9
	M	73	1.1	0.3	0.1			
						(2)		
Leather								
	Chamois dust	—	8.3	—	—	—	—	1
	Hide meal, ground leather	19–68	0.4–7.5	0.1–0.2	0.02–0.3	—	—	4
	M	52	6.1	0.1	0.15			
			(7)					
Malt fiber		7	4.4–5.2	1.6–2.1	1.1–2.2	—	—	2
		(1)						
Paper waste		85	0.1	—	—	—	—	1
Phosphoric acid waste		—	—	14.6	—	—	—	1

Polyurethane foam			—	—	4.4	—	—	—	1
Sawdust			4–66	0.1–0.9	0.01–0.5	0.04–1.4	—	—	4
		M	48	0.2	0.06	0.07			
Soap residue			—	—	—	17.5	—	—	1
Tannery waste			6–85	0.1–14.1	—	—	—	—	10
Tobacco waste		M	55	5.3					
	Fresh		26–61	0.5–3.9	0.1–1.0	0.001–0.1	—	—	5
		M	54	0.8	0.6	0.005			
	Stored in open		26–70	0.4–0.8	0.1–0.6	0.001–0.1	—	—	5
		M	55	0.5	0.3	0.03			
Town refuse									
	Composted		20–49	0.5–0.6	0.4–0.8	0.2	—	—	2
	Screened		4–91	0.01–3.1	0.1–2.1	0.04–1.6	—	—	14
		M	35	0.5	0.4	0.35			
	Municipal		4–78	0.3–1.0	0.1–2.1	0.2–1.6	—	—	10
		M	35	0.5	0.4	0.4			
Urea formaldehyde			—	21.4	—	—	—	—	1
Vermiculite			7	0.3	0.2	3.3	—	—	1
Woodash			1–42	0.1–0.4	0.1–3.4	0.1–9.4	—	0.6	6
		M	34	0.1	0.4	1.1		(1)	
Woodshavings			71	0.4	0.1	0.9	—	—	1

[a]Percent in material as received.

[b]The range in composition and the median value M are given where possible; where the number of determinations differs from the total number of samples, the number is given in parentheses below the analysis.

Table A-47

Organic Matter, Heavy Metals, and Other Nutrients in Industrial and Miscellaneous Wastes[a,b]

Material	Details	NH_4-N (ppm)	NO_3-N (ppm)	% OM	pH	% Ash	Total ppm Cd	Cr	Cu	Pb	Ni	Zn	B	Available ppm Cd	Cr	Cu	Pb	Ni	Zn	B	% $CaCO_3$
Brewers waste				84		0.2															
				(1)		(1)															
Carbon (activated)						17.9		5	100–		4–20	24–									
waste									169			79									
						(1)		(1)	(2)		(2)	(2)									
Cocoa waste				27	12.5	15.7–														32–	
						23.9														54	
	M																			38	
				(1)	(1)	(3)														(3)	
Coconut fiber/matting																					
waste																					
	Dust			76		2.2–															
						5.7															
				(1)		(2)															
Coffee waste		11–78	6–29	28	5.1–	0.4–			30			60									
					5.9	5.9															
	M		9		5.3	1.0															
		(2)	(4)	(1)	(8)	(3)			(1)			(1)									
Coffee-chicory residue					4.6																
					(1)																
Flax																					
	Cavings and cleanings			28																	
				(1)																	
	Residues from retting tank (dried)			81																	
				(1)																	
Fly ash				8–10	3.3–			4	66–		86	81–	71–							18–	0.3–
					12.2				116			1,294	1,950							156	0.5
	M				10.3				112			303	210							85	
				(2)	(5)			(1)	(3)		(1)	(4)	(3)							(3)	(2)

Glue factory waste			0.6	0.26	32–46		10.7										
			(1)	(1)	(2)		(1)										
Hemp waste					20												
					(1)												
Hops	Spent				64–84	4.6–											
						5.5											
					(2)	(2)											
Leather																	
	Hide meal, ground leather				28	5.1	6.0	40–	14–	5	2–5	121					
								12,800	15								
		M						1,650									
					(1)	(1)	(1)	(8)	(2)	(1)	(2)	(1)					
Malt fiber					83												
					(1)												
Sawdust					33–44		3.0										3.1–
																	5.6
		M															3.5
					(2)		(1)										(4)
Soap residue					3.5												
					(1)												
Tobacco waste	Fresh				28.2												
Town refuse	Composted				15						102		772				
					(1)						(1)		(1)				
	Screened				12–51	7.3–	39.8										
						8.5											
		M			26												
					(6)	(2)	(1)										
	Municipal									90–		55–	250–	7	2–78	245–	
										570		79	2,200			500	
		M								335			984			372	
										(6)		(2)	(8)	(1)	(2)	(2)	
Wood ash					6–42												
					17												
					(4)												

[a] Data refer to materials as received.

[b] The range in composition and the median value M are given where possible; where the number of determinations differs from the total number of samples, the number is given in parentheses below the analysis.

AUTHOR INDEX

Numbers in parentheses are reference numbers and indicate that an author's work is referred to although his name is not cited in the text. Numbers in italics show the page on which the complete reference is listed.

A

B

C

F

G

H

I

J

K

L

Q

R

S

T

Y

Z

SUBJECT INDEX

L

M

N

O

P